AF606774

SAP® SuccessFactors® Employee Central Payroll

SAP PRESS is a joint initiative of SAP and Rheinwerk Publishing. The know-how offered by SAP specialists combined with the expertise of Rheinwerk Publishing offers the reader expert books in the field. SAP PRESS features first-hand information and expert advice, and provides useful skills for professional decision-making.

SAP PRESS offers a variety of books on technical and business-related topics for the SAP user. For further information, please visit our website: *www.sap-press.com*.

Marson, Murray, Toombs
SAP SuccessFactors Employee Central: The Comprehensive Guide (3rd Edition)
2021, 712 pages, hardcover and e-book
www.sap-press.com/5352

Leong-Cohen, Kandi, Karthik, Gan
Integrating SAP SuccessFactors
2020, 746 pages, hardcover and e-book
www.sap-press.com/5019

Smith, Yang, Churin
SAP SuccessFactors Learning: The Comprehensive Guide
2018, 680 pages, hardcover and e-book
www.sap-press.com/4577

Amy Grubb, Kim Lessley
SAP SuccessFactors Recruiting and Onboarding: The Comprehensive Guide
2017, 672 pages, hardcover and e-book
www.sap-press.com/4290

Wanless, Villet, McGough
SAP Fieldglass: Functionality and Implementation
2022, 587 pages, hardcover and e-book
www.sap-press.com/5239

Deepankar Maitra, Stacy Wilkins

SAP® SuccessFactors® Employee Central Payroll

Editor Meagan White
Acquisitions Editor Emily Nicholls
Copyeditor Yvette Chin
Cover Design Graham Geary
Photo Credit iStockphoto.com: 524283637/© malerapaso
Layout Design Vera Brauner
Production Kelly O'Callaghan
Typesetting III-satz, Husby (Germany)
Printed and bound in Canada, on paper from sustainable sources

ISBN 978-1-4932-2170-7

1st edition 2022

Library of Congress Cataloging-in-Publication Data
Names: Maitra, Deepankar, author. | Wilkins, Stacy, author.
Title: SAP SuccessFactors Employee Central Payroll / by Deepankar Maitra, Stacy Wilkins.
Description: 1st edition. | Bonn ; Boston : Rheinwerk Publishing, 2021. | Includes index.
Identifiers: LCCN 2021046225 | ISBN 9781493221707 (hardcover) | ISBN 9781493221714 (ebook)
Subjects: LCSH: SAP SuccessFactors Employee Central Payroll. | Wages--Accounting--Computer programs. | Wages--Accounting--Data processing.
Classification: LCC HF5681.W3 M35 2021 | DDC 657/.742--dc23/eng/20211130
LC record available at https://lccn.loc.gov/2021046225

Contents at a Glance

Dear Reader,

Most people I know have encountered some sort of error with their paycheck at some point in their lives.

For me, it was at my very first job, working for minimum wage at the local library. The state minimum wage was raised by $.25 an hour, and it was a few weeks before I realized that my pay had not gone up with it! Sorting out the backpay (which, if I recall correctly, amounted to less than $10) took months, as the payroll system was inflexible and the process unclear.

Even when things are going right, payroll can be difficult. At a previous employer, my husband used to earn wildly different rates of pay depending on which job he was performing at the moment. Additionally, there were "extra" industry-specific rules about overtime pay, which could kick in after a certain number of hours worked in a single day or if he worked during certain hours of the night (and, of course, any work over 40 hours per week). The company's HR system could barely handle its own complex policies.

In short, payroll is complicated and essential. Getting it right requires clear processes and a well-designed, *properly configured* system. Between these pages, you'll find instructions for just that!

What did you think about *SAP SuccessFactors Employee Central Payroll*? Your comments and suggestions are the most useful tools to help us make our books the best they can be. Please feel free to contact me and share any praise or criticism you may have.

Thank you for purchasing a book from SAP PRESS!

Meagan White
Editor, SAP PRESS

meaganw@rheinwerk-publishing.com
www.sap-press.com
Rheinwerk Publishing · Boston, MA

Contents

Preface

SAP ERP HCM Payroll has evolved into a user-friendly, cloud-based, modern global payroll product that is now branded as SAP SuccessFactors Employee Central Payroll. This book is a comprehensive guide for Employee Central Payroll that gives human resources (HR) consultants, payroll managers, and administrators step-by-step instructions for configuring and running Employee Central Payroll and integrating it with Employee Central.

Introducing fundamental payroll concepts, this book explores the configuration of the standard payroll engine as well as new developments for integrating with SAP's flagship HR product, Employee Central. We'll cover the Payroll Control Center in detail, from both usability and configuration viewpoints.

Step-by-step configuration instructions are provided that follow the sequence of an Employee Central Payroll project: payroll elements, provisioning, Employee Central, Employee Central Payroll, master data integration, master data replication, and the Payroll Control Center.

The book also covers optional steps covering important new topics in postprocessing with the Employee Central user interface (UI): setting up pay statements, transferring pay results back to Employee Central, and configuring single sign-on (SSO). Our book ends with instructions on monitoring and troubleshooting and a practical discussion on implementation techniques and methods that are essential to succeed in implementing Employee Central Payroll.

While this book covers the steps required to run payroll, our emphasis is on the configuration—how to turn on the features that make Employee Central Payroll one of the best payroll systems in the world. Our discussion is applicable to all countries, but we'll use US examples whenever the topics touch upon country-specific items.

Objective

The main purpose of writing this book is to provide both new and experienced Employee Central Payroll users tasked with configuration an easy-to-use guide that is structurally intuitive and procedurally complete. Additionally, this book is designed to achieve the following goals:

- **Provide a general understanding of the complex payroll process**
  Payroll is a complex process with many steps and many possible points of failure. It requires many players to successfully configure it. Team members often play diverse roles—technical, functional, support, processor, etc. Executing one's specific

role in this complex chain of processes does not provide them the holistic view that is sometimes required to identify issues and resolve problems. This book is designed to provide both the "big picture" and the details.

- **Bridge the gap between SAP ERP HCM Payroll and Employee Central Payroll**
  SAP ERP HCM Payroll practitioners have spent many years learning how to configure SAP ERP HCM Payroll. While the payroll engine is the same, SAP has made some significant additions to their original product and rebranded it as Employee Central Payroll. The original SAP ERP HCM Payroll product is also included in the HR system. With the acquisition of SuccessFactors, SAP moved HR processes over to Employee Central, which is a completely new product that operates in the cloud and requires a different skillset to configure. Connecting the Employee Central Payroll product with Employee Central requires some of those skills. This book will address those skills specifically for components that impact payroll.
- **Reveal the mystery of new features that make up Employee Central Payroll**
  SAP has introduced the Payroll Control Center—a cockpit-like UI that helps drive payroll processes with an end goal in mind and superior visibility into processing progress. This book explains the nuances of the Payroll Control Center, including how to best configure it and how to derive the most value out of it.
- **Present the configuration in a logical sequence**
  Configuration in Employee Central Payroll is based on many tables and parameters that are technical in nature. The layout of configuration in the IMG (Implementation Guide) configuration settings is not in the order of the configuration sequence, which makes it harder for payroll professionals to understand the steps and ensure that settings are not missed. This book lays out the configuration in a logical sequence that provides continuity and helps explain the techniques for each feature and how these help the overall process.
- **Highlight opportunities and pitfalls**
  SAP documentation does not instruct the user on which features should be enabled—in fact, the strength of Employee Central Payroll is that it offers a multitude of features that can resolve many problems in multiple countries. While a great strength, this flexibility makes it harder for a user to determine which features are essential and which should be delayed until basic processes are mastered. Overwhelming a payroll team with the rich capability of Employee Central Payroll is an easy mistake. This book helps sort out the priorities and identifies scenarios that can cause trouble during implementation.
- **Provide a clear picture of the complex configuration required across multiple systems (Employee Central and Employee Central Payroll)**
  The documentation for Employee Central Payroll and Employee Central are in different repositories. To turn on the integration between the two systems to leverage their full potential, one must be an expert in both systems. This book carefully isolates

those features in Employee Central that are relevant for payroll and discusses step-by-step techniques for configuring them, thus bringing knowledge about configuring both systems into one place.

- **Prepare the reader for the implementation journey**
  Implementing Employee Central Payroll involves careful planning and risk assessment. For a large organization, this goal requires a large team with specific tasks that must be completed in carefully developed timeframes. The process must follow a methodology that can be articulated and audited as well as tracked over the duration of the project to assure successful completion and acceptance. Many methodologies are available to deliver information technology implementation projects, which naturally can be confusing users of the product. This book provides a short but informative guideline for users to understand the key aspects of any methodology to help make informed choices about the management framework for the implementation project.

Target Audience

The book is intended for readers with the following characteristics:

- Eager to gain an understanding of the payroll process in SAP
- Eager to learn about configuring Employee Central Payroll
- Seeking guidance to connect the dots between the different technical components that constitute Employee Central Payroll
- Struggling to identify the right sequence for the many configuration steps required for Employee Central Payroll
- Understands that Employee Central is a key component of payroll processes in SAP but struggles to start learning it
- Eager to implement Employee Central Payroll for their organization
- Eager to migrate their current SAP ERP HCM Payroll system to the cloud using Employee Central Payroll
- Anxious to keep their on-premise SAP ERP HCM Payroll system but would like to explore the use of some new features like the Payroll Control Center

This book is also helpful for project managers, implementation partners, configuration analysts, payroll managers, payroll administrators, integration architects, and technology auditors.

Familiarity with SAP systems, with standard navigation for configuration and operation using transaction codes, and with standard table-based data structures in a multiple-environment landscape are required to get the maximum value out of this book.

Structure of This Book

This book is organized so you can follow along while configuring Employee Central Payroll. We'll begin with an introduction to the payroll process and then explore the fundamental concepts for configuring the basic payroll engine that comes with Employee Central Payroll, which has not changed for the last 20 years. This book then covers integration concepts and discusses how these concepts can be extended to leverage Employee Central. After a detailed discussion on the Payroll Control Center, this book highlights various solutions for pay statements and discusses how to tackle common issues associated with a split landscape. This book ends with a concise and lucid explanation of project implementation methodology.

Each chapter covers the following topics:

- **Chapter 1: Introduction to Payroll with SAP SuccessFactors**
  This introduction explains payroll processes and introduces the payroll solutions available from SAP: We'll describe the different components of the solution, introduce the key terminology, and provide a good foundation for understanding how Employee Central Payroll is configured in subsequent chapters. We'll list the key differences between on-premise and cloud payroll systems. This chapter introduces the key components of the Employee Central Payroll system and identifies the countries that are supported. This chapter closes by introducing an entity relationship diagram for configuration that we'll use throughout the book to explain the interrelationships among the different configuration components.
- **Chapter 2: Configuring Payroll Elements**
  This chapter provides step-by-step instructions for configuring the standard payroll elements that originated in the on-premise SAP ERP HCM Payroll engine. We'll cover wage types, infotypes, rules and schemas, and the Data Medium Exchange (DME) process to transfer money to employee bank accounts. We'll also cover financial postings and third-party remittances and end with a short discussion on processing benefits in the US. The content is global; however, since each country has uniquely defined their own regulated, local processes for taxes, garnishments, and benefits, the examples provided in this chapter are for one country—the United States.
- **Chapter 3: System Provisioning**
  This chapter shows you how to perform provisioning: turning on a series of switches to initiate configuration for Employee Central Payroll. Provisioning is a concept unique to Employee Central and accessible to SAP and their partners only. This chapter describes the specific provisioning settings that are required to turn on Employee Central Payroll.
- **Chapter 4: Configuring Employee Central**
  Quite a bit of configuration is required in Employee Central to ensure your users can interact with this UI to work with Employee Central Payroll. This chapter provides step-by-step instructions for activating these items. This discussion includes the

rules engine and its application for validating data that is important for payroll but originates from an HR system.

- **Chapter 5: Configuring Employee Central Payroll**
  This chapter provides step-by-step instructions for configuring the basic features of Employee Central Payroll. Our discussion begins with the switches that you must turn on in Employee Central Payroll. Enabling the HR system through personnel administration configuration tasks is discussed as a precursor to replication—the technique that brings in HR data from Employee Central and holds it in Employee Central Payroll for payroll processing. This topic is followed by an overview of pay statements and a short discussion on concurrent employment. This chapter ends with an analysis of security and the roles required in the Employee Central Payroll system.
- **Chapter 6: Integration**
  This chapter introduces the integration concept in Employee Central Payroll, explaining at a high level how it integrates with related solutions for time management, benefits, finance, and the human resources information system (HRIS), that is, for Employee Central. We'll cover modern concepts of connectivity and describe the options available for connecting the various systems that make up the payroll galaxy. Three different technologies are discussed in this chapter, highlighting their applicability for various payroll functions.
- **Chapter 7: Master Data Integration**
  SAP has enabled *point-to-point replication* as the integration between Employee Central and Employee Central Payroll. This chapter explains how to set up the first component (master data integration) with step-by-step instructions. This configurable web service is explained in detail including concepts like country filtering and master data push replication. Our discussion in this chapter includes the mapping of field values from Employee Central to Employee Central Payroll.
- **Chapter 8: Master Data Replication**
  This chapter explains how to set up the second phase of point-to-point integration (master data replication) with step-by-step instructions. Each infotype brought over from Employee Central is discussed with details about the fields that are critical for payroll processing, Both pay and non-pay infotypes are broken down into their component fields, and elements inherited from Employee Central are identified with country-specific details where applicable. We include a special discussion on personnel numbers in this chapter as well as how to handle no-shows.
- **Chapter 9: Using the Payroll Control Center**
  After introducing key features of the Payroll Control Center, this chapter provides users (especially payroll managers) instructions for running payroll, from simulation through postprocessing. These features are described through examples and screenshots for easy understanding of what these features deliver. A real-life example takes you through all the steps that typically converge into a successful payroll run, with the goal of money for employees deposited in their banks.

- **Chapter 10: Configuring the Payroll Control Center**
  This chapter expands upon the concept of the Payroll Control Center and describes its configuration. Components of the Payroll Control Center, such as validations, alerts, and key performance indicators (KPIs), are discussed with step-by-step instructions on configuration. Concepts like step templates, analytics charts, policy types, and process types are covered in detail. The setup involves steps in Employee Central as well as Employee Central Payroll. The latest development, the *simplified configuration technique*, is described, including instructions on how to correlate policies with validations and correlate processes with alerts. Collaboration is essential for completing the payroll for large employee bases on time, and this chapter explains how the Payroll Control Center helps achieve the required efficiency by sharing work.
- **Chapter 11: Setting Up Pay Statements**
  Pay statements, or paystubs, show an employee what they are being paid for. This chapter provides step-by-step instructions for configuring pay statements that have already been designed in HR Forms. Our discussion of the different types of pay statements include recommendations on which are more suitable for specific situations.
- **Chapter 12: Using Payroll Results in Employee Central**
  In specific scenarios (e.g., SAP Fiori-like pay statements, mobile pay statements, and Employee Central Advanced Reporting), the payroll result from Employee Central Payroll should be carried over to Employee Central. This technique is somewhat the reverse of a replication because the data is going from Employee Central Payroll to Employee Central. This integration enables analytics reports that combine HR data with details about the payments made to employees under various classifications. This chapter describes these scenarios and their configuration.
- **Chapter 13: Setting Up Single Sign-On and Logout**
  Working on two systems ordinarily requires the user to log on to both systems and remember multiple user names and passwords. This chapter provides step-by-step instructions for setting up the SSO functionality for employees and managers to enjoy a smooth UI experience between Employee Central and Employee Central Payroll. SSO enables an Employee Central logon to be used for authenticating to Employee Central Payroll, thus avoiding the need to remember extra credentials and navigating seamlessly from one system to another. Our discussion also includes the SSO mechanism from Employee Central Payroll to Employee Central using OAuth 2.0, which is required for using payroll results in Employee Central.
- **Chapter 14: Monitoring Tools**
  This chapter teaches you how to use the available tools to audit and report on Employee Central Payroll system operations. These tools help you monitor the pulse and the health of the system by ensuring any data dropout is discovered early and thus resolved before payroll processing.

- **Chapter 15: Troubleshooting**
  Out of all things that can go wrong, a few issues are difficult to fix. This chapter covers advanced topics for Employee Central Payroll, including troubleshooting and changing hire dates in the system. These problems can be challenging to resolve due to a split system landscape where HR and payroll are independent environments.
- **Chapter 16: Implementation Project Planning**
  This chapter discusses the steps you must take to implement Employee Central Payroll. We'll provide high-level guidelines on the strategies required to perform a successful implementation. We'll explain the discover, design, build, deploy, and hypercare phases. In addition, we'll highlight the importance of maintaining a deliverables tracker. Finally, testing and conversion considerations are discussed, especially as they are applicable for payroll implementations.

Acknowledgments

This book is dedicated to my parents for being my everyday heroes spontaneously inspiring me and selflessly supporting me throughout my academic and professional journey. As I spent long hours developing the material through the challenges of the pandemic, I realize that I could not have completed this book without the constant support from my family.

Thanks to Chumki, Olie, Kakoli, Lahiru, and Anika for being the pillars of support that I leaned on while completing 14 of the 16 chapters in this book. Thanks to Subba Rao Inaganti for helping with reviews, Santhosha Edukulla for her help with screenshots, and Wendy Rooney for supporting the initiative to write this book. Last but not the least, thanks to Satish Badgi for his reviews and invaluable advice.

Deepankar Maitra
November 2021

This book is dedicated to my mom and dad who always pushed that education and 100% effort were the keys to success in work and life. A special thank you to Rich and McKenna for being patient with me as I worked on this book at night and on weekends. Thanks to Satish Badgi for recommending I join this project as well as for providing insight into all things payroll. And a final thanks to Nina Oddo and David Zhou for assistance in reviewing and providing valuable input into this book.

Stacy Wilkins
November 2021

Chapter 1
Introduction to Payroll with SAP SuccessFactors

Configuring SAP SuccessFactors Employee Central Payroll is a complex and demanding task that requires a high level of collaboration between the technical and business teams. Configured correctly, Employee Central Payroll can leverage an SAP SuccessFactors Employee Central system effectively. Standard features should be used, but components can be tailored to the specific needs of your business. The configuration is also dependent on the country in which your business operates.

This chapter is an introduction to Employee Central Payroll, what it does, how it is organized, and why it is necessary to configure it.

When we buy a new appliance, we look for the instruction manual on how to operate it. We skip all the complicated pages with circuit diagrams and find the *Quick Start Guide*. Then, we start using the appliance. We have no difficulty using it until we encounter a situation that demands a feature that we desperately need. We look for information online as well as in the manual and then try to learn how to use it. In our rapid learning process, we start with identifying whether the appliance can do the specific task we need help with. With minimal investigation, we can form conclusions about whether a task can be done. Once we've made up our mind, we happily read the pages that we skipped to discover how we can put this feature to good use by configuring the appliance, to make most of what we paid for. If we cannot achieve this task, we feel that the value of the device is diminished.

SAP works in a similar way. You can do a lot with the startup guide. But a lot of documentation online and countless YouTube videos describe specific features that users have found helpful. Consider the first two chapters of this book as the quick start guide for Employee Central Payroll. A lot of information must be learned, but the material covered in these first chapters will set you up for understanding what Employee Central Payroll is capable of and introduce you to the overall scope of this product.

Just as a quick start guide provides us an introduction into what's inside the box, we'll begin with an introduction to payroll with SAP (mostly focusing on Employee Central Payroll) in Section 1.1. Next, we'll walk you through the major payroll processes and how they function in Employee Central Payroll in Section 1.2. Finally, in Section 1.3, we'll introduce you to entity relationship diagrams, which we'll make use of throughout this book.

1.1 Payroll with SAP

Let's first look at the components of an Employee Central Payroll system, as shown in Figure 1.1.

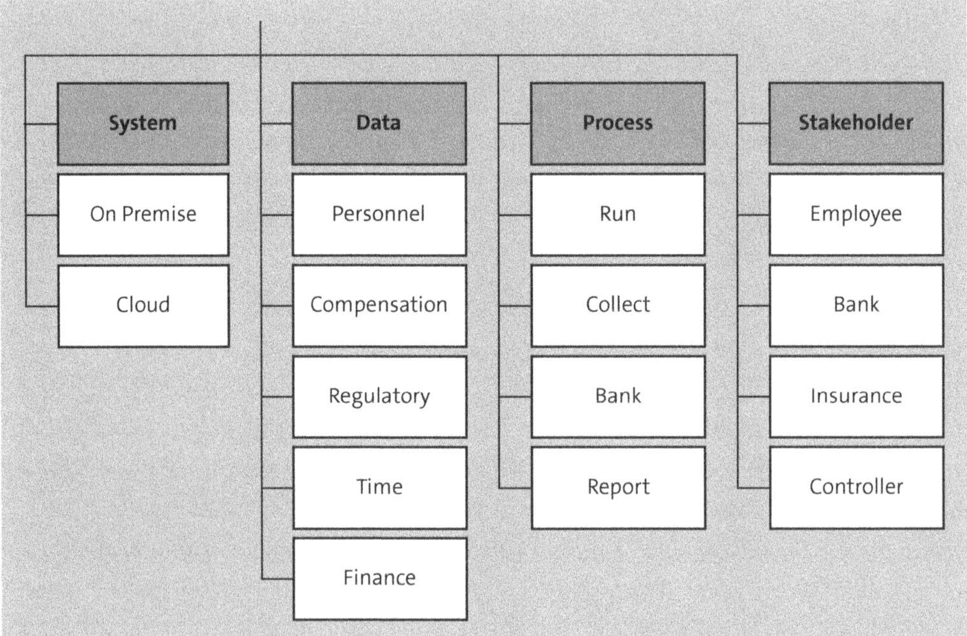

Figure 1.1 Components of the Employee Central Payroll System

In the broad universe of Employee Central Payroll, components can be organized into the following major classifications:

- Systems
- Data
- Processes
- Stakeholders

An Employee Central Payroll system is one central component that we discuss in this book. But other systems play an integral part in reliable payroll processing. While an Employee Central Payroll system is in the cloud, other supporting systems may be on-premise or cloud based. These systems process the data that comes in and goes out of them in an orchestrated rhythm. The data is generated through the execution of key business processes that are tightly linked with the employee lifecycle and involve several stakeholders who execute these processes and either generate or consume this data. This view of the payroll universe is further explained in the next few pages.

If you're running payroll with SAP, your system can be on-premise or cloud based. The system is where the software actually runs.

An on-premise system is a system that is fully available in the client's technology infrastructure and is surrounded by a firewall. This system is quite secure because no one else outside the business has access to it. SAP provides the software installed on the purchaser's own hardware. SAP will provide assistance to install it, as will other partners who deal with implementing this software. The on-premise version can go out of date because of new versions that SAP releases. SAP provides an update every time there is a legislation change as well as at half-yearly intervals. The SAP software release page provides detailed guidance on all patches and service packs that have been released and are planned for release in the near future. When a new release needs to be installed, SAP makes it available on their servers. A purchaser of an on-premise system will need to download these updates and apply the releases themselves to their version of the software. Thus, the system requires regular maintenance from both a hardware and software perspective. On the hardware side, a server has disk drives, CPU chips, and memory components that are subject to normal wear and tear like any other electronic component. This degradation will need to be managed by the buyer. Similarly, some software tasks, such as emptying log tables and ensuring that the product is being used according to corporate guidelines, will also fall under the purview of tasks that the buyer needs to perform—any assistance from SAP or its partners is on a request basis and typically at an additional cost. The on-premise solution for payroll that SAP offers is SAP ERP HCM Payroll.

A cloud-based system, on the other hand, can be fully provided by SAP. The technology infrastructure is SAP owned and SAP managed. This system is protected so that only the buyer and SAP have access to it. SAP installs the software for the purchaser on SAP's own hardware. This version can also go out of date because of new versions that SAP releases. SAP provides an update every time a legislation change occurs as well as in half-yearly intervals. Since this system is the same code set as installed by on-premise purchasers (with some additional cloud-enabled web integration capabilities) the SAP software release page provides the same detailed guidance of all patches and service packs that have been released and are planned for release in the near future. When a new release needs to be installed, SAP makes it available on their servers. The purchaser of a cloud-based system doesn't need to download it or apply the release themselves to their version of the software. They will need to request the application of the software through a ticket to SAP. However, testing- and data-related activities are the responsibility of the buyer. System maintenance from a hardware perspective, such as server disk drives, CPU chips, and memory components, are all SAP's responsibility. Most software tasks, such as emptying log tables and ensuring that the product is being used according to corporate guidelines, are still the responsibility of the buyer—any assistance from SAP or its partners is available on request, typically at an additional cost.

Employee Central Payroll is entirely cloud based. The role that SAP plays is clearly defined in a document called the Employee Central Payroll – Service Catalog, available at *http://s-prs.co/v538600*. The full document runs many pages, but a short excerpt is shown in Figure 1.2.

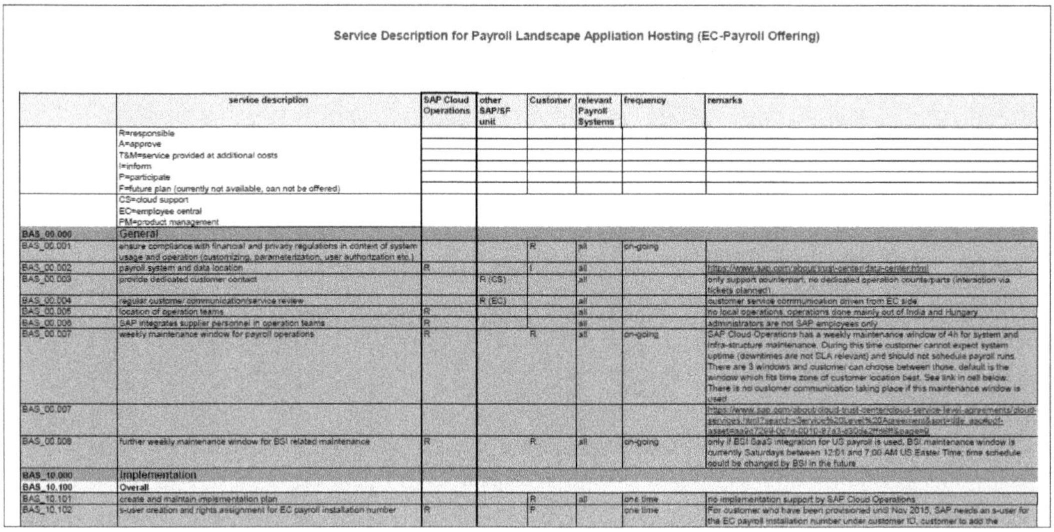

Service Description for Payroll Landscape Appliation Hosting (EC-Payroll Offering)

	service description	SAP Cloud Operations	other SAP/SF unit	Customer	relevant Payroll Systems	frequency	remarks
	R=responsible A=approve T&M=service provided at additional costs I=inform P=participate F=future plan (currently not available, can not be offered)						
	CS=cloud support EC=employee central PM=product management						
BAS_00.000	General						
BAS_00.001	ensure compliance with financial and privacy regulations in context of system usage and operation (customizing, parameterization, user authorization etc.)			R	all	on-going	
BAS_00.002	payroll system and data location	R		I	all		https://www.sap.com/about/trust-center/data-center.html
BAS_00.003	provide dedicated customer contact		R (CS)		all		only support counterpart, no dedicated operation counterparts (interaction via tickets planned)
BAS_00.004	regular customer communication/service review		R (EC)		all		customer service communication driven from EC side
BAS_00.005	location of operation teams	R			all		no local operations, operations done mainly out of India and Hungary
BAS_00.006	SAP integrates supplier personnel in operation teams	R			all		administrators are not SAP employees only
BAS_00.007	weekly maintenance window for payroll operations	R		R	all	on-going	SAP Cloud Operations has a weekly maintenance window of 4h for system and infra-structure maintenance. During this time customer cannot expect system uptime (downtimes are not SLA relevant) and should not schedule payroll runs. There are 3 windows and customer can choose between those, default is the window which fits time zone of customer location best. See link in cell below. There is no customer communication taking place if this maintenance window is used
BAS_00.007							https://www.sap.com/about/cloud-trust-center/cloud-service-level-agreements/cloud-services.html?search=Service%20Level%20Agreement&sort=title_asc#pdf-asset=aa9c7299-0c7d-0010-87a3-c30de2ffd8ff&page=0
BAS_00.008	further weekly mainenance window for BSI related maintenance	R		R	all	on-going	only if BSI SaaS integration for US payroll is used, BSI maintenance window is currently Saturdays between 12:01 and 7:00 AM US Easter Time; time schedule could be changed by BSI in the future
BAS_10.000	Implementation						
BAS_10.100	Overall						
BAS_10.101	create and maintain implementation plan			R	all	one time	no implementation support by SAP Cloud Operations
BAS_10.102	s-user creation and rights assignment for EC payroll installation number	R		P		one time	For customer who have been provisioned until Nov 2015, SAP needs an s-user for the EC payroll installation number under customer ID, customer to add the

Figure 1.2 SAP's Responsibilities for Employee Central Payroll

Employee Central Payroll is a fully developed system available and configurable for 46 countries. In general, the currency and language for each available country are also incorporated into the solution. However, Employee Central is available in many more countries (99 in total). The list of countries for both systems is shown in Table 1.1. Note, however, that this list is subject to change, and for the latest list, always refer to SAP's official documentation.

Countries	Employee Central	Employee Central Payroll
Angola	√	
Argentina	√	√
Australia	√	√
Austria	√	√
Bangladesh	√	
Belarus	√	
Belgium	√	√
Botswana	√	
Brazil	√	√
Bulgaria	√	
Cambodia	√	

Table 1.1 Countries in Scope for Employee Central and Employee Central Payroll

Countries	Employee Central	Employee Central Payroll
Canada	√	√
Chile	√	√
China	√	√
Colombia	√	√
Costa Rica	√	
Croatia	√	
Czech Republic	√	√
Denmark	√	
Dominican Republic	√	
Ecuador	√	
Egypt	√	√
El Salvador	√	
Estonia	√	
Finland	√	√
France	√	√
Germany	√	√
Greece	√	
Guatemala	√	
Honduras	√	
Hong Kong	√	√
Hungary	√	√
Iceland	√	
India	√	√
Indonesia	√	√
Iraq	√	
Ireland	√	√

Table 1.1 Countries in Scope for Employee Central and Employee Central Payroll (Cont.)

Countries	Employee Central	Employee Central Payroll
Israel	√	
Italy	√	√
Japan	√	√
Jordan	√	
Kazakhstan	√	
Kenya	√	
Kosovo	√	
Kuwait	√	√
Latvia	√	
Lebanon	√	
Lithuania	√	
Luxembourg	√	
Malawi	√	
Malaysia	√	√
Malta	√	
Mexico	√	√
Morocco	√	
Mozambique	√	
Myanmar	√	
Namibia	√	
Netherlands	√	√
New Zealand	√	√
Nicaragua	√	
Nigeria	√	
North Macedonia	√	
Norway	√	√

Table 1.1 Countries in Scope for Employee Central and Employee Central Payroll (Cont.)

Countries	Employee Central	Employee Central Payroll
Oman	√	√
Pakistan	√	
Panama	√	
Peru	√	
Philippines	√	√
Poland	√	√
Portugal	√	√
Puerto Rico	√	
Qatar	√	√
Romania	√	
Russia	√	√
Saudi Arabia	√	√
Serbia	√	
Singapore	√	√
Slovakia	√	
Slovenia	√	
South Africa	√	√
South Korea	√	√
South Sudan	√	
Spain	√	√
Sri Lanka	√	
Sweden	√	√
Switzerland	√	√
Taiwan	√	√
Tanzania	√	
Thailand	√	√

Table 1.1 Countries in Scope for Employee Central and Employee Central Payroll (Cont.)

Countries	Employee Central	Employee Central Payroll
Turkey	√	
Uganda	√	
Ukraine	√	
United Arab Emirates	√	√
United Kingdom	√	√
United States	√	√
Uruguay	√	
Venezuela	√	√
Vietnam	√	
Zimbabwe	√	

Table 1.1 Countries in Scope for Employee Central and Employee Central Payroll (Cont.)

A payroll system relies heavily on data. Payroll happens to be one of the first systems digitalized after the invention of the computer. The reason for this early focus is that the core of payroll is data processing. Calculating pay for a multitude of employees is a perfect task for a computer as the successor of the calculator—both performing quick calculations with data that can be input and output in electronic devices. The following five types of data can be processed by Employee Central Payroll:

- **Personal data**
  Personal data is data about the employee, including where the employee lives, what the employee's full name is, and how many hours the person is supposed to work each week. Other information includes the type of contract (hourly or salaried), the location assigned for work, and the employee's supervisor—all part of the information required to correctly place an employee in your organization. This data is vital for calculating and reporting on employee pay and is typically stored in a human resources (HR) system. For Employee Central Payroll to work, that HR system must be Employee Central.
- **Compensation data**
  Compensation data is all about the contracted rate of pay—the hourly rate for employees paid by the hour or a monthly or annual salary for salaried employees paid a contracted amount regardless of hours worked. Compensation may also include certain incomes that are not periodic. Some companies provide a bonus at the end of the year that may depend on the company's performance as well as the employee's performance. Other business like to provide frequent incentives to motivate their workers and recognize their special efforts at the beginning or the

end of a significant project. Sometimes, these incentives are deferred for a period of time with the condition that they will be paid out only if the employee continues to work for this company for a specified number of years. These incentives are also called *retention bonuses* because they are aimed at retaining talent. Determining the rate of pay or the portion of annual or ad-hoc bonus that should be paid to an employee is part of a compensation process. However, once determined, the actual money owed to the employee is calculated by the payroll system. In practice, for Employee Central Payroll to work, your compensation planning system doesn't have to be SAP SuccessFactors. However, the final rates of pay and bonus amounts must be available in Employee Central or in Employee Central Payroll.

- **Regulatory data**
  Regulatory constraints are placed on employers in the form of taxes and garnishments. For example, in the US, a company must pay unemployment insurance for every person employed. The income tax withheld from the employee's pay often depends on the place of work. These calculations can be complex and sometimes require the identification of the locality and state in which the person is working in as well as the locality and state in which the person resides, which goes beyond just the postal address. This regulatory data must be captured for each employee to derive the appropriate tax deductions and thus derive the net pay accurately. Most of this regulatory data is directly entered into the payroll system since this data is not really relevant for HR processes. However, some features in Employee Central allow for the entry of this data directly into the payroll system from the Employee Central user interface (UI). These features are called *mashups*.

- **Time data**
  Time data is essential for hourly employees to be paid accurately because they are paid by the hours of work they have completed. Time data is typically handled through a different system such as WorkForce Software or Kronos. Complex rules are associated with the evaluation of time data. For example, if an employee works over 8 hours, the employee may be eligible for overtime pay, perhaps time and a half of their normal pay. In some places, if they work over 12 hours, they may be eligible for twice their regular pay. These rules are generally located in the time system (whichever system that may be). The evaluated results are sent across to the payroll system through an interface. The Employee Central Payroll system then uses this time data to work out what the pay should be by applying the rate of pay. Similarly, data related to paid leave, both for hourly and salaried employees, must flow through to Employee Central Payroll if leave data is required to work out the employee's pay. For some companies, the leave data for salaried employees does not affect pay—this data, however, may still be required to be sent to the payroll system so that it can be included on the employee pay slip along with other pay-related data.

- **Finance data**
  Finance data is required to keep the accounting of the payments correct in the company's financial books. This data exists in the form of company codes, cost centers,

profit centers, vendor codes, and general ledger codes. A cost center is a key piece of information at the employee level that determines the profit center. A general ledger code is not at the associate level and is instead derived through configuration, based on the type of employee and the type of payment or deduction. Liabilities are automatically transferred to the finance system, from which the accounts payable system can receive liability information and thus pay the appropriate tax authorities or garnishment agencies. If the finance system is an SAP system (like SAP ERP Financials or SAP S/4HANA Finance), direct interfaces are available that provide the necessary financial master data to the payroll system and refresh this data periodically to keep it current.

1.2 Payroll Processes

The processes that payroll uses can be summarized by the diagram shown in Figure 1.3. This simplified diagram shows four groups of stakeholders—HR, information technology (IT), payroll, and finance. HR is generally responsible for maintaining and stewarding the employee's personal, compensation, and time data. IT is involved in ensuring the system functions correctly and that data flows seamlessly to and from payroll partners from outside the organization for whom payroll is being processed. Payroll is responsible for the periodic processing of payroll, ensuring its accuracy. Finance ensures the financial ledgers are updated accurately to reflect the movement of funds and costs in the company's financial books. More discussion on the high-level steps follows in this section.

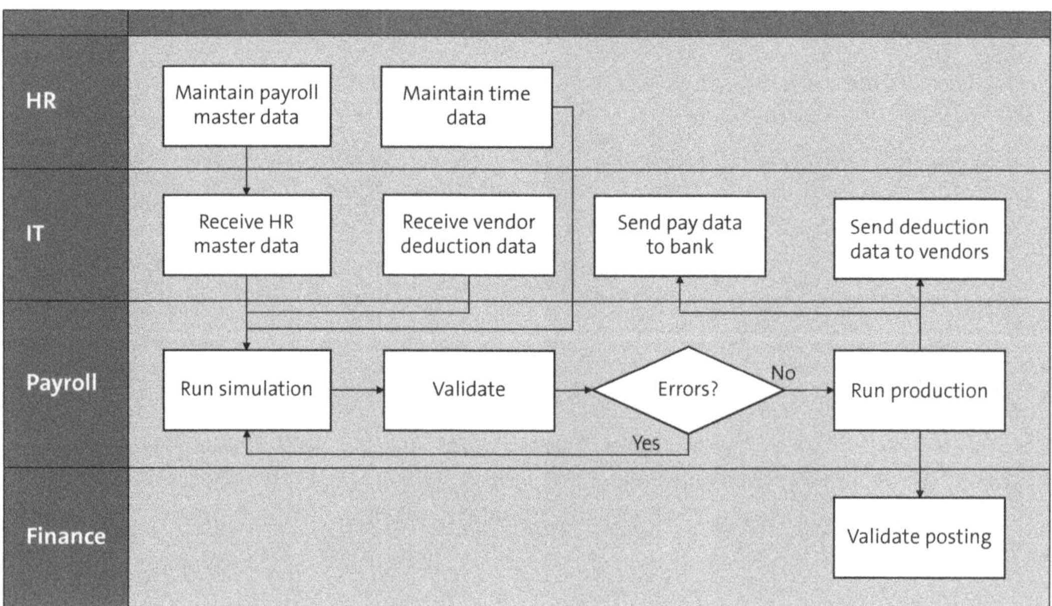

Figure 1.3 The Payroll Process

1.2.1 Receiving Human Resources Master Data

Receiving HR master data is the first step in the chain of payroll processes. A fundamental ingredient for pay computation, this step ensures that all the personal and compensation data captured in Employee Central is transferred to the payroll environment as accurately as possible.

We'll discuss the methods utilized in bringing Employee Central data over to Employee Central Payroll in Chapter 7 and Chapter 8. However, in this section, we want to highlight the concepts of accuracy and timeliness as they relate to this type of data.

An employee's situation in a company changes with time. Salaries go up. Duties change. Supervisors resign and get reassigned. All these events in the employee lifecycle are tracked accurately in the HR system (Employee Central) using a technique called *effective dating*. This technique ensures that any piece of data about an employee, current or historical, has a start and an end date. The history is required for reporting as well as for the determination of various employee benefits and pay components.

This data is sent along to payroll, and the technique of effective dating must be faithfully represented in the payroll system as well. The start and end dates are necessary to calculate retroactive pay adjustments. For example, let's say an employee got a raise from $10 to $15 an hour on March 1, but the approval process takes some time and arrives a month late. If the company is on a biweekly pay cycle, two cycles of pay have been processed in the interim. When this data arrives, the system must be able to calculate the additional money owed to the employee for the 2 months during which the employee was paid at $10 an hour. For this task, the payroll system must retain both the current rate and the past rate of pay.

Employee Central has an "implicit" end date for its data, as shown in Figure 1.4. The end date for the previous record is 1 day less than the start date of the current record. Employee Central Payroll expresses explicit end dates in infotypes, as shown in Figure 1.5.

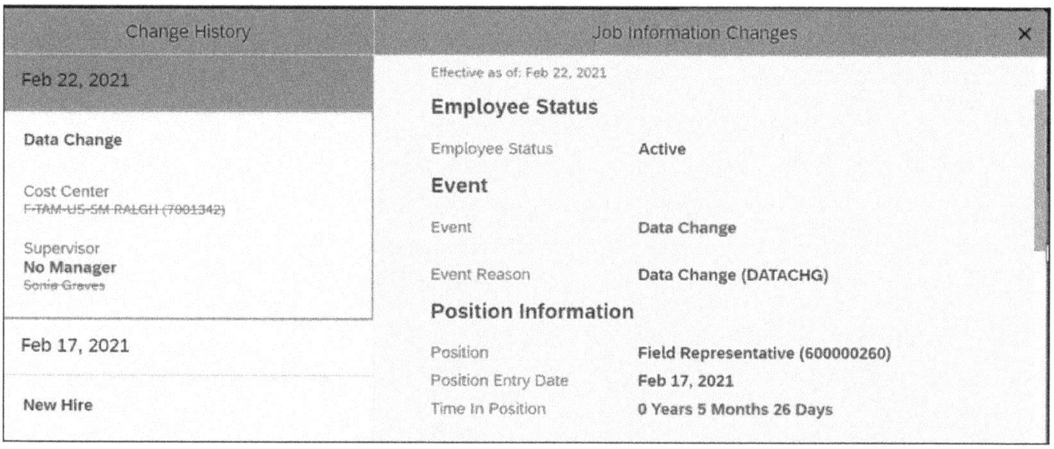

Figure 1.4 Implicit End Date of February 21 in the Employee Central Job Information Portlet

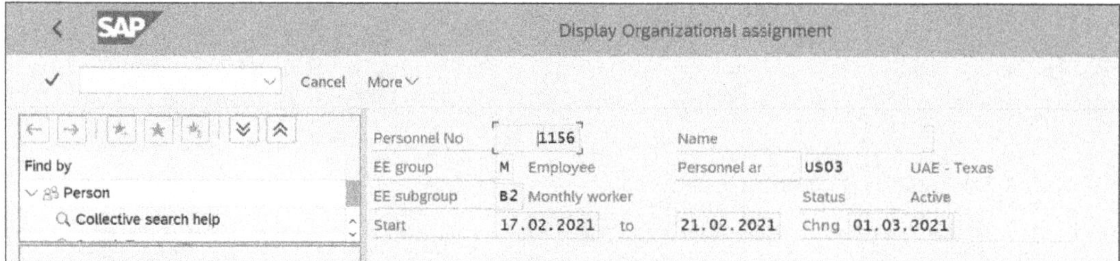

Figure 1.5 Explicit End Date of February 21 in Employee Central Payroll: Infotype 1 Record

When a change is made in Employee Central, the previous record remains, and its end date is implicitly derived from the new record's start date (1 day less than the start date).

Transaction HRSFEC_EE_PTP_REPL allows you to receive HR master data, as shown in Figure 1.6.

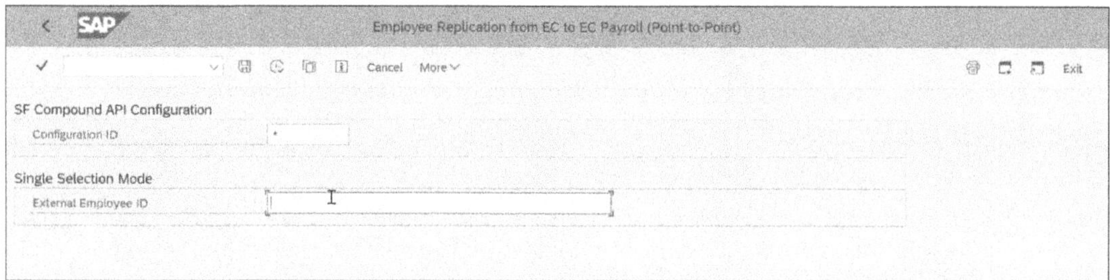

Figure 1.6 Transaction HRSFEC_EE_PTP_REPL in Employee Central Payroll for Receiving HR Data from Employee Central

More information on Transaction HRSFEC_EE_PTP_REPL is available in Chapter 7.

1.2.2 Maintaining Payroll Data

Some data is not usually maintained in Employee Central. Some typical reasons for this difference include the following:

- Tax-related data is not required to complete hiring but is required for payroll.
- Regulations for tax, garnishment, etc. are location specific and can be derived automatically once the employee starts work and the work location is confirmed.
- Regulatory data is not required for HR processes, making it hard to incentivize Employee Central system owners to store and maintain this data.

SAP provides specific infotypes for each country that deals with regulatory data. Table 1.2 provides a list of country-specific infotypes for the United States.

Additional Resources

SAP Help Portal will have the most current list of country-specific infotypes at *http://s-prs.co/v538601*. A list for other countries is also provided at the publisher's website for this book at *https://www.sap-press.com/5386/*.

Country	Infotype	Infotype Description
USA	0048	Residence Status
USA	0057	Membership Fees
USA	0076	Workers' Comp. NA
USA	0082	Additional Abs. Data
USA	0094	Residence Status
USA	0102	Grievances NA
USA	0103	Bond Purchases
USA	0104	Bond Denominations
USA	0161	IRS Limits USA
USA	0170	Flexible Spending Accounts
USA	0172	Flexible Spending Account Claims
USA	0194	Garnishment Document
USA	0195	Garnishment Order
USA	0207	Residence Tax Area
USA	0208	Work Tax Area
USA	0209	Unemployment State
USA	0210	Withholding Info W4/W5 US
USA	0211	COBRA-Qualified Beneficiary
USA	0212	COBRA Health Plans
USA	0216	Garnish. Adjustment
USA	0221	Payroll Results Adjustment
USA	0234	Add. Withh. Info. US
USA	0235	Other Taxes US

Table 1.2 Country-Specific Infotypes for the United States

Country	Infotype	Infotype Description
USA	0269	Not used at present
USA	0270	COBRA Payments
USA	0375	HCE Information
USA	0446	Payroll US Fed Taxes
USA	0447	Payroll US Fed Taxes MTD
USA	0448	Payroll US Fed Taxes QTD
USA	0449	Payroll US Fed Taxes YTD
USA	0450	Payroll US State Taxes
USA	0451	Payroll US State Taxes MTD
USA	0452	Payroll US State Taxes QTD
USA	0453	Payroll US State Taxes YTD
USA	0454	Payroll US Local Taxes
USA	0455	Payroll US Local Taxes MTD
USA	0456	Payroll US Local Taxes QTD
USA	0457	Payroll US Local Taxes YTD
USA	0496	Payroll US Benefits data
USA	0497	Payroll US Benefits data MTD
USA	0498	Payroll US Benefits data QTD
USA	0499	Payroll US Benefits data YTD
USA	0500	Statistical Data
USA	0501	Other Social Insurance Data
USA	0506	Tip Indicators
USA	0509	Activity with Higher Rate of Pay
USA	0510	Tax-Sheltered Pension (US)
USA	0554	Hourly Rate per Assignment
USA	0556	Tax Treaty
USA	0566	US Pension Plan QDRO Information

Table 1.2 Country-Specific Infotypes for the United States (Cont.)

Country	Infotype	Infotype Description
USA	0613	Absence Donation Administration US
USA	0633	EEO Complaint
USA	0656	Nature of Actions
USA	0697	Drug Screening

Table 1.2 Country-Specific Infotypes for the United States (Cont.)

In an Employee Central Payroll system, data maintenance is simple once you know which infotype to use. The transaction code for maintaining an infotype is always Transaction PA30, as shown in Figure 1.7. Use Transaction PA20 for displaying an infotype.

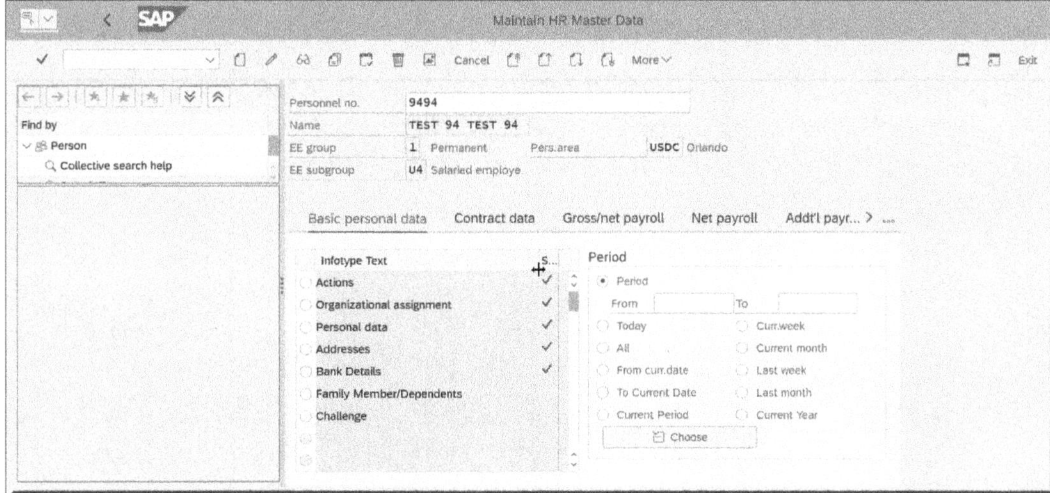

Figure 1.7 Transaction PA30 for Maintaining Infotypes

1.2.3 Maintaining Time Data

In the previous two sections, you learned about the need to set up employees. Although this data may change during the lifecycle of the employee, it does not change regularly. However, another kind of data, time data, is needed on a regular basis to compute pay—this data can be different every day. Time data reflects how long an employee worked. Pay for work follows this simple formula:

$$Net\ Pay = Rate \times Hours - Deductions$$

The rate is part of master data, but the hours consist of time data.

The need for time data to calculate pay varies according to the employment contract. The following three types of employment contracts are common in the business world:

- Hourly employee contracts where pay is based on hours worked
- Salaried employment where the pay is a fixed amount per pay period regardless of hours worked
- A combination of hourly and salaried where a salary is provided for work up to a certain number of hours, but the employee is paid by the hour when a threshold is crossed during a time period

Time data is needed for the first and third scenarios. This time data can be entered in Employee Central but is usually captured outside the SAP system.

Employee Central Payroll allows the use of its own time management functionality. The time management function has two components:

- Time acquisition
- Time evaluation

The standard time acquisition function is not modern in design and lacks a GUI, but the time evaluation function is sophisticated and is capable of evaluating complex rules.

Most businesses using Employee Central Payroll prefer to use third-party time acquisition and evaluation systems. The most popular ones include the following:

- SAP SuccessFactors Employee Central Time Management
- WorkForce Software
- Kronos

Table 1.3 provides a quick comparison of the features available in these three systems.

Features	SAP SuccessFactors Employee Central Time Management	WorkForce Software	Kronos
Evaluation of straight time versus overtime	Yes	Yes	Yes
Complex rules	No	Yes	Yes
Leave tracking	Yes	Yes	Yes
Overtime equalization	No	Yes	Yes
Fatigue management	No	Yes	Yes
Scheduling	Not available	Yes	Yes
HR data import	Not required	Required	Required
Standard integration with Employee Central Payroll	Yes	Yes	Yes

Table 1.3 Quick Comparison of the Features Available in Three Time Management Systems

The time data imported from the time system goes into infotype 2010, as shown in Figure 1.8. The **Wage type** field determines the type of calculation required (straight time, overtime, etc.). The hours are already evaluated and reflected in the wage type that comes through to Employee Central Payroll. The **Date** field is an important field in this infotype—this date is the date the work was done. Employee Central Payroll will use this date to identify the rate of pay that is effective for that date.

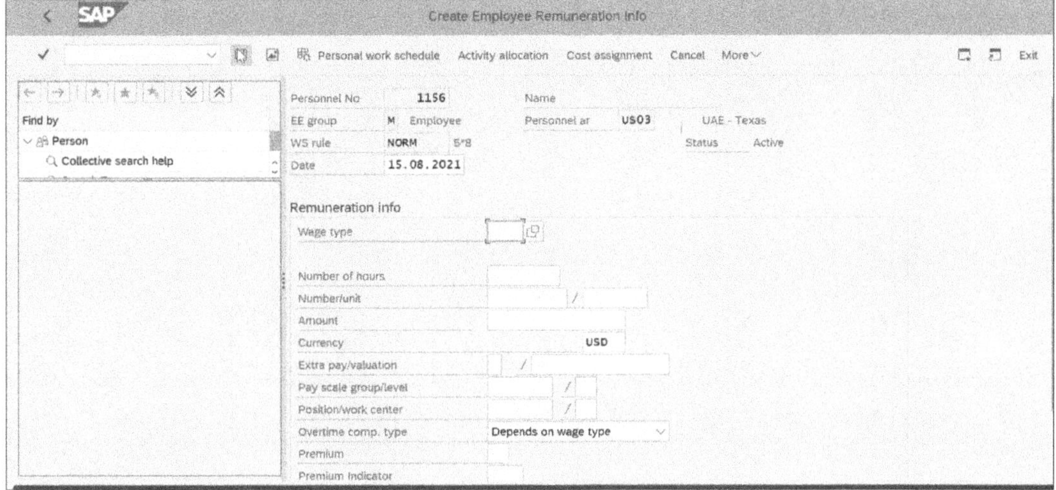

Figure 1.8 Infotype 2010

The integration for time data uses IDocs—a standard messaging function that SAP provides for all external systems. To bring this data into Employee Central Payroll, typically two steps are performed:

- Export of the data from the time system after all data is entered and approved
- Import of the data into SAP using a middleware such as SAP Cloud Integration (more information is available on this topic in Chapter 6)

1.2.4 Receiving Vendor Deduction Data

This step is for retrieving deduction data from various vendors that provide benefits and other services to the employee. A good example of this scenario is healthcare. Many employers offer health insurance to their employees at a reduced rate. This rate is dependent on the specific choices the employee makes, which is controlled through a vendor portal that employees can access. But the deduction amount is not known to the payroll system and therefore must be communicated in advance of running payroll.

The interfaces that bring in vendor deduction data are usually custom built, since each vendor can have a different system that processes these amounts. However, on the Employee Central Payroll side, this data is typically imported into infotype 0014

(Recurring Deduction) or infotype 0015 (One-Time Deduction), as shown in Figure 1.9 and Figure 1.10, respectively. Chapter 6, Section 6.3 on integration discusses a potential strategy where custom development may be avoided for processing benefits.

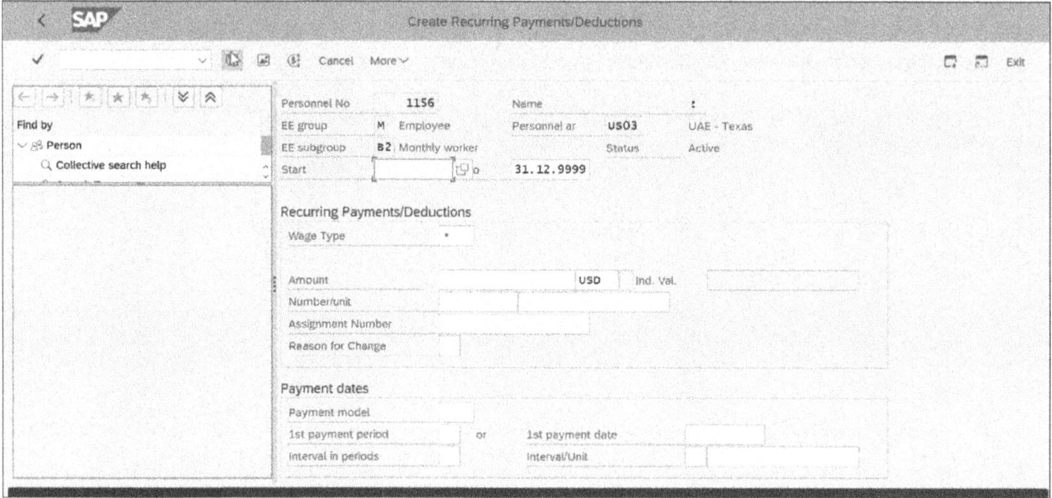

Figure 1.9 Infotype 0014

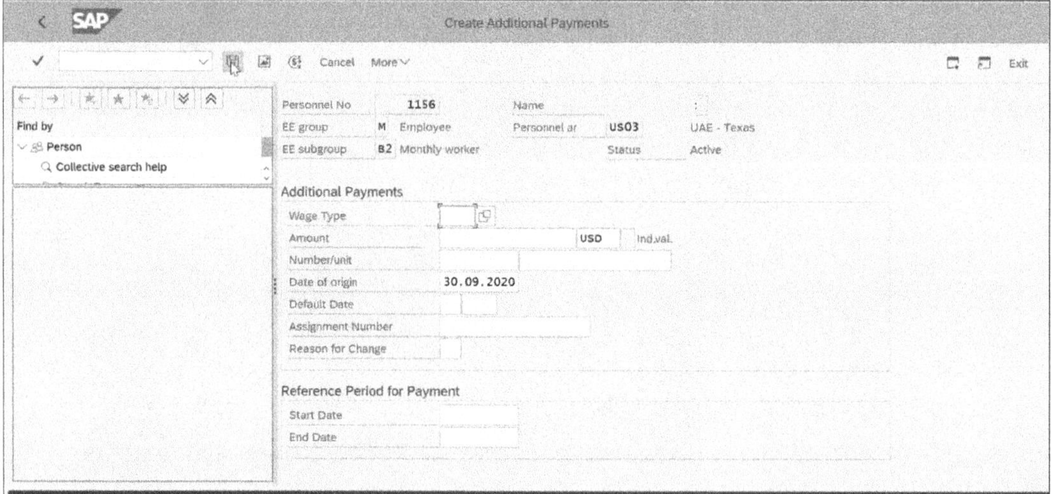

Figure 1.10 Infotype 0015

1.2.5 Running Simulations

In this step, the "rubber meets the road" for the first time. Using a car manufacturing analogy, we've assembled all the parts including the gear box and engine, painted the doors, put on the tires, and set it on the road for the first time for a "trial" run. A simulation is a trial run of the payroll engine. We run simulations to determine if any errors need to be fixed before the final execution. A simulation payroll can be run anytime—before or after the final pay run. You might wonder what's the point of running a simulation after running the actual payroll. In some testing situations, it is helpful to see what the system would have calculated if the configuration was corrected, to confirm that the system will work correctly going forward. Of course, no change can be made to a previously run payroll—these are protected and can be only reported on, not recalculated. Changes that are required to be made for an employee's past pay, for instance, due to inaccuracy or untimeliness of data, are known as *retroactive changes*. Employee Central Payroll is extremely proficient in sensing and recalculating a retroactive payroll situation, generally by conducting these steps:

1. Calculating the past pay period with the corrected data
2. Retrieving the data that was generated in the original pay period
3. Calculating the difference between the two amounts
4. Paying the difference in the next pay period

1.2.6 Validating the Simulation Results

The validation step is about reviewing the payroll simulation results as well as running other checks in the system in preparation for the final run.

The simulation run discussed in the previous subsection can be scheduled to run every day so that the payroll team has ample time to review its results. Any errors or warnings are displayed in the run. While a user typically views this result in the Payroll Control Center (discussed in Chapter 9), you can also run a simulation on your own using Transaction PC00_MXX_CALC_SIMU, where XX is the country code (10 for the USA), as shown in Figure 1.11.

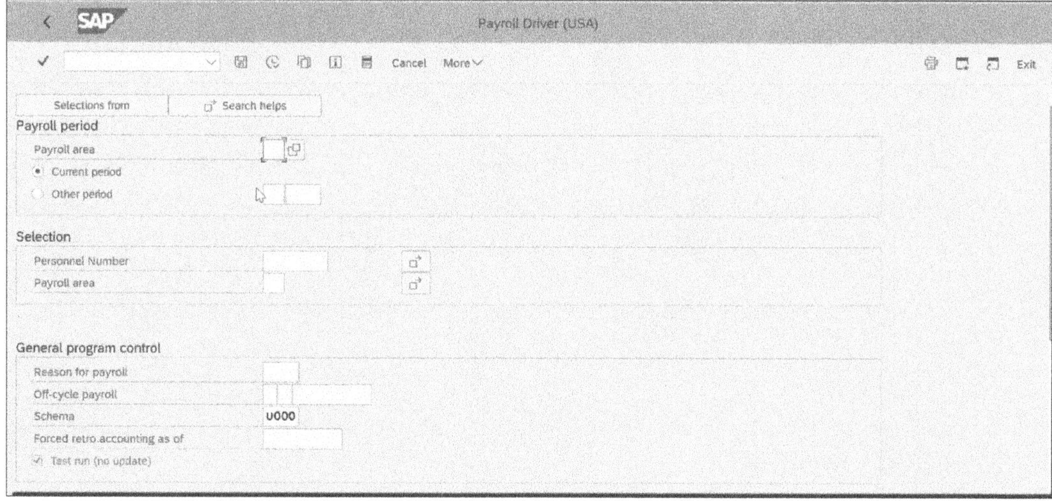

Figure 1.11 Payroll Simulation

The errors produced will highlight configuration settings that might be missing or special situations for data that came in from Employee Central or for data from time or vendor systems that have not yet been integrated with Employee Central Payroll.

The validation process also extends to reviewing errors that may have been produced by the interfaces that bring in Employee Central data, time data, or vendor data. If the data encountered a fatal error in one of those systems, that data would not have entered into Employee Central Payroll—therefore, you cannot expect to see these errors in the error log of the simulation program.

1.2.7 Running Production

In this step, payroll is processed for real. A prerequisite for this step is for the release of the payroll control record for this payroll area. This release is achieved through Transaction PCOO_M99_PA03_RELEA or simply Transaction PA03, as shown in Figure 1.12.

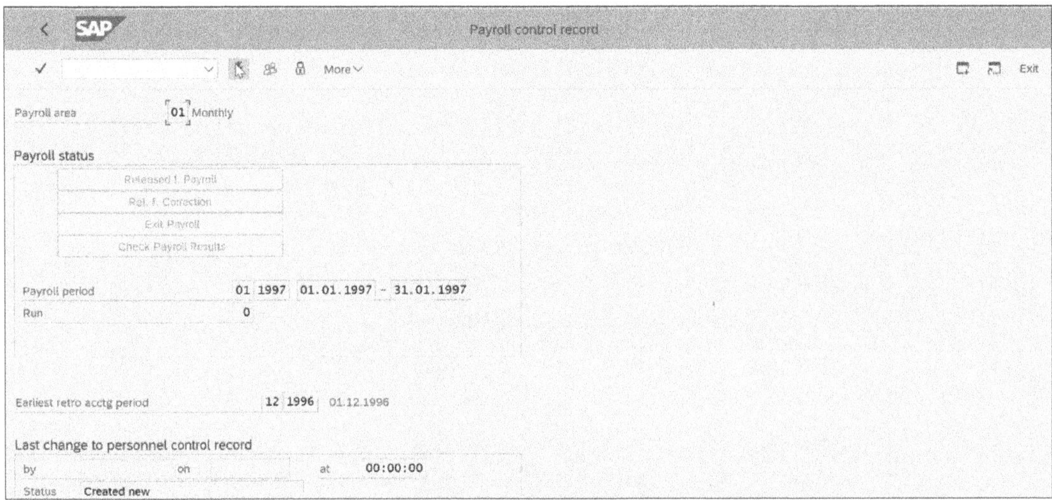

Figure 1.12 Transaction PA03 to Release the Payroll Area for Payroll

Payroll areas are groups of employees whose payroll follows the same process cycle timing. These individuals may have totally different employment contracts (salaried versus hourly, for example), but they all have one thing in common—they get paid for the same time period on the same date. The date of pay is usually referred to as the "check date"—regardless of whether the payment is made by bank transfer or by check.

Payroll is run using Transaction PCOO_MXX_CALC—similar to the simulation transaction, but without "SIMU" in the transaction code. As shown in Figure 1.13, note that this program is the same program as for simulation, but with the **Test run (no update)** checkbox turned off.

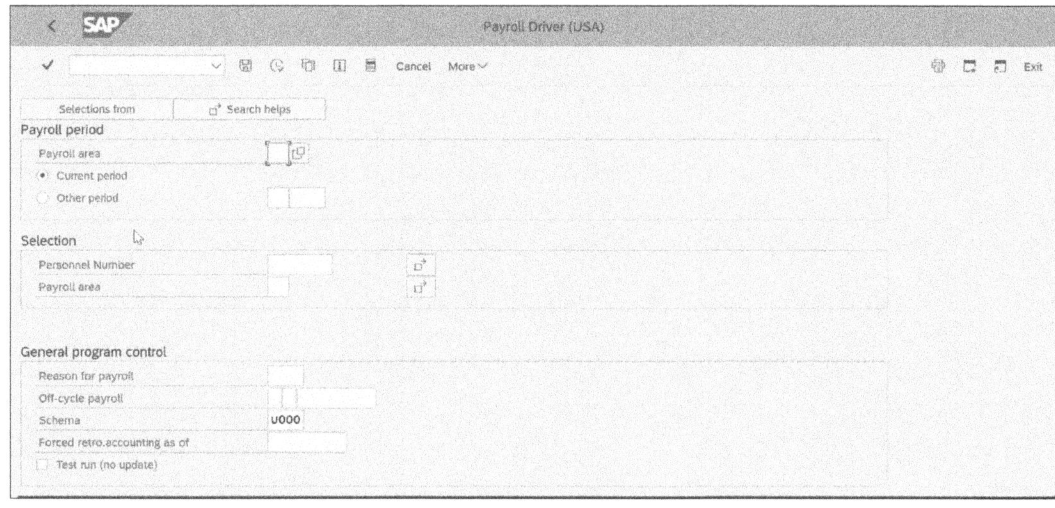

Figure 1.13 Production Payroll Run: Part I

Scroll down (see Figure 1.14) and you will see the **Display log** checkbox, which produces a log of how the computation was performed. You can use the log to investigate specific employees if you suspect an error in calculation. The log shows details about the calculation at every step—gross, taxes, net, etc. However, we recommend turning this log off when running in production or simulation mode in bulk for multiple employees since this log file can become exceptionally large. Overly large log files can potentially delay or even abort the payroll process due to lack of disk space.

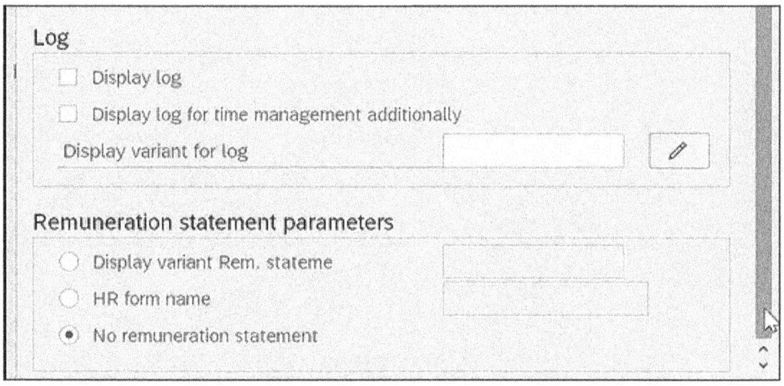

Figure 1.14 Production Payroll Run: Part II

1.2.8 Sending Pay Data to Banks

Two programs are run to prepare data for the bank, namely, Transactions PC00_MXX_CDTC and PC00_M99_FPAYM, as shown in Figure 1.15 and Figure 1.16, which prepare and then conduct Data Medium Exchange (DME) bank payments.

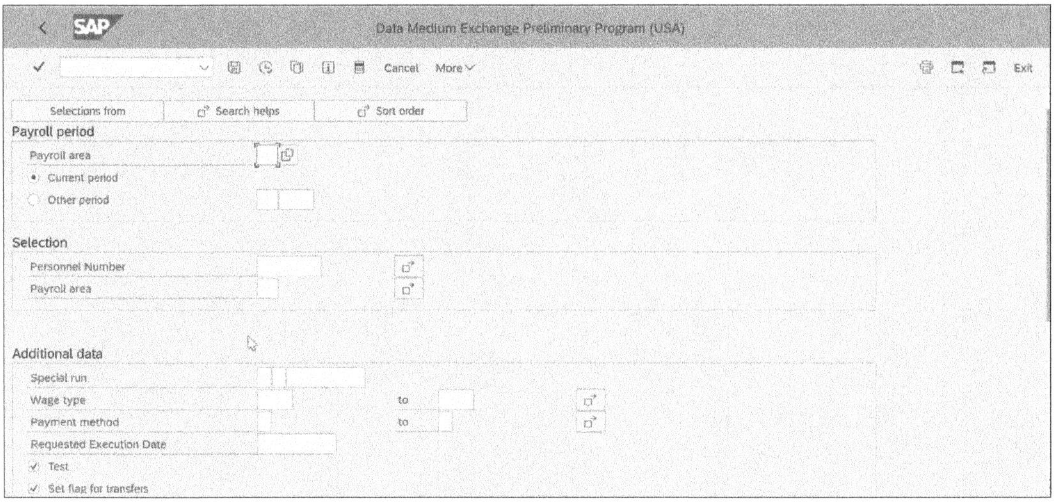

Figure 1.15 Transaction PC00_M10_CDTC before DME Bank Payments

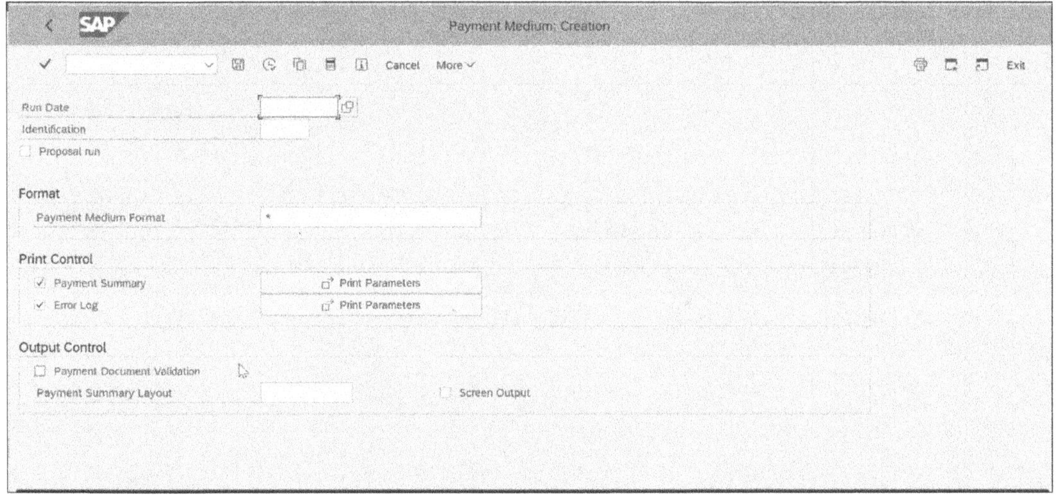

Figure 1.16 Transaction PC00_M99_FPAYM for DME Bank Payments

The final step produces a bank file that can then be transported to the bank servers using standard messaging protocols such as electronic data interchange (EDI) or Secure File Transfer Protocol (SFTP).

The process also generates a payment summary that can be saved for auditing purposes later.

1.2.9 Sending Deduction Data to Vendors

This deduction data is different from the deduction data that was received from vendors before the payroll was run, as described earlier in Section 1.2.4.

A deduction taken in a pay period may not be the same as what was intended. Perhaps, earnings fluctuate from period to period for various reasons, such as hours of work, unpaid leave situations, or other deduction priorities that may reduce the available gross pay to an amount less than the planned deduction. In Employee Central Payroll, during configuration, you can specify whether to take the deduction fully or partially through Transaction OH13, as shown in Figure 1.17. Setting the **Arrear characteristics** field to **2: Deduct what is possible** will deduct what is available.

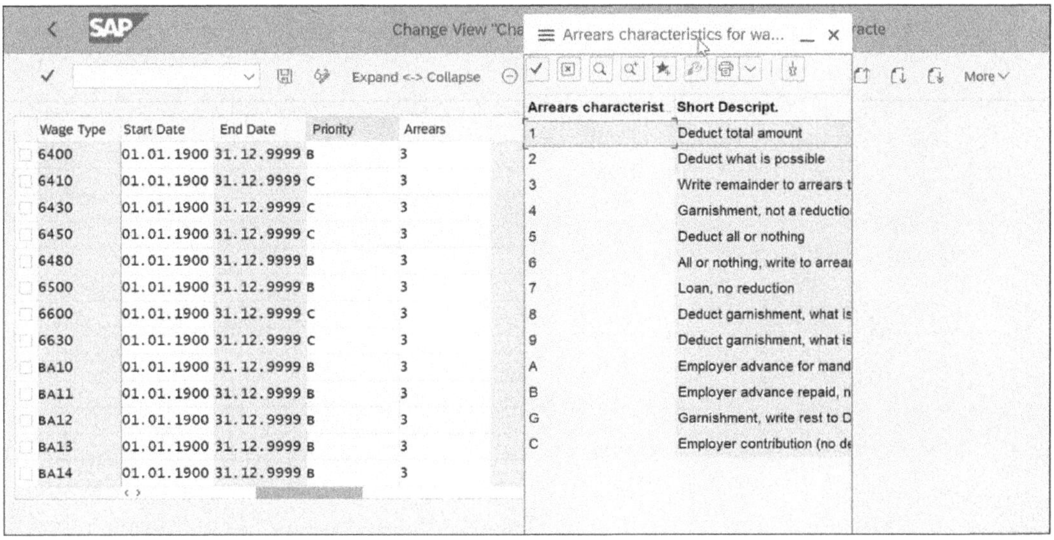

Figure 1.17 Setting Deduction Characteristics in Employee Central Payroll Configuration

1.2.10 Validating Postings

After payroll has been processed and the money has been sent to the bank, the final step to close the payroll cycle is called the *financial posting*. The finance system may or may not be the same system in which payroll is processed. More discussion on integration with finance systems is available in Chapter 6. But if you're connecting to another SAP finance system using Application Link Enabling (ALE), all the validations for account codes, vendor codes, and cost centers are performed in the finance system.

Finance postings are required to achieve two important objectives:

- Reflect the money paid to the associate in the profit and loss (P&L), balance sheet, and costing entities. This process is known as *financial posting*.
- Prepare the amounts collected for payment for remittance to the respective agencies through accounts payable. This process is known as *third-party remittance*.

Financial postings use Transaction PC00_M99_CIPE, as shown in Figure 1.18.

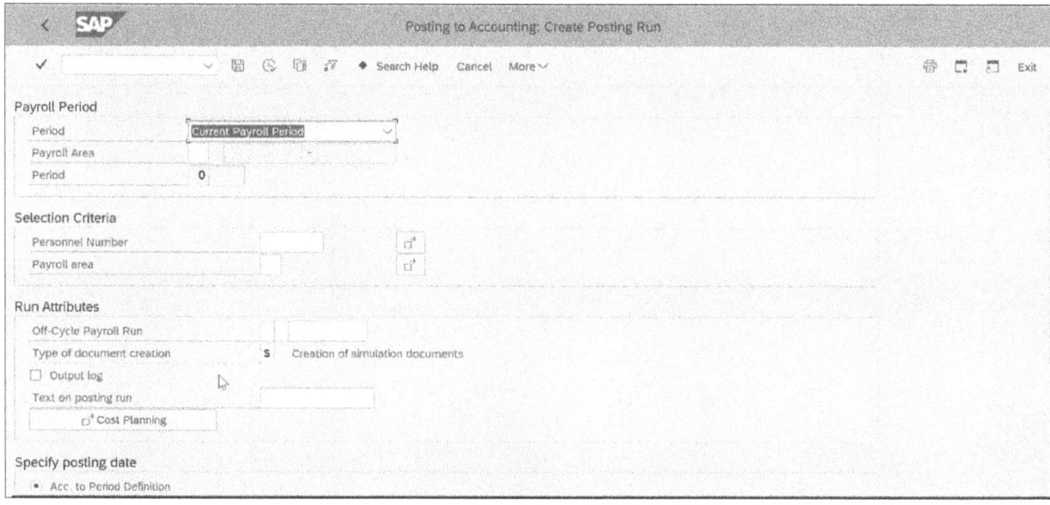

Figure 1.18 Posting to Financial Accounting

A financial posting moves a cost, as shown in Figure 1.19, for instance, for gross wages, which are posted to an expense account. Employee taxes are posted to a liability account, ready to be moved into accounts payable for payment.

Employer taxes have both expense and liability account postings. Furthermore, the net pay is posted to the cash liability account ready for payment—this amount will be later reconciled with the bank transactions that arise out of treasury to fund the payroll payment.

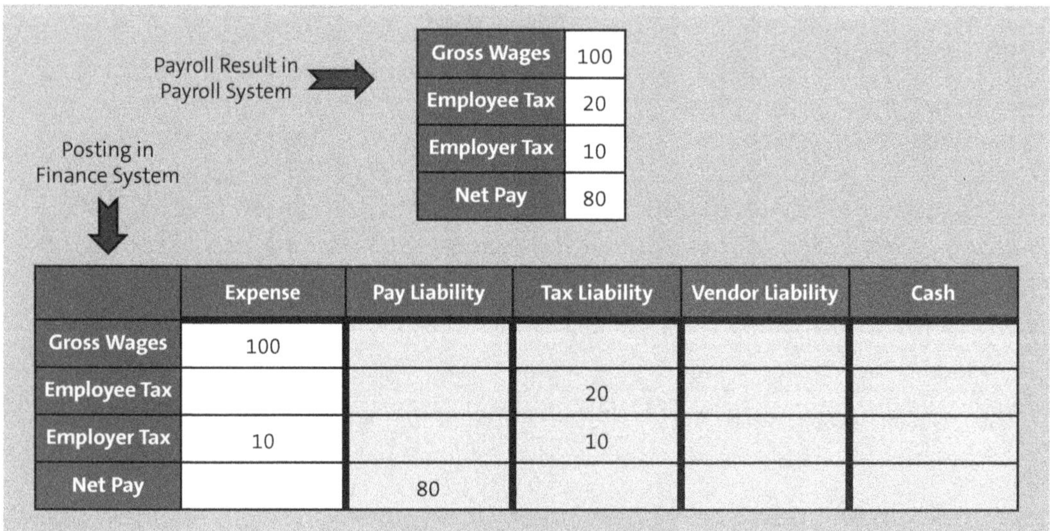

Gross Wages	100
Employee Tax	20
Employer Tax	10
Net Pay	80

	Expense	Pay Liability	Tax Liability	Vendor Liability	Cash
Gross Wages	100				
Employee Tax			20		
Employer Tax	10		10		
Net Pay		80			

Figure 1.19 Posting Payroll Results to Financial Accounting

Third-party remittance is executed using Transactions PC00_M99_URME (Evaluation) and PC00_M99_URMP (Posting), as shown in Figure 1.20 and Figure 1.21, respectively.

Create Third-Party Remittance Evaluation Run
Cancel More
Exit
Payroll Period
Period Current Payroll Period
Payroll Area
Period 0
Selection Criteria
Personnel Number
Payroll area
HR payee type to be posted
HR payee type
Control data
ALV control
Grid view
Tree view
No ALV view
ALV Layout

Figure 1.20 Third-Party Remittance Evaluation Run

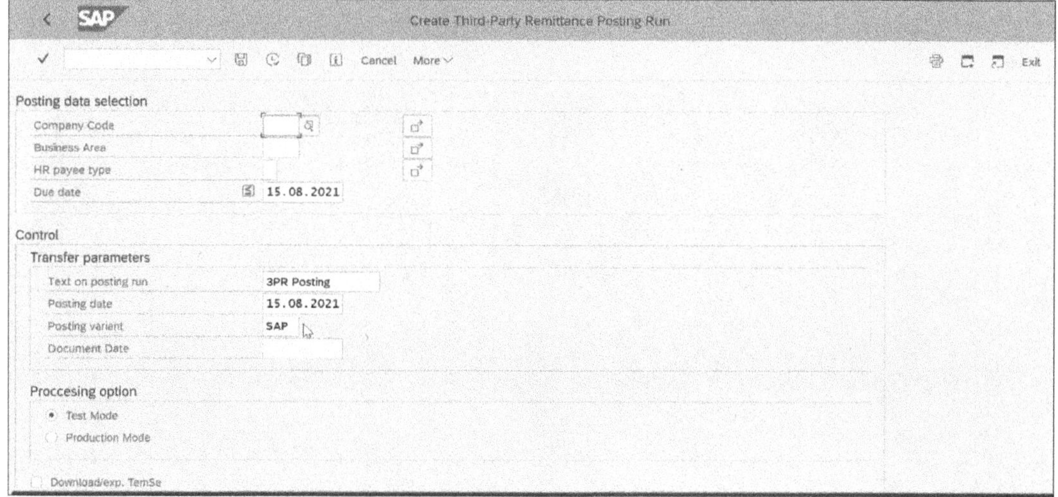

Figure 1.21 Third-Party Remittance Posting Run

Third-party remittance moves the required amounts to vendor accounts, as shown in Figure 1.22. Our simple example only involved one tax vendor—you may deal with many, depending on the country, state, and city where the work is done. Using this program, a posting is made to vendor liability account and removed from the liability account.

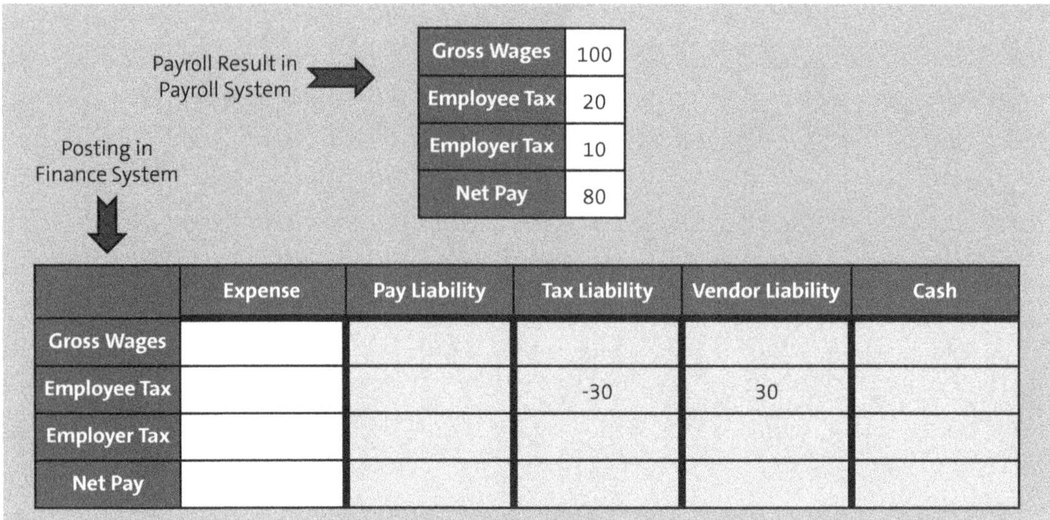

Figure 1.22 Posting Third-Party Remittances

Finally, when accounts payable makes a payment, the vendor accounts are cleared, as shown in Figure 1.23. Accounts payable payments in SAP automatically relieve the vendor liability and debit the cash account.

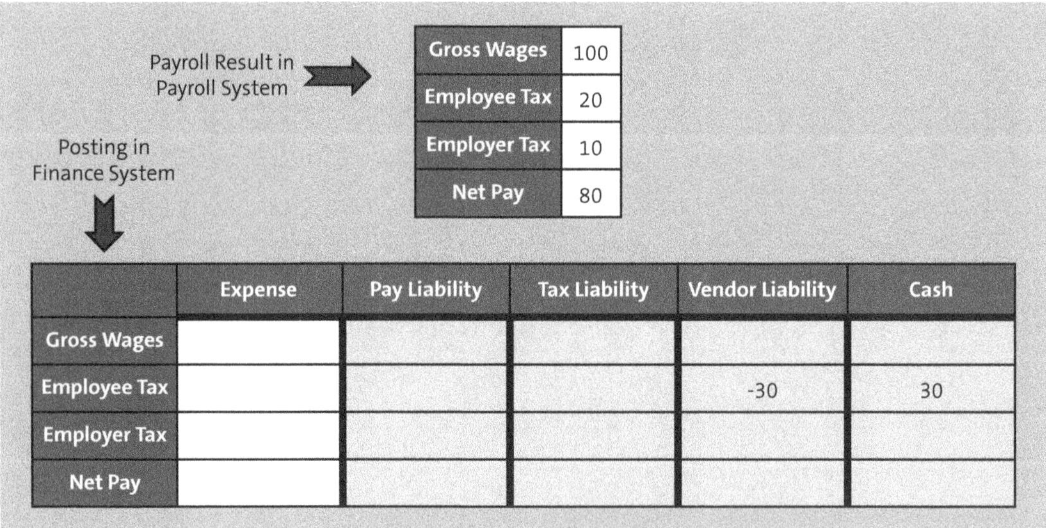

Figure 1.23 Posting Vendor Payments in Accounts Payable

The validation task involves checking the error and warning messages that these programs may produce. A common error is a cost center blocked against direct postings. This block may occur if the cost center is locked from direct postings in the finance system—the system ignores the status of this cost center in Employee Central Payroll.

1.2.11 Reporting Wages

Two popular reports in Employee Central Payroll report wages. The first report is the Wage Type Reporter report, which is initiated with Transaction PC00_M99_CWTR, as shown in Figure 1.24 and Figure 1.25.

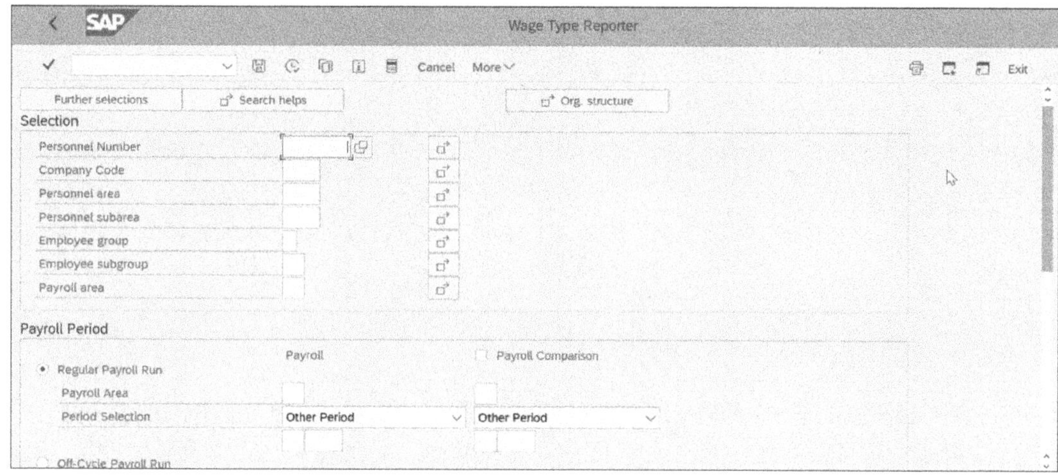

Figure 1.24 Selection Screen for Wage Type Reporter: Part I

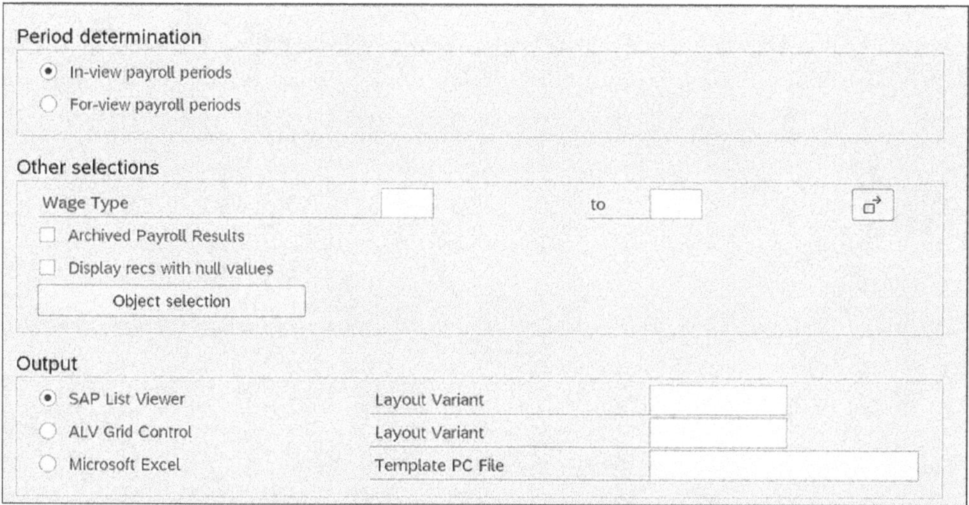

Figure 1.25 Selection Screen for Wage Type Reporter: Part II

The selection screen contains the usual organizational data for identifying the group of employees you want to report on. The fields required to be reported can be selected with the **Object selection** button. The report shows current period as well as year-to-date (YTD), month-to-date (MTD), and quarter-to-date values for each wage type, if the appropriate checkboxes are selected. A sample output is shown in Figure 1.26.

Wage Type Reporter

Pers.No.	Last name First name	ID number	CoCd	PA	PY Area, IP	In-Period	WT	Amount	Crcy
982	TEST, AR OK	100050140	UAE	UAE1	US	202109	/101	3.500,00	USD
982	TEST, AR OK	100050140	UAE	UAE1	US	202109	/102	3.500,00	USD
982	TEST, AR OK	100050140	UAE	UAE1	US	202109	/114	3.500,00	USD
982	TEST, AR OK	100050140	UAE	UAE1	US	202109	/301	10.500,00	USD
982	TEST, AR OK	100050140	UAE	UAE1	US	202109	/303	3.500,00	USD
982	TEST, AR OK	100050140	UAE	UAE1	US	202109	/304	3.500,00	USD
982	TEST, AR OK	100050140	UAE	UAE1	US	202109	/305	3.500,00	USD
982	TEST, AR OK	100050140	UAE	UAE1	US	202109	/306	3.500,00	USD
982	TEST, AR OK	100050140	UAE	UAE1	US	202109	/310	7.000,00	USD
982	TEST, AR OK	100050140	UAE	UAE1	US	202109	/401	599,07	USD
982	TEST, AR OK	100050140	UAE	UAE1	US	202109	/403	217,00	USD
982	TEST, AR OK	100050140	UAE	UAE1	US	202109	/404	217,00	USD
982	TEST, AR OK	100050140	UAE	UAE1	US	202109	/405	50,75	USD
982	TEST, AR OK	100050140	UAE	UAE1	US	202109	/406	50,75	USD
982	TEST, AR OK	100050140	UAE	UAE1	US	202109	/550	2.633,18	USD
982	TEST, AR OK	100050140	UAE	UAE1	US	202109	/559	2.633,18	USD
982	TEST, AR OK	100050140	UAE	UAE1	US	202109	/560	2.633,18	USD
982	TEST, AR OK	100050140	UAE	UAE1	US	202109	/5PY	3.500,00	USD
982	TEST, AR OK	100050140	UAE	UAE1	US	202109	/5U0	866,82	USD
982	TEST, AR OK	100050140	UAE	UAE1	US	202109	/5U1	267,75	USD
982	TEST, AR OK	100050140	UAE	UAE1	US	202109	/5U3	0,00	USD
982	TEST, AR OK	100050140	UAE	UAE1	US	202109	/5UA	3.500,00	USD

Figure 1.26 Sample Wage Type Report

The second report is the Payroll Reconciliation report, accessed via Transaction PC00_MXX_REC. (XX is the country code—e.g., 10 for the USA), as shown in Figure 1.27.

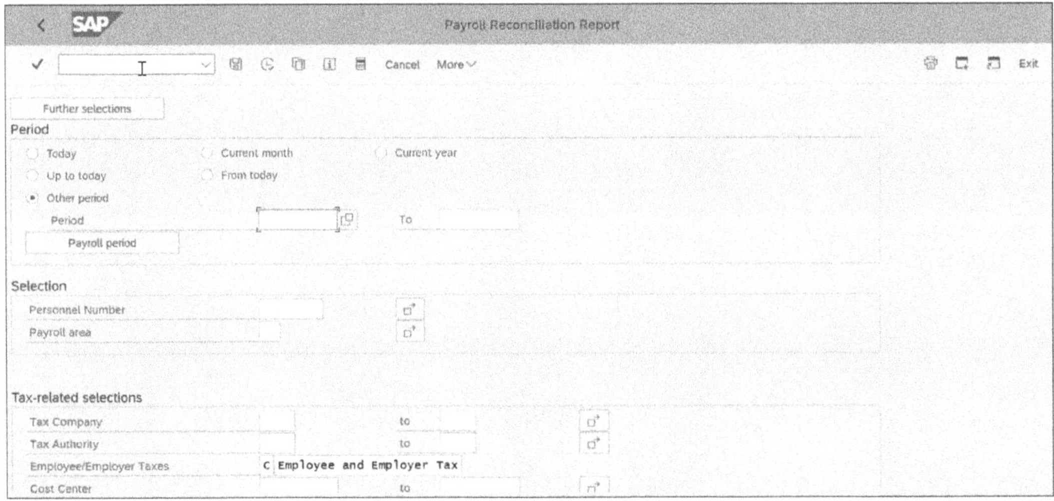

Figure 1.27 Selection Screen for the Payroll Reconciliation Report

This report is similar to the Wage Type Reporter report but with one key difference: This report can identify the tax authorities for which taxes were collected, as shown in Figure 1.28.

Payroll Reconciliation Report

Cancel More Exit

Payroll Reconciliation Report

Selection Period 02.2021 (01.01.2021 - 15.01.2021)
Run Date/Time 15.08.2021 / 20:33:44
Wage Type Selections Wage Type

Tx	Auth	PersNo	Payroll Results	WT	Wage Type Text	Current Pe	Current Pe	Current Pe	Current Pe	M-T-D Numb	M-T-D Amou	Q-T-D Numb	Q-T-D Amou	Y-T-D Numb	Y-T
UAE		970	Normal Period Results	/101	Total gross	02.21-02.21	17.01.2021		3.000,00						3
UAE		970	Normal Period Results	/102	401(k) Wages	02.21-02.21	17.01.2021		3.000,00						3
UAE		970	Normal Period Results	/110	Net payments/Deductions	02.21-02.21	17.01.2021		788,97-						
UAE		970	Normal Period Results	/114	Base wage for BSI	02.21-02.21	17.01.2021		3.000,00						3
UAE	FED	970	Normal Period Results	/301	TG Withholding Tax	02.21-02.21	17.01.2021		3.000,00						3
UAE	IL	970	Normal Period Results	/301	TG Withholding Tax	02.21-02.21	17.01.2021		3.000,00						3
UAE	FED	970	Normal Period Results	/303	TG EE Social Security Tax	02.21-02.21	17.01.2021		3.000,00						3
UAE	FED	970	Normal Period Results	/304	TG ER Social Security Tax	02.21-02.21	17.01.2021		3.000,00						3
UAE	FED	970	Normal Period Results	/305	TG EE Medicare Tax	02.21-02.21	17.01.2021		3.000,00						3
UAE	FED	970	Normal Period Results	/306	TG ER Medicare Tax	02.21-02.21	17.01.2021		3.000,00						3
UAE	FED	970	Normal Period Results	/310	TG ER Unemployment Tax	02.21-02.21	17.01.2021		3.000,00						3

Figure 1.28 Example Payroll Reconciliation Report

1.3 Entity Relationship Diagrams

Throughout this book, we've used entity relationship diagrams to clarify some intricate concepts. Before we jump into the rest of the book, we want to briefly describe these diagrams in case you're not familiar with this kind of diagramming technique:

- An entity is a data item, a programming object, or a process.
- A relationship is the cardinal representation of the association between two "entities."

The entity relationship diagram shown in Figure 1.29 illustrates the associations between different pay data components for an employee to help you understand each item better, as follows:

- An *employee* has many *pay runs*.
- Each *pay run* is linked to a *pay period*.

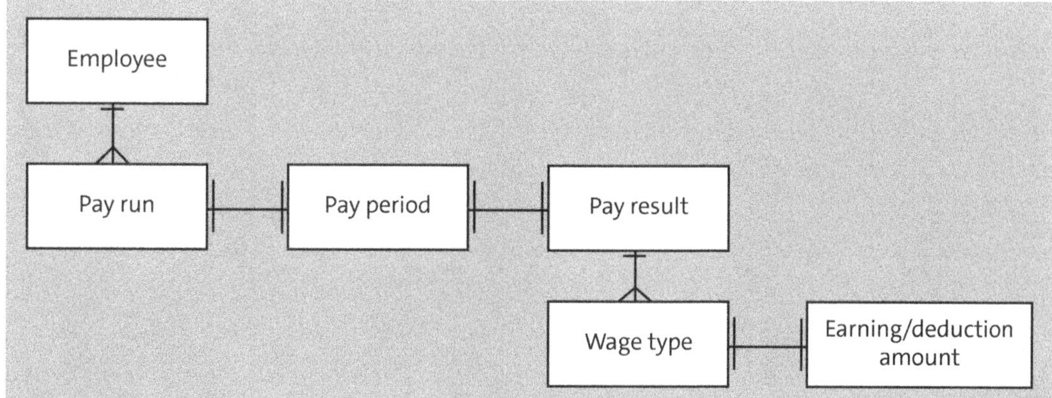

Figure 1.29 Example Entity Relationship Diagram

- A *pay run* produces one *pay result* for an employee.
- A *pay result* has many *wage types*.
- Each *wage type* in a *pay result* has one and only one *earning* or *deduction amount*.

1.4 Summary

This chapter discusses the payroll process and its numerous data and process components. After reading this chapter, you're now ready to discover how these features are configured in the next few chapters.

The simulation process requires special mention. A payroll can be run as many times as needed before finalization. So, in that sense, all the runs before we "exit" payroll are also simulation runs. You might wonder, then, why have two different ways to run a trial payroll? The two methods have their own advantages.

The key difference between a *simulation run* and a *non-final payroll run* is that a simulation run does not store payroll results, while the non-final pay run does. This behavior enables the user to run reports (a reconciliation report or a wage type report) on the intermediate results, which is not possible with a simulation run. On the other hand, a real payroll run requires the payroll control record to be in a **Released** or **Correction** mode for that period. Once exited, the payroll run cannot be run again. However, a simulation run can be performed anytime—regardless of the control record status. Note that the recently released business function (HCM_LOC_CI_110) allows test payroll results to be stored when processed in the Payroll Control Center.

In the next chapter, we'll review how to configure some of the basic payroll features that we discussed here.

Chapter 2
Configuring Payroll Elements

SAP's payroll engine is powerful, but like any good engine, parts and fuel are required to function. The purpose of this chapter is to provide you a good understanding of its different components. You'll learn how to configure the system and set up the parts required for the engine to run. Many of the components we discuss in this chapter are global; however, we'll focus on defining US-specific elements.

In Chapter 1, we introduced you to the payroll process in SAP including the ability to run payroll simulations, conduct live payroll runs, and post payroll activities. For payroll to process, key fundamental configuration items tailored to each business are required. In this chapter, you'll learn about the key payroll elements needed to process payroll—from data elements to the payroll driver. The flexibility and robustness of Employee Central Payroll is what has made it the global leader in payroll products.

This chapter introduces the configuration steps required to process payroll. We'll cover the fundamental elements for building wage types and explore payroll infotypes and how they impact payroll processing. You'll also gain an understanding of how to configure the "engine," including maintaining its rules and schemas as well as other key elements, such as benefits, finance postings, third-party remittances, and Data Medium Exchange (DME). All these elements work together for the payroll process.

2.1 Wage Types

Every payroll system needs earnings and deductions to calculate pay for an employee, and these elements are the foundation of everything in payroll. Earnings are wages earned or provided by the company (salary or bonus) to the employee, and deductions are the income or wages withheld from the employee (taxes or benefits). In this section, we'll delve into wage types and how to configure them to meet your requirements. We'll start with an introduction of the key concepts behind and categories of wage types, but we'll also cover certain attributes of wage types, including processing classes and cumulations. The lifecycle of a wage type and its documentation are also important topics covered in this chapter.

2.1.1 Concepts

For a better understanding of wage types, let's briefly discuss a few basic concepts, as follows:

- **Model wage types**
  SAP-delivered wage types that can be used to copy and create a new wage type. You should never use a standard, predelivered wage type for customer use.
- **Earnings**
  Wages earned or provided by the company for labor or services provided.
- **Post-tax deductions**
  Amount withheld from earnings for purposes of benefits, garnishments, and other purposes after taxes are taken.
- **Pre-tax deductions**
  Amount withheld from earnings for purposes of benefits, and other purposes before taxes are taken. For example, 401k deductions are pre-tax deductions.
- **Taxes**
  A withholding of an amount of money from the employee for the purpose of paying taxes to a tax authority. Could also be employer-paid taxes.
- **Net pay**
  The amount of earnings remaining after all deductions have been taken.
- **Permissibility**
  Controlling access of a wage type by employee type. For example, you may not want an hourly employee to access some benefit plans.

The three elements of a wage type, listed in Table 2.1, work together.

Name	Description
RTE (rate)	Rate (for example, 10.00 dollars per hour)
NUM (number)	Number of hours worked (for example, 40 hours)
AMT (amount)	Amount of pay (for example, 400 dollars)

Table 2.1 Elements of a Wage Type

2.1.2 Categories of Wage Types

Two categories of wage types exist: primary and secondary/technical. Primary wage types are further broken into the following two groups:

- *Dialog wage types* are typically used by infotypes (more information found in Section 2.2) to enter data, for example, hourly rates or benefit deductions. These amounts can be earnings or deductions.
- *Time wage types* are created when a time evaluation is run. These wage types are not maintained using infotypes, but in special tables.

Secondary wage types are system generated and not maintained using infotypes. Instead, these wage types are created during the payroll run.

Technical wage types are delivered by SAP and start with a forward slash (/). Examples include total gross wages (/101) or net pay (/559). Most of these wage types are delivered as standard, but you can also create your own technical wage types, for example, a wage type for total hours worked.

2.1.3 Attributes of Wage Types

To create new wage types, you'll copy model wage types or existing wage types. Then, you'll want to validate/update the attributes of the wage type to meet your business requirements. In this step, you'll set the following parameters:

- Deduction wage type: Select this indicator to specify whether this wage type should be considered a deduction or an earning.
- Amount or number: Specifies if this wage type stores a dollar amount or a number.
- Minimum or maximum values allowed: If needed, you can specify the minimum and maximum allowed values for a wage type.
- Indirect valuation allowed: If applicable, this setting would allow values for this wage type to be calculated dynamically.
- Priority: Define the order in which deductions are taken.
- Arrears: This setting is responsible for determining how payroll will be processed if deductions exceed the net pay.

2.1.4 The Lifecycle

Figure 2.1 shows the end-to-end process of a dialog wage type from wage type creation to after payroll. This diagram will help structure our discussion of wage types and how they are used in payroll.

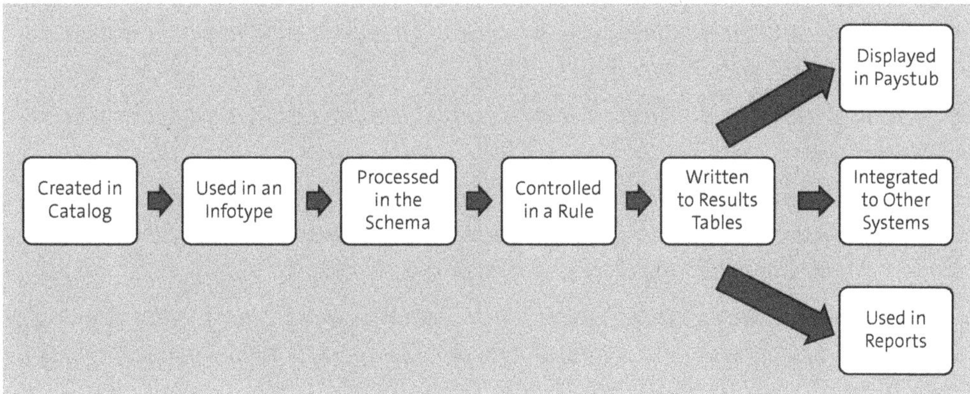

Figure 2.1 Lifecycle of a Wage Type

This lifecycle starts with the documentation of requirements (Section 2.1.8), the copying of a wage type (Section 2.1.7), and either standard processing or updating a schema or rule (for special processing). The result of the payroll processing is a wage type written to the payroll results table, which can then be used on paystubs and reports or sent to third-party systems or vendors.

2.1.5 Processing Classes

Processing classes determine how wage types will be processed during a payroll run. SAP has provided processing classes 1 through 89, which can be set for each wage type, but custom processing classes can be created between processing classes 90 and 99.

Each processing class has a set of specifications that define how a wage type should be handled. Figure 2.2 shows processing class 68 (Payment Type for Tax Calculation), which determines how an earning should be handled in tax processing—regular payment, supplemental, etc. By setting this processing class for all earnings, the payroll processor can tax that earning appropriately.

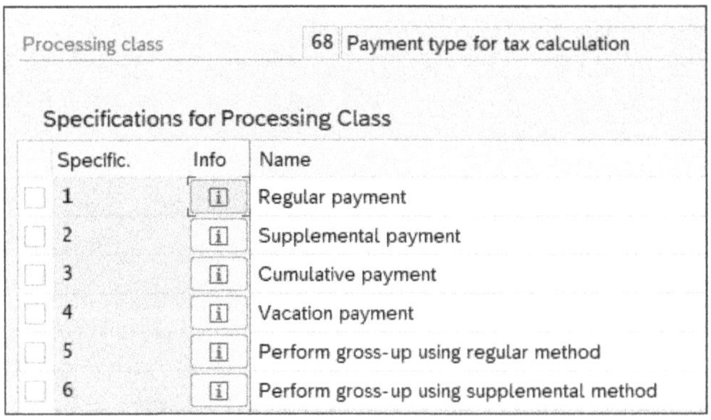

Processing class 68 Payment type for tax calculation

Specifications for Processing Class

Specific.	Info	Name
1	i	Regular payment
2	i	Supplemental payment
3	i	Cumulative payment
4	i	Vacation payment
5	i	Perform gross-up using regular method
6	i	Perform gross-up using supplemental method

Figure 2.2 Processing Class 68

A good way to view all the processing classes set for a wage type is Transaction SE16 and view V_512W_D, as shown in Figure 2.3. Using this report, you can also view the appropriate cumulations and evaluation classes.

In some situations, you may need to create a custom processing class and will need to create one within the range from 90 to 99.

To create a custom processing class, follow the menu path **Payroll • Payroll: USA • Basic Settings • Environment for Maintaining Wage Types • Processing Classes and Evaluation Classes • Maintain processing classes and their specifications.**

After you select **New Entries**, define the number and description for your processing class. Then, click **Save**. In our example shown in Figure 2.4, we've created processing class 99 for total hours worked.

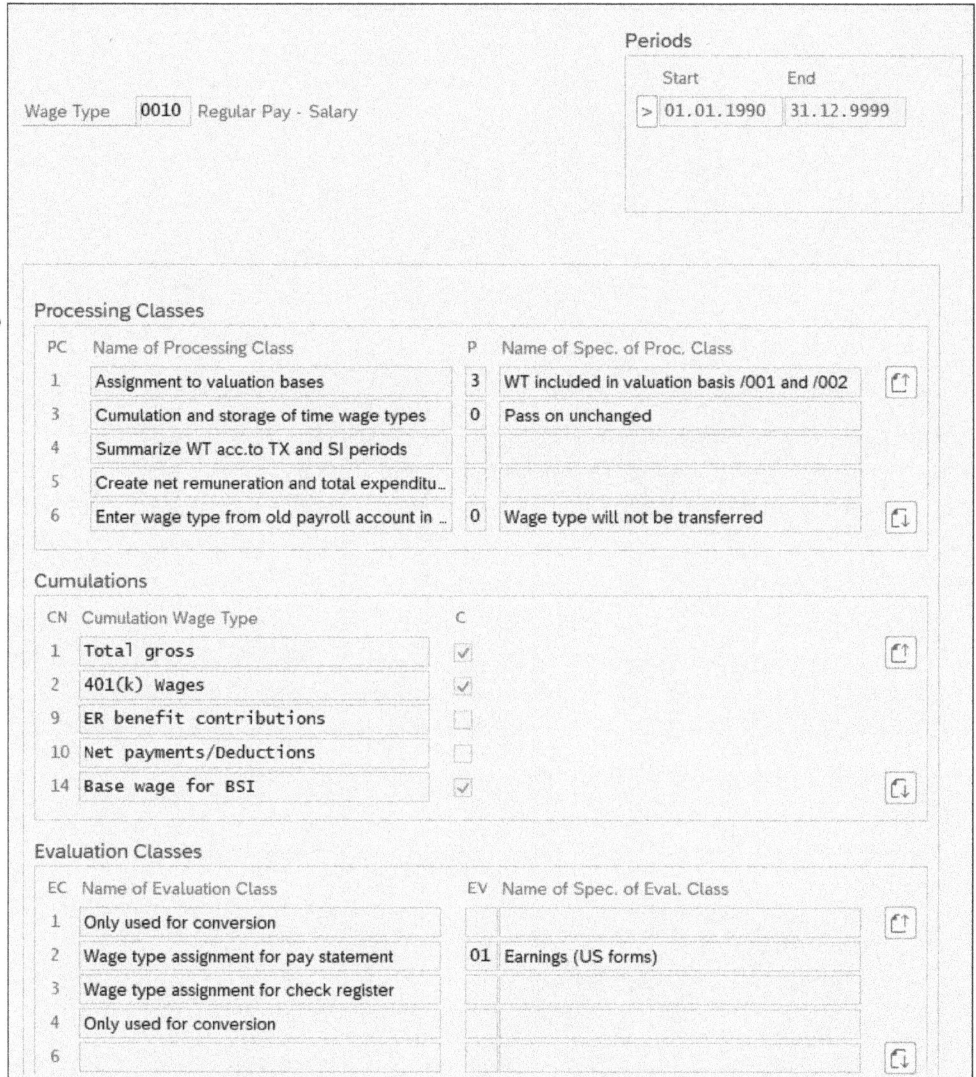

Figure 2.3 View V_512W_D

PCl	Info	Description
99		Total Hours Worked

Figure 2.4 New Processing Class 99

Select your processing class specifications and processing class and then select **New Entries**. You can now define the specifications needed for the new processing class. In our example shown in Figure 2.5, our new processing class has two specifications: **1** for the number of hours as basic worked hours and **2** for the number of hours worked overtime.

Processing class 99

Specifications for Processing Class

Specific.	Info	Name
1		Cumulate the number of hours as basic worked hours
2		Cumulate the number of hours as overtime

Figure 2.5 Processing Class 99: Specifications

Wage types that make up the total hours worked can now use this processing class.

2.1.6 Cumulations

SAP provides the cumulation functionality to sum up several wage types into another wage type. Cumulation wage types are technical wage types and start with a /1. Figure 2.6 shows the cumulation for an earning wage type that should be included in the total gross (/101), 401k wages (/102), and base wage for Business Software Incorporated (BSI) wage types (/114).

A common wage type is /101 for total gross wages. All wage types that should be added to a total gross wage should check for processing class 1. When payroll is processed, /101 is created as a summation of all wage types with this check and can be used for other calculations in schemas or rules or by reports or interfaces.

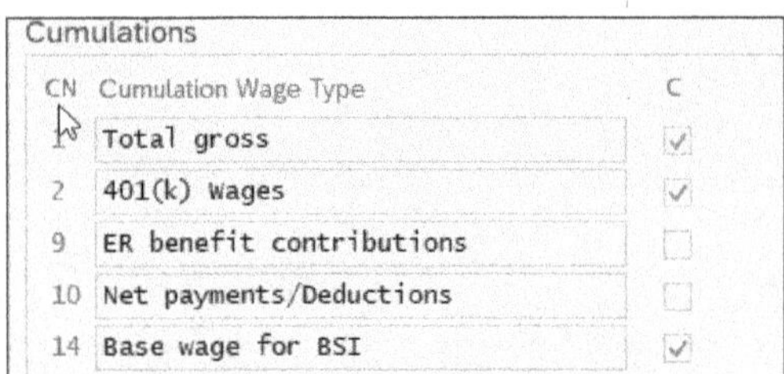

Cumulations

CN	Cumulation Wage Type	C
1	Total gross	✓
2	401(k) wages	✓
9	ER benefit contributions	
10	Net payments/Deductions	
14	Base wage for BSI	✓

Figure 2.6 Sample Wage Type Cumulation Settings

2.1.7 Creating a New Wage Type

Creating wage types is a task you'll need to perform for every project and then as the business changes. Previously, in this section, we discussed several different aspects of a wage type—now let's create a wage type.

As mentioned earlier, you should never use a standard-delivered wage type, nor should you create one from scratch. Instead, copy either a model wage type or another existing wage type, which copies all its attributes. Update this copy as needed by following menu path **Payroll • Payroll: USA • Basic Settings • Environment for Maintaining Wage Types • Create Wage Type Catalog • Copy**.

As shown in Figure 2.7, for example, we're using wage type 0010 (Regular Pay - Salary) to copy to a new wage type (0045), and this new wage type will be used for a different group.

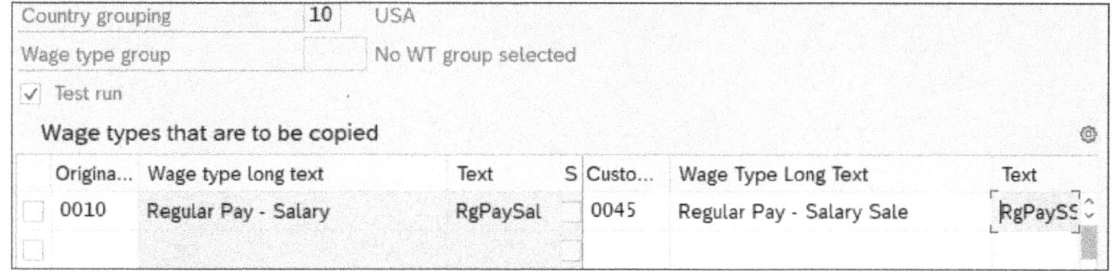

Figure 2.7 Copying a Wage Type

If needed, you can update attributes, processing classes, and permissibility, as needed, but ideally, you should find the closest wage type to what you want to create so only a few updates are required.

2.1.8 Documentation and Numbering

As mentioned earlier, wage types are critical in payroll calculations and require detailed design and documentation to maintain consistency and integrity. Most companies end up with hundreds, even thousands, of custom wage types to meet specific business requirements. Thus, defining and developing great documentation are crucial to the success of the implementation and ongoing maintenance of your Employee Central Payroll system.

The first step in this goal is defining good number ranges. While smart numbering for codes (i.e., having meaningful numbering) is not always recommended, having defined ranges increases clarity. Before you start creating/configuring wage types, you should define your wage type number ranges.

Standard ranges that are used include the following:

- 1*** for earnings
- 2*** for employee deductions
- 3*** for employer contributions
- 8*** for loans and balances
- 9*** for system generated or internal wage types

Table 2.2 shows an example of how you can define number ranges for wage types.

Wage Type Category	Description	Ranges
Earnings	Basic Pay Wage Types on IT 0008	1000 to 1099
Earnings	Bonus Wage Types	1100 to 1199
Earnings	Gross Up	1200 to 1299
Earnings	Time (2010) Related	1400 to 1499

Table 2.2 Sample Ranges for Wage Types

Wage Type Category	Description	Ranges
Earnings	Bonus and Miscellaneous	1500 to 1699
Earnings	Recurring/One Time Earning	1800 to 1899
Earnings	Imputed Income	1900 to 1999
Deductions	Pre-tax Benefit Deductions	2000 to 2099
Deductions	Pre-tax Deductions	2100 to 2199
Deductions	Post-tax Benefit Deductions	2200 to 2399
Deductions	Post-tax Deductions	2400 to 2799
Deductions	Employee Loan	2800 to 2899
Deductions	Garnishments	2900 to 2999
Deductions	Employer Contributions	3000 to 3299
Loans and balances	Balances	8A00 to 8A99
Loans and balances	Total	8B00 to 8B99
Loans and balances	Limits	8L00 to 8L99

Table 2.2 Sample Ranges for Wage Types (Cont.)

Another key form of documentation is the *wage type catalog*. This documentation provides a clear view of the wage types that have been created, where and how they are used, and how they are built. A good catalog includes the following attributes:

- Wage type number, short text, and long text
- Attributes (processing class, evaluation class, cumulation, priority, arrears, deduction flag, min/max, amount/number)
- Date of last update
- Infotype permissibility

Tax wage types are defined by SAP and delivered with a specific structure and format, /XYY, as follows:

- /X describes what type of tax wage type. Following are some of the most common ones:
  - /3XX stands for taxable gross (TG) and represents all wages eligible for that tax type and authority.
  - /6XX stands for taxable base (TB) and represents TG minus all pretax deductions.
  - /7XX stands for reportable earnings (RD) and represents TB with limits applied.
  - /4XX stands for tax (TX) and represents the actual tax.
  - /QXX stands for taxable but not taxed (TNTD) and represents wages eligible for taxation but not actually taxed.

- YY describes the tax, for example, in the following format:
  - 01 for withholding
  - 02 for EE (employee) Social Security
  - 04 for ER (employer) Social Security
  - 10 for unemployment

2.2 Payroll Infotypes

SAP stores data about an employee in infotypes. Infotypes are a group of related data that can then be used in payroll processing. Infotypes can store master data or transactional data and can be created within Employee Central Payroll. Alternatively, infotypes can be created outside and replicated or interfaced from a third-party system. Infotypes are grouped into the following categories:

- Employee personal data, which includes names, addresses, and Social Security Numbers (SSNs)
- Payroll/compensation, including salaries or hourly rates
- Benefits information for health, insurance, and savings plans
- Garnishment documents and deduction information
- Taxes including residence, work, unemployment, and IRS limits
- Time data needed for payroll processing, which can be calculated and stored in SAP or captured outside and interfaced into Employee Central Payroll
- Other data, such as cost overrides

Some infotypes are required for processing US payroll; others are optional. For example, all employees must have an hourly rate or a salary wage type for processing; however, employees are not required to have benefits infotypes. For US payroll to be calculated, Employee Central Payroll requires the following infotypes:

- Infotype 0000 (Actions)
- Infotype 0001 (Organizational Data)
- Infotype 0002 (Personal Data)
- Infotype 0006, subtype 1 (Addresses)
- Infotype 0007 (Planned Working Time)
- Infotype 0008 (Basic Pay)
- Infotype 0009, subtype 0 (Bank Details – Main Bank)
- Infotype 0207 (Residence Tax Area)
- Infotype 0209 (Unemployment Tax Area)
- Infotype 0210 (Fed - Withholding Information (W4))

To display/edit infotypes, you can access Transactions PA20 or PA30 through the command field, as shown in Figure 2.8.

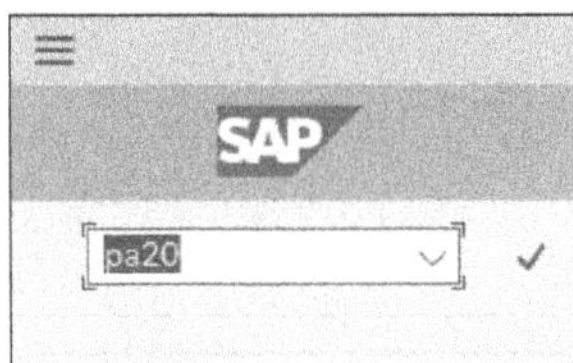

Figure 2.8 Command Field

Through Transaction PA20, you can select a personnel number (the **Personnel no.** field) and the infotype you're interested in viewing (the **Infotype** field at the bottom), as shown in Figure 2.9. Alternatively, you can use Transaction PA30 to make changes.

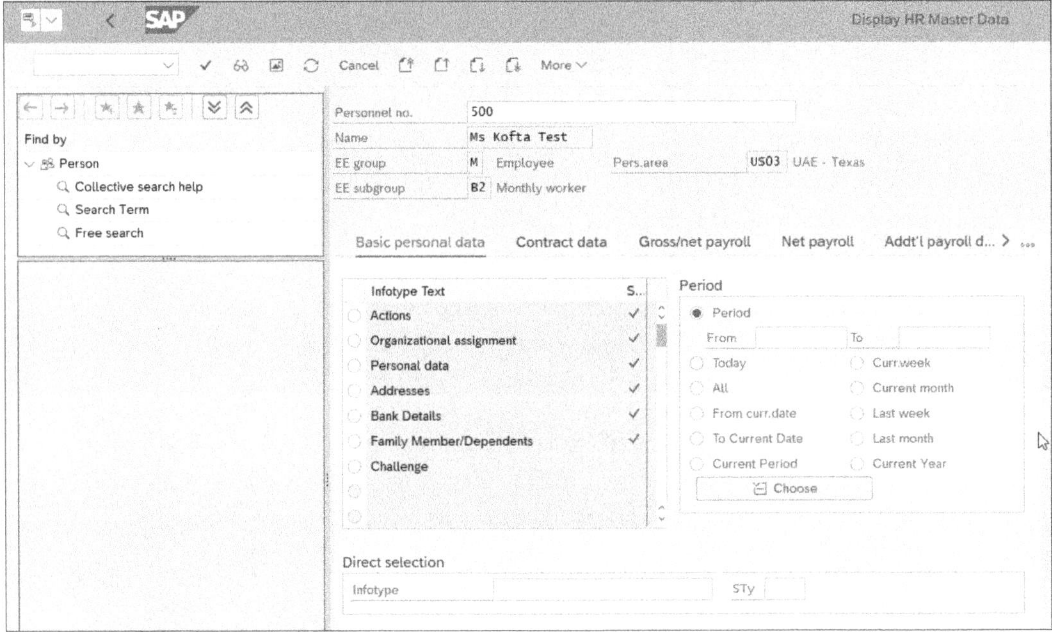

Figure 2.9 Transaction PA20

In this section, we'll cover each category of infotype (personal data, payroll/compensation data, tax data, and benefits data).

2.2.1 Personal Data

Personal data includes data about an employee—name, address, SSN, etc. If the human resource (HR) system is Employee Central, then this data is replicated from Employee Central to be used by Employee Central Payroll. Table 2.3 lists the infotypes typically considered personal data, each of which we'll discuss in the following sections.

Infotype Number	Infotype Description
0000	Actions
0001	Organizational Assignment
0002	Personal Data
0006	Addresses

Table 2.3 Personal Data Infotypes

Infotype 0000: Actions

Let's begin with infotype 0000 (Actions), which is used to store the details about an employee's status with the company and is updated when the employee goes through a personnel change. This infotype is required for processing payroll and is updated whenever a person changes status—a new hire, a change in the organization, a change in pay, a separation, etc. An employee can only have one active infotype 0000 at a time, its status is updated when an applicable change occurs, with an appropriate effective date, as shown in Figure 2.10.

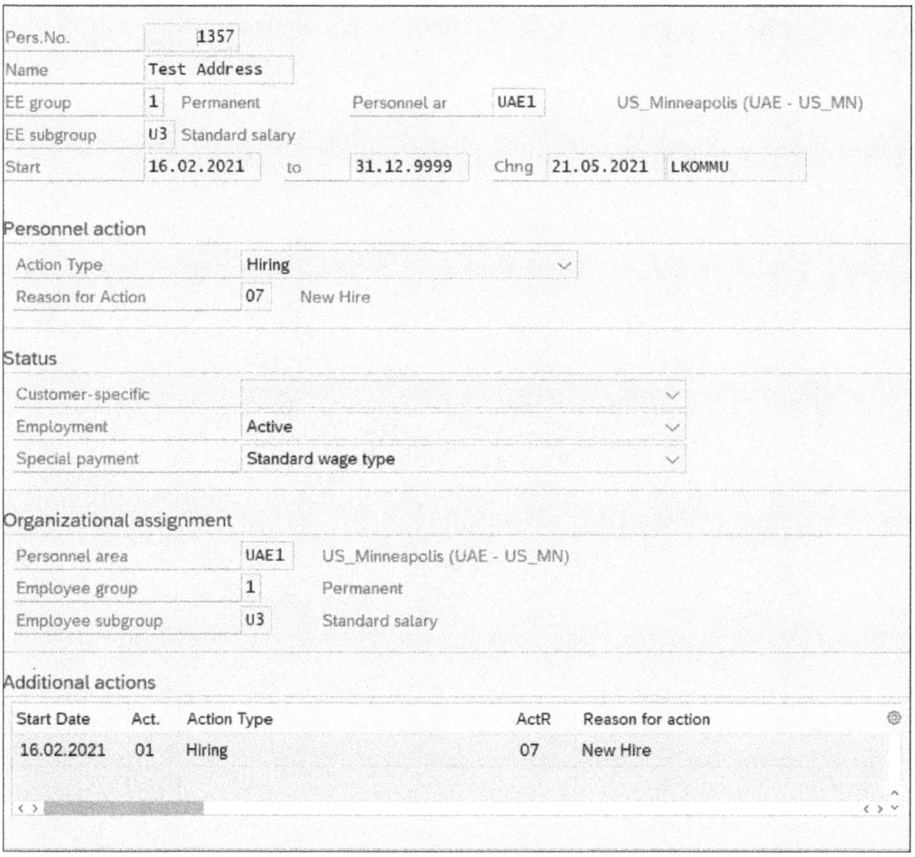

Pers.No. 1357
Name Test Address
EE group 1 Permanent Personnel ar UAE1 US_Minneapolis (UAE - US_MN)
EE subgroup U3 Standard salary
Start 16.02.2021 to 31.12.9999 Chng 21.05.2021 LKOMMU

Personnel action
Action Type Hiring
Reason for Action 07 New Hire

Status
Customer-specific
Employment Active
Special payment Standard wage type

Organizational assignment
Personnel area UAE1 US_Minneapolis (UAE - US_MN)
Employee group 1 Permanent
Employee subgroup U3 Standard salary

Additional actions

Start Date	Act.	Action Type	ActR	Reason for action
16.02.2021	01	Hiring	07	New Hire

Figure 2.10 Infotype 0000: Actions

When updating this infotype, you'll specify the **Action Type** (for example, **Hiring** or **Separation**) and **Reason for Action** (for example, **New Hire** or **Voluntary**). Both fields are customer specific and drive the employee's **Employment** status (**Active**, **Inactive**, **Retiree**, or **Withdrawn**).

This status drives payroll, especially whether this employee should be considered part of payroll processing or not.

Infotype 0001: Organizational Assignment

Infotype 0001 (Organizational Assignment) stores an employee's position, personnel area, employee group, and employee subgroup. This infotype is required infotype for processing payroll and is updated whenever a change in a person's place in the organizational or personnel structure. An employee can only have one active infotype 0001 record at a time, which is updated whenever an applicable change, with the appropriate effective date occurs, as shown in Figure 2.11.

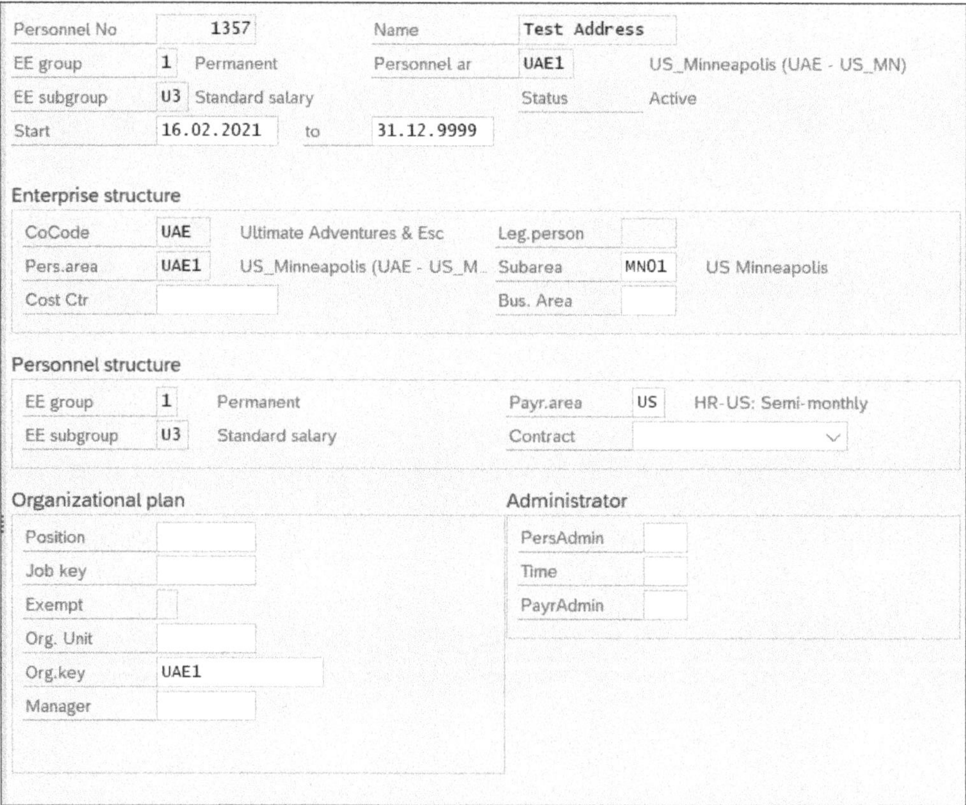

Figure 2.11 Infotype 0001: Organizational Assignment

The **Enterprise structure** settings define the key components of your organization and the employee's place in that structure and include the **CoCode** (company code) field, the **Cost Ctr** (cost center), and the **Pers. Area** (personnel area) field, which is typically a location. The **Personnel structure** settings define the type of employee (e.g., permanent, part time, retiree, salary, hourly). These settings are also where the employee's payroll area is stored. The **Payr.area** (payroll area) field is for grouping employees that run through payroll processing together based on same calendar and payment date.

Infotype 0002: Personal Data

Infotype 0002 (Personal Data) holds an employee's personal information and is an infotype containing personal identifiable information (PII) data. One difference between this infotype and others is its effective date is set as the employee's date of birth, rather than based on the action because this data is all about the employee, and the data is effective since day that person was born. Thus, you'll only find one infotype 0002 record per employee, and this record is required to run payroll. Anytime this infotype is changed, for example, to reflect a formal name change, the record would be updated. This infotype, shown in Figure 2.12, stores a formal name (**First name** and **Last name**), a Social Security Number (**SSN**) or another country identifier, a **Date of Birth**, and a **Gender**. This infotype can also store other fields, such as **Nationality** or **Marital Status**.

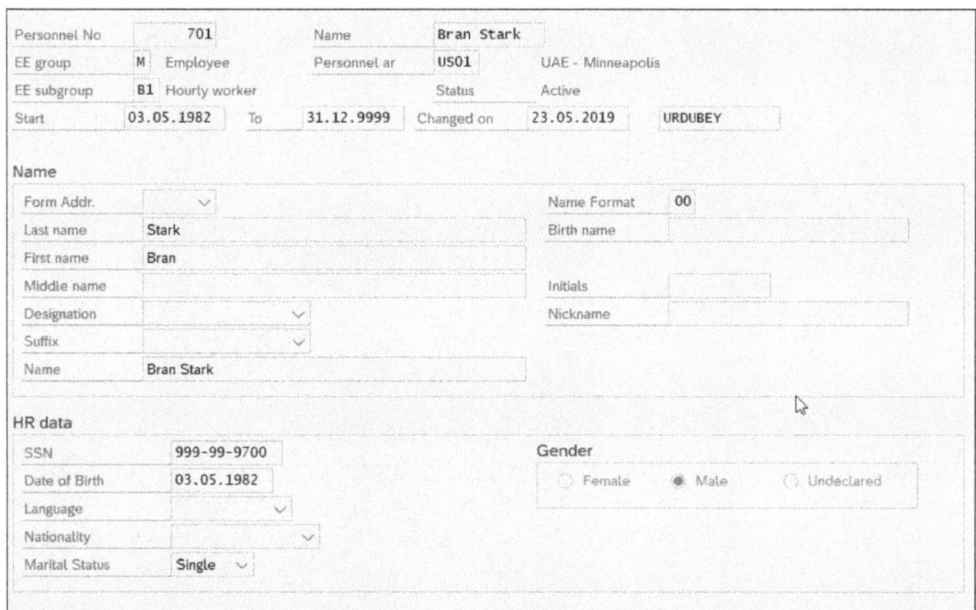

Figure 2.12 Infotype 0002: Personal Data

Infotype 0006: Address

Infotype 0006 (Address) is used to store an employee's address. This infotype can store multiple addresses to differentiate among permanent, temporary, and mailing addresses depending on your company's policies. However, a permanent address is required for payroll processing. As shown in Figure 2.13, you would select **Permanent residence** from the **Address type** dropdown list.

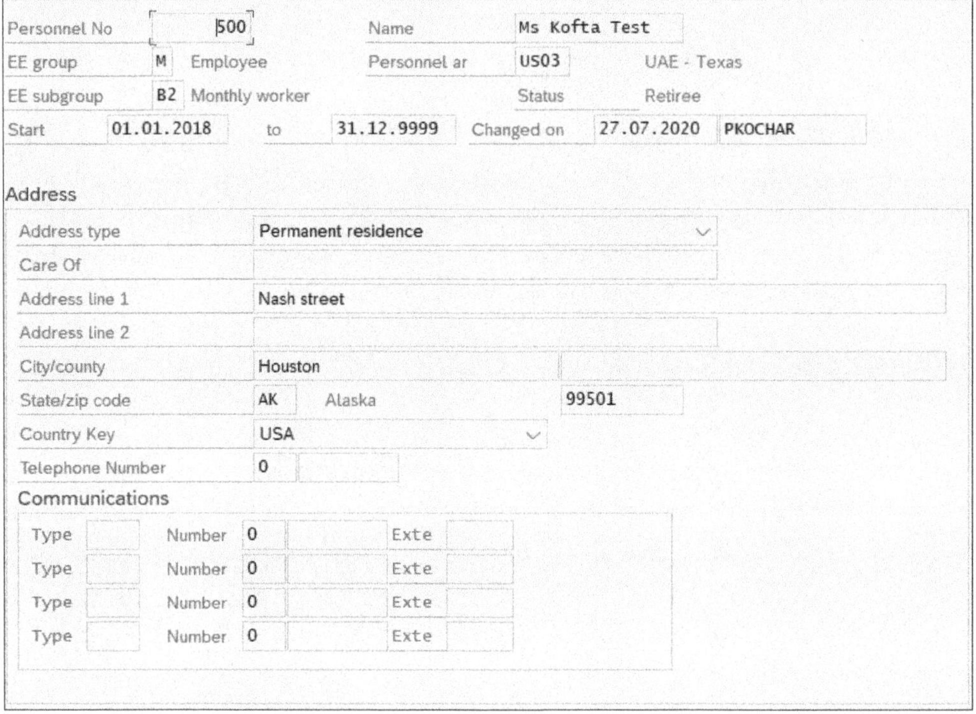

Figure 2.13 Infotype 0006: Address

This address is also used to define residence tax state, which is stored in infotype 0207, discussed in Section 2.2.3.

2.2.2 Payroll/Compensation Data

An employee's payroll or compensation data is stored in various infotypes and drive key payroll processing calculations. Many of these infotypes are created outside of Employee Central Payroll and are interfaced into Employee Central Payroll for processing, for example, earnings and deductions from benefit providers or timesheets from a time-tracking solution.

Table 2.4 lists payroll/compensation infotypes that are typically considered payroll/compensation data. We'll discuss each of these infotypes in the following sections.

Infotype Number	Infotype Description
0003	Payroll Status
0007	Planned Working Time

Table 2.4 Payroll/Compensation Infotypes

Infotype Number	Infotype Description
0008	Basic Pay
0009	Bank Details
0014	Recurring Payments/Deductions
0015	Additional Payments
0221	Payroll Results Adjustment
0267	Additional Off-Cycle Payment
2010	Employee Remuneration
0027	Cost Distribution

Table 2.4 Payroll/Compensation Infotypes (Cont.)

Infotype 0003: Payroll Status

Infotype 0003 (Payroll Status), shown in Figure 2.14, is automatically created by the system in response to a hiring action. This infotype stores data that controls the employee's payroll and time evaluation.

Figure 2.14 Infotype 0003: Payroll Status

The data on this infotype is updated automatically when time is evaluated, payroll is run, or changes are made to past payroll for an employee. This infotype is also unique in that it has its own transaction, Transaction PU03. This infotype can also be updated manually but with caution.

Some key fields stored in this infotype include the following:

- **Earl.pers. RA date (earliest personnel retroactive date)**
  The function of this date is to stop retroactive payroll calculations prior to this date. Defines the earliest date on which master data and time data changes are allowed—thus no changes can be made prior to this date.
- **Accounted to**
  This date is set during the payroll processing and specifies the date up to which payroll accounting has been run for the employee. You should never update this field.
- **Run payroll up to**
  This field is left blank unless an employee has left the company. In that case, this field is used to force payroll to be run for the employee. An example of the use of this field is for a severance payment.
- **Earliest MD change (earliest master data change)**
  When a change occurs in master data, this date is the earliest date from which the master data change was changed. If this date falls in a prior period, a retroactive calculation will be triggered. After payroll is calculated, this field reset to blank, ready for the next change.
- **Do not account after**
  This date is the final date up to which payroll is run for the employee.
- **Mast.data chng.bonus (master data change bonus)**
  This field is similar to the **Earliest MD Change** field and is set when a master data change occurs. This field is cleared after processing bonus payments or by unauthorized manual checks.
- **Pers.no.locked (personnel number locked)**
  If you set this flag, the employee is excluded from payroll. This flag is especially useful when trying to exit the payroll run, but you are unable to successfully process an employee (and therefore cannot exit). If you lock this employee, you can exit payroll and then correct the employee afterwards.
- **Payroll correction**
  Automatically flagged if a personnel number was rejected in the payroll run or if changes were made to payroll-relevant data in the correction phase. This flag is sometimes called matchcode W and can be useful to identify employees that will need to be reprocessed before a payroll run can be successful. You cannot exit payroll if any employees have this flag.

Warning!

Infotype 0003 is extremely powerful and if not managed properly can negatively impact payroll.

Infotype 0007: Planned Working Time

Infotype 0007 (Planned Working Time), shown in Figure 2.15, defines an employee's work schedule, how time should be evaluated, and the working time for an employee. Some key fields stored in this infotype include the following:

- **Work schedule rule**
  The working hours for a specific period and how hours are distributed.
- **Time Mgmt status (time management status)**
  Select this indicator if a person is processed using SAP ERP HCM Time Evaluation. Enter "0" or leave this field blank when you do not want to process time using time evaluation.
- **Working time**
  The fields in this section are derived from the work schedule rule selected earlier.

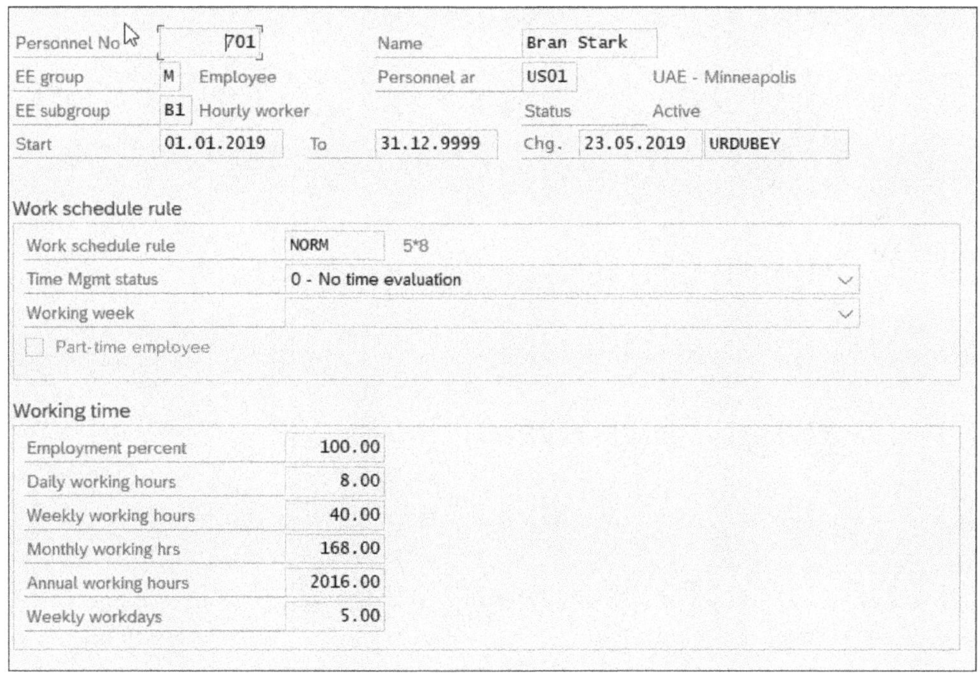

Figure 2.15 Infotype 0007: Planned Working Time

Infotype 0008: Basic Pay

Infotype 0008 (Basic Pay) stores an hourly rate or a salary for an employee. The pay structure data is typically defined by the job or the position. For salaried employees, the annual salary is defined by pay frequency. Hourly rates can be driven by a union-mandaded pay structure. Figure 2.16 shows a record from an hourly employee. Some key fields stored in this infotype include the following:

- **PS type, PS Area, PS group, and Level fields**
  Pay structure that is typically defaulted from the position or job.
- **WkHrs/period**
  Defaulted from the hours defined in infotype 0007.
- **Ann.salary (annual salary)**
  Defines the annual salary for a salaried employee.
- **WageType**
  Salary wage type or hourly wage type, dependent on the employee type.
- **Amount**
  Either the period salary or hourly rate, dependent on the employee type.

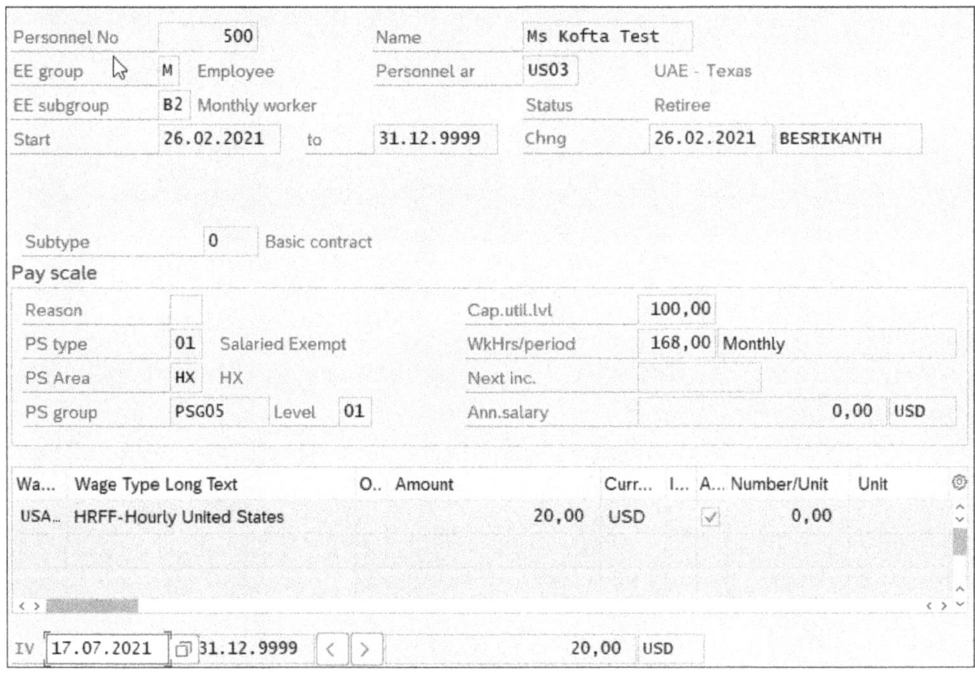

Figure 2.16 Infotype 0008: Basic Pay

Infotype 0009: Bank Details

Infotype 0009 (Bank Details) stores the payee and bank details for net pay or travel expenses. Every employee must have a main bank, but other banks can be stored. As shown in Figure 2.17, **Main bank** has been selected from the **Bank details type** dropdown list. The usage of the other subtypes depends on your business requirements, in the following ways:

- Infotype 0009, subtype 0 (Main Bank): Every employee must have this subtype defined. All money goes to this bank unless directed to another bank.
- Infotype 0009, subtype 1 (Other Bank): Specify a flat dollar amount or a percentage that should go to another bank. Any money over this dollar amount or the rest of the percentage would go to main bank. An employee can have multiple infotype 0009, subtype 1 records.
- Infotype 0009, subtype 2 (Travel Expense): Specify where you want travel expenses to go.
- Infotype 0009, subtype 5 (Main Bank Details for Off-Cycle): Main account to use when receiving an off-cycle payment. All off-cycle payments would go into this subtype unless directed to another bank. If this subtype does not exist, the main bank is used by default.
- Infotype 0009, subtype 6 (Other Bank Details for Off-Cycle): Another account to use when receiving an off-cycle payment. Specify a flat dollar amount or a percentage for a portion of the off-cycle amount to go to another bank. Any money over this dollar amount or the rest of the percentage would go to the main bank (the bank specified in infotype 0009, subtype 5) for the off-cycle payment.

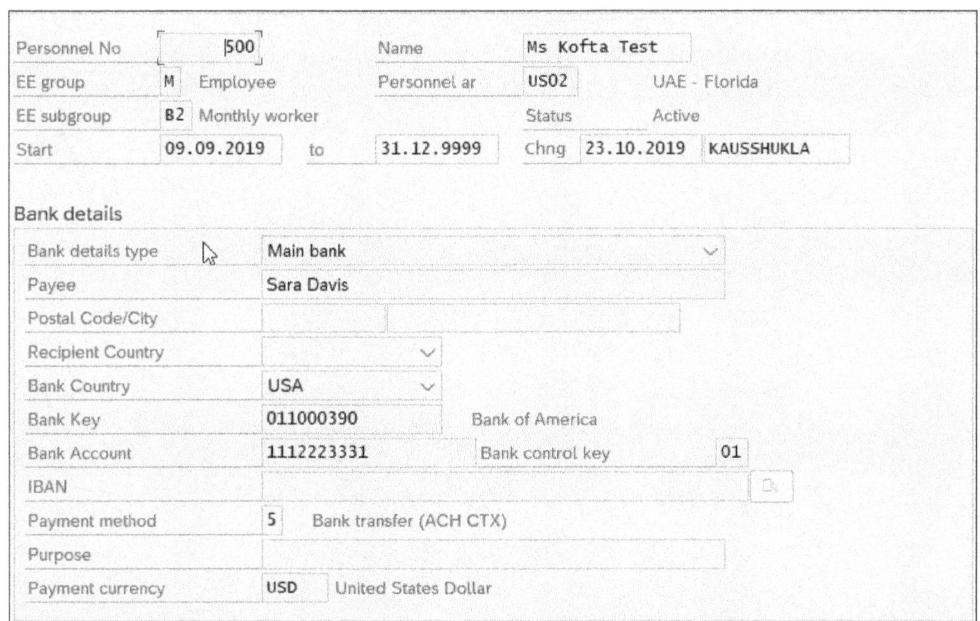

Figure 2.17 Infotype 0009: Bank Details

Key fields in this infotype include the following:

- **Payee**
  The name is defaulted from the employee name from infotype 0002, discussed in detail in Section 2.2.1. when we created the infotype. Only time this name would be different is after the death of the employee.
- **Bank Country**
  Identifies the country where the bank is located and the banking rules for bank account numbers, which are used for validation.
- **Bank Key**
  This key is the specific bank number for your bank (for the US, this is the routing number).
- **Bank Account**
  Employee's specific bank account number.
- **Bank control key**
  Determines if an account is a checking account (01) or a savings account (02).
- **Payment method**
  This field determines how the payment will be made, check, Automated Clearing House (ACH), etc.
- **Payment currency**
  Currency in which the payment will be made.

Infotype 0014: Recurring Payments/Deductions

Infotype 0014 (Recurring Payments/Deductions) is for wage elements that are paid or deducted on a defined recurring basis. Recurring payments and deductions only need to be entered into the system once with a defined start date and end date, and these payments/deductions will be processed every pay period during that time period. The amount to be paid or deducted can either be entered, looked up in a table, or calculated. As shown in Figure 2.18, some key fields of this infotype include the following:

- **Wage Type**
  The earning or deduction code to be processed by payroll.
- **Amount**
  Stores the amount to be deducted or paid during the payroll processing. If a deduction wage type, you'll see an **A** next to the field. In this case, do not enter a negative amount since the system knows the amount is for a deduction (based on the configuration of the wage type).
- **Number/Unit**
  Used if the wage type should be processed as hours or units rather than as an amount.
- **Payment Model**
  If needed, you can define a frequency. If this field is left blank, every pay period is the default payment model.

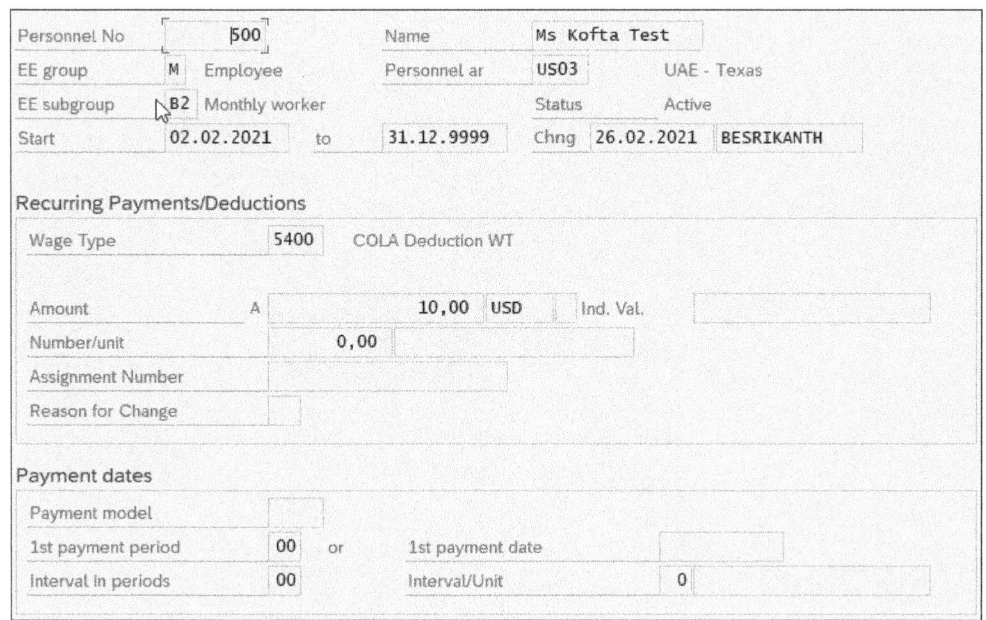

Figure 2.18 Infotype 0014: Recurring Payments/Deductions

Infotype 0015: Additional Payments

Infotype 0015 (Additional Payments) are wage elements that are not paid or deducted on a defined recurring basis, which is the case with infotype 0014, but instead in a single payroll period. This infotype uses the **Date of origin** field to define an effective date. As shown in Figure 2.19, notice how this infotype lacks effective dates or start and end dates. The important fields for this infotype include the following:

- **Wage Type**
  The earning or deduction code to be processed by payroll.
- **Amount**
  Stores the amount to be deducted or paid during the payroll processing. If a deduction wage type, you'll see an **A** next to the field. In this case, do not enter a negative amount since the system knows the amount is for a deduction (based on the configuration of the wage type).
- **Ind.val. (indirect valuation)**
  Indicates whether the wage type is to be valuated indirectly.
- **Number/unit**
  Used if the wage type should be processed as hours or units rather than as an amount.
- **Date of origin**
  Effective date of the change.

Figure 2.19 Infotype 0015: Additional Payments

Infotype 0221: Payroll Results Adjustment

Infotype 0221 (Payroll Results Adjustment) allows adjustments to be made to payroll tables, including payments made to employees outside of Employee Central Payroll. These changes are entered into Employee Central Payroll, and the employee's results tables are updated. Different than how most infotypes are updated, this infotype uses Transactions PAUX or PAUY. Transaction PAUX is for updating the subtypes NAMC, ADJT, and TWEG. Subtypes YANA and YAWA use Transaction PAUY. These transactions should not be used during a normal payroll process but instead should be used during a special payroll run. Table 2.5 lists the subtypes available.

Subtype	Description
Blank or subtype NAMC (Non-Authorized Manual Check)	Enter gross amounts and employee taxes, and the system will calculate the employer taxes and other technical wage types.
Subtype ADJT (Adjustment)	Enter gross amounts, employee taxes, and employer taxes, and the system will calculate the technical wage types.
Subtype TWEG (Prior Period Gross Adjustment)	Enter gross amounts, and the system will calculate both employee taxes and employer taxes as well as calculate technical wage types.

Table 2.5 Infotype 0221: Payroll Results Adjustment Subtypes

Subtype	Description
Subtype YANA (Adjustments, No Tax Calculation)	All updates to payroll results are entered (gross, technical, and taxes); nothing is generated from the system. This subtype allows you to make non-cash adjustments to an employee's results without triggering a retroactive calculation.
Subtype YAWA (Adjustments with Tax Calculation)	Gross amounts are entered as well as employee taxes (optional), and the system will generate the employee taxes and employer taxes as well as the technical wage types. This subtype allows you to make non-cash adjustments to an employee's results without triggering a retroactive calculation.

Table 2.5 Infotype 0221: Payroll Results Adjustment Subtypes (Cont.)

Infotype 0267: Additional Off-Cycle Payments

Infotype 0267 (Additional Off-Cycle Payments) is used when making payments outside of the regular payroll, for example, for the payment of bonuses. As shown in Figure 2.20, and similar to infotype 0015, this infotype lacks a start date and an end date. Instead, this infotype has a **Payment date** field for the date of an off-cycle payment. This date *must* be dated after the pay date of the regular payroll run. Key fields in this infotype include the following:

- **Wage Type**
  The earning code for the payment.
- **Amount**
  Amount to be paid during the off-cycle payroll.
- **Payment date**
  Date of payment that is used for off-cycle payments.
- **Reason for Change**
  Enter the reason for change in the data in the infotype, for instance, to reflect a bonus payment.
- **Off-cycle reas. (off-cycle reason)**
  Specifies the reason why an off-cycle payroll run is performed.
- **Payroll type**
  Feature used by the system to recognize which of the following payroll runs should be executed:
  - Regular payroll run (type <blank>)
  - Payroll run for bonus payments or absences (type A)
  - Correction run due to a change in master data (type B)
  - Payroll run for payments from infotype 0221, subtype C (Payroll Results Adjustment)

- **Payroll Identifier**
  Indicator used to distinguish between different off-cycle payroll runs created on the same day.

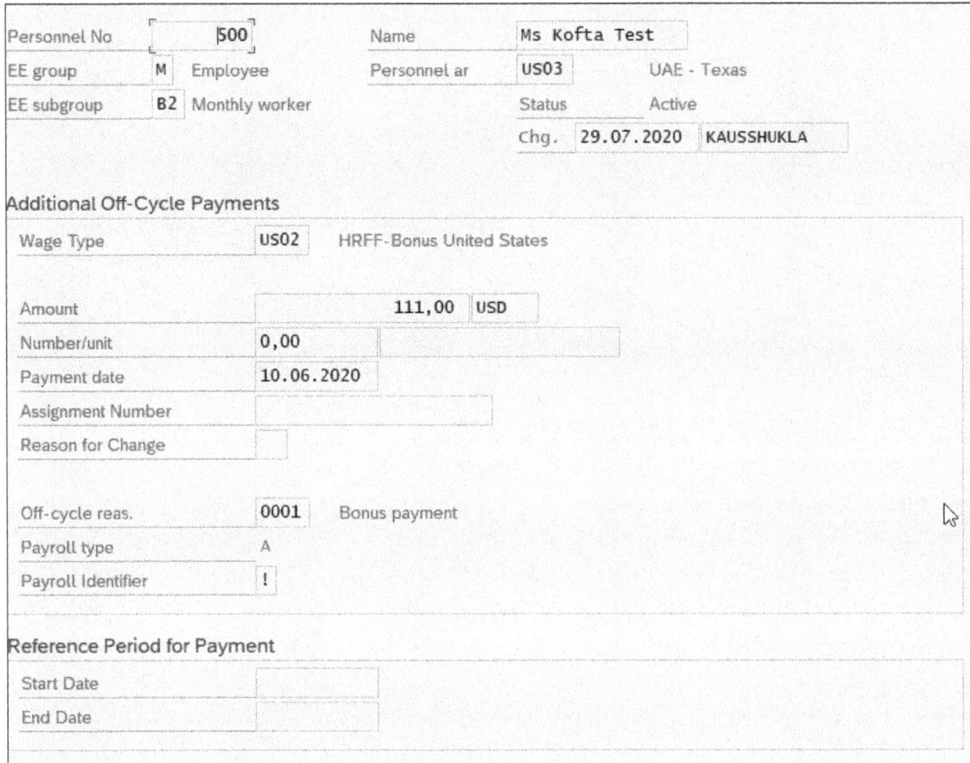

Figure 2.20 Infotype 0267: Additional Off-Cycle Payments

Infotype 2010: Employee Remuneration Info

Infotype 2010 (Employee Remuneration Info) is used to load time into Employee Central Payroll for processing. In many situations, employee time is managed and evaluated outside the system, and this infotype is used to receive this information for processing payment. Figure 2.21 is an example of an infotype 2010 record. Key fields in this infotype include the following:

- **Wage Type**
  The earning code for the payment.
- **Number of hours**
  Dependent on the wage type configuration; used for entering the number of hours.
- **Number/unit**
  Dependent on the wage type configuration; used for entering numbers and units. Can be the number of hours or can be used to pay based off of some other number (such as miles).

- **Amount and Currency**
  Dependent on the wage type configuration, but for entering amounts and currencies.
- **Position/work center**
  Can override the rate of payment from the infotype with data from the position.
- **Premium and Premium Indicator**
  Can set a premium to be added to the payment. For example, if you enter a premium for 10 hours of regular pay, the calculation might result in an additional $1 per hour.

Personnel No	500	Name	Ms Kofta Test
EE group	M Employee	Personnel ar	US03 UAE - Texas
WS rule	REG 5*8	Status	Retiree
Date	29.08.2021		

Remuneration info

Wage type	1600 Regular working time
Number of hours	
Number/unit	40,00 / Hours
Amount	
Currency	USD
Extra pay/valuation	/
Pay scale group/level	/
Position/work center	/
Overtime comp. type	Depends on wage type
Premium	
Premium Indicator	
Tax area record type	
External document number	

Figure 2.21 Infotype 2010: Employee Remuneration Info

Infotype 0027: Cost Distribution

Infotype 0027 (Cost Distribution) is used to override the home cost center. Without an override, the payroll costs will post to the home cost center and company code stored on infotype 0001. However, in some situations, you might need to override the usual cost center for a specific time period, which is possible using this infotype. For example, let's say an employee is working 50% for another cost center for 3 months; in this case, you would create an infotype to represent that time period, cost assignment, and percentage, as shown, for example, in Figure 2.22. Key fields in this infotype include the following:

- **Distrib. (distribution type)**
  Category of the type of costs to be distributed.
- **CoCode**
  Company code of the overriding cost center.
- **Cost ctr (cost center)**
  Cost center you want to use for the override.
- **Pct (percentage)**
  Percentage of the costs that you want to override.

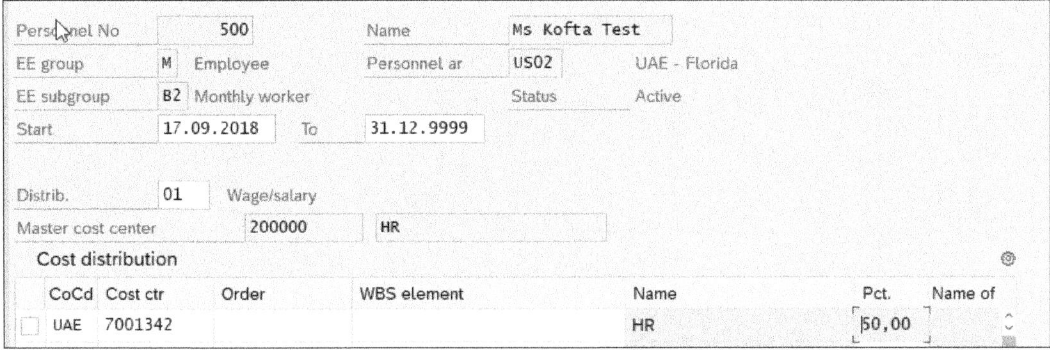

Figure 2.22 Infotype 0027: Cost Distribution

2.2.3 Tax Data

The tax infotypes provide the payroll driver the information required to accurately calculate employee taxes and employer taxes for all levels, from the federal level down to the school district. These infotypes provide information on where the employee lives and works, for which state unemployment should be collected, and the employee's W-4 elections. Table 2.6 lists the tax infotypes, each of which we'll cover in the following sections.

Infotype Number	Infotype Description
0207	Residence Tax Area
0208	Work Tax Area
0209	Unemployment State
0210	W-4 Withholding Information
0235	Other Taxes US

Table 2.6 Tax Infotypes

2

Infotype 0207: Residence Tax Area

Infotype 0207 (Residence Tax Area) is used to store an employee's resident tax area, that is, where the employee lives. This infotype should indicate the same state as defined in infotype 0006, discussed in detail in Section 2.2.1, except in situations where a local tax exists. You should always select the lowest tax authority for this infotype. Figure 2.23 shows an infotype 207 record for someone living in the state of Alabama. Once you select a tax area from the available options in the dropdown menu, all appropriate tax authority areas are populated.

Figure 2.23 Infotype 0207: Residence Tax Area

Infotype 0208: Work Tax Area

Infotype 0208 (Work Tax Area) takes into consideration how an employee is subject to taxes in the location where they work. Infotype 0208 records the amount of time an employee spends in each locality, as shown in Figure 2.24. You cannot have records totaling over 100%, so before creating a new record, you'll need to update the end date of the existing record, prior to creating a new record.

Key fields in the infotype include the following:

- **Tax Area**
  Lowest-level tax authority for where an employee works.
- **Allocation**
  Percentage of the employee's working time spent in the tax area.

- **Predominant Tax Area**
  Defines if a work tax area is predominant. This field is only relevant for employees who reside and work in the Commonwealth of Pennsylvania (PA) and do not work in their resident municipality. This field is not relevant for employees who reside or work in Philadelphia.

  Similar to infotype 0207, when selecting a tax authority, an area populates. In this example, a city tax was selected (**CA01: San Francisco**), so the state tax (**CA: California**) was also populated.

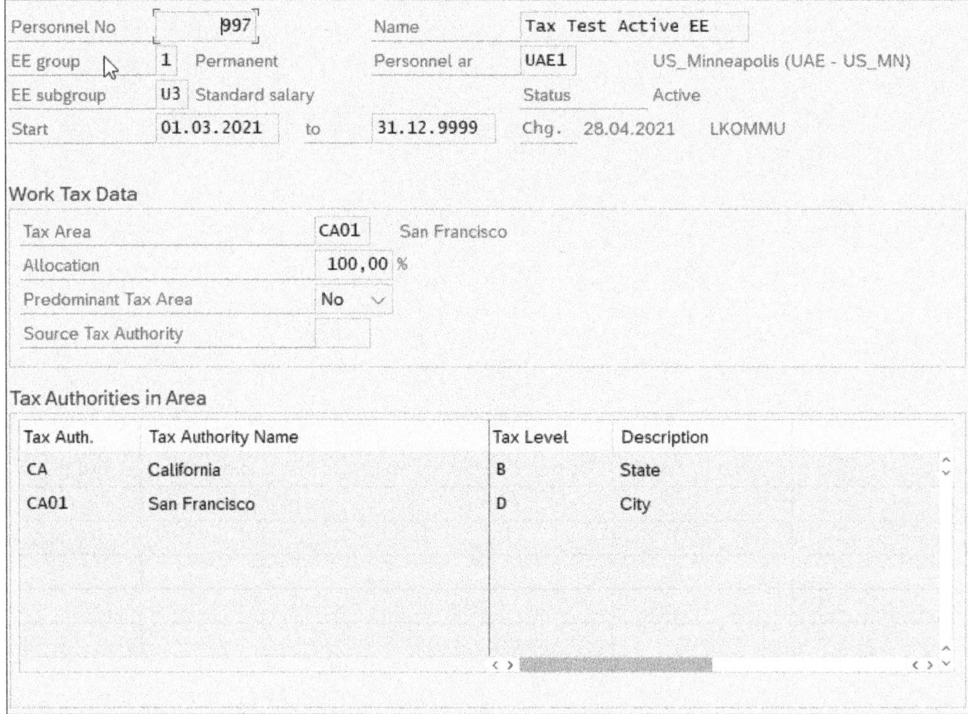

Figure 2.24 Infotype 0208: Work Tax Area

Infotype 0209: Unemployment State

Infotype 0209 (Unemployment State) is used for the payment of unemployment insurance for an employee. As shown in Figure 2.25, this infotype collects unemployment-related state information as well as provides statistical information to unemployment authorities requiring multiple work site reports. This infotype is required for payroll processing and is based on the work location. Each employee only has one active record. Key fields in this infotype include the following:

- **Tax authority**
  This field identifies the government agency authorized to levy certain taxes.

- **Worksite**
  Worksites are used in certain states to provide statistical information to the tax authority for governing unemployment.

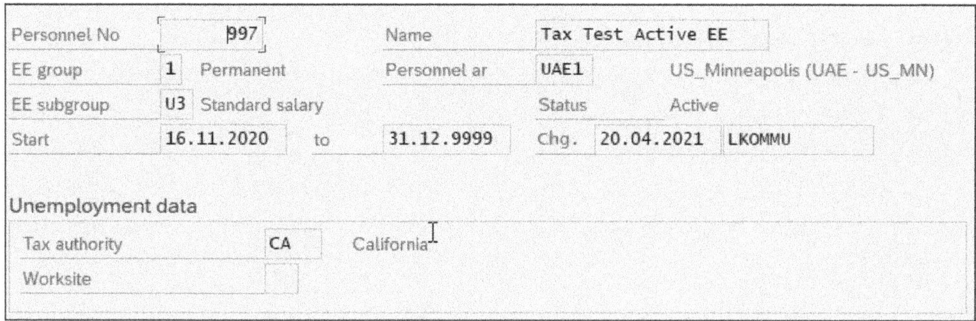

Figure 2.25 Infotype 0209: Unemployment State

Infotype 0210: Withholding Info W4/W5 US

Infotype 0210 (Withholding Info W4/W5 US) is used to store the details from an employee's Federal Employee Withholding Allowance Certificates (Form W-4) and from the employee's equivalent state and local withholding allowance certificates, if applicable. Every employee needs an infotype 0210 record for the federal withholding to process payroll. Some states and localities also require an infotype 0210 record. State forms can be different than the federal form.

When you create a new infotype 0210 record, the system displays a dialog box listing the available subtypes for various federal, state, and local authorities. Figure 2.26 shows the federal tax form with the following fields:

- **Tax authority**
  Tax authority levying withholding tax.
- **Tax level**
  This field defines the hierarchy level to which the tax authority belongs. Federal is level A. This level is defaulted from the tax authority.
- **Filing Status**
  The employee's filing status for tax purposes.
- **Allowances**
  Number of allowances claimed by the employee.
- **Tax Exempt Ind. (tax-exempt indicator)**
  Indicator that specifies whether an employee's earnings are exempt from taxation. R is the tax-exempt indicator.
- **Add.withholding (additional withholding)**
  Additional amount to be withheld, at the employee's request, for the payment of taxes.

- **Default formula**
  Alternative formula assigned to calculate withholding tax. Only some authorities allow employees to define an alternative method for calculating their withholding tax.
- **Non-resident tax calculation**
  Used to indicate that an employee has been designated as a non-resident for payroll tax calculation purposes.
- **Alternative formula**
  Alternative formula assigned to calculate withholding tax. Only some authorities allow employees to define an alternative method for calculating their withholding tax.

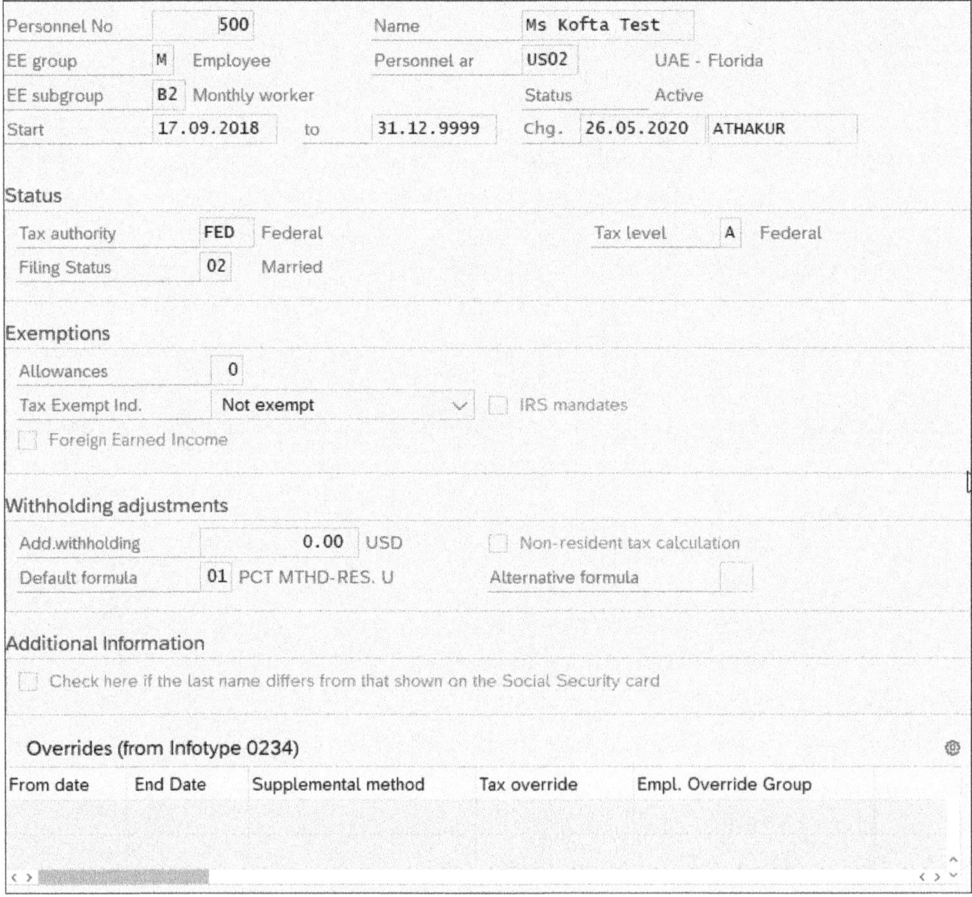

Figure 2.26 Infotype 0210: Withholding Info W-4/W-5 US

Infotype 0235: Other Taxes US

Infotype 0235 (Other Taxes US) is used to exempt an employee from certain taxes, for example, a student or a non-resident. Figure 2.27 shows exemptions related to Earned

Income Credit Payment and Social Security Tax. The field you use for these exemptions is the **Exempt** column on the right, where you would enter "Y" for those tax types you want to exempt from tax calculations.

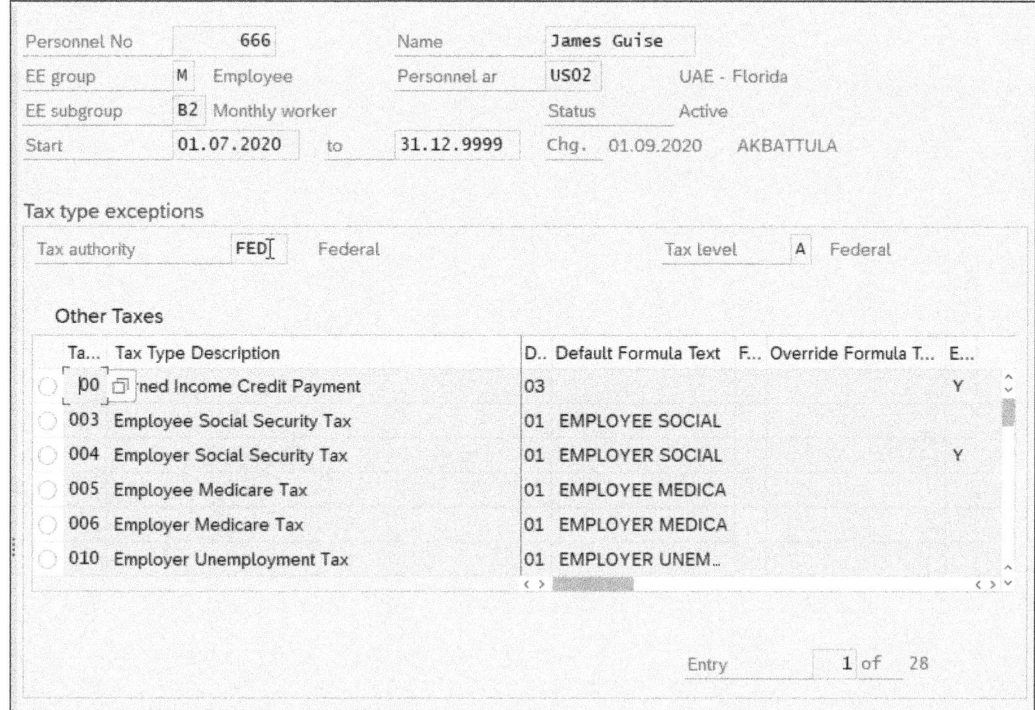

Figure 2.27 Infotype 0235: Other Taxes US

2.2.4 Benefits Data

Benefits vary significantly by country and can be handled in numerous ways by different organizations. Table 2.7 shows the various types of benefits-related infotypes to cover health insurance, insurance plans, savings plans, and Flexible Spending Accounts (FSAs), each of which will be discussed in the following sections.

Infotype Number	Infotype Description
0167	Health Plans
0168	Insurance Plans
0169	Savings Plans
0170	Flexible Spending Accounts
0171	General Benefits Information

Table 2.7 Benefits Infotypes

Infotype 0167: Health Plans

Infotype 0167 (Health Plans) stores information on the health plans an employee is enrolled in. For each health plan, a different record exists, distinguished by the plan name, which is also a subtype of the infotype. The structure of this infotype is listed in Table 2.8, and the infotype itself is shown in Figure 2.28.

Tab	Description
Plan data	Defines the plan's attributes for the specific benefit plan as well as the cost rule variant.
Administration	Documents the eligibility date, evidence of insurability, and the type of enrollment.
Costs	Defines the calculated costs of the benefit plan.
Dependents	Eligible dependents are stored under this tab with indicators to specify coverage.
Additional data	If the processing of claims must be coordinated, the additional information needed is stored under this tab.

Table 2.8 Structure of Infotype 0167

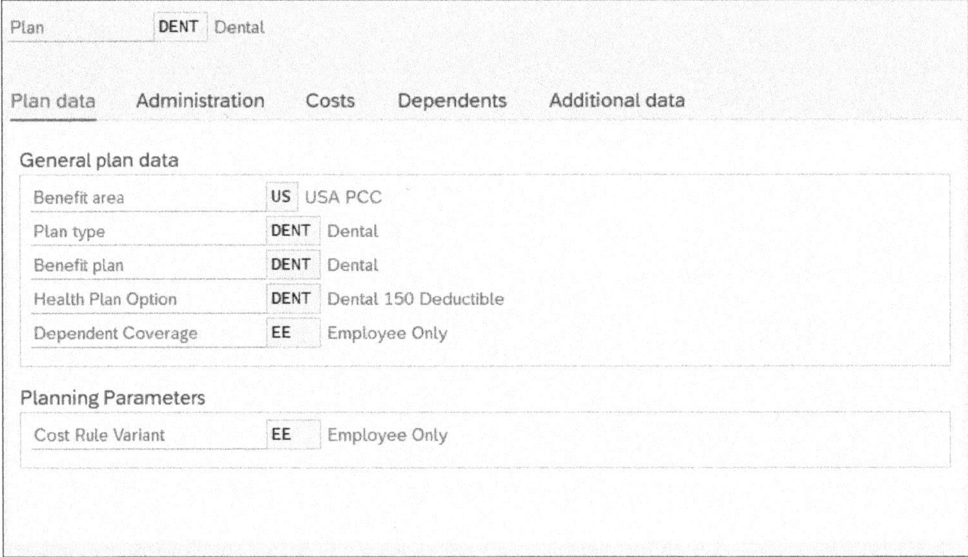

Figure 2.28 Infotype 0167: Health Plans

Infotype 0168: Insurance Plans

Infotype 0168 (Insurance Plans) is used to store information on the insurance plans an employee is enrolled in. For each insurance plan, a different record exists, distinguished by the plan name, which is also a subtype of the infotype. The structure of the infotype record is listed in Table 2.9, and the infotype itself is shown in Figure 2.29.

Tab Name	Description
Plan data	Defines the plan's attributes for the specific benefit plan as well as the cost rule variant.
Administration	Documents the eligibility date, evidence of insurability, and type of enrollment.
Insurance cov.	Defines the coverage related to that plan.
Costs	Defines the calculated costs of the benefit plan.
Beneficiaries	Documents the beneficiaries identified.

Table 2.9 Structure of Infotype 0168

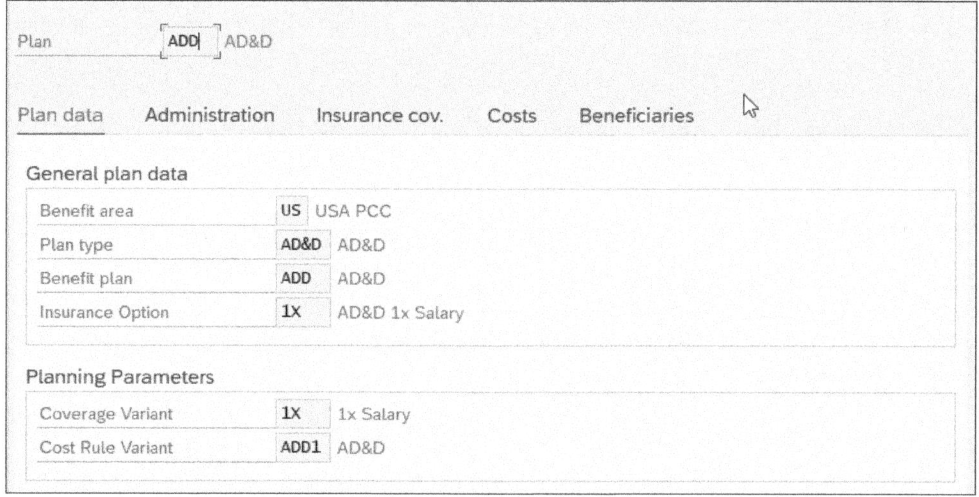

Figure 2.29 Infotype 0168: Insurance Plans

Infotype 0169: Savings Plans

Infotype 0169 (Savings Plans) is used to store information on the savings plans an employee is enrolled in. For each savings plan, a different record exists, distinguished by the plan name, which is also a subtype of the infotype. The structure of the infotype record is listed in Table 2.10, and the infotype itself is shown in Figure 2.30.

Tab Name	Description
Plan data	Defines the plan's attributes for the specific benefit plan as well as the cost rule variant.

Table 2.10 Structure of Infotype 0169

Tab Name	Description
Administration	Documents the eligibility date, evidence of insurability, and type of enrollment.
Regular contrib.	Stores the contribution percentage or amount to be deducted and can be split between pre- and post-tax contributions.
Bonus contribution	Stores the contribution percentage or amount to be deducted and can be split between pre- and post-tax contributions for a bonus payment. If blank, the default is to use the regular contribution.
Beneficiaries	Documents the beneficiaries identified.
Investments	Stores investments related to the plan.
Additional data	Documents the account number and employee pension contribution amount.

Table 2.10 Structure of Infotype 0169 (Cont.)

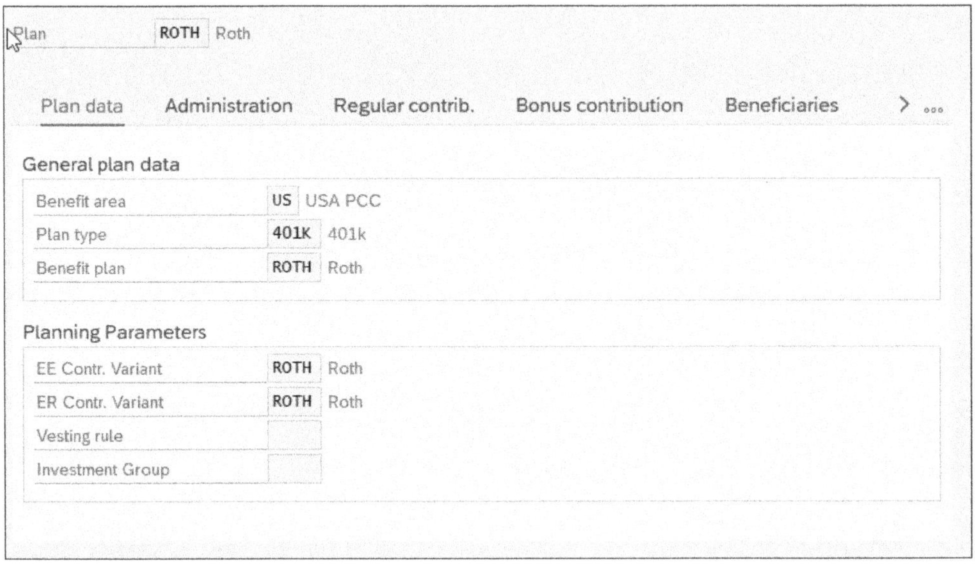

Figure 2.30 Infotype 0169: Savings Plans

Infotype 0170: Flexible Spending Accounts

Infotype 0170 (Flexible Spending Accounts) is used to store information on any flexible spending account an employee may be enrolled in. If enrolled in either type of plan—health or dependent, a different record exists, distinguished by the plan name, which is also a subtype of the infotype. Two records will exist if an employee is enrolled in both kinds of flexible spending accounts. The structure of the infotype record is listed in Table 2.11, and the infotype itself is shown in Figure 2.31.

Tab Name	Description
Plan data	Defines the plan's attributes for the specific benefit plan as well any employer contribution.
Administration	Documents the eligibility date and type of enrollment.
Employee contribution	Stores the annual pre-tax employee contribution.

Table 2.11 Structure of Infotype 0170

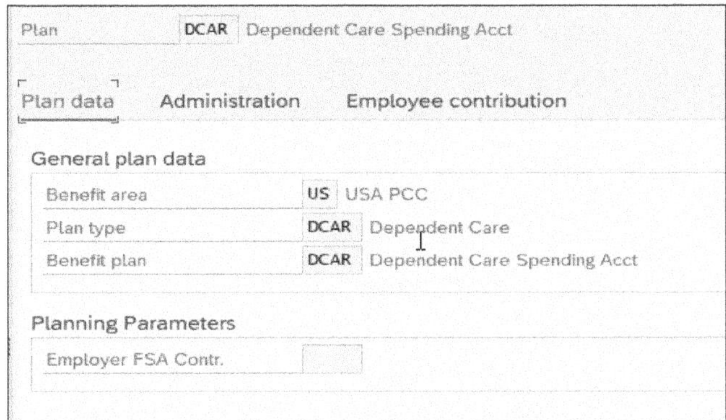

Figure 2.31 Infotype 0170: Flexible Spending Accounts

2.2.5 Garnishments

Garnishments are set up when a court orders that a deduction be taken from an employee's check. Three infotypes are used in the US, listed in Table 2.12, for documenting the garnishment information needed for payroll processing and garnishment remittance. We'll discuss each type in the following sections.

Infotype Number	Infotype Description
0194	Garnishment Document
0195	Garnishment Order
0216	Garnishment Adjustment

Table 2.12 Garnishment Infotypes

Infotype 0194: Garnishment Document

Infotype 0194 (Garnishment Document) stores all the relevant information from a garnishment order, as shown in Figure 2.32.

When updating this infotype, you'll maintain the **Case no**; **Status** (**Active**, **Pending**, **Released**); **Received**; **Priority**; **Jurisdiction**; **Category** (**Creditor**, **Support**, **Student Loan**);

Vendor; **Remittance** (**Immediate – ACH**, **Immediate – Check**, **Quarterly – Check**); and so on, depending on individual requirements. Other fields, such as **Svc.Charge** (service charge), can be used if your company charges a service charge for garnishments.

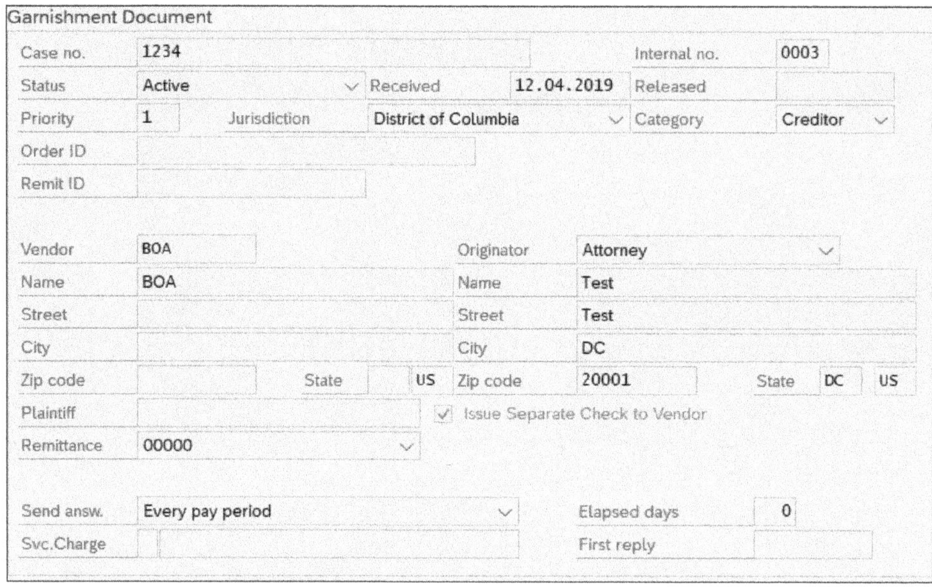

Figure 2.32 Infotype 0194: Garnishment Document

Infotype 0195: Garnishment Order

Infotype 0195 (Garnishment Order) is created once an infotype 0194 record is saved, at which point the infotype 0195 record will be available for updating. As shown in Figure 2.33, the **Case no.**, **Internal no**, **Sequence no.**, and **Order Type** fields are all defaulted from the garnishment document (infotype 0194).

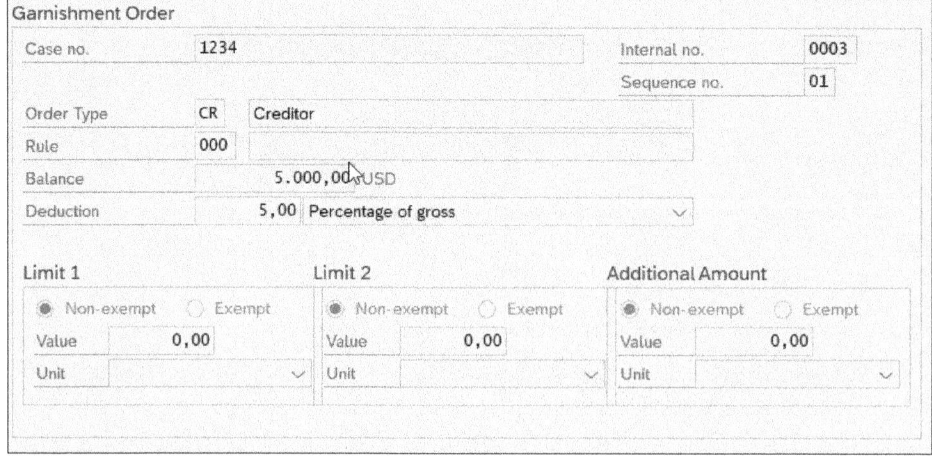

Figure 2.33 Infotype 0195: Garnishment Order

When updating this infotype, you'll maintain the **Rule**, **Balance**, and **Deduction** fields and (if applicable) maintain the fields in the **Limit 1**, **Limit 2**, and/or **Additional Amount** areas.

Infotype 0216: Garnishment Adjustment

Infotype 0216 (Garnishment Adjustment) is used to refund an employee for an earlier garnishment, as shown in Figure 2.34. You create this adjustment through infotype 0194 by navigating to **More • Garnishment • Adjustment • Refund**. The **Case no.**, **Internal Number**, **SequenceNo**, and **Order Type** fields are all defaulted, but you must maintain the **Adjustment Amount**, **Date of Adjustment**, and **Adjustment Wage Type** fields. You can also use this infotype to change a balance if needed, by selecting **More • Garnishment • Adjustment • New Balance**.

Figure 2.34 Infotype 216: Garnishment Adjustment

2.3 Schemas and Rules

The US payroll driver RPCALCU0 is an ABAP program that processes a series of schemas and rules. In a concurrent employment environment, program RPCALCU0_CE is used, but this scenario is beyond the scope of this book.

A *schema* is a code that drives or controls the logic for payroll calculations and processes. A schema is made up of a series of *subschemas* and *rules* that define the actions it should take. This code is the heart of payroll and is what controls everything—calculating taxes and net pay, creating technical wage types, and more. A schema is also quite configurable but can be tricky, and some people find it difficult to configure. In Chapter 1, we provided an overview of the main steps of the payroll process; in this section, you'll learn how the payroll driver works and learn how to update or create the main payroll schema, including its subschemas, tables, rules, and operations.

2.3.1 Payroll Driver

The payroll driver RPCALCU0 runs the payroll process using a main schema (by standard, schema U000). The U standards for US, and each country has a unique payroll driver. The payroll driver is controlled by the payroll control record, which sets the payroll period, the status of the payroll, and earliest retroactive accounting period. The job of the payroll driver is to process payroll as outlined in the schemas, rules, and functions specified. Use Transaction PC00_M10_CALC to access the US payroll driver.

Figure 2.35 shows the payroll driver screen. Key fields to run payroll include the following:

- **Payroll area**
  Select the payroll area for which you want to calculate payroll.
- **Current period/Other period**
  When running payroll in production, the system will run payroll for current period. The system will read the payroll area's control record to select the appropriate period to run. You can select the **Other period** option for simulation or testing purposes.
- **Personnel Number**
  When running payroll in production, leaving this field blank will select all employees in the selected payroll area. Alternatively, you can enter one or more employee numbers in this field.
- **Payroll area**
  You must enter a payroll area for the employee (or set of employees) you want included in payroll.
- **Schema**
  Specify the main schema defined for your company.
- **Forced retro.accounting as of**
  To force payroll to calculate retroactive back to a specific date for all employees, enter a specific date in this field.
- **Test run (no update)**
  Set this flag if you want to run payroll in simulation mode. This mode will not lock records and doesn't save results, but this option does allow you to test the payroll calculation for an employee or for a set of employees.
- **Display log**
  Provides a detailed payroll log, which is useful if running simulations on a small number of employees. You would not use this field for a large population, which would slow down performance and could cause failures.

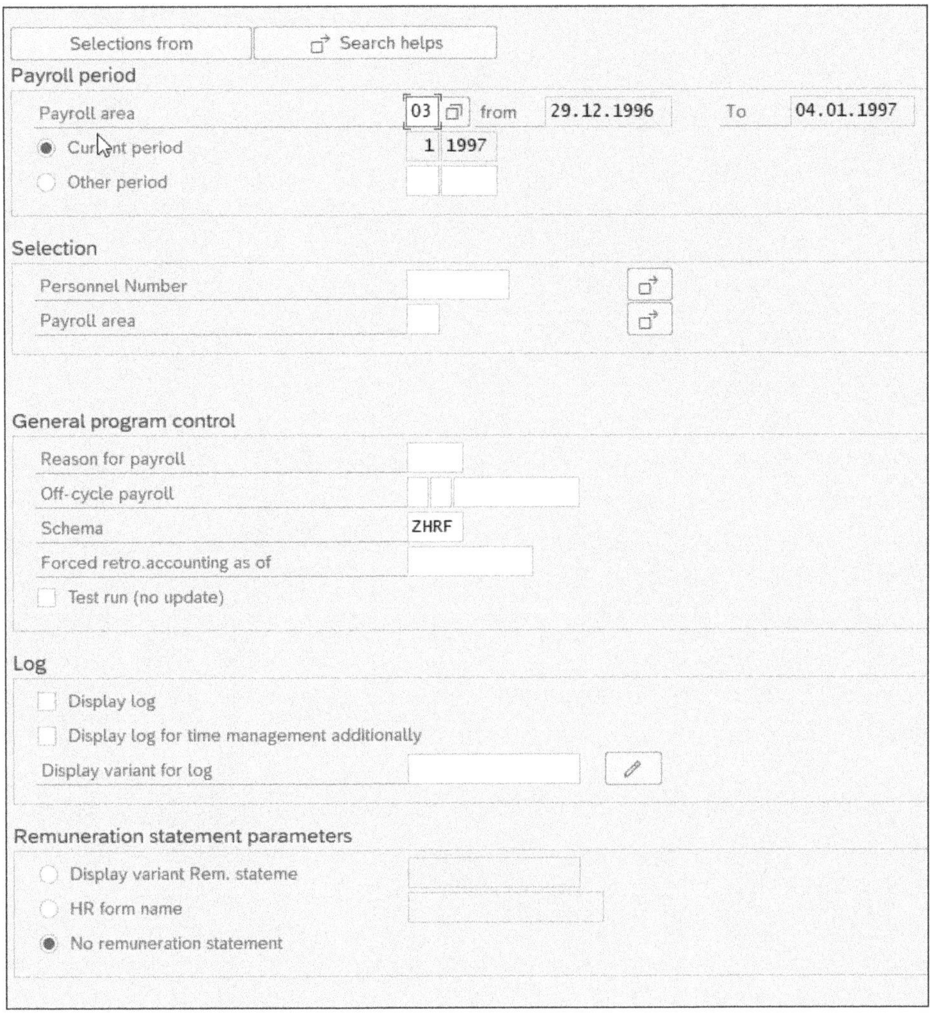

Selections from | Search helps

Payroll period

Payroll area 03 from 29.12.1996 To 04.01.1997

Current period 1 1997

Other period

Selection

Personnel Number

Payroll area

General program control

Reason for payroll

Off-cycle payroll

Schema ZHRF

Forced retro.accounting as of

Test run (no update)

Log

Display log

Display log for time management additionally

Display variant for log

Remuneration statement parameters

Display variant Rem. stateme

HR form name

No remuneration statement

Figure 2.35 Payroll Driver

2.3.2 Main Schema and Its Subschemas

The main schema used by the payroll driver during each payroll run is driver RPCALCU0. The standard schema that is delivered for the US is U000. You should never update this schema, which exists in SAP's namespace and could be overwritten with SAP Notes and other updates. If you must make changes to this schema, you should copy schema U000 and create a Z version of your own to make any updates as needed to meet your business requirements. In the payroll driver, you can specify the primary schema to drive the payroll process.

Let's look at the main schema in more detail. To view or edit a schema, open Transaction PE01 and enter the schema you would like to copy, review, or update, as shown in Figure 2.36. To create a new schema, you should copy an existing one.

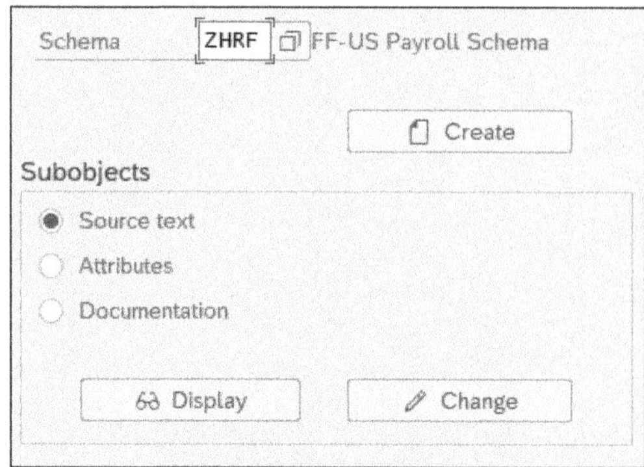

Figure 2.36 Transaction Code PE01

Let's look into schema ZHRF to see how it's been configured so you can see best way to make and document the changes needed. Make sure you select the **Source text** radio button. Before we jump into looking at the actual schema, note a few important aspects of updating a schema, such as the following:

- Custom schemas are copied from other existing schemas and created using the same name, but replacing the U with a Z. In this way, you can always clearly see what schema was copied. The example shown in Figure 2.37 is **ZINO**, which is a copy of **UINO**.
- The original schema is prefixed by an asterisk (*), which means this schema is commented out and not in use. Do not delete the row, which helps you understand changes. As shown in Figure 2.37, **UINO** has an asterisk (*), indicating that it will not be processed.
- Not all schemas should be changed. If you don't need to change a schema to meet your business requirements, use the standard schemas. As shown in Figure 2.37, **UPRO** was not copied or updated, and is currently being used.
- Schemas are highly structured and must follow the format listed in Table 2.13.

Column Names	Column Description
Line	Sequential. You can add or delete rows.
Func.	Function. A function is high-level logic built in ABAP code that can be used for calculations and executed as part of the schema.
Par 1, **Par 2**, **Par 3**, and **Par 4**	Up to 4 parameters can exist per line.
D	Used for commenting out lines that won't be executed. Insert an asterisk (*) in this column.
Text	Freeform text to document the purpose of the line.

Table 2.13 Schema Structure

Figure 2.37 shows a copy of the schema, where you can see examples of copied and renamed schemas using the rules we just described.

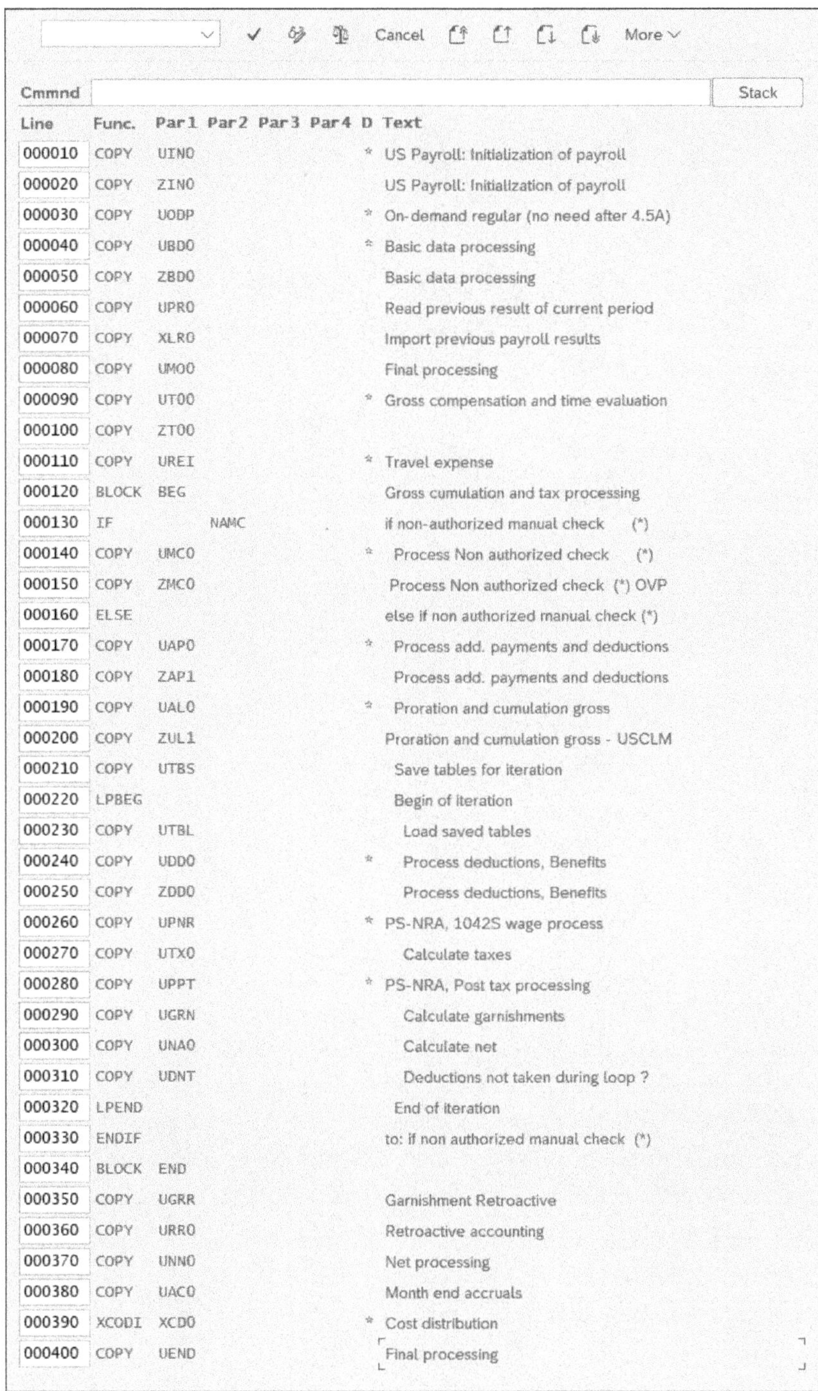

Cancel More

Cmmnd Stack

Line	Func.	Par1	Par2	Par3	Par4	D	Text
000010	COPY	UINO				*	US Payroll: Initialization of payroll
000020	COPY	ZINO					US Payroll: Initialization of payroll
000030	COPY	UODP				*	On-demand regular (no need after 4.5A)
000040	COPY	UBDO				*	Basic data processing
000050	COPY	ZBDO					Basic data processing
000060	COPY	UPRO					Read previous result of current period
000070	COPY	XLRO					Import previous payroll results
000080	COPY	UMOO					Final processing
000090	COPY	UTOO				*	Gross compensation and time evaluation
000100	COPY	ZTOO					
000110	COPY	UREI				*	Travel expense
000120	BLOCK	BEG					Gross cumulation and tax processing
000130	IF		NAMC				if non-authorized manual check (*)
000140	COPY	UMCO				*	Process Non authorized check (*)
000150	COPY	ZMCO					Process Non authorized check (*) OVP
000160	ELSE						else if non authorized manual check (*)
000170	COPY	UAPO				*	Process add. payments and deductions
000180	COPY	ZAP1					Process add. payments and deductions
000190	COPY	UALO				*	Proration and cumulation gross
000200	COPY	ZUL1					Proration and cumulation gross - USCLM
000210	COPY	UTBS					Save tables for iteration
000220	LPBEG						Begin of iteration
000230	COPY	UTBL					Load saved tables
000240	COPY	UDDO				*	Process deductions, Benefits
000250	COPY	ZDDO					Process deductions, Benefits
000260	COPY	UPNR				*	PS-NRA, 1042S wage process
000270	COPY	UTXO					Calculate taxes
000280	COPY	UPPT				*	PS-NRA, Post tax processing
000290	COPY	UGRN					Calculate garnishments
000300	COPY	UNAO					Calculate net
000310	COPY	UDNT					Deductions not taken during loop ?
000320	LPEND						End of iteration
000330	ENDIF						to: if non authorized manual check (*)
000340	BLOCK	END					
000350	COPY	UGRR					Garnishment Retroactive
000360	COPY	URRO					Retroactive accounting
000370	COPY	UNNO					Net processing
000380	COPY	UACO					Month end accruals
000390	XCODI	XCDO				*	Cost distribution
000400	COPY	UEND					Final processing

Figure 2.37 Schema ZHRF

2.3.3 Tables

Tables are used throughout SAP for storing different types of data. In Section 2.2, we discussed how all master data (employee data) is stored in tables, but for this section, we'll focus specifically on tables related to payroll processing. Table 2.14 lists a sample of some key payroll tables.

Table	Description
Table RT (Results Table)	Stores data after the payroll calculation.
Table CRT (Cumulative Tax Results)	Stores cumulative results as well as month-to-date (MTD) and year-to-date (YTD) accumulations.
Table TCRT (Tax Cumulative Tax Results)	Stores cumulative tax results as well as MTD and YTD accumulations.
Table IT (Input)	Used during processing only; passes its results to table RT.
Table DDNTK (Deductions Not Taken)	Stores the amount of deductions not able to be taken during processing.
Table ARRRS (Arrears)	Stores the deduction amounts not able to be taken during the pay period, which are to be carried over to the next pay period.
Table BT (Bank Table)	Payment information that includes bank information for /559 records.
Table WPBP (Work Place Basic Pay)	If changes for infotypes 0001, 0007, 0008, or 0027 have occurred, this table is created and is used to support payroll processing.

Table 2.14 Payroll Tables

2.3.4 Rules and Operations

Sometimes referred to as *personnel calculation rules (PCRs)*, a rule is a sequence of statements for calculating values in the payroll process. Each step in a rule corresponds to one operation. Rules are what gives Employee Central Payroll the flexibility and the capability to configure complex payroll calculations to meet your company's requirements. For a better understanding of what a rule looks like, let's look at rule **UW14**, which is called within schema **UBDO** by function **P0014**, as shown in Figure 2.38.

By double-clicking on **UW14** (line **000080**), you'll open rule UW14, as shown in Figure 2.39. If you need to update this rule, copy and create a new rule called ZW14 since you should never modify an SAP-delivered rule.

Cmmnd | Stack

Line	Func.	Par1	Par2	Par3	Par4	D	Text
000010	BLOCK	BEG					Basic data processing
000020	ENAME						Retrieve employee name
000030	WPBP						Read org. assignment / basic pay
000040	P0002						Read personal data
000050	P0006						Read address
000060	P0207	2					Read tax data
000070	GON						Continue with complete data
000080	P0014	UW14	GEN	NOAB			Split WPBP and set APZNR for P0014
000090	PRINT	NP	NAM				Print employee name
000100	PRINT		WPBP				Print org. assignment / basic pay
000110	PRINT		TAXR				Print resident authority
000120	PRINT		PERM				Print personal data
000130	IF		SPRN				If special run
000140	RFRSH		IT				Clear internal table IT
000150	ENDIF						Endif
000160	BLOCK	END					

Figure 2.38 Schema UBDO

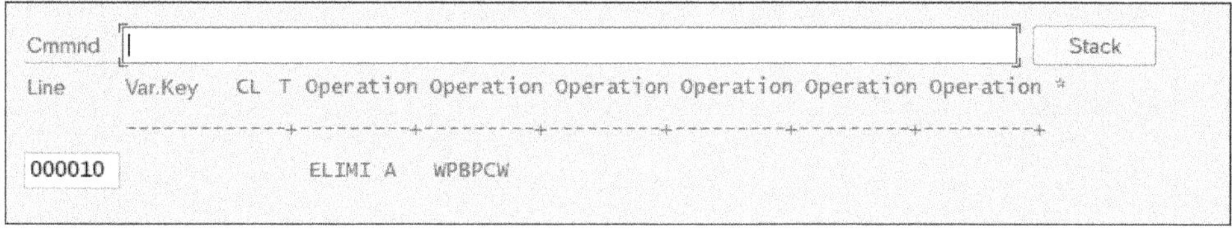

Cmmnd | Stack

Line	Var.Key	CL	T	Operation	Operation	Operation	Operation	Operation	Operation	*
000010				ELIMI A	WPBPCW					

Figure 2.39 Rule UW14

To copy or create a new rule, use Transaction PE02 to open the Rules Editor, as shown in Figure 2.40. From this screen, you can create or update a rule. Remember, never update a standard rule; instead, update a copy of a predelivered rule.

Rule UW14 …itting WPBP and assigning the split to IT

Create

Subobjects

- (●) Source text
  - ESGrp grouping: *
  - Wage/time type: ****
- ( ) Attribute
- ( ) Documentation

Display | Change

Figure 2.40 Transaction PE02 Rules Editor

Operations are used within a rule to represent actions. A large number of operations are available. Table 2.15 lists some common operations, but this list is not exhaustive.

Operation	Description
ADDWT*	Add the values of the current wage types to the cumulative results table.
AMT?0	Compare the value in the amount field to 0
ELIMI A	Eliminates splits for a work center period (WPBP)
NUM?0	Compare the value in the number field to 0
NUM=0	Set the number field to zero
PRINT	Prints the IT
RTE*100	Multiply the percentage in the rate field by 100
RTE/100	Divide the rate by 100
ZERO= AN	Sets the AMT and NUM to zero

Table 2.15 Payroll Operations

In this section, you learned about the standard US payroll driver RPCALCU0 and how to view and update schemas and rules and learned about operations. Next, let's look at some additional components, such as benefits.

2.4 Benefits

Benefits are a key component of US payroll systems. Some organizations outsource their benefits so that their administration is performed outside of the system, and deductions are sent to payroll to process (typically to infotype 0014 or infotype 0015, but sometimes to infotype 0169 as well). Other organizations may choose to implement a benefit module in SAP, and still other organizations will adopt a hybrid of the two models. Whatever the scenario, Employee Central Payroll can handle benefits processing. For more information on what data is stored and managed with benefits infotypes, see Section 2.2.4.

2.4.1 Types of Benefits

Four types of benefits can be managed within SAP, listed in Table 2.16. Employee-specific data for benefits are stored in infotypes, as discussed in Section 2.2.4.

Type of Benefit	Description	Infotype
Health: medical, vision, dental	Each plan will have plan types, coverage options, costs, and eligibility requirements.	0167
Insurance: long term disability (LTD), short term disability (STD), accidental death and dismemberment (AD&D), life	Each plan will have plan types, coverage options, costs, and eligibility requirements.	0168
Savings: 401k, Roth, deferred comp	Each plan can set eligibility and plan details.	0169
Flexible spending plans: health and dependent	Each plan will set eligibility and plan details.	0170

Table 2.16 Types of Benefits

2.4.2 Benefits Processing

Benefits are calculated in the payroll driver by calling various subschemas for calculation. They read the employee data stored in the benefits infotypes and the payroll configuration to produce benefit deductions. Table 2.17 lists the subschemas involved.

Subschema Name	Description
UBE1	Process Benefits (first time)
UBE2	Process Benefits (second time)
UBEN	Process Benefits Infotypes

Table 2.17 Benefit Subschemas

Depending on the employee selection as well as the configuration, employee and employer deductions will be calculated and stored as part of the payroll run. These selections will be stored in custom-defined wage types that you can build as part of your wage type catalog, as discussed in Section 2.1.7. Two additional standard, predelivered wage types will be calculated:

- Wage type /102 (401k Wages)
- Wage type /109 (ER Benefit Contributions)

These wage types can be used in payroll rules or calculations. For example, wage type /102 can be used to calculate a 401k deduction from the percentage stored in infotype 0169.

After payroll is processed, you must transmit the deductions and employee data to the benefit providers. A standard IDoc can be used where you'll download enrollment data and map and send that data to a provider. Some companies follow this approach, but other providers may have their own requirements. A company could also utilize middleware for this process.

2.5 Subsequent Processing

After payroll processing is completed, a set of activities generally must occur, frequently referred to as *subsequent processing*. The key activities that make up subsequent processing include finance postings, third-party remittances, and electronic funds transfers, including DME. Important aspects of each of these processes are covered in this section.

2.5.1 Finance Postings

Integration to a finance system is a crucial part of the overall payroll process. Employee Central Payroll is strongly integrated with SAP ERP Financials and SAP S/4HANA Finance but can also be integrated with another financial system. The integration points between Employee Central Payroll and either finance system from SAP are listed in Table 2.18.

Financial Elements	Payroll
Company codes	Infotype 0001 (Home Company Code)
Cost centers	Infotype 0001 (Home Cost Center) Infotype 0027 (Overriding Cost Center) Infotypes 0014, 0015, and 2010, which override cost centers for specific wage types
General ledger mapping	Mapping tables to map general ledger accounts to financial general ledger accounts

Table 2.18 Financial Integration with Payroll

As payroll is processed, results are stored based on the cost elements for an employee. After payroll is completed, these results will be posted to the financial system. Figure 2.41 shows an overview of the financial posting process, which is triggered by payroll being executed and payroll results being stored. Next, the program reads the results and the mapping tables and generates posting documents. Once successful and validated, these documents are sent to a financial system to update the general ledger for payroll.

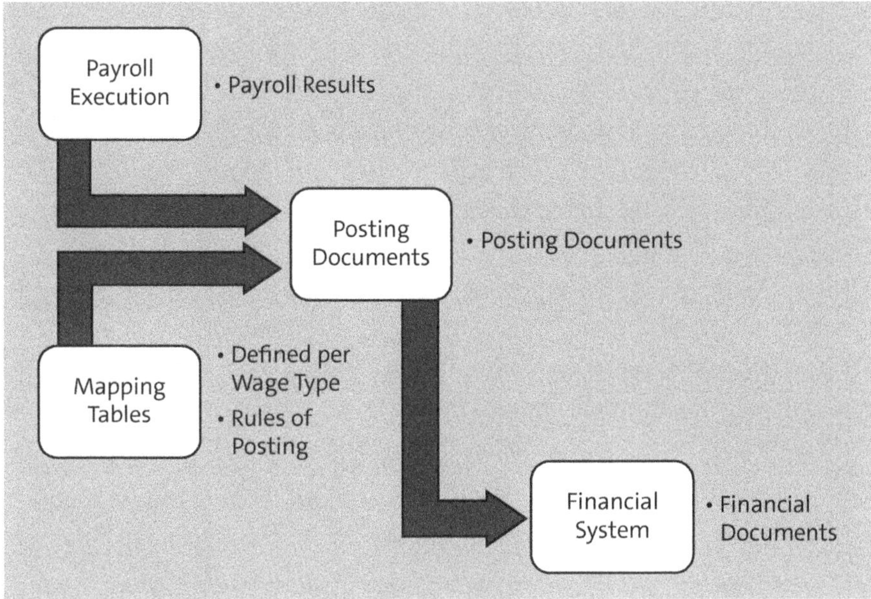

Figure 2.41 Financial Posting Process Overview

In this section, we'll discuss how to configure postings to both SAP ERP Financials and SAP S/4HANA Finance. The program for posting to finance is RPCIPE00 or RPCIPE01 (which allows for parallel processing in posting runs). However, program RPCIPE01CE is used for concurrent employment.

We'll start by setting up symbolic accounts. Wage types are not directly assigned to accounts in SAP ERP Financials and SAP S/4HANA Finance; instead, they are mapped to symbolic accounts. Each symbolic account can be mapped to multiple wage types and is also assigned to an account assignment, which specifies the type of account it is. This symbolic account will be assigned to a general ledger account. Figure 2.42 shows some sample symbolic accounts.

Symbolic Accounts

SymAc	Description	AATyp	MOMAG	Fixed MOM	Negative Postg	Neg. Posting Debit
		C				
0732	Vacation payment (SE)	C	✓			
1100	Wages and salaries	C	✓			
1110	Direct labor costs	C	✓			
1120	NBI (Public Sector France)	C	✓			
1190	Other Wage and Salary Expenses	C	✓			
1210	Emloyer Tax expense	C	✓			
1220	Shift bonus	C	✓			
1221	Work center bonus	C	✓			
1230	Remuneration for idle time	C	✓			

Figure 2.42 Symbolic Accounts

Key fields that must be maintained to configure symbolic accounts include the following:

- **SymAc (symbolic account)**
  Unique number used to map wage types to symbolic accounts.
- **Description**
  Define a description that best indicates the type of account.
- **AATyp (account type)**
  Defines the account type ("C" is for posting to an expense account, "F" for posting to a balance sheet account).
- **MOMAG (grouping for account determination)**
  Select this flag if the mapping will require the use of an employee grouping account to determine the specific account to use for the symbolic account.

If requirements are too complex to use the configuration tables for the mapping, you may need to use feature PPMOD, as shown in Figure 2.43.

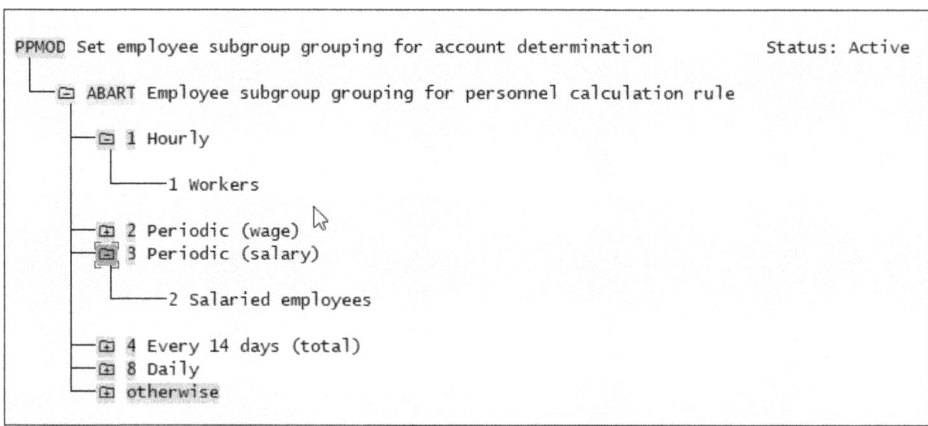

Figure 2.43 Feature PPMOD

With your symbolic accounts defined, you can now update the wage types by assigning the applicable symbolic accounts. Figure 2.44 shows an earning wage type that has a mapping based on whether it's a positive or negative value. If needed, a month-end accruals mapping can also be stored.

Dialog Structure
Wage Type Overview
Posting a Wage Type
Posting US Tax

Wage Type 0010 Regular Pay - Salary
End Date 31.12.9999

Posting a Wage Type

No	+/- sign	Proc.Type	SymAcctAss	Description of SymAcctAss	AATyp	Ignore Cost ...
1	Positive amounts in debit, negative amounts in cred...	Normal...	1100	Wages and salaries	C	
9	Positive amounts in debit, negative amounts in cred...	Month en...	1100	Wages and salaries	C	
10	Positive amounts in credit, negative amounts in deb...	Month ...	2110	Wages and salaries payable	FO	

Figure 2.44 Wage Type Mapping to Symbolic Earnings Account

Key fields must be maintained to configure the postings of an earning wage type, including the following:

- **No (sequence number)**
  Unique number for each posting type.
- **Proc.Type (processing type)**
  To distinguish between normal versus month-end accrual processing.
- **SymAcctAss (symbolic account assigned)**
  Defines the symbolic account to which you want to map the wage type.

Mapping for taxes is different, as shown in Figure 2.45. For taxes, you must include all the appropriate tax authorities in this table. If not already found in the table, you can copy or create a new entry.

Dialog Structure
- Wage Type Overview
  - Posting a Wage Type
  - Posting US Tax

Posting US Tax

Wa...	Wage Type Long Text	TxCo	Description	Auth	Description	Start Date	End Date	No	A...	Sy...	Description	+/- sign
/40..	TX Withholding Tax			AK	Alaska	01.01.1990	31.12.9999	1	FC	AK		- Positive amounts in cr
/40..	TX Withholding Tax			AL	Alabama	01.01.1990	31.12.9999	1	FC	AL		- Positive amounts in cr
/40..	TX Withholding Tax			AL0..	Attalla	01.01.1990	31.12.9999	1	FC	AL0..		- Positive amounts in cr
/40..	TX Withholding Tax			AL0..	Bessemer	01.01.1990	31.12.9999	1	FC	AL0..		- Positive amounts in cr
/40..	TX Withholding Tax			AL0..	Brilliant Town	01.01.1990	31.12.9999	1	FC	AL0..		- Positive amounts in cr
/40..	TX Withholding Tax			AL0..	Gadsen	01.01.1990	31.12.9999	1	FC	AL0..		- Positive amounts in cr

Figure 2.45 Mapping from Wage Type to Symbolic Account for Tax Wage Type

In the next step, you'll generate the posting date for each payroll period of each payroll area which we'll discuss further in Section 4.2.6. Once generated, you can update these dates as needed. This process is similar to when generating payroll periods. A posting variant must also be created, as shown in Figure 2.46.

Posting variant: ZUS
Name: ECP Posting Variant

Parameters for document creation

Country Grouping: 10
Run type: PP
Document Type: SB
Language for line item text:
Text key:
Prefix for reference: XXXXX
☐ Post to co.code clearing accounts

Document split

Doc. split: Document Split Based on Company Code
Additional document split acc.to no.of lines:
☐ Separate summarized items
☐ Fixed cost posting to summarized items
Doc.Type Customers:
Creditors Doc. Type:
DocTyp MEA:

Figure 2.46 Posting Variant

The key fields that must be maintained define a posting variant include the following:

- **Posting variant**
  Unique name for variant.
- **Description**
  Description defining the variant.
- **Country Grouping**
  Define for the country you are processing payroll for. The country grouping for the US is 10.
- **Document Type**
  This classification represents the document type for the posting to accounting. All documents posted with this variant will use this defined document type.

We've mapped wage types to symbolic accounts, and now, we need to assign symbolic accounts to general ledger accounts as part of the account assignment. This assignment is performed for each account type: financial, expense, technical, customer, and vendor, as shown in Figure 2.47.

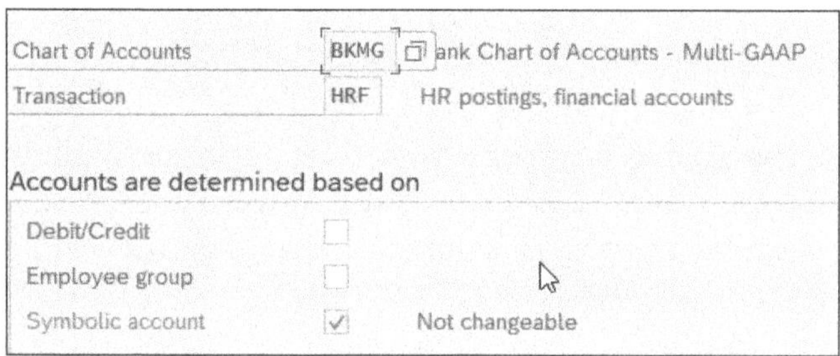

Figure 2.47 Account Assignment

The following key fields must be maintained to determine the account assignment for financial accounts:

- **Chart of Accounts**
  Select the chart of accounts for the company.
- **Transaction**
  This field always defaults based on the type of account—financial, expense, technical, customer, or vendor.
- **Employee group**
  Select if any of the accounts must be driven by the type of employee, for example, for postings for salaried employees versus hourly.

Once this information is saved, the system will then assign symbolic accounts that are aligned to the financial accounts (see Figure 2.48).

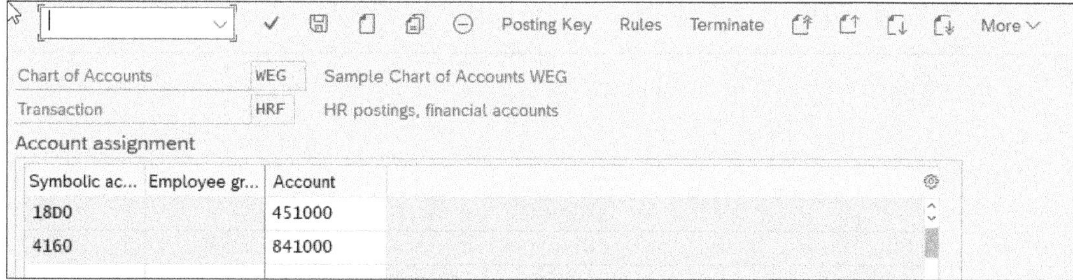

Figure 2.48 Symbolic Accounts Assigned

The following key fields must be maintained to determine the account assignment for financial accounts:

- **Symbolic account**
  List all the symbolic accounts that aligned to the financial accounts.
- **Account**
  Enter the account that is aligned to the appropriate symbolic account.

2.5.2 Third-Party Remittances

Third-party remittance (also sometimes referred to as a third-party process or 3PR) is the process to create payment requests with corresponding financial postings for vendor payments due from payroll deductions. The four primary types are taxes, benefits, garnishments, and "other." This standard process reads the payroll results, posts the data to SAP ERP Financials or SAP S/4HANA Finance, and generates the bank transfer and check payment requests due to third parties. As part of setting up wage types, you should have configured processing classes 73 and 78, which process deductions for third-party remittances and tell the system how to treat the sign (positive or negative) of the wage type. Figure 2.49 shows how this process begins by reading the payroll results from Employee Central Payroll and the rules configured for the remittance. Then, the posting documents are created, which are then reviewed and released, and finally posts to finance are made with vendor vouchers and posting documents.

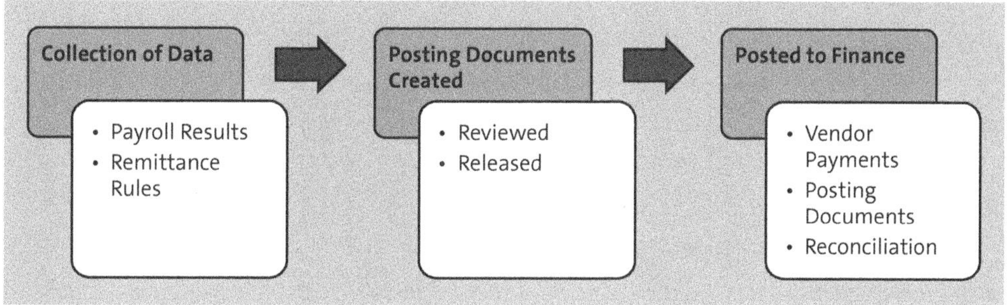

Figure 2.49 Third-Party Process Overview

We'll start by defining the remittance rules, as needed for the various payment requests, by following menu path **Payroll • Payroll: USA • Third-Party Remittance • Remittance Rule and Accumulator Maintenance • Define Rules**. This step is a prerequisite for defining the HR payee later.

When setting up the remittance rule, you'll need to identify "when" the payment should be sent to the vendor and "how" or what payment method should be used. Each remittance rule is a unique set of rules that can be used across HR payees, as shown in Figure 2.50.

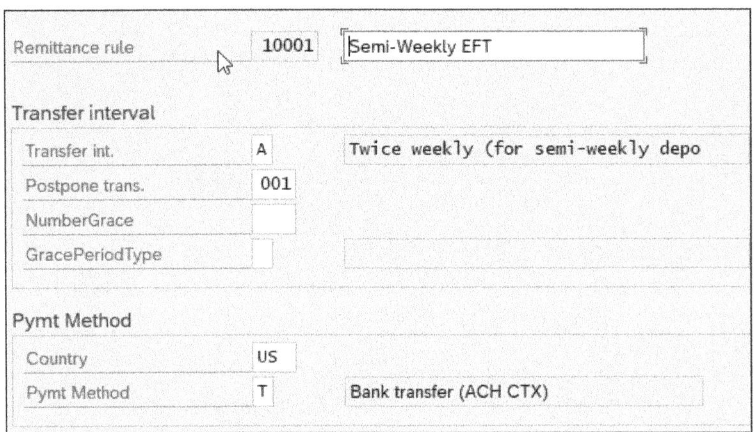

Figure 2.50 A Remittance Rule

The following key fields must be maintained to define a remittance rule:

- **Remittance rule**
  Unique rule identifier and rule name. Make the rule name descriptive.
- **Transfer int. (transfer interval)**
  Select the appropriate indicator for the transfer rules (e.g., weekly, monthly).
- **Postpone trans. (postpone transfer)**
  This field represents the period of time required for processing, which will be subtracted from the due date. For example, if accounts payable needs 4 days to process a check, enter the number "4."
- **NumberGrace**
  The number of days to add to the due date if needed.
- **GracePeriodType**
  Select the period of time for the grace period (in calendar days, workdays, weeks, etc.).
- **Country**
  Country code for the paying company.
- **Payment Method**
  Defines the how the payment will be made (e.g., via bank transfer, check, etc.).

Now that we've defined the remittance rule, we're ready to define the HR payees that will be part of the third-party process by following the menu path **Payroll • Payroll: USA • Third-Party Remittance • HR Payee Maintenance • Set Up HR Payees for Tax Remittance.** This step will open configuration table shown in Figure 2.51.

Dialog Structure
- HR payee assignment
  - Remittance rule details for
  - Dependent HR payees
  - Tax Wage Types for HR pa

HR payee type: 2 Tax payee(s)

HR payee assignment

HR payee	Text	Co...	Vendor	Name
90000001	Federal Withholding Tax	USO...		
90000003	EE OASDI	USO...		
90000004	ER OASDI	USO...		
90000005	EE Medicare	USO...		
90000006	ER Medicare	USO...		
90000010	Federal Unemployment	USO...		
90010001	Alabama income Tax	USO...		
90010010	Alabama unemployment tax	USO...		
90020001	Alaska income tax	USO...		

Figure 2.51 HR Payee Tax Configuration Table

This step will display all previously defined HR payees for taxes, with HR payee type set to **2: Tax Payments**. You'll need to create an HR payee assignment for each type of tax payment you would like included in the process. Figure 2.52 shows an existing record for a tax payee.

HR payee type	Tax payee(s)	
HR payee	90000001	Federal Withholding Tax
Payee		
Vendor		
Company Code	US01	
Vendor subgrp.	1	
HR trans.accumulator		
HR accumulator	1	100,000 counter for US01
☐ Data at employee level		

Figure 2.52 Creating a New Tax HR Payee

The key fields needed to define an HR payee type rule include the following:

- **HR payee type**
  Defaulted to **2: Tax payee(s)**.

- **HR Payee**
  Defined number given for the unique payee as well as defined name.
- **Vendor**
  This number represents the vendor for which you want to create a payment voucher. Each vendor has a unique vendor number and can be selected from a drop-down or entered in this field.
- **Company Code**
  Company code of the vendor.
- **Vendor subgroup**
  Used if multiple HR payees are to be processed to one vendor, and you want the documents kept separately, grouped by HR payee.
- **HR Accumulator**
  If remittances require a special rule to override standard processing if specific thresholds are set. Example, for the US, according to the $100,000 next-day deposit rule, if you collect over $100,000 in taxes in one day, you must immediately deposit the tax collected by the next day. Rather than follow the standard process, would want to set this requirement up in this field.
- **Data at employee level**
  Select this flag if you need to store detailed data for each employee for the remittance, rather than consolidated within the HR creditor.

Next, we want to select the remittance rule details for the HR payee to set the remittance rule, as shown in Figure 2.53.

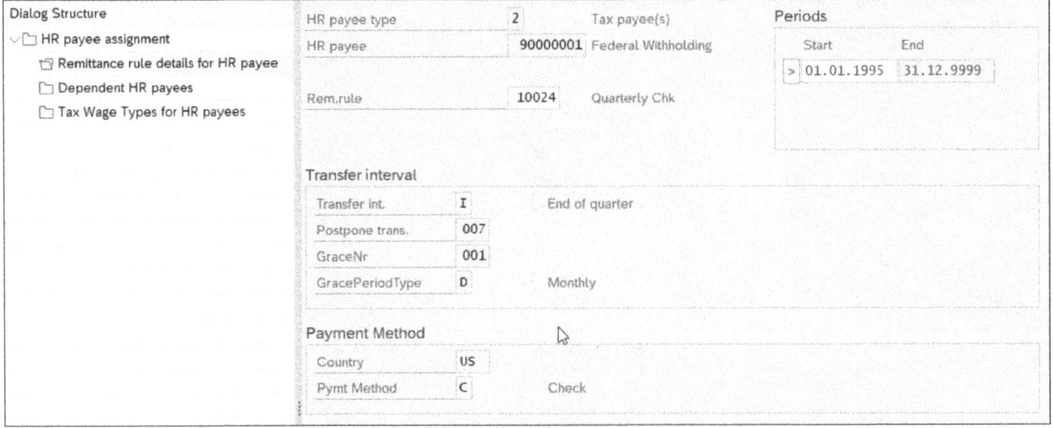

Figure 2.53 Remittance Rule

The final step is to align the wage types to the HR payee, as shown in Figure 2.54.

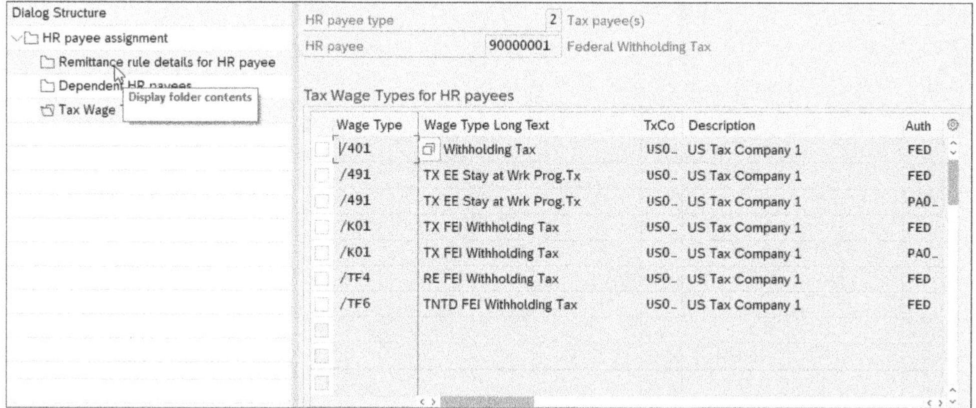

Figure 2.54 Tax Types for HR Payees

Create an entry for each combination of wage type, tax company, and authority that is aligned to the HR payee in the configuration table. Consider the following guidance for maintaining these fields:

- **Wage Type**
  Enter the wage type number aligned to the HR payee.
- **Tax Company**
  Legal entity for the tax company.
- **Authority**
  The tax authority for the HR payee.

To set up HR payees for regular remittance (HR payee 1), follow the menu path **Payroll • Payroll: USA • Third-Party Remittance • HR Payee Maintenance • Set Up HR Payees for Regular Remittance**. To maintain HR payees for benefits (HR payee 3), follow the menu path **Third-Party Remittance • HR Payee Maintenance • Maintain HR Payees for Benefits**. Then, follow the same process as setting up tax remittance HR payees, but with a different HR payee type. Figure 2.55 shows the different types of HR payees available.

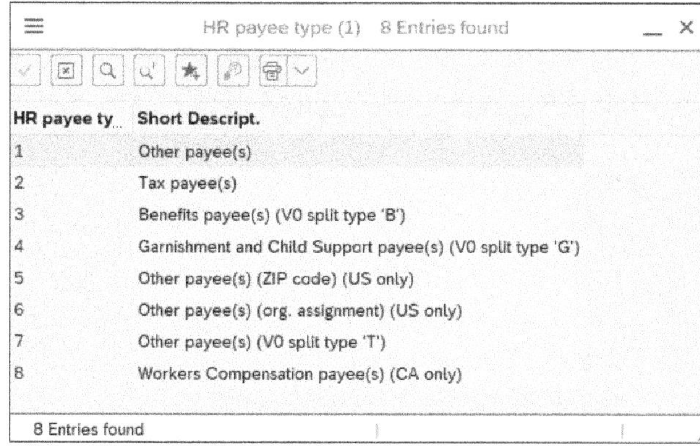

Figure 2.55 HR Payee Types

Now that the wage types have been updated, remittance rules have been created, and HR payees have been set up, the final step is setting up number ranges. You must set up separate number ranges up for remittance originators, vendor subgroups, and garnishment payees. To set up number ranges, follow the menu path **Payroll • Payroll: USA • Third-Party Remittance • Number Ranges • Maintain Remittance Originator Number Ranges**. On the screen shown in Figure 2.56, create a number range for the remittance process.

Note that you can create a single range to cover all remittance payees; there's no need to split this number range, as earlier when creating number ranges for vendor subgroup and garnishment payees.

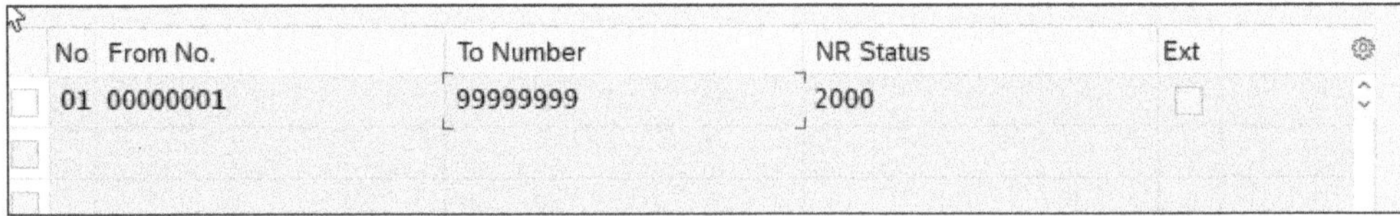

Figure 2.56 Number Range for Remittance

2.5.3 Electronic Funds Transfers, Including Data Medium Exchange

The SAP process to create checks or ACH payments after payroll is run is DME. Several key processes are included in DME. To set up prenotifications, follow the IMG path **Payroll • Payroll: USA • Electronic Funds Transfer (EFT) • Prenotification • Set Up Prenotification**.

Prenotification is a process where a test transfer is sent to the bank, prior to money being transferred for the first time. No money is sent during this test; it simply validates that the bank information is correct. Not all companies choose to turn this functionality on.

Another process is the ability to print a zero net check. In this situation, an employee does not have wage type /559 (Bank Transfer) or /558 (Payment of Balance), but your company would like to produce a remuneration statement for the employee anyway.

To process payments through DME, you must set up banks with your corresponding accounts from which payments will be made. In this scenario, we assume that house banks and associated accounts have already been configured—a finance task, not a payroll task. As shown in Figure 2.57, we've assigned house bank Z004 to company code ZA01.

Feature DTAKT assigns the house bank and account number for the employee's company code to the bank transfers. Figure 2.58 shows an example of feature DTAKT for a company. The first 5 characters—CHEMM are the code for the bank, followed by a /, and then 27 stands for the bank account code, which could be up to 5 digits.

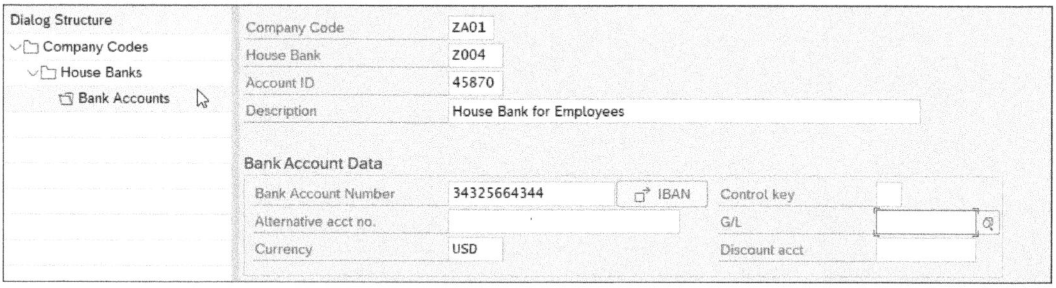

Figure 2.57 House Bank

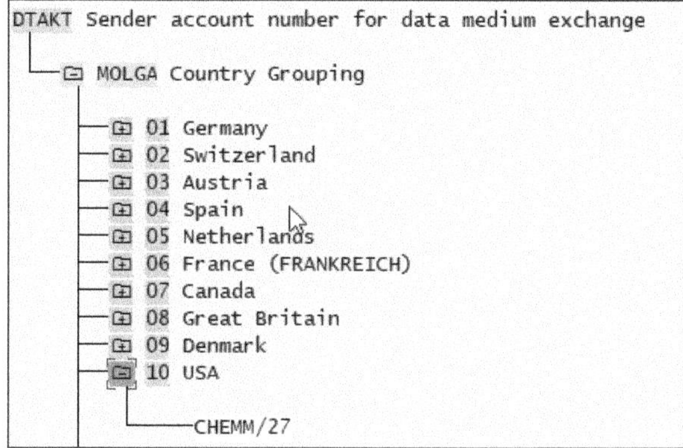

Figure 2.58 Feature DTAKT

2.6 Summary

In this chapter, we covered key elements needed to process payroll—from infotypes to payroll drivers, including schemas, rules, and operations. Employee Central Payroll is powerful and flexible enough to meet your company's unique business requirements. Understanding how to update the system to meet your specific needs is key. Chapter 7 and Chapter 8 will teach you how infotypes are updated and replicated from Employee Central. Chapter 1 included the overall payroll process, and in this chapter, we walked through how you can set up the post-payroll processes to process payments, financial postings, and third-party processes.

Chapter 3
System Provisioning

Employee Central is required for Employee Central Payroll to function. This book is written with the viewpoint that Employee Central is your human resources (HR) system. While enabling an Employee Central system for HR use, the settings discussed in this chapter may have already been configured. However, a good idea is to check since these settings are related to payroll and may not have been tested before your Employee Central Payroll journey begins. After all, the start of a good payroll system is a good HR system. Thus, most companies implement Employee Central at least a year ahead of implementing Employee Central Payroll.

This chapter is a precursor to enabling Employee Central for payroll. Although Employee Central is mostly configurable from the admin screens in the main Employee Central application, some required settings must be configured in a separate enabling area called *provisioning*. If you compare this with the traditional SAP ERP HCM Payroll, this step would probably be classified as a Basis task that technical support staff perform before configuration activities can begin.

Think of provisioning as the backend of Employee Central. In this chapter, we'll walk you through how to use provisioning to activate the Employee Central Payroll integration in Section 3.1. Then, you'll learn how to set up a succession data model in Section 3.2, a corporate data model in Section 3.3, and a country-specific data model in Section 3.4. We'll set up the web service in Section 3.5 and enable push replication in Section 3.6.

Note that provisioning tasks are not directly available to most organizations. To access the provisioning system for your implementation environment for Employee Central, you need to be an SAP partner certified in Employee Central.

3.1 Activating Employee Central Payroll Integration

To activate Employee Central Payroll, follow these steps:

1. Log on to provisioning for your Company_ID, which identifies your unique area in the SAP data center where your Employee Central data and applications are housed. Different Company_IDs are assigned to different environments. For example, for a company called xTravel, the development system could have xTravelD as its Company_ID, with xTravelT for the quality system, and xTravelP for the production system.

2. Once there, go to **Edit Company Settings** and then to **Company Settings**.
3. Enable Employee Central Payroll and the Payroll Control Center by selecting the following options, as shown at the bottom of Figure 3.1:
   - **Enable Employee Central Payroll**
   - **Enable Payroll Control Center**

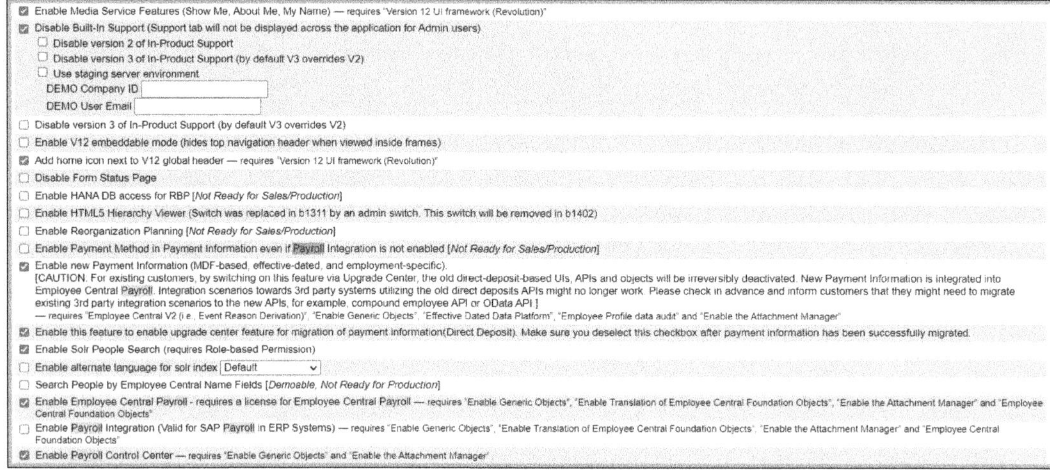

Figure 3.1 Enabling the Employee Central Payroll Module in Provisioning

4. Enable cost distribution by selecting the **Enable Cost Distribution** checkbox; in our example shown in Figure 3.2, this option is the second checkbox that has been selected.

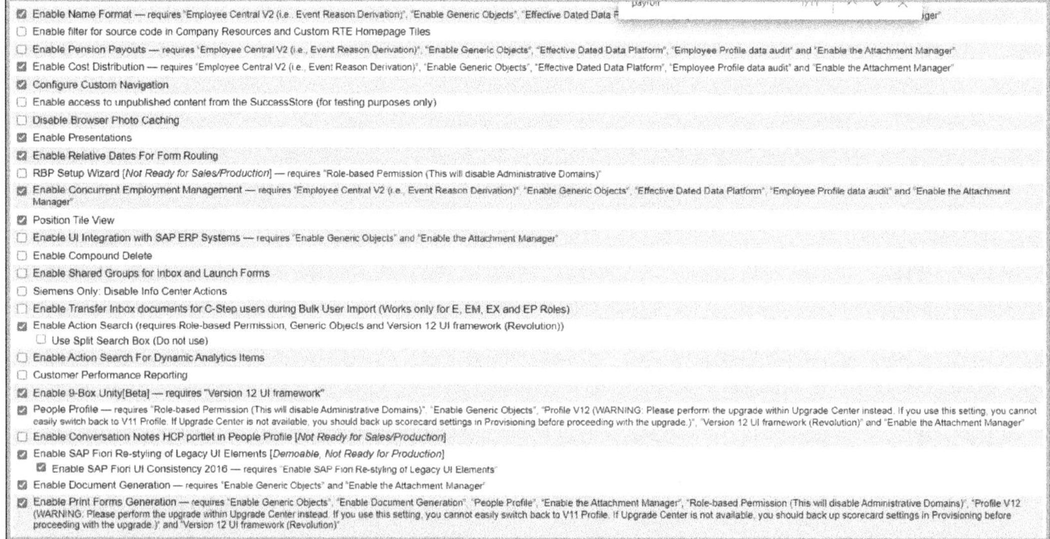

Figure 3.2 Enabling Cost Distribution in Provisioning

3.2 Setting Up the Succession Data Model for Master Data Replication

A succession data model in Employee Central holds the structure and rules of the data about employees, in XML files. The model consists of human resources information system (HRIS) elements that require maintenance in provisioning. Many of these elements have moved to the **Admin** screen in Employee Central so that clients can have access to modify them.

Certain fields must be enabled in this model. In the succession data model's XML file, characteristics of the fields and their properties are identified by tags. These models are extremely sensitive and are not available for customers to make changes. SAP has been transferring these model fields into Metadata Framework (MDF) structures that you can configure from the Employee Central application itself.

All the fields discussed in this section are available from the **Admin** screen in Employee Central by typing "Manage Business Configuration" in the search bar. Once on this screen, the important settings you must maintain require the following steps:

1. Select **Employee Central • HRIS Elements • personInfo** from the menu on the left, as shown in Figure 3.3. On this screen, fill out the following information:
   - In the **Identifier** column, locate **place-of-birth** and change **Enabled** to **Yes** and **Mandatory** to **No**.
   - In the **Identifier** column, locate **birth-name** and change **Enabled** to **No** and **Mandatory** to **No**.

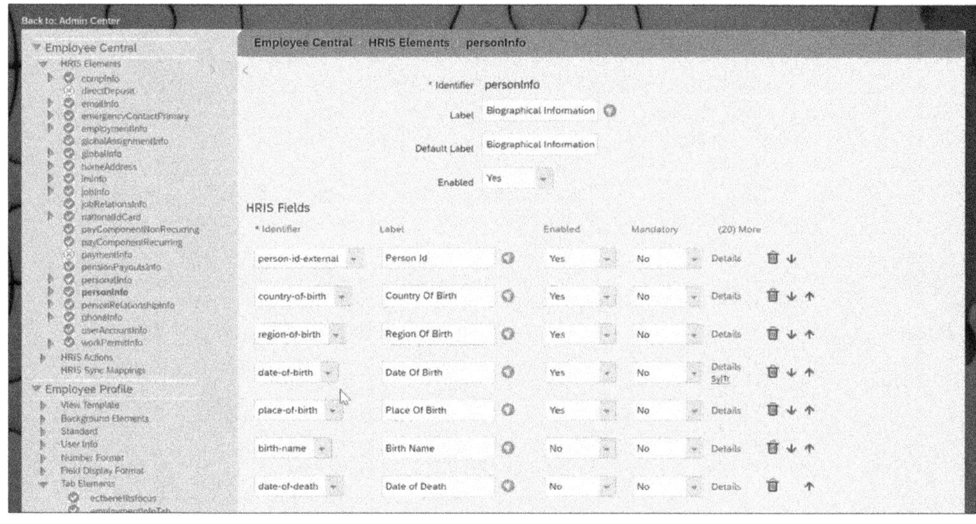

Figure 3.3 Changing the Settings for the personInfo Object

2. Next, select **personalInfo** from the menu on the left. In the **Identifier** column, locate **nationality** and change **Enabled** to **Yes** and **Mandatory** to **No**, as shown in Figure 3.4.

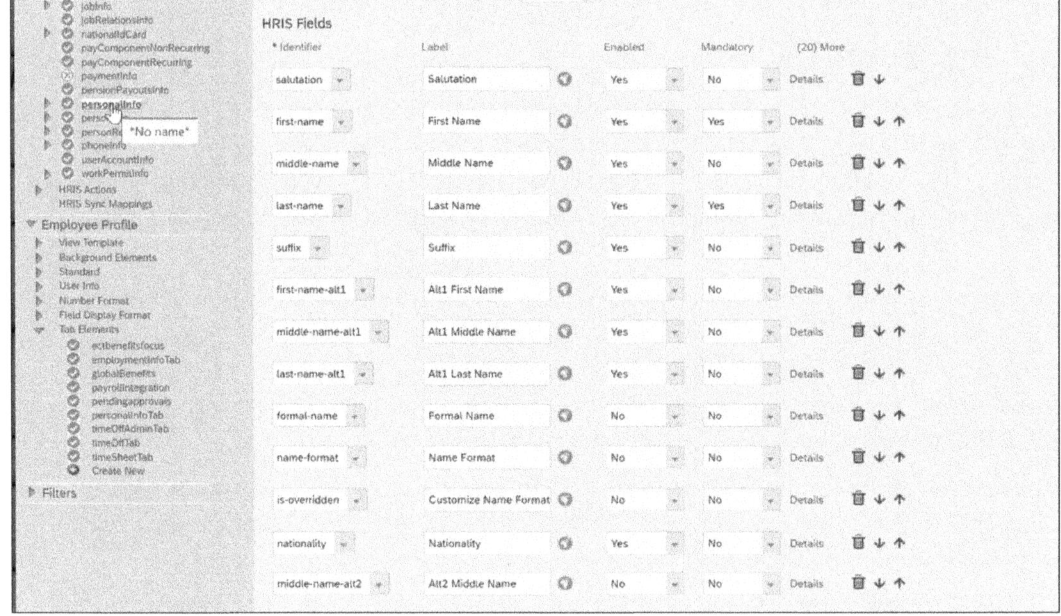

Figure 3.4 Changing the Settings for the personalInfo Object

3. Then, select **employmentInfo** from the menu on the left. In the **Identifier** column, locate **firstDateWorked** (US only) and change **Enabled** to **Yes** and **Mandatory** to **No**, as shown in Figure 3.5.

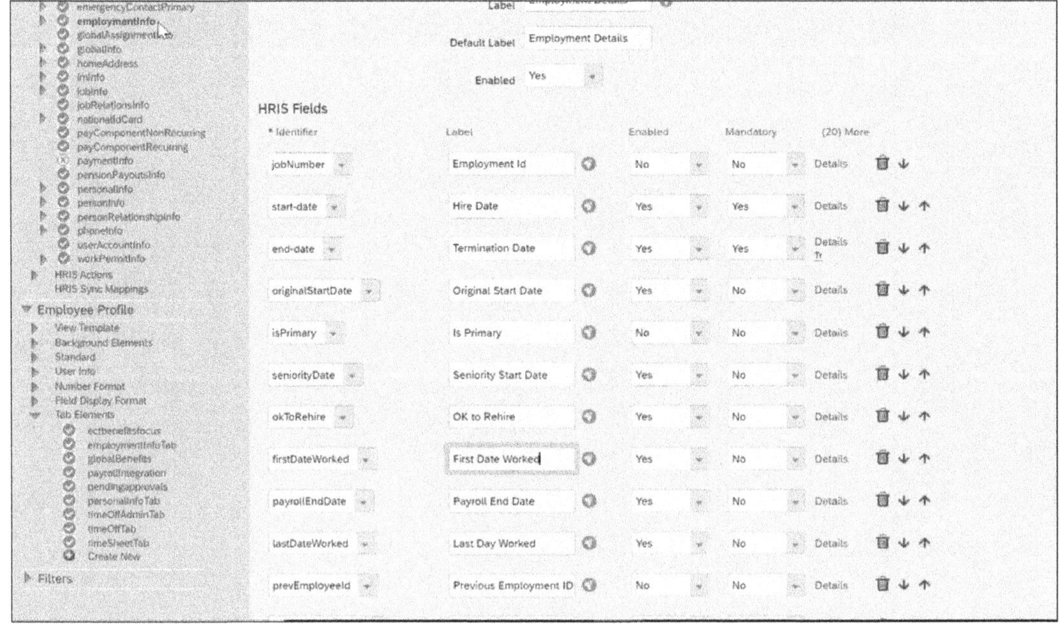

Figure 3.5 Changing the Settings for the employmentInfo Object

4. Select **jobInfo** from the menu on the left. Fill out the following information, as shown in Figure 3.6:
   - In the **Identifier** column, locate **workingDaysPerWeek** and change **Enabled** to **Yes**.
   - In the **Identifier** column, locate **employment-type** and change **Enabled** to **Yes**.

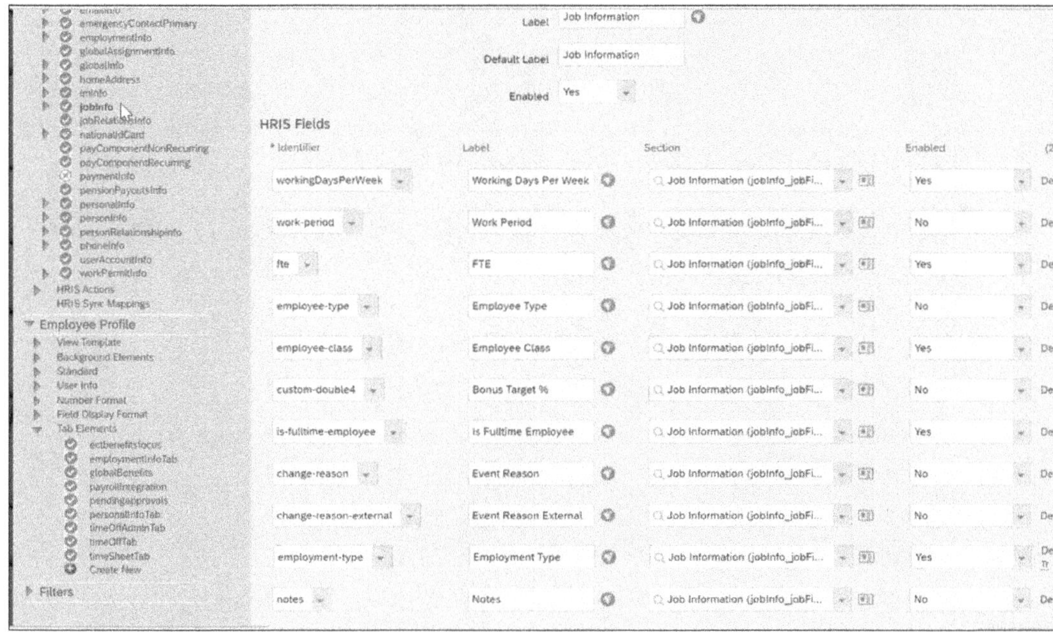

Figure 3.6 Settings for the jobInfo Object

5. Select **Employee Central • HRIS Elements • Employee Profile**. For the element **payroll-Integration**, create the label "Payroll Information," as shown in Figure 3.7.

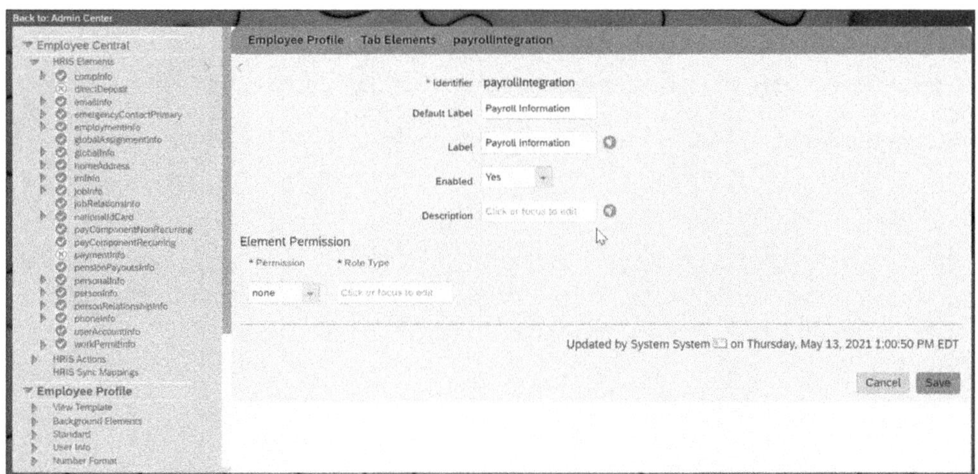

Figure 3.7 Changing the Settings for payrollIntegration under Employee Profile

3.3 Setting Up the Corporate Data Model

Data elements in Employee Central that are essential but not dependent on an employee are called *foundation objects*. These objects are defined in a data model using XML that is similar to the succession data model described in the previous section.

A corporate data model might have been already set up with the necessary fields during the Employee Central implementation. If any were not set up, you'll need to configure them for use with Employee Central Payroll. However, limitations in field length in Employee Central Payroll may cause problems for the payroll implementation. Table 3.1 lists typical fields that may cause issues.

Field	Employee Central	MDF Object	Employee Central Payroll
Company (Legal Entity)	32 characters	Yes	4 characters
Location	32 characters		4 characters
Cost Center	32 characters	Yes	10 characters
Job Classification	32 characters	Yes	10 characters
Pay Group	32 characters	Yes	4 characters
Pay Component	32 characters		4 characters
Frequency	32 characters		2 characters

Table 3.1 Differences in Field Lengths for Critical Fields

If, for example, company codes are set up with 6 characters during the Employee Central implementation, they cannot be replicated as they are into Employee Central Payroll where this field is only 4 characters. You have two options to resolve this issue:

- Reduce the number of characters in Employee Central
- Use a mapping in the Employee Central Payroll system during replication

Having the same code for these important fields in both systems increases visibility and operational efficiency, and thus, the first choice preferable, although you can choose either option for any field.

Changing field lengths can be performed either by a partner using the corporate data model, as shown in Figure 3.8, or by a client using the **Configure Object Definitions** screen in Employee Central, as shown in Figure 3.9.

```
<hris-element id="frequency">
  <label>Frequency</label>
  <hris-field max-length="32" id="externalCode" visibility="both" required="true">
    <label>Frequency ID</label>
  </hris-field>
  <hris-field max-length="32" id="name" visibility="both">
    <label>Name</label>
  </hris-field>
  <hris-field max-length="128" id="description" visibility="both">
    <label>Description</label>
  </hris-field>
  <hris-field id="annualizationFactor" visibility="both" required="true">
    <label>Annualization Factor</label>
  </hris-field>
  <hris-field max-length="256" id="custom-string1" visibility="none">
    <label>Custom String1</label>
  </hris-field>
```

Figure 3.8 Setting the Length of Frequency Using the Corporate Data Model

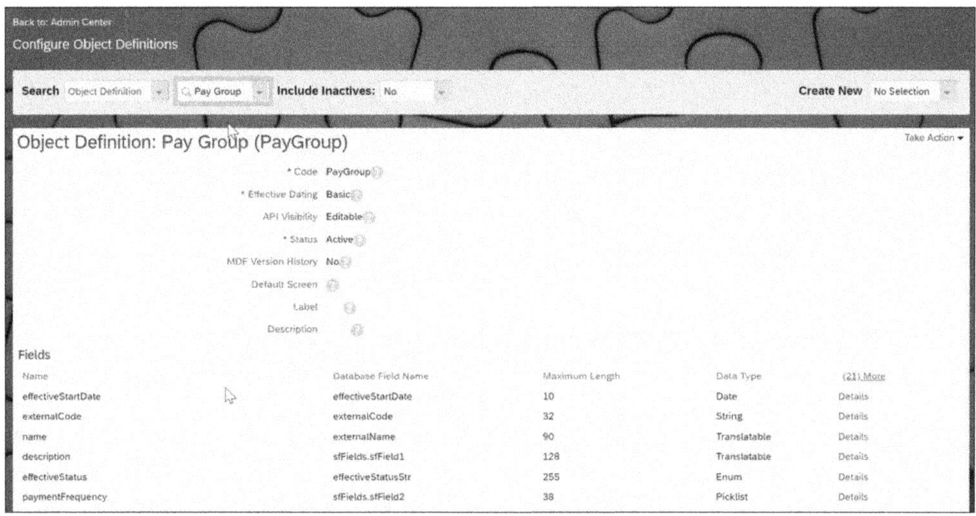

Figure 3.9 Setting the Field Length for Pay Group (externalCode)

3.4 Setting Up the Country-Specific Succession Data Model

The next phase of system provisioning is to set up country-specific succession data models. The following fields that record basic pay in Employee Central Payroll must be enabled in Employee Central and replicated back to Employee Central Payroll:

- Pay scale type (**payScaleType**)
- Pay scale area (**payScaleArea**)
- Pay scale group (**payScaleGroup**)
- Pay scale level (**payScaleLevel**)

These fields uniquely define an hourly rate and sets it up for progression if required. These fields reside in the country-specific corporate data model for the jobInfo object.

Similar to the succession data model fields set up in Section 3.2, these fields can be activated by partners using the country-specific data model or by clients by toggling the **Enabled** setting to **Yes** through the **Manage Business Configuration** screen shown in Figure 3.10.

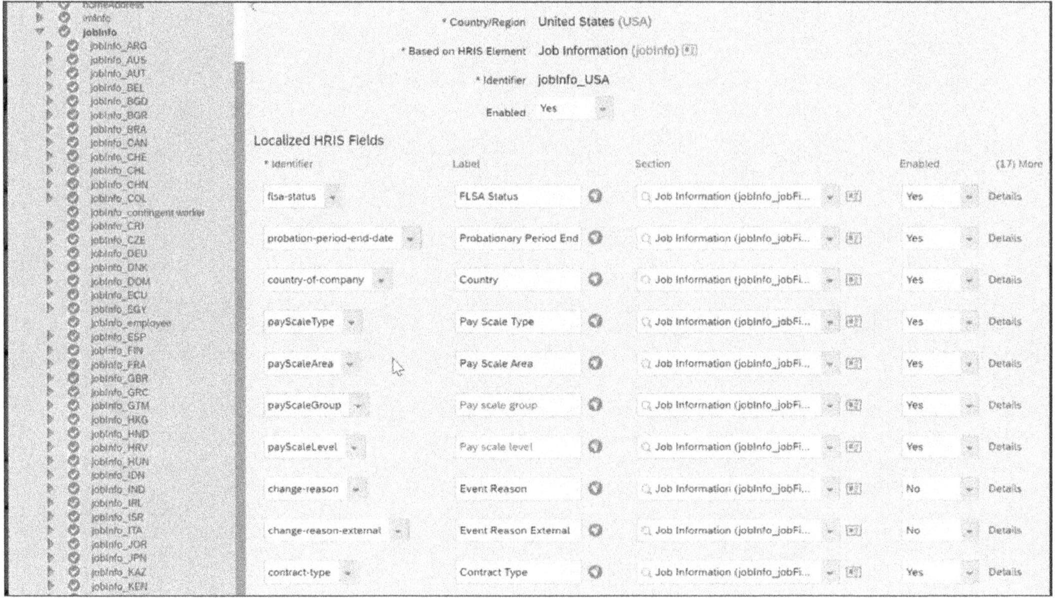

Figure 3.10 Enabling Pay Scale Fields Using Manage Business Configuration

3.5 Setting Up the Web Service

To set up web services for Employee Central Payroll, go to **Edit Company Settings • Company Settings** in provisioning. In the **Web Services** section, select the following checkboxes, as shown in Figure 3.11:

- **SFAPI**
- **Employee Central SOAP API**

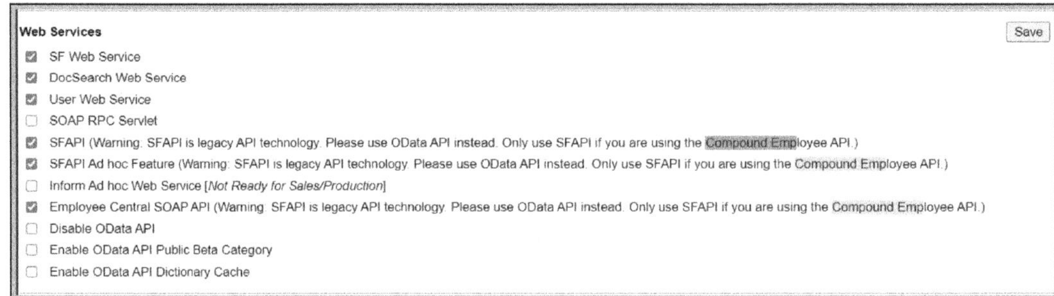

Figure 3.11 Enabling Web Services in Provisioning

3.6 Enabling Push Replication from Provisioning

To enable push replication from provisioning, go to **Edit Company Settings** and then to **Company Settings**. Mark the following checkboxes, as shown at the top of Figure 3.12:

- **Enable Intelligent Services**
- **Enable Intelligent Services Center (ISC)**

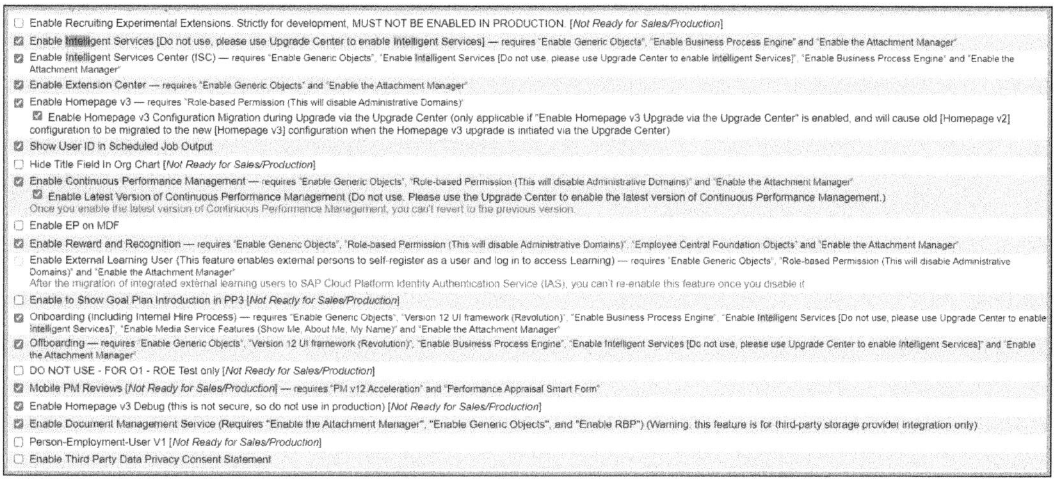

Figure 3.12 Enabling Push Replication Using Provisioning

3.7 Summary

This chapter prepares you for implementing Employee Central in the next chapter. The features discussed in the next chapter will not work until the settings in this chapter are completed. But these settings may need to be done by SAP or your implementation partner since you may lack access to provisioning as an end-user client. As SAP continues to migrate these settings to MDF, more and more of these fields will be available in the **Admin** screen, accessible to the client.

Chapter 4
Configuring Employee Central

Employee Central Payroll depends on Employee Central for its sleek user interface (UI) and thus, unsurprisingly, some configuration is required in Employee Central to enable the features and capabilities of Employee Central Payroll. The best part of Employee Central Payroll is that it is fully integrated with Employee Central, which enables SAP to offer a high level of automation and improved usability.

This chapter is about enabling Employee Central for payroll. Employee Central is a feature-rich human resources (HR) system that specializes in employee processes starting from hiring to termination and retirement. This solution does not perform all these tasks this alone. Employee Central integrates with many other systems to cover two important aspects of systems that manage people in an organization:

- Taking control of all employee data
- Seamlessly integrating with other systems so that the user experiences a unified process with no boundaries

In this chapter, we'll walk through establishing permissions for administrators and general users to use Employee Central Payroll in Section 4.1. We'll then review the basic configuration of the UI in Employee Central Payroll in Section 4.2. In Section 4.3, we'll discuss how to set up payroll tasks, and we'll close the chapter in Section 4.4 by building the payroll information block.

4.1 Permissions for Employee Central Payroll

Before you can perform any task in Employee Central, you'll need permission. Unlike other systems, where you start using the system with full access and then restrict each user to what a user is allowed to use, Employee Central requires that every feature be enabled individually before it can be accessed.

Setting up permissions is not as complicated as it sounds. In Employee Central, permissions are set up through *roles*, as shown in Figure 4.1. You can assign both a role and a target for that role. In this way, two users with the same role may have completely different targets. This distinction enables the same system function for two users who do similar jobs in the company. However, the target feature protects the visibility of certain data to each user and segregates duties when required.

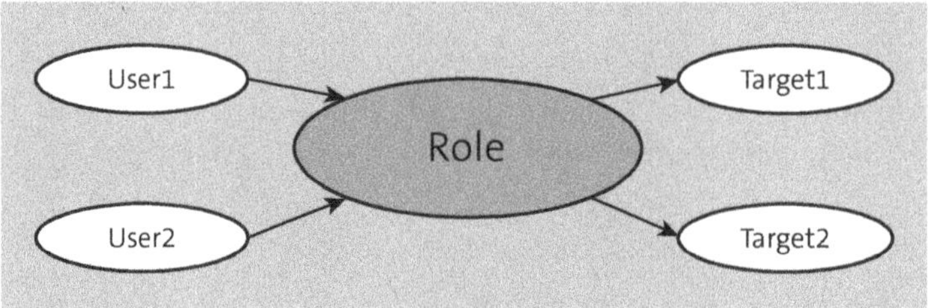

Figure 4.1 Role-Based Permissions in Employee Central

Two roles need to be set up in Employee Central:

- Administrator
- General user

4.1.1 Administrator Role

To set up any role, go to the Admin Center, open the **Tools** menu, and select **Manage Permission Roles**. Click **Take action** next to the role that you have designed for administrators and select **Edit**. In our example shown in Figure 4.2, we're editing the **System Admin** role.

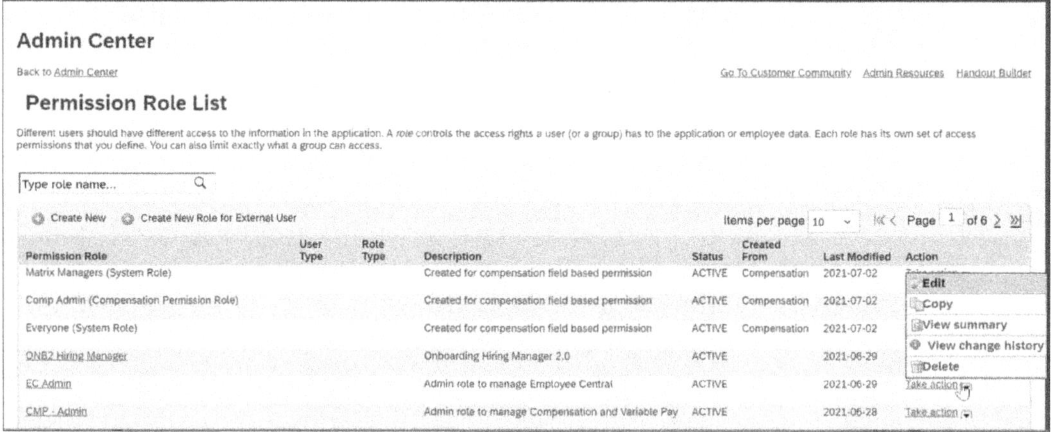

Figure 4.2 Editing Permission Roles

From the **Permission** view, click on the **Permission** button. Table 4.1 shows the groups you need to set up and their relevant permissions (a *group* is a collection of required functions). Settings can be maintained for each group by clicking on the group in the menu on the left and then selecting all the checkboxes on the right, next to the required permissions.

Functionality Required	Permission Setup
Administrator access to payroll UI mash-ups, customizing, and admin services for maintaining employee data required by Employee Central Payroll	Turn on permission for **Payroll Administration** under **Payroll Permissions**.
Payroll information for maintaining bank and other payment-related information on behalf of employees	Turn on **Edit** access for **Payment Information** under **Employee Data**.
Payment information	Turn on **Edit** access for **Payment Information involving pay components**.
Access to payroll-related job information fields for employees	Turn on permission for job information fields under **Employee Central Effective Dated Entities**.
Access to payroll-related compensation information fields for employees	Turn on permission for compensation information fields under **Employee Central Effective Dated Entities**.
Access check tool, allow configuration export, and use the check tool quick fix feature	Turn on permission to **Access check tool, Allow Check Tool Quick Fix**, and **Allow configuration Export** under **Check Tool**.
Show system dashboard for payroll	Turn on **Check Payroll System Dashboard** under **Payroll Permissions**.

Table 4.1 Permission Settings

Our example Figure 4.3 shows an example group called **General User Permission**.

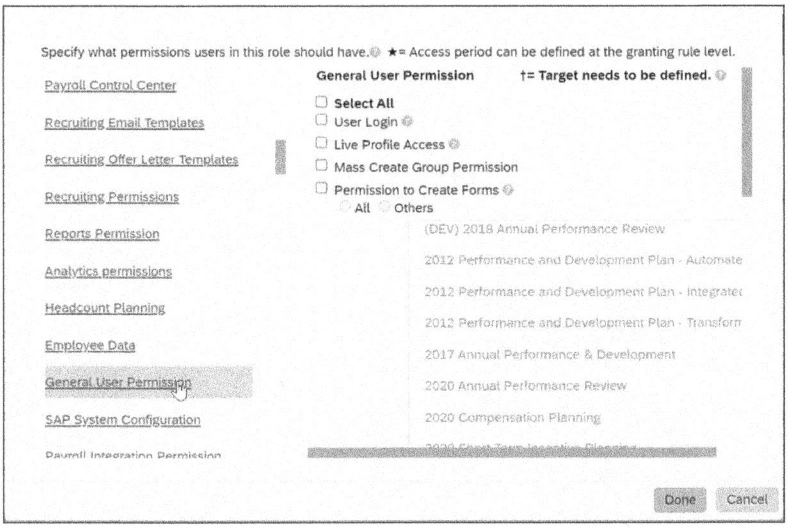

Figure 4.3 Permission Role Settings

Once you've selected the checkbox next to the required permissions, click on **Done**. Scroll down to the **Grant this role to...** section, click **Add**, and select the user to whom you want to grant this role in the **Grant role to** dropdown. Decide whether the target population this role can access includes everyone or a subset of this population. If the latter, select from the **Target population of** dropdown menu shown in Figure 4.4.

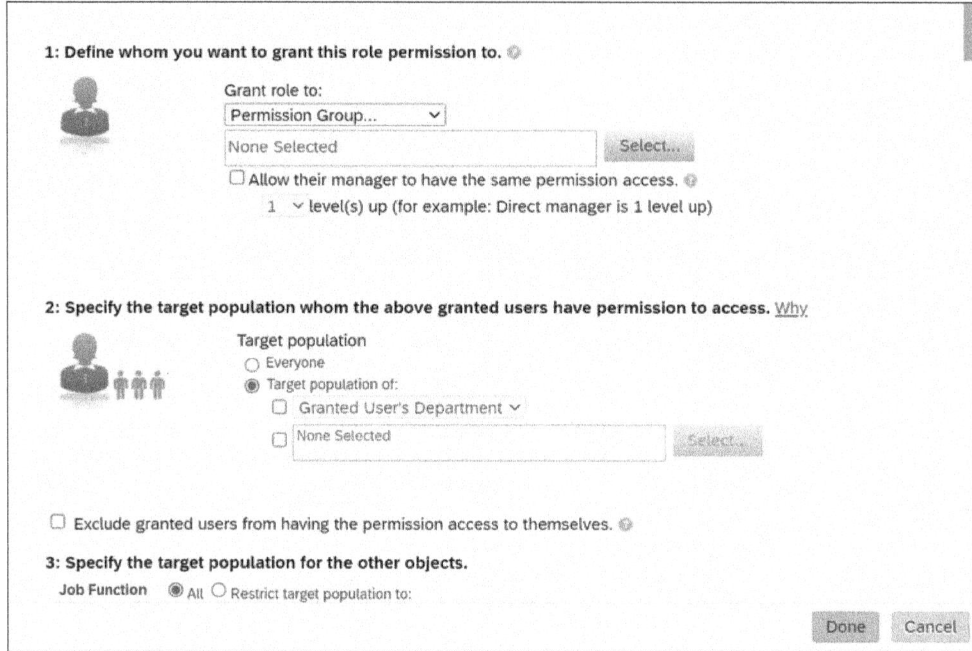

Figure 4.4 Granting Permission

Depending on whether you chose any additional features, you may need to select the checkboxes for the following:

- **Mobile pay statements**
- **Payroll run results**
- **Complete payroll tasks**
- **Check tool**

4.1.2 General User Role

To set up permissions for general users, follow the same process as for administrators, except this time select a general user role that you've earmarked for this purpose for editing. Click on **Permissions** and set up the groups listed in Table 4.2 for which you need to set up relevant permissions. The general user role will be assigned to everyone in the organization, thus enabling every employee to view their pay slips and interact with their own payroll-related data.

Functionality Required	Permission Setup
Payroll self-service for employee self-services, such as access to the pay statement	Turn on the **Payroll Self-Service** permission under **Payroll Permissions**.
Access to payroll information and the payroll information block	Turn on permission to **Payroll Information** under **Employee Views**.
Review status of employment	Turn on permission to **STATUS** under **Employee Data**.
Review pay data such as gross pay in the payroll block	Turn on view permission for **Employee Payroll Run Results** under **Payroll Integration Permission**. Turn on view permission for **Employee Payroll Run Results ITEMS** under **Payroll Integration Permission**. Turn on permission for **Allow Admin to access the OData application programming interface (API) through basic authentication** under **Manage Integration Tools**. Turn on permission for **Manage Data** under **Metadata Framework**.

Table 4.2 Additional Permission Settings

Once complete, click on **Done**. Scroll down to the **Grant this role to...** section, click **Add**, and select the user to whom you want to grant this role. Then, choose the target population.

4.2 Configuration

In this section, we'll review the different configuration activities required to set up Employee Central so that it can work with Employee Central Payroll. We'll begin with the check tool, which provides a synopsis of missing configurations. Then, we'll look at setting up Employee Central Payroll, its configuration, and its many links to payroll infotypes. We'll then look at the block where the payroll information will be displayed on the whole page. Finally, we'll review the picklists required for subsequent settings and explore how to share the payroll calendar between the Employee Central Payroll and Employee Central systems.

4.2.1 Using the Check Tool

To use the check tool, go to the Admin Center and search for "Check Tool" in the search bar.

On the screen that appears, from the **Application** dropdown list, select the required application. In our example, we'll select **Data Replication**, as shown in Figure 4.5.

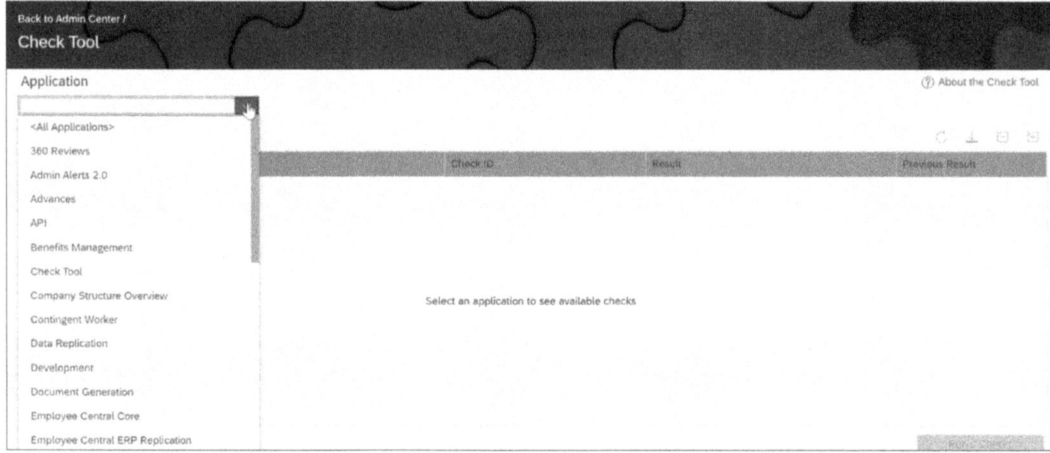

Figure 4.5 Choosing an Application in the Check Tool

The resulting screen will show you a list of available checks. Select the checkboxes next to the checks you wish to run, as shown in Figure 4.6, and then click the **Run Check** button. In our example, we have chosen to run the **Data Replication Configuration is filtered by employee class.** check.

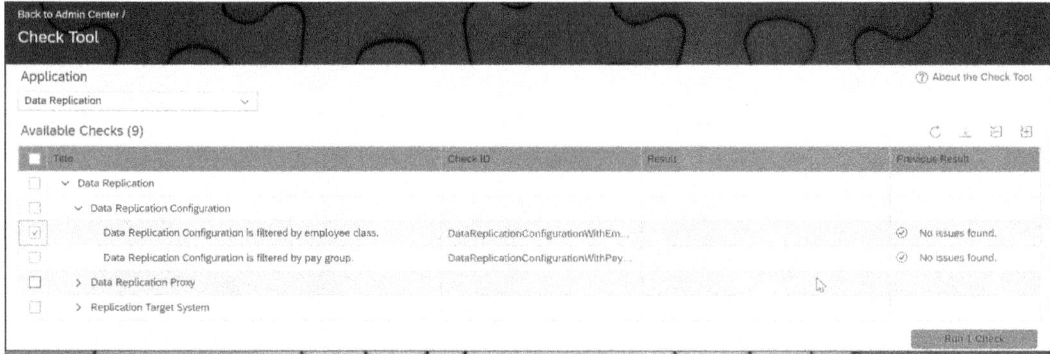

Figure 4.6 Running the Check Tool

A good time to run the check tool is if the system is behaving in an unexpected way. After running the tool, not only will you see the issue (if one exists), but also a resolution will be recommended. Current and previous results can be compared.

You can see the errors on the **Error** page. Click an error to see its details. You can also click on the **Need Assistance** column to directly log an SAP Support ticket if required. If a quick fix exists, a four-step process to conduct the fix will automatically appear in the window. Step 3 allows you to skip out if you change your mind. Some additional discussion of this tool can be found in Section 4.4.

4.2.2 Using Employee Central Payroll System Information

The next configuration phase is for using Employee Central Payroll system information, which is retrieved and displayed in a dashboard. Since this connection is a reverse connection from Employee Central Payroll to Employee Central, an OAuth 2.0 connection is required (discussed in more detail in Chapter 13). You'll also need the service `HRSFEC_ECP_INFO_SRV` activated in Employee Central Payroll.

To start, enter "Payroll System Information Dashboard" in the search bar. This screen uses color coding indicate the statuses of all the support packages in your Employee Central Payroll system(s), as follows:

- Green for applied packages
- Blue for planned packages
- Red for packages that should have been applied but have not been yet
- Orange for available packages that are not mandatory

This color coding is another example of how the Employee Central and Employee Central Payroll systems can work together and provide administrators the opportunity to monitor both solutions from a single UI.

4.2.3 Setting Up Payroll Configuration

This activity is probably the most important configuration that you must perform in Employee Central to enable payroll.

In this step, you'll identify all the necessary system parameters and infotypes in advance, so that you can use the Employee Central UI to navigate directly into the Employee Central Payroll system and review an employee's payroll-related data. This access is created through the **Payroll Information** block in the People Profile, as shown in Figure 4.7. You can navigate to this screen by clicking on **My Employee File** in the menu and scrolling down to find the block.

Figure 4.7 Payroll Information in the People Profile

To configure payroll in Employee Central, follow these steps:

1. Enter "Payroll Unified Configuration" into the search bar in Employee Central.
2. Select countries from the panel on the left by clicking **Add Country/Region.** For each country, click the **SAP System Configuration** links or the payroll tasks that are applicable, as shown in Figure 4.8.

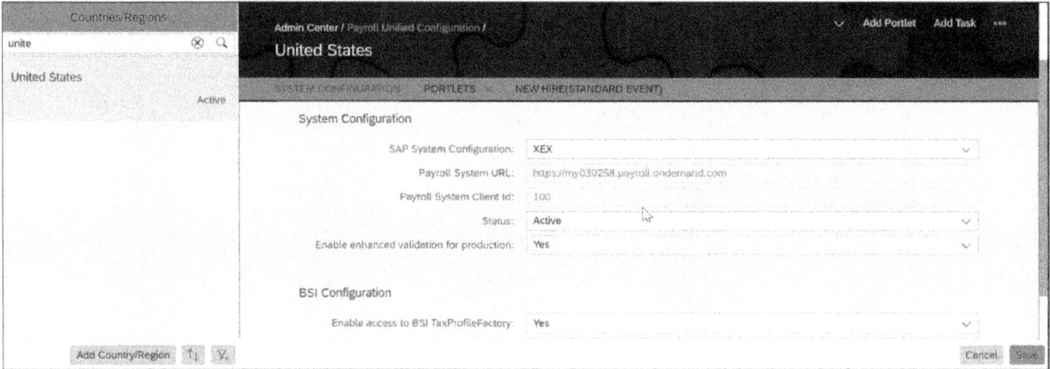

Figure 4.8 Unified Configuration Screen

3. Choose an SAP system configuration from the **SAP System Configuration** dropdown menu. This system configuration should have already been created. If not, go to **Manage Data** page by entering "Manage Data" into the search bar and create the configuration, as shown in Figure 4.9.

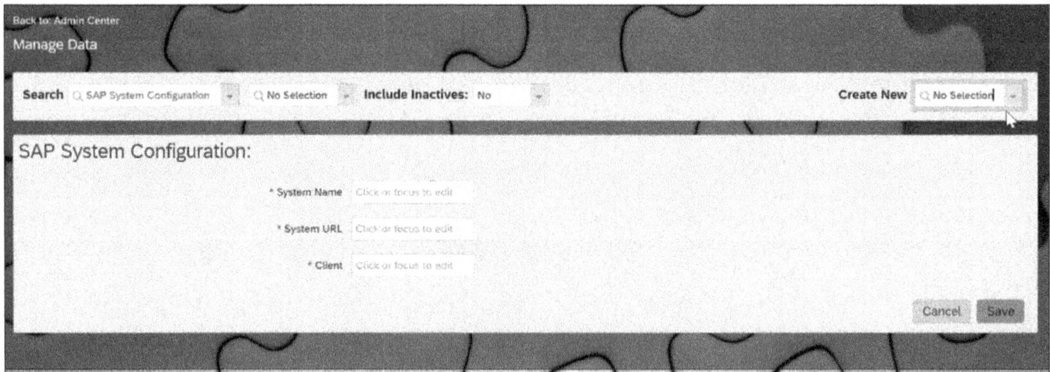

Figure 4.9 Creating an SAP System Configuration

4. Once you enter the **SAP System Configuration** screen, the **URL** and **Client** fields automatically populate.
5. Back in the unified configuration screen, shown earlier in Figure 4.8, the **Status** field indicates whether the payroll for that country is **Active** (this status is automatically set).

6. The **Enable enhanced validation for production** field is automatically set to **Yes** for all countries as soon as this feature is selected for a country. Enhanced validation is not country specific. One example of enhanced validation is the prevention of importing event reasons that are already active in Employee Central.
7. The next section of the screen is for Business Software Incorporated (BSI) configuration. The BSI EFormsFactory link provides a connection to BSI's software for managing tax profiles for countries that have a BSI solution, such as the United States. This solution provides a cloud-based repository for all tax forms that the employee needs to fill in (e.g., the W-4 form in the US, which allows a taxpayer to indicate tax withholding exemptions and filing status for accurate tax calculations in payroll). The use of BSI eForms is optional; SAP provides similar screens in Employee Central Payroll. Tax calculations in Employee Central Payroll uses a different software called BSI Tax Factory, separate from BSI eForms Factory. Enable the use of BSI by selecting **Yes** in the **Enable access to BSI TaxProfileFactory** field, as shown earlier in Figure 4.8.
8. Keep the **Third-Party Identity Provider URL** field blank, which is the default. Note that BSI will automatically use Employee Central as its identity provider (IdP) for single sign-on (SSO), but setting this up is beyond the scope of this book.
9. From the **Portlets** dropdown list, select **Tax • Add Link.**
10. Add the service **BSI EFormsFactory – Employee** with the following values, as shown in Figure 4.10:
    - In the **Link Name** field, enter a meaningful name so employees can easily recognize what this link will do (e.g., "Fill tax forms").
    - For the **Link Type** field, select the **Service** radio button.
    - From the **Service** dropdown list, select the **BSI EFormsFactory – Employee** option.
    - Under **Permission Category**, select the **Self-Service** checkbox.
11. Click **Save**.

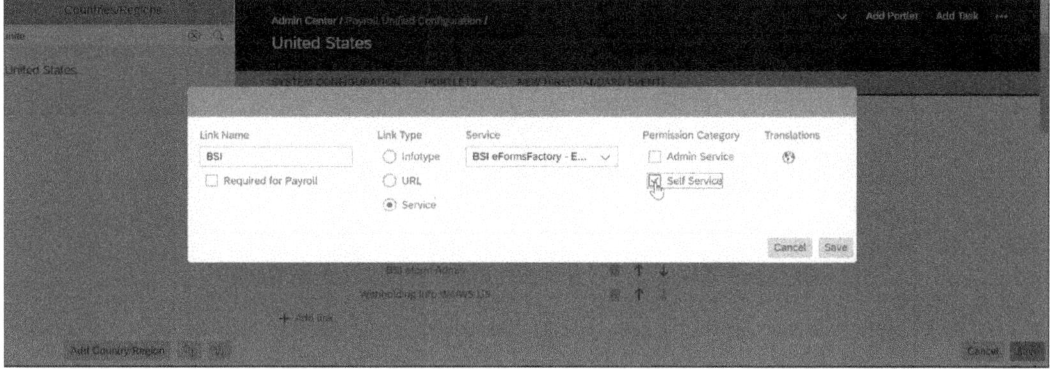

Figure 4.10 Setting Up the BSI Link in Unified Tax Configuration

Now, the link to BSI will be available on the **My Employee File** page automatically.

Note

Note that some technical configuration steps for connecting BSI are not covered in this book. Refer to *http://s-prs.co/v538604* for full details.

A *portlet* is a screen of related data in the Employee Central system. For example, the **Personal Information** portlet shows an employee's first name, last name, nationality, and marital status. Now, we'll choose all the portlets relevant for your Employee Central Payroll system, by clicking on the **Portlets** tab, as shown in Figure 4.11. Set up the portlets you need by choosing the relevant fields under the **Add Link** button.

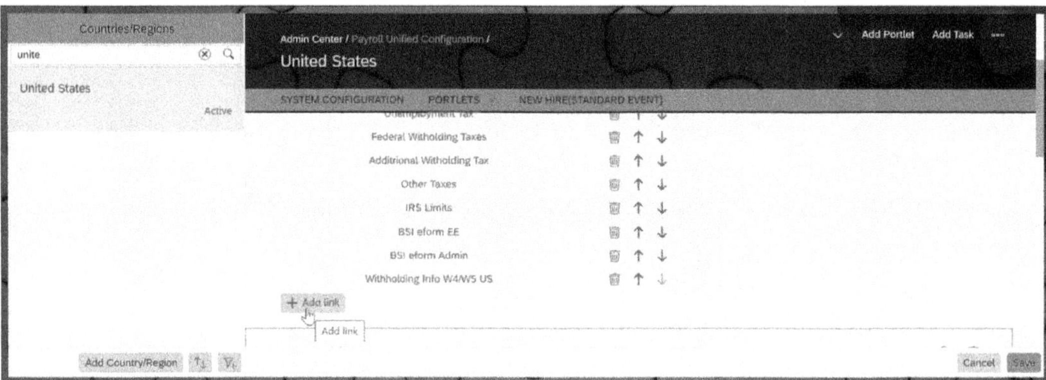

Figure 4.11 Adding a New Portlet Configuration

Three types of portlets are available: infotypes, link, and service.

An infotypes portlet is the easiest to configure. Simply select the available infotypes from the dropdown list. You'll must ensure the Web Dynpro application HRPAO_PAOM_MASTERDATA is activated in payroll. (The activation steps are discussed in Chapter 7.)

Clicking on **Required for Payroll** makes this link mandatory for infotypes—the system provides a warning if no data has been maintained for that infotype. The **URL** setting allows you to enter a web URL, such as a benefit provider's homepage, so that the employee can go to that website from the same screen.

The **Service** portlet has a fixed number of services that SAP offers that you'll need to choose from:

- **Direct Pay Statement**
- **Pay Statement**
- **Legacy Pay Statement**
- **Payment Summary** (Australia only)

- **Superannuation** (Australia only)
- **Federal Withholding Taxes - Employee Self-Service** (US only)
- **BSI EFormsFactory - Admin** (US only)
- **BSI EFormsFactory - Employee** (US only)
- **Reporting of Online W-2 - Employee Self-Service** (US only)
- **Election for Online W-2 - Employee Self-Service** (US only)
- **Form 16** (India only)

The pay statement service is discussed in more detail in Chapter 11.

The **Admin Service** and **Self-Service** checkboxes within the portlet determine whether the portlet is available only for administrators, only for employees, or available to both. The following defaults apply for each choice under **Link Type**:

- **URL**: **Admin Service** will be selected. If you deselect this option, then you must select the **Self Service** checkbox. You cannot select both checkboxes.
- **Infotype**: **Admin Service**.
- **Service** (with the **Pay Statements and Forms** chosen): both **Admin Service** and **Self Service**.

The last configuration step in this section involves tasks, as follows:

1. Click on **Add Task** in the top-right corner of the screen to add a payroll task.
2. Under the task, choose the infotype.
3. Save these settings.

You should now have the tasks ready for action when required. We'll look at task configuration later in Section 4.3.

4.2.4 Setting Up Pay Data in Payroll Block for Employee Self-Service

Setting up pay data in a block is helpful for the employee, but you must have the **People Profile** checkbox enabled in provisioning. Employees can then see gross pay, taxes, deductions, and net pay data from the **Payroll Information** block.

To set up pay data in block, follow these steps:

1. Enable the **Payroll Information** block from the **Configure People Profile** page, as shown in Figure 4.12, by dragging and dropping the block into the required position. In our example, we've dragged the block to the spot below **Job Relationship** details.
2. Make sure that the **Enable Payroll Integration** checkbox has been selected in the **Enable Payroll Integration** section in provisioning, that the People Profile is enabled in the Upgrade Center, and that Employee Central Payroll results are present in Employee Central for the employee (see Chapter 16).

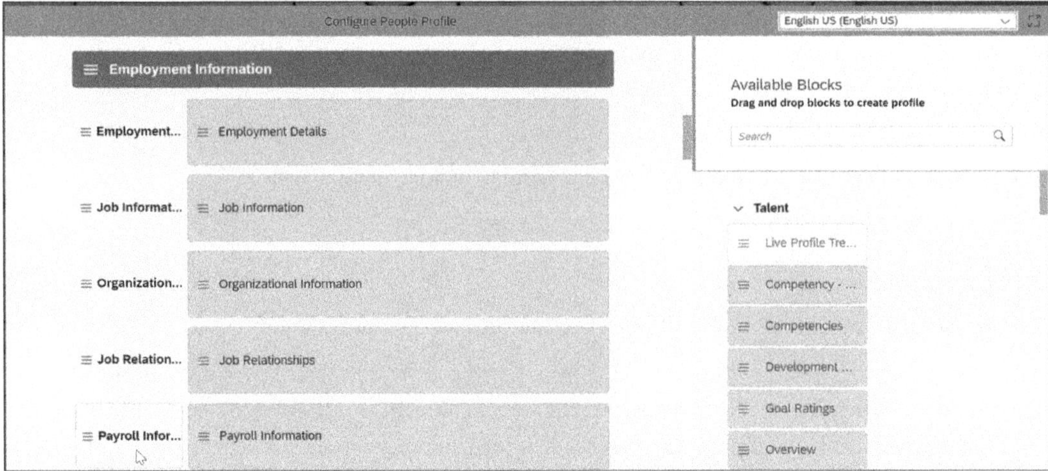

Figure 4.12 Configuring the Payroll Block on the People Profile

3. Ensure access is available to employees for the following permissions: **WageTypeAccumulationConfig**, **Employee Payroll Run Results**, and **Allow Admin to Access OData API through the Basic Authentication Group**.

4.2.5 Uploading the Picklists

A *picklist* is a list of valid values. You can upload picklists by entering "Picklist Center" into the search bar, which opens the screen shown in Figure 4.13.

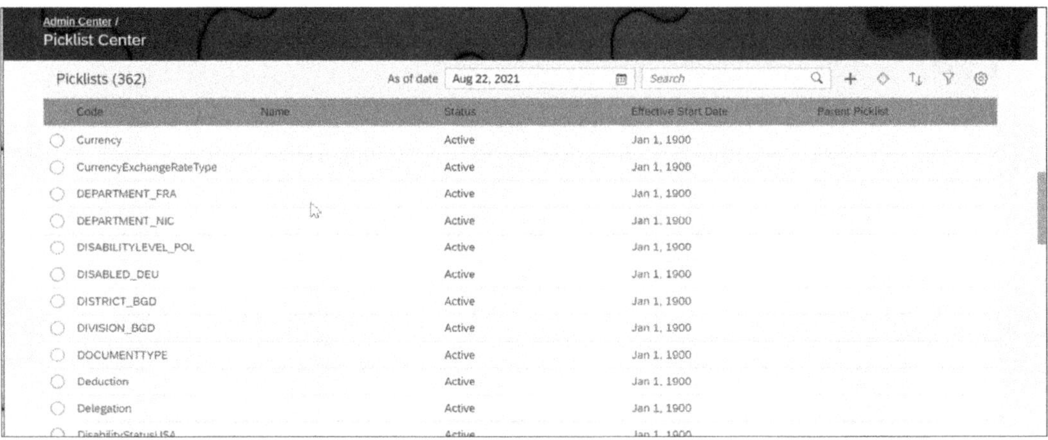

Figure 4.13 Managing Picklists

The following picklists are required:

- **EmploymentType**
- **EmployeeClass**

- **AddressType**
- **STATE_XXX** where XXX is the ISO country/region code for the country (e.g., USA for the United States). For certain countries, more than one code may be available, for regional differentiation.

To find a picklist, enter its name in the search bar. Click on the picklist once found, as shown in Figure 4.14. Clicking on the right arrow displays details about the picklist, as shown in Figure 4.15.

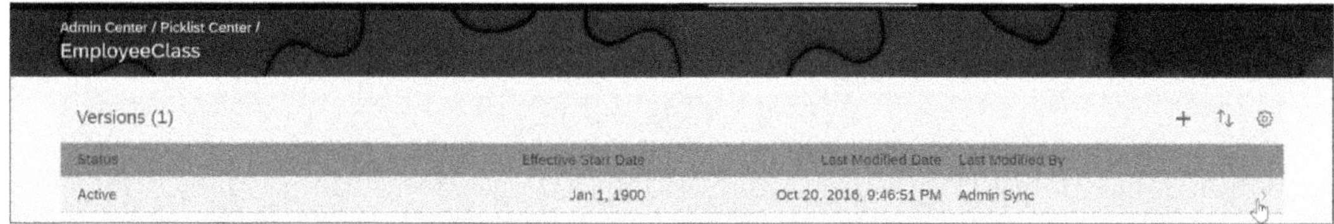

Figure 4.14 Identifying a Picklist

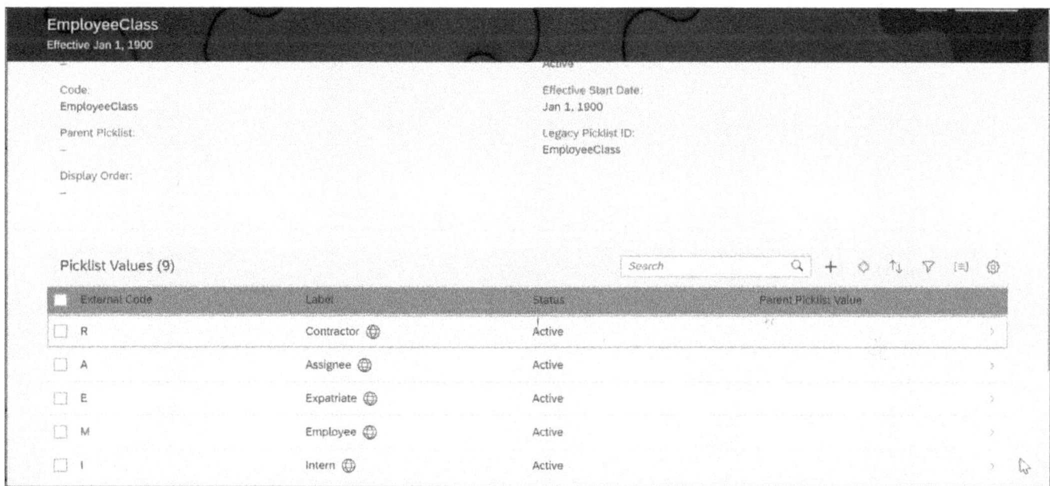

Figure 4.15 Details of Picklist Values

You can click on any value to change it. To add new values, click on the plus sign next to the search field.

4.2.6 Exporting and Importing Payroll Calendar Date Information

A *payroll calendar* is a list of dates indicating when payroll periods start and end as well as when the pay for that period will be made available to employees. The dates are important to drive the schedule for payroll processing as well as to ensure that all the data needed for processing has arrived on time.

Payroll periods are defined in Employee Central Payroll. Defining payroll periods requires that you configure pay periods and check dates using report RPCSHS00 in Employee Central Payroll. You can execute this program using Transaction SE38, as shown in Figure 4.16. In our example, our biweekly pay periods go from a Monday to the next Sunday. July 25th is the end of a pay period that started on July 12th, and the pay date for this period is July 16th, a Friday.

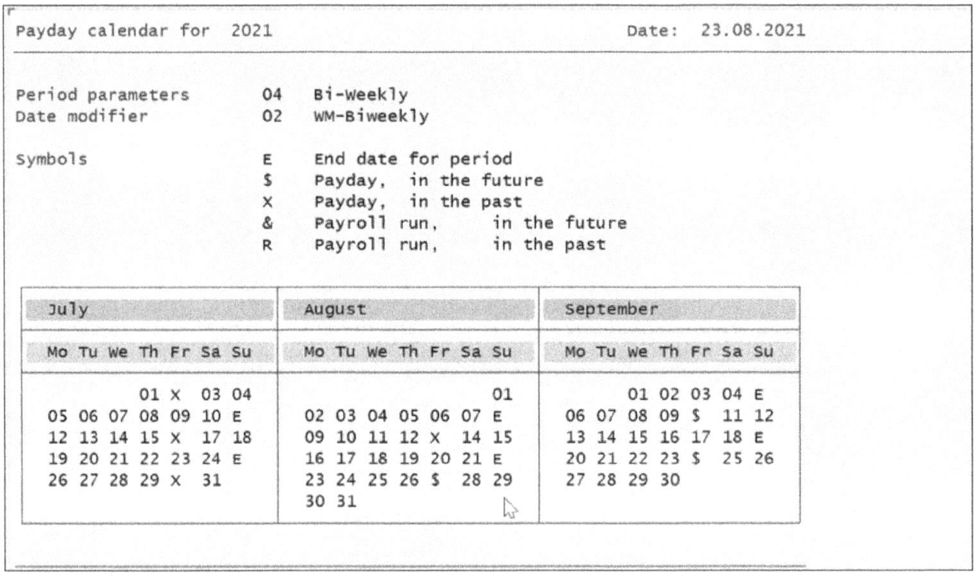

```
Payday calendar for  2021                                   Date:  23.08.2021

Period parameters          04    Bi-Weekly
Date modifier              02    WM-Biweekly

Symbols                    E     End date for period
                           $     Payday,  in the future
                           X     Payday,  in the past
                           &     Payroll run,     in the future
                           R     Payroll run,     in the past

 July                    August                  September
 Mo Tu We Th Fr Sa Su    Mo Tu We Th Fr Sa Su    Mo Tu We Th Fr Sa Su
          01 X  03 04                      01          01 02 03 04 E
 05 06 07 08 09 10 E     02 03 04 05 06 07 E     06 07 08 09 $  11 12
 12 13 14 15 X  17 18    09 10 11 12 X  14 15    13 14 15 16 17 18 E
 19 20 21 22 23 24 E     16 17 18 19 20 21 E     20 21 22 23 $  25 26
 26 27 28 29 X  31       23 24 25 26 $  28 29    27 28 29 30
                         30 31
```

Figure 4.16 Payroll Calendar Report RPCSHS00

To configure a payroll calendar, in Transaction SPRO, go to the payroll for the country you want to configure (for example, **Payroll • Germany**) and select **Basic Settings • Payroll Organization**. Then, as shown in Figure 4.17, follow these steps to define the period of pay (monthly, weekly, etc.):

1. Click on **New Entries**.
2. Enter a 2-digit code in **Period Parameters** column.
3. Enter a name in the **Name** field.
4. Choose a time unit (e.g., weekly) from the **Time unit** field.
5. Enter a start date.
6. Save these settings.
7. Define date modifiers if multiple dates of pay exist for the same period (e.g., pensioners are paid on the first day of month; hourly workers, on the 15th of the month; etc.), as shown in Figure 4.18.
8. Click on **New Entries**.
9. Add a 2-character code in the **Date modifier** field.

10. Add a name in the **Name** field.
11. Save these settings.

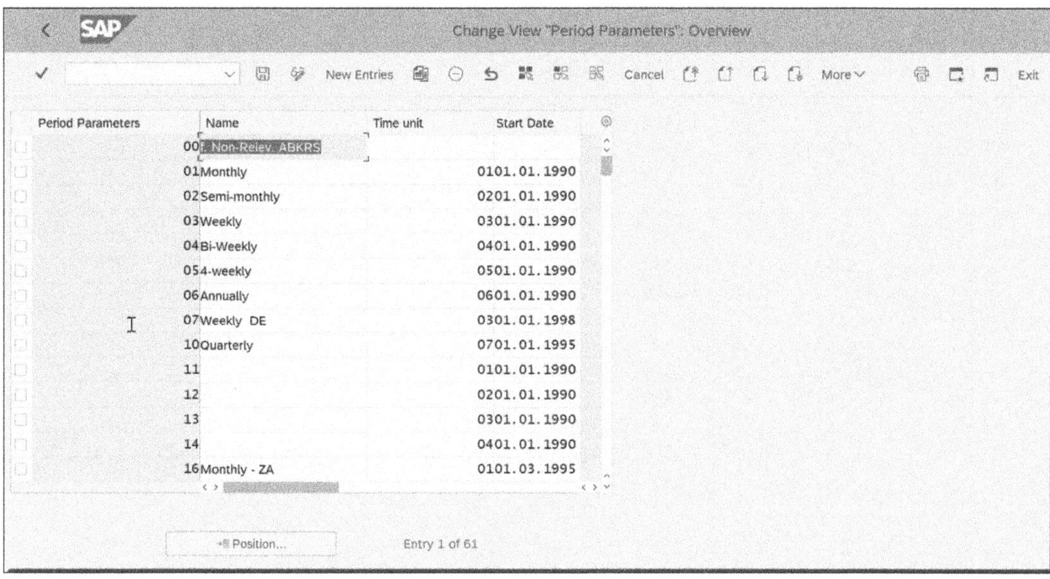

Figure 4.17 Period Parameter Setting

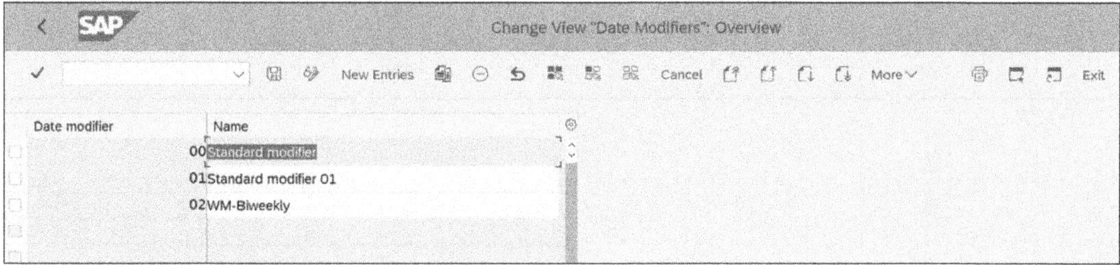

Figure 4.18 Date Modifiers

Next, as shown in Figure 4.19, you then need to assign the new payroll accounting areas to a period parameter:

1. Click on **New Entries**.
2. Add a 2-character code in the **Payroll Area** field.
3. Add a name in the **Payroll Area Text** field.
4. Choose the period parameter we just set up from the **Period Parameter** field. Select the **Run payroll** checkbox to make this payroll area active.
5. Choose the date modifier we just created from the **Date Modifier** field.
6. Save these settings.

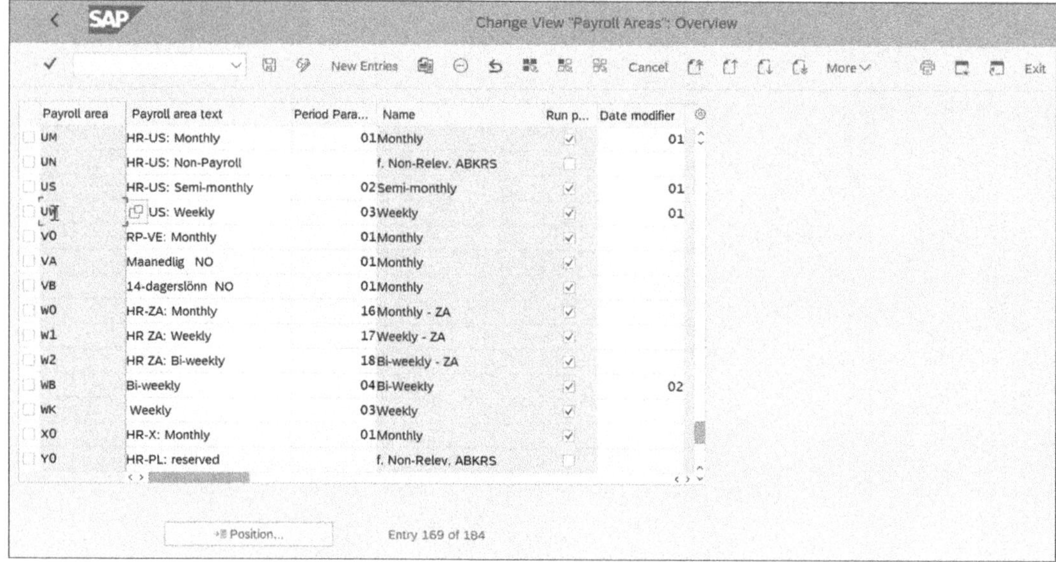

Figure 4.19 Assigning a Period Modifier to a Payroll Area

Finally, you will generate the payroll periods, as shown in Figure 4.20.

1. Maintain the **Period Parameters**, **Date modifier**, **Start Date**, and **Final year** fields.
2. Choose a payday rule (e.g., to add the number of days to the period start date).
3. Enter the days required for the rule in the **Number of days** field.
4. The **Determine period number** field allows you to set a period number for each pay period based on period start end dates or pay dates.

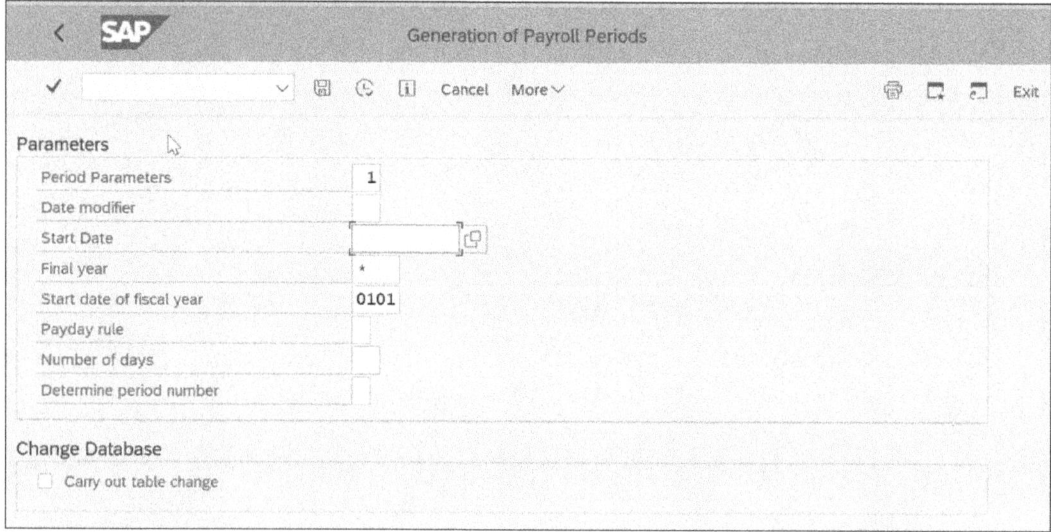

Figure 4.20 Generating Payroll Periods

Once designed, this payroll calendar information can be easily transferred to Employee Central by following these steps:

1. Run Transaction HRSFEC_PAY_CALENDAR in Employee Central Payroll. The screen shown in Figure 4.21 will appear, where you'll maintain the following information:
   - **Country Grouping**
   - **Employee Central Pay Group** (payroll area)
   - **From Payroll Year** and **To Payroll Year** (start and end years for payroll)
2. Execute, and the system will generate a CSV file.

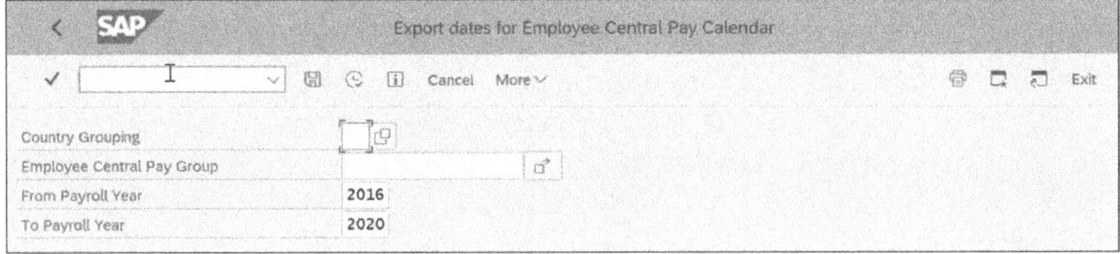

Figure 4.21 Downloading the Payroll Calendar

3. Next, go to the Admin Center in Employee Central and select **Import and Export Data**, as shown in Figure 4.22. On this screen, perform the following actions:
   - Select **Pay Calendar** from the **Select Generic Object** dropdown list.
   - Select **Incremental Load** in the **Purge Type** dropdown list.
   - Choose the file to import by clicking the **Choose File** button and selecting the CSV file we downloaded in the previous step.
4. Click on **Import**.

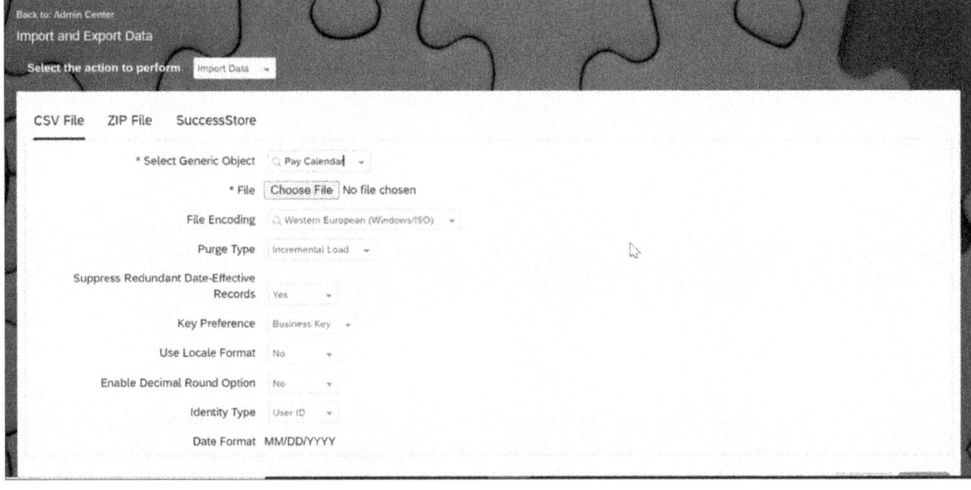

Figure 4.22 Importing a Pay Calendar

> **Note**
>
> Note that if the pay group in Employee Central is not named the same as the payroll area in Employee Central Payroll, then table T77SFEC_CVMAPC must be maintained for mapping.

The pay calendar in Employee Central is not only important for visibility. Another rather important use is when certain dates are captured for employees in Employee Central that must align with pay dates. For example, some companies do not enter a salary change or a transfer action in the middle of a pay period; others may not allow retroactive changes in previous years to avoid tax filing corrections and penalties.

Rules can be created for such validations using the Employee Central rule engine, but rules will be dependent on the payroll calendar dates in Employee Central. For example, a rule that validates an effective date of change is never before the pay period start date, as shown in Figure 4.23. Choose **1** for the **Date Type** field, which corresponds to pay period start date.

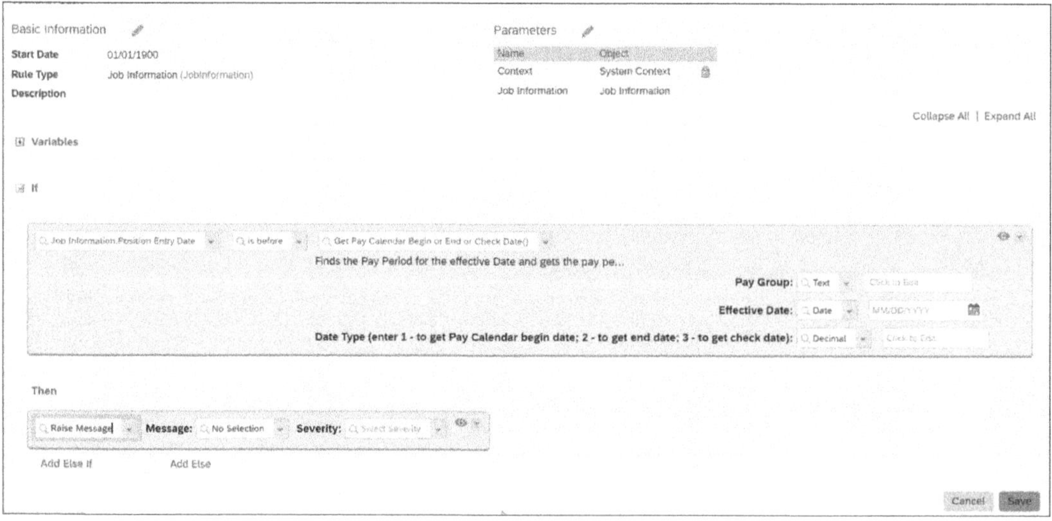

Figure 4.23 Example Rule for Validating with Pay Calendar Date

4.3 Completing Payroll Tasks (Optional)

A *payroll task* is typically an infotype that is maintained for a payroll data item that is not stored in Employee Central. All data items stored in Employee Central are moved promptly to Employee Central Payroll. Maintaining these data items are part of data administration in Employee Central.

However, maintaining payroll data is a duty of the payroll administrator. Employee Central Payroll keeps this data separate in the Employee Central Payroll system. When a new hire joins the company or an employee transfers to another branch of the company, an urgent need exists to enter or update this data. Having these tasks populate into the to-do list of an administrator ensures that tasks are not missed.

To configure payroll tasks, follow these steps:

1. The necessary permissions must be set in role-based permissions under **Payroll Integration Permission**, as described in Section 4.1. Ensure that the administrator has access to the **Payroll Data Maintenance Task** settings, as shown in Figure 4.24.

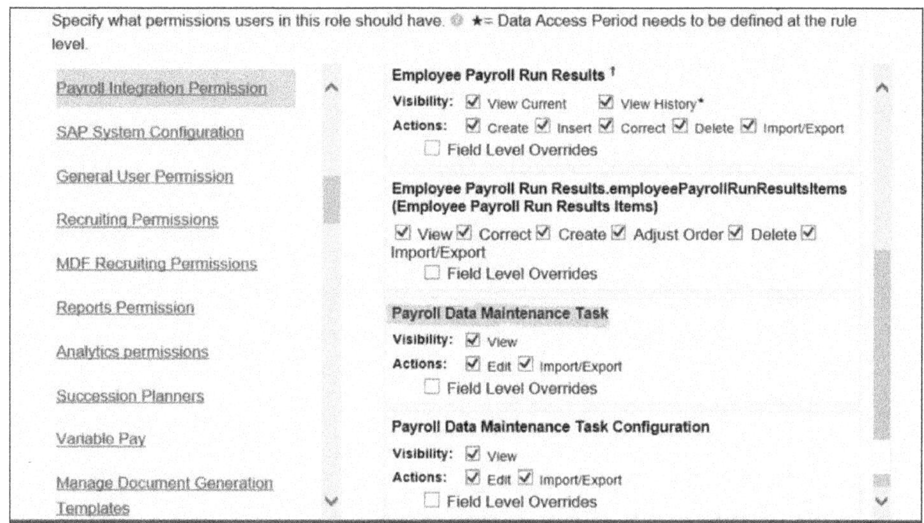

Figure 4.24 Role-Based Permissions for Payroll Tasks

   You can restrict this access by pay group (payroll area in Employee Central Payroll). Simply set up the **Target Population** field, as described in Section 4.1. In this context, you'll need to choose the **Pay Group** field.

2. Next, set up the event notification for payroll tasks. In the search bar, enter "Event notification subscription" to reach the screen shown in Figure 4.25.

Event Notification Subscription

Subscriber | External Event | SEB External Event

+ Add Subscriber | Save | Cancel

Category	Subscriber Id	Name	Group	Client Id	Created On	Last Modified	Last Modifi...	Deleted
Customized	Subscriber Id	Name	Group	Client Id				☐
Customized	Subscriber Id	Name	Group	Client Id				☐
Customized	Subscriber Id	Name	Group	Client Id				☐
Customized	Subscriber Id	Name	Group	Client Id				☐
Customized	Subscriber Id	Name	Group	Client Id				☐
Customized	Subscriber Id	Name	Group	Client Id				☐

Figure 4.25 Event Notification Subscription

3. Under the **Subscriber** tab, create a subscriber by clicking on **Add Subscriber** and then maintaining the **Subscriber ID**, **Name**, and **Group** fields, which are free format fields.
4. Under the **SEB External Event** tab, shown in Figure 4.26, click on **Add Topic**.
5. Select **Payroll Data Maintenance Task Configuration Change**.
6. Select **Add Subscription** and enter the following:
   - Under **SEB Event Type**, select the **External Event** alert type.
   - Under **Endpoint URL**, enter the payroll access URL.
   - Under **Subscriber**, enter the subscriber we created earlier.
   - Under **Protocol**, choose **SOAP over HTTP/HTTPS**.
   - Under **Authentication**, select basic authentication (**Authentication**).

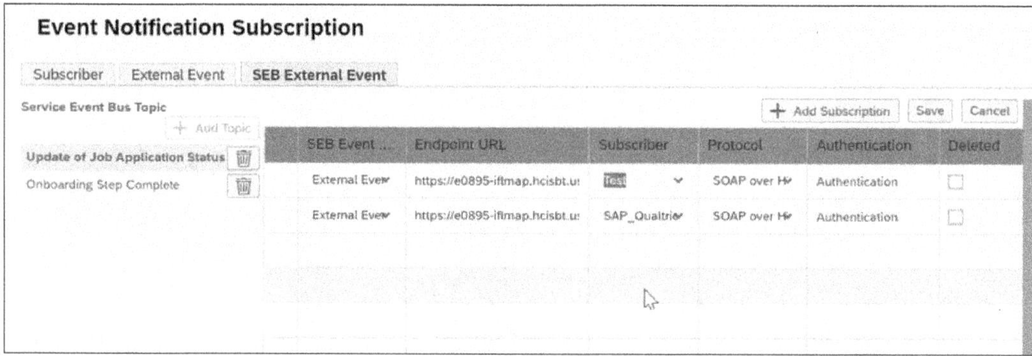

Figure 4.26 External Event Settings

To find the access URL, in the Employee Central Payroll system, use Transaction SOAMANAGER, as shown in Figure 4.27. Under the **Service Administration** tab, select **Simplified Web Service Configuration.** Enter "*payrolldata*" into search bar and click on **Go**.

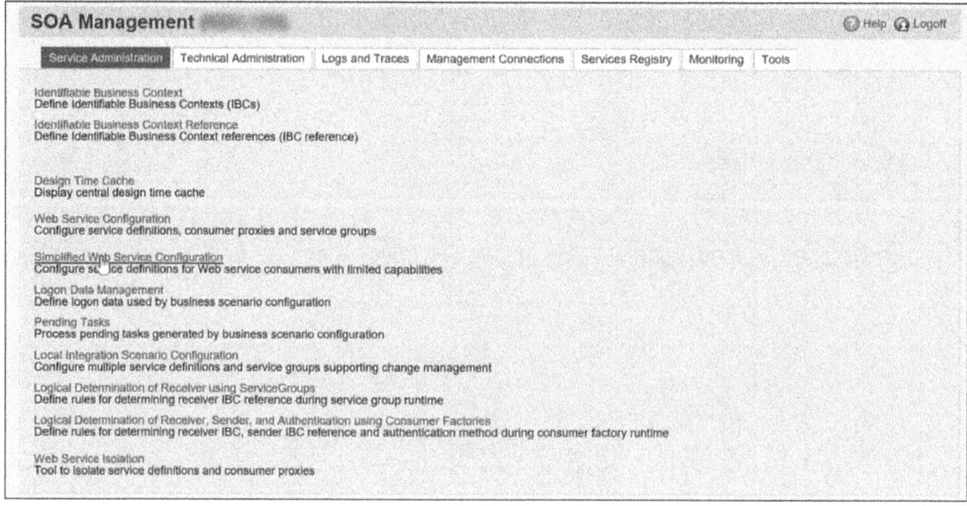

Figure 4.27 Finding the URL Using Transaction SOAMANAGER

In the results, choose the **PAYROLLDATAMAINTENANCE TASKCONF.** Select the user and password and click **Save.** Then, select **Show Details.** As shown in Figure 4.28, the configuration details screen includes the URL, which you can copy for use in the configuration.

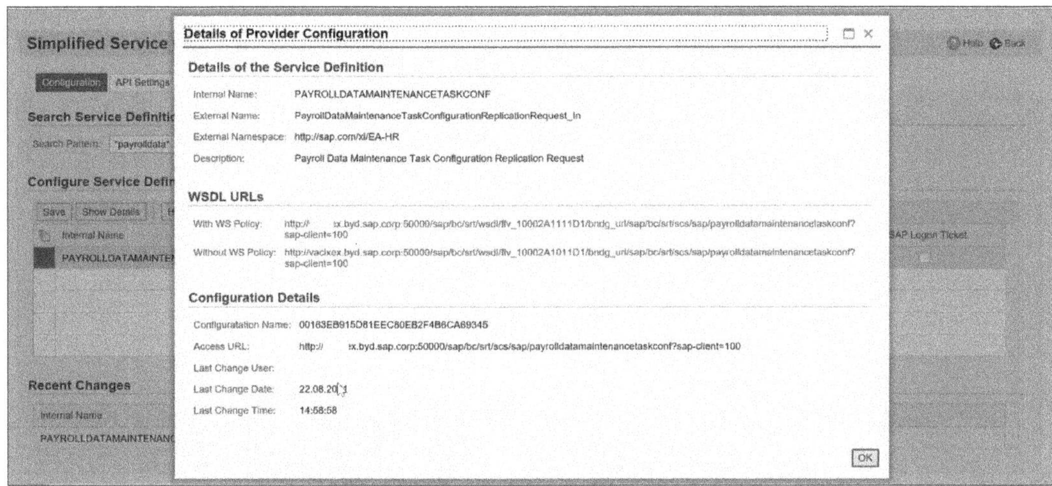

Figure 4.28 Accessing the URL in Transaction SOAMANAGER

Note

Note that the URLs are only accessible internally. To access a URL externally, copy the first part of the URL from the Transaction SOAMANAGER browser window, for example, *http://xxx.sap.corp:44322/*. Then, replace the first part of the Web Services Description Language (WSDL) and endpoint URLs displayed in the **Details of Provider Configuration** screen with the external part you copied earlier from the SOA Management browser window.

If configured correctly, you should see the event notification audit log, as shown in Figure 4.29.

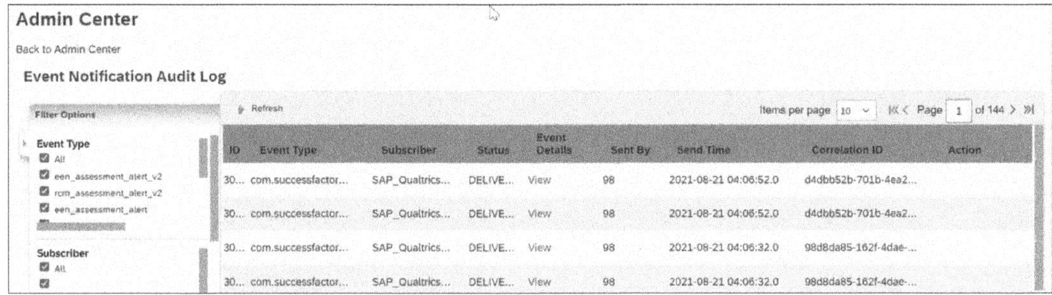

Figure 4.29 Event Notification Audit Log

Next, update the settings for switch T77SO by following these steps:

1. Go to Employee Central Payroll and bring up Transaction SM30 and table T77S0.
2. Click **New Entries**.
3. Maintain the following fields:
   - **Group**: SFEC
   - **Semantic abbreviation**: ECTSK
   - **Value**: X
4. Save these settings.

As mentioned earlier in Section 4.2.3, you'll need to set up the task configuration, as shown in Figure 4.30, with the payroll infotypes required for that country. In our example company running payroll in the United States, the required infotypes are the resident tax area (infotype 207), the work tax area (infotype 208), and unemployment tax (infotype 209).

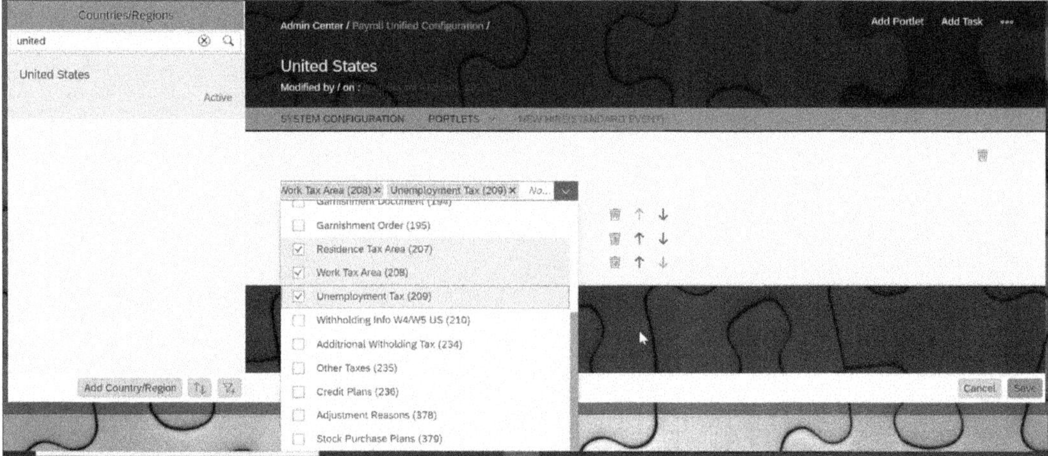

Figure 4.30 Task Configuration on the Unified Configuration Screen

This next step is to run report RP_HRSFEC_PTP_TASK_CFG_REPL in Employee Central Payroll using Transaction SE38, as shown in Figure 4.31.

Once run, the system indicates whether replication was successful, as shown in Figure 4.32.

Note that you must run this replication task every time changes are made to the configuration of Employee Central (Figure 4.33).

The final step is to check that the **Complete Payroll Tasks** tile is available for the payroll administration group. As shown in Figure 4.33, under **To-Do Settings**, click on **Show/ Hide To-Do Tiles**. In the pop-up that appears, find the **Complete Payroll Tasks** tile, and ensure the activation button is set to **Yes**, then save these settings.

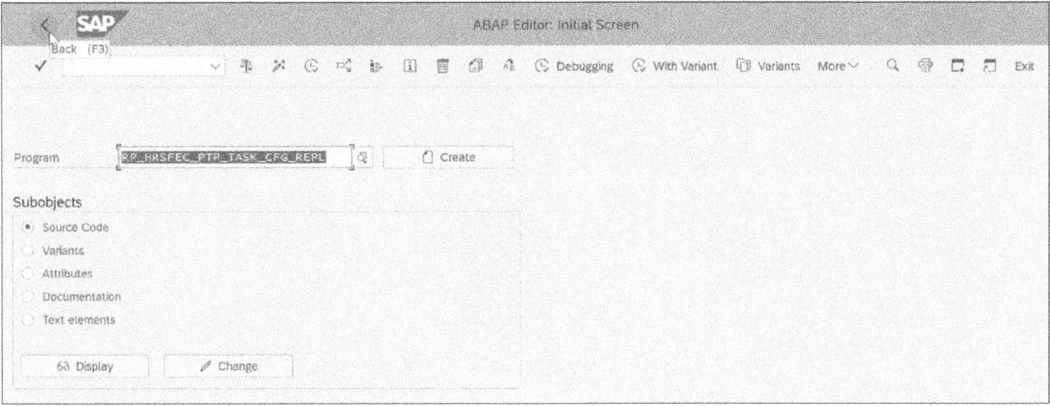

Figure 4.31 Replication Report in Payroll

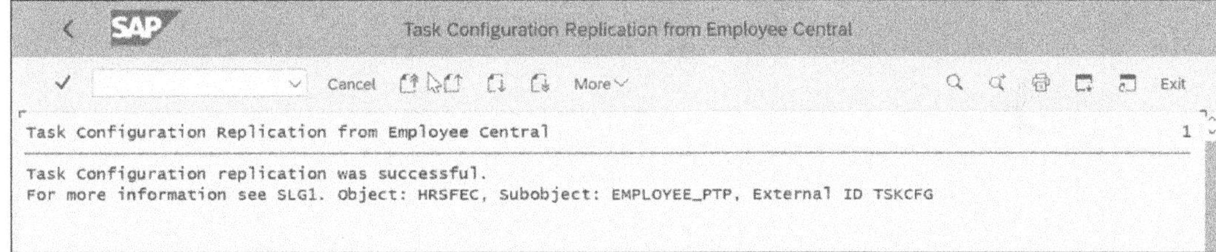

Figure 4.32 Replication of Task Configuration Success Message

Figure 4.33 Manage Home Page Screen to Confirm the Payroll Task Tile

As a result of this configuration, the payroll administrator will receive a notification to complete these infotypes (the tax infotypes 0207, 0208 and 0209 in our US example) through a special tile called **Complete Payroll Tasks**, as shown in Figure 4.34.

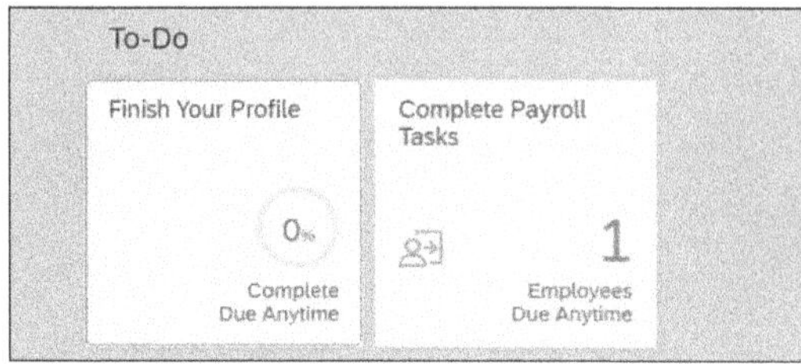

Figure 4.34 Complete Payroll Tasks Tile

4.4 Payroll Information Application in Employee Central (Optional)

The **Payroll Information** application shows the health of the configuration you've maintained specifically for payroll. In this way, you can ensure that all the configuration is in order and identify potential mistakes that may prevent the system from functioning well. You can configure the **Payroll Information** application using the Check Tool, by following these steps:

1. Go to the search bar and enter "Check Tool."
2. A list will display different components, as shown in Figure 4.35. Choose **Payroll Information** to open the following checks:
   - **Data Consistency**
   - **Integration to BSI TaxProfileFactory**
   - **Payroll Results Consumption**
   - **Replication Relevant Data Model**
   - **SAML Configuration**

Figure 4.35 Check Tool for Payroll Information

3. Click on the **Run Checks** button. All the checks you've selected will run at once.
4. Once you run the check, the outcome will be one of the following:
   - No inconsistencies found.
   - *N* inconsistencies found (*N* being the number of inconsistencies).

For each inconsistency, a yellow or red icon will be displayed. Yellow means the problem is not so serious, while red is a serious issue. The system proposes a solution, which you'll need to review and accept to complete the configuration.

4.5 Summary

This chapter prepared you for the configuration activities required in the HR system. From permissions to events, a few important components in Employee Central must be configured to enable the seamless continuity from your HR system to your Employee Central Payroll system. Some technical connectivity functions depend on this configuration, while business features like payroll tasks integrate with HR events to make keeping track of important employee activities easy.

In the next chapter, we review the settings necessary in Employee Central Payroll. After completing the basic configuration in these two systems, you'll be ready to turn on the features that Employee Central Payroll is built for: efficient payroll processing through the Payroll Control Center and effective control of master data through replication.

Chapter 5
Configuring Employee Central Payroll

Now that you've configured Employee Central, you're ready to prepare your Employee Central Payroll system. To unlock the rich features of Employee Central Payroll, you'll first need to lay the foundation by configuring the system.

This chapter is about configuring Employee Central Payroll. We'll review a few techniques, such as turning on switches, defining settings in personal administration, setting up pay statements, identifying Business Add-Ins (BAdIs) that may be necessary, and setting up concurrent employment.

You must fulfill the following two prerequisites before the instructions in this chapter can be performed:

- Employee Central Payroll has been turned on in provisioning (Chapter 3)
- Employee Central has been configured (Chapter 4)

Additionally, business function HCM_SFEC_MDEC2HR must be turned on for different features in Employee Central Payroll. You can turn this function on in Transaction SFW5 by opening the **ENTERPRISE_BUSINESS_FUNCTIONS** folder, as shown in Figure 5.1.

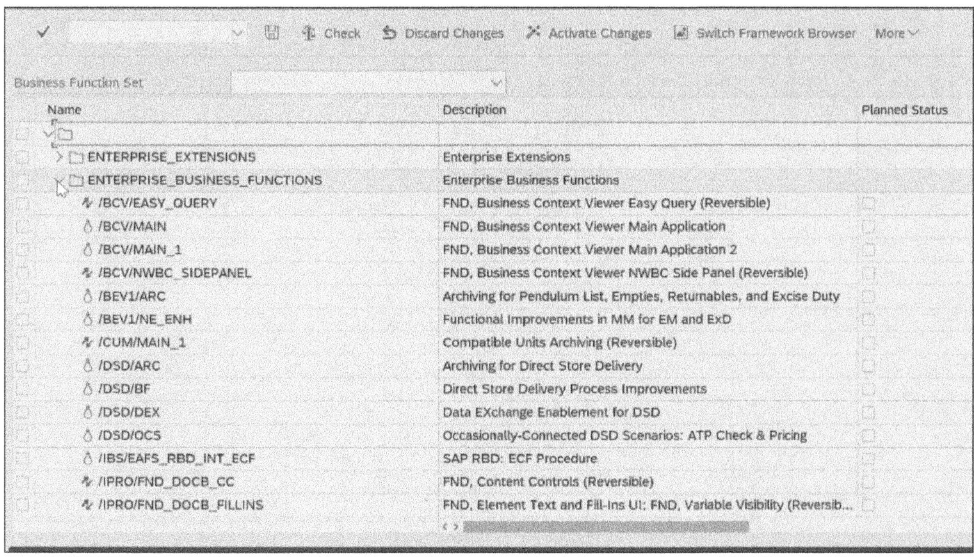

Figure 5.1 Transaction SFW5 Switch Framework

Now, scroll down to the required business functions and activate them by selecting the **Planned Status** checkbox, as shown in Figure 5.2.

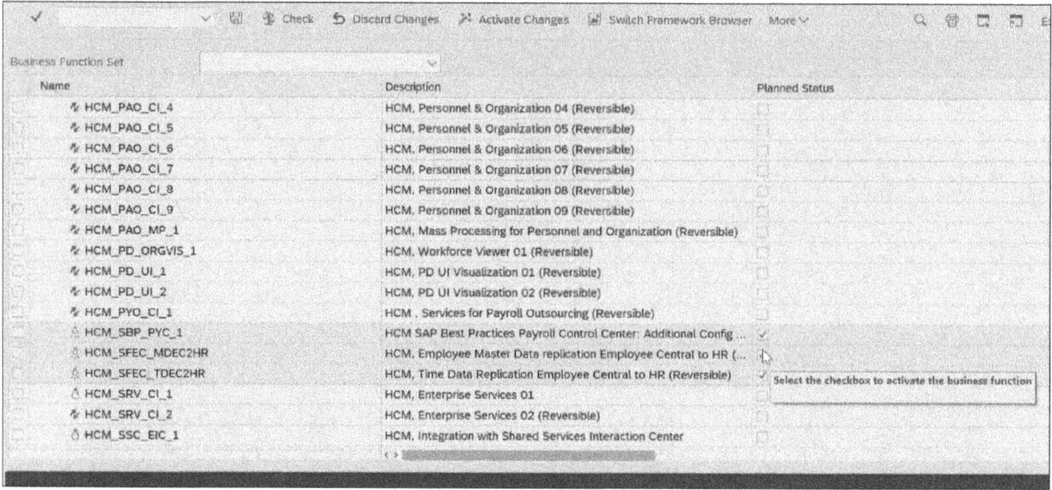

Figure 5.2 Activating the Function HCM_SFEC_MDEC2HR

> **Tip**
>
> Your Employee Central settings must match exactly with the settings chosen in Employee Central Payroll. If you have a separate SAP financial system (SAP ERP Financials or SAP S/4HANA Finance), that configuration must be repeated in Employee Central Payroll for the following objects:
>
> - Company codes
> - Cost centers
> - Profit centers
> - General ledger accounts
> - Controlling areas
> - Vendor master records

5.1 Switches for Payroll

Turning on switches for Employee Central Payroll is achieved in Transaction SM30. Once you've executed the transaction, enter table name "T77SO" in the **Table/View** field. We'll discuss this screen in more detail later in Section 5.2.2.

Click the **Position** button at the bottom of the next screen, which will open a dialog box. In the field called **Group Name**, enter the following groups individually in the **Group** field, one at a time:

- SFEC
- ADMIN
- CCURE

Various switches will appear, which can be turned on by entering, in the **Value abbr.** field (shown later in Figure 5.20), the values listed in Table 5.1.

Switch Group	Switch Name	Description	Variable Choices	Use Case
SFEC	CCMM	Cost center mapping mode	Blank = Standard cost center mapping mode 1 = Extended cost center mapping mode	Mapping cost center keys
SFEC	DRTM	Evaluation of SAP SuccessFactors compound application programming interface (API) purge objects	Blank = Default X = Employee Central Data retention management	Purging employee master data replicated to Employee Central Payroll
SFEC	ECTSK	Creation of payroll data maintenance tasks	Blank = No creation of payroll data maintenance tasks X = Creation of payroll data maintenance tasks	Enabling the creation of payroll data maintenance tasks
SFEC	EECOH	Cut-off history when rehiring employees inactive at full transmission start date (FTSD)	Blank = No cut-off X = Cut-off enabled	While using FTSD and configuring employment filtering
SFEC	EXJPO	External numbering for replicated job, position, and organization	Blank = Don't use external numbering X = Use external numbering	Checking and creating jobs, positions, and organizational units

Table 5.1 Switches for Employee Central Payroll

Switch Group	Switch Name	Description	Variable Choices	Use Case
SFEC	HDREP	Switch on higher duty replication	X = Enable the replication of higher duties	Enabling the replication of higher duty or temporary assignment
SFEC	JBPS	Replication of jobs and positions	Blank = No replication of jobs and positions X = Replication of jobs and positions	Job and position
SFEC	LGLVE	Log level for replication of employee master data	0 = Don't write to application log 1 = Log errors and success messages only 2 = Log warnings, errors, and success messages only 3 = Log all messages	Application log: setting the log level
SFEC	LGLVP	Log level (technical point to point logs)	0 = No log 1 = Standard log 2 = Detailed log	Application log: setting the log level
SFEC	OCHAT	Default personnel action for organizational change	Blank = Default X = Define personnel action for organizational change	Artificial actions
SFEC	OCRSN	Support OCRSN (off-cycle payroll run reason) and PAYID of infotype 0267	Blank = Default X = Support OCRSN and PAYID of infotype 0267	Off-cycle reason
SFEC	ONGO	Ongoing payroll (ignore correction phase)	Blank = Default X = Correction phase indicator enabled	Enabling the correction phase indicator

Table 5.1 Switches for Employee Central Payroll (Cont.)

Switch Group	Switch Name	Description	Variable Choices	Use Case
SFEC	OURPL	Define source object for replication of organization	Blank = Don't replicate organizational units 1 = Use business unit as source for organizational unit 2 = Use department as source for organizational unit 3 = Use division as source for organizational unit	Checking and creating jobs, positions, and organizational units
SFEC	PERNR	Enable the creation of personnel numbers using the corresponding Assignment ID	Blank = Default 1 = Creation of PERNR using Assignment ID	Using the Employee Central Assignment ID as the PERNR (optional)
SFEC	SPYSL	Activate secure pay slip	Blank = Default 1 = Additional security layer enabled	Enabling additional security for pay statements
SFEC	WKWDY	Weekly workday information	Blank = PERNR-based infotype 0007, Filter: Switch ON X =PERNR-based infotype 0007, Filter: Switch OFF 1 = PERNR-based infotype 0007, Filter: Switch OFF 2 = PERNR-based infotype 0007, Filter: Replicate-IT7 when WKWDY = 0	Blank does not replicate infotype 0007 from Employee Central. X or 1 replicates infotype 0007 from Employee Central, but throws an error if the **Weekly Work Days** are zero in Employee Central. 2 replicates infotype 0007 from Employee Central, but does not throw an error if the **Weekly Work Days** are zero in Employee Central

Table 5.1 Switches for Employee Central Payroll (Cont.)

Switch Group	Switch Name	Description	Variable Choices	Use Case
ADMIN	EVSUP	Additional actions	Enabled only by running report RPUEVSUP	Multiple events per day (Normally, SAP has a restriction of only 1 event a day.)
CCURE	PC_UI (Several scenarios exists when using concurrent employment in Employee Central Payroll.)	PC UI main switch	Blank = No data sharing for Transaction PA30 X = Data sharing for Transaction PA30	Basic settings enabling concurrent employment in Employee Central Payroll

Table 5.1 Switches for Employee Central Payroll (Cont.)

5.2 Defining Settings in Personnel Administration

The settings we'll describe in this section are required to match the key characteristics of personnel records between Employee Central and Employee Central Payroll. The items that you must configure include following:

- Key fields
- Basic settings
- Bank details
- Organizational data
- Basic pay
- Recurring payments and deductions
- Additional payments
- Evaluation basis (dates)
- Communication
- Actions
- Disabling the transport connection for personnel planning
- Integration between personnel data and organizational data

Master data items in Employee Central Payroll are often tightly coupled with an equivalent data item in Employee Central, as shown in Figure 5.3. (Dark gray is Employee Central, and white is Employee Central Payroll.)

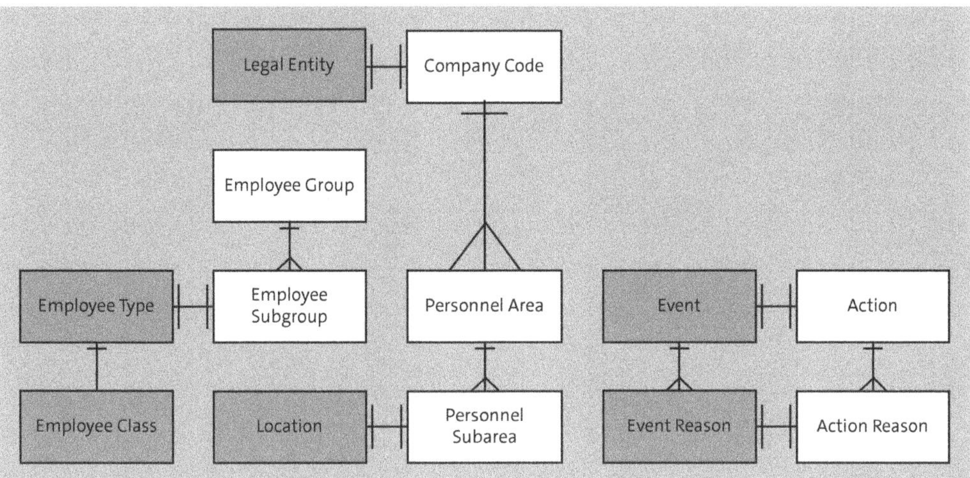

Figure 5.3 Entity Relationship Diagram Comparing Personnel Administration Fields

5.2.1 Key Fields

In the following sections, we'll look at two types of key fields in Employee Central Payroll: personnel areas (and subareas) and employee groups (and subgroups).

Personnel Areas and Subareas

Personnel areas are based on work locations and are directly linked to a company code. This information is essential for determining several key components, primarily based on the location of work, about an employee. Bring up the configuration screens for personnel areas in Transaction SPRO by following the menu path **Enterprise Structure • Definition • Human Resources Management • Personnel Areas.**

Click on **New Entries**, and then in the next screen, maintain the **Personnel area** field and other location data, as shown in Figure 5.4.

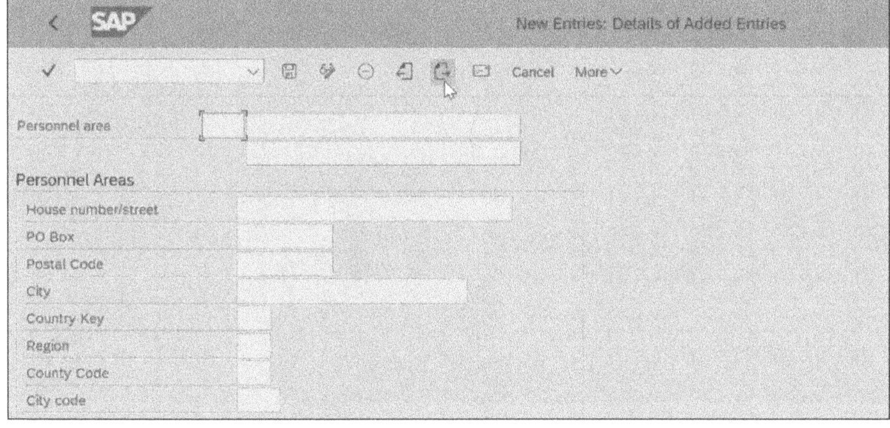

Figure 5.4 Creating New Personnel Areas

After you've saved your settings by clicking the **Save** icon, your screen should resemble the screen shown in Figure 5.5.

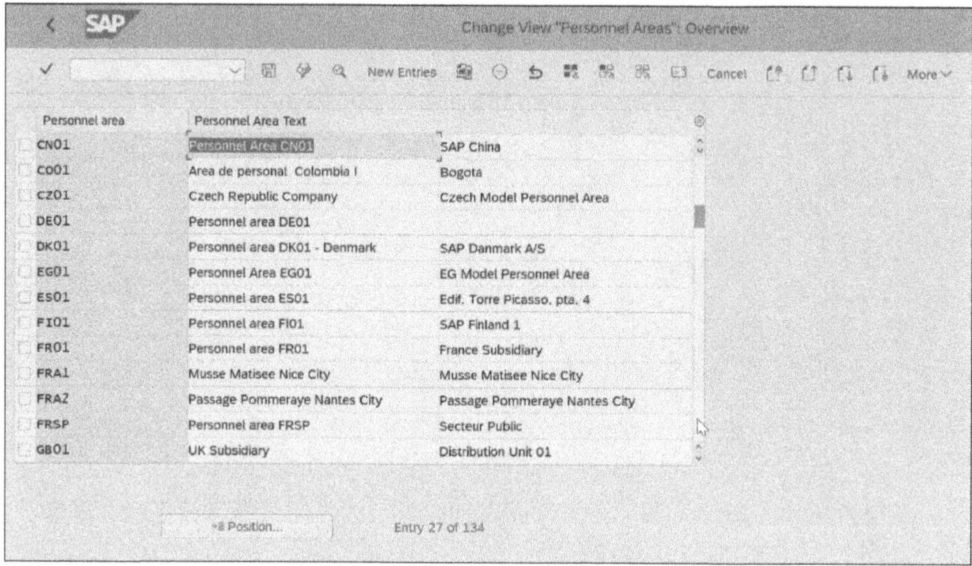

Figure 5.5 List of Personnel Areas

A *personnel area* is a structural unit that represents a location group such as a city or a state. A personnel area may have several subareas, which would represent, for example, zones or suburbs within a city. Bring up the configuration screens for personnel subareas in Transaction SPRO by following the menu path **Enterprise Structure • Definition • Human Resources Management • Personnel Subareas**. Click on **Create Personnel Subareas** to open the popup window shown in Figure 5.6. Maintain the **Personnel area** field first and then click on the green checkmark to continue.

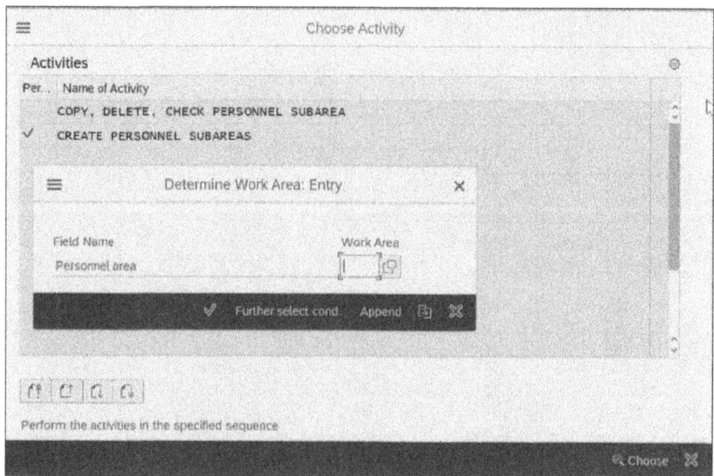

Figure 5.6 Creating New Personnel Subareas

Next, click on **New Entries** and maintain the **Personnel subarea** and **Pers. subarea text** fields, as shown in Figure 5.7. Both fields derive their values from the **Location** field in Employee Central, as in the entity relationship diagram shown earlier in Figure 5.3.

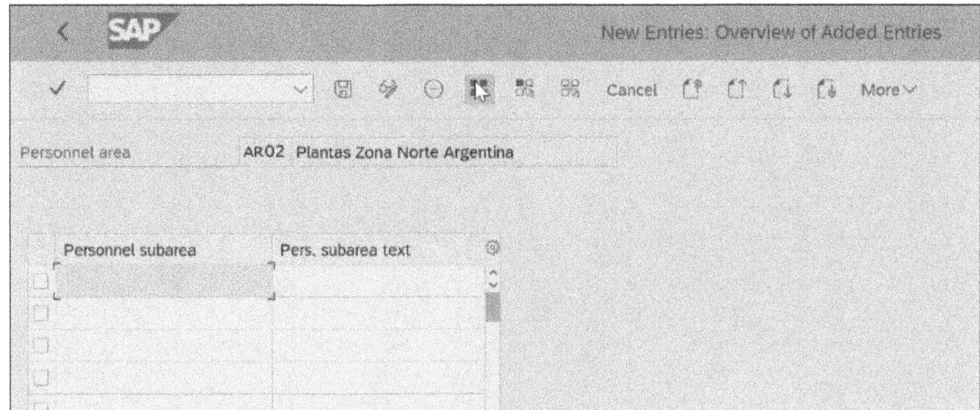

Figure 5.7 Creating New Personnel Subareas: Entering Values

Saving will result in the screen shown in Figure 5.8.

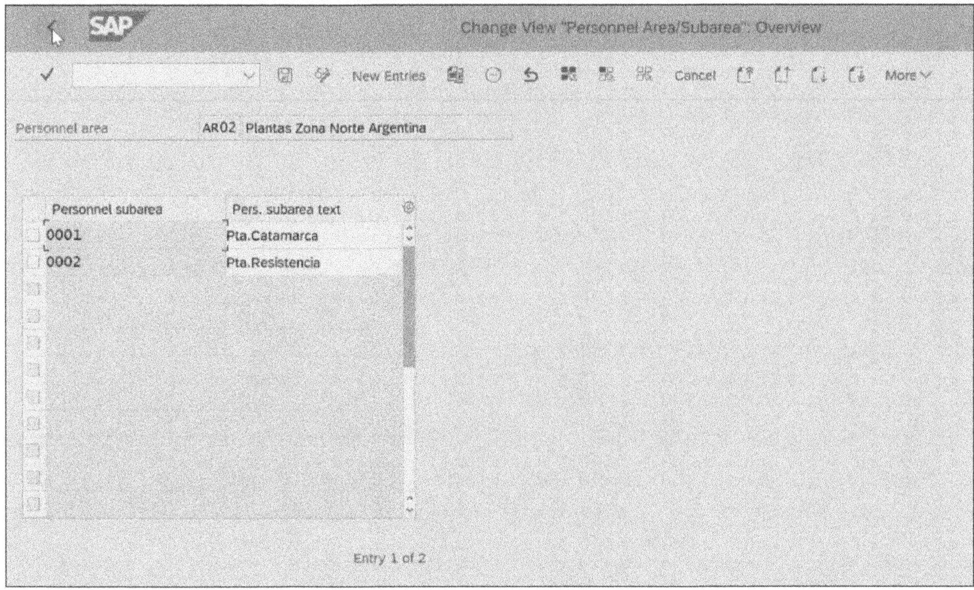

Figure 5.8 Personnel Subarea List

One additional step is required for configuring personnel areas—you must link them to the correct company codes. For this step, bring up the configuration screens in Transaction SPRO by following the menu path **Enterprise Structure • Assignment • Human Resources Management • Assignment Of Personnel Area To Company Code.**

In the row for the relevant personnel area, as shown in Figure 5.9, click on the **Company Code** field and choose the correct company code from the dropdown list.

Change View "Assignment of Personnel Area to Company Code": Overview

Pers.area	Personnel Area Text	Company Code	Company Name	Ctry Grpg
CN01	Personnel Area CN01	CN01	Country Template CN	28
CO01	Area de personal Colombia I	CO01	Country Template CO	38
CZ01	Czech Republic Company	CZ01	Country Template CZ	18
DE01	Personnel area DE01	DE01	Country Template DE	01
DK01	Personnel area DK01 - Denmar	DK01	Country Template DK	09
EG01	Personnel Area EG01	EG01	Country Template Egypt	EG
ES01	Personnel area ES01	ES01	Country Template ES	04
FI01	Personnel area FI01	FI01	Country Template FI	44
FR01	Personnel area FR01	FR01	Country Template FR	06
FRA1	Musse Matisee Nice City	UAEF	Ultimate Adventures & Esc	06
FRA2	Passage Pommeraye Nantes C	UAEF	Ultimate Adventures & Esc	06
FRSP	Personnel area FRSP	FR01	Country Template FR	06
GB01	UK Subsidiary	GB01	Country Template GB	08

Position... Entry 27 of 134

Figure 5.9 Linking a Personnel Area to a Company Code

Employee Groups and Subgroups

Employee groups represent basic personnel categories such as employees, pensioners, retirees, etc. Employee subgroups further break down these categories. Some companies distinguish their employees as hourly versus salaried depending on how the employee is paid. Employee groups and subgroups typically derive their values from the **Employee Type** and **Employee Class** fields in Employee Central.

To configure these fields, follow these steps:

1. Bring up the configuration screens in Transaction SPRO by following the menu path **Enterprise Structure • Definition • Human Resources Management • Employee Group**.
2. Now, click on the **New Entries** button and then add a new employee group by maintaining the **Employee group** and **Name of employee grp** fields, as shown in Figure 5.10.

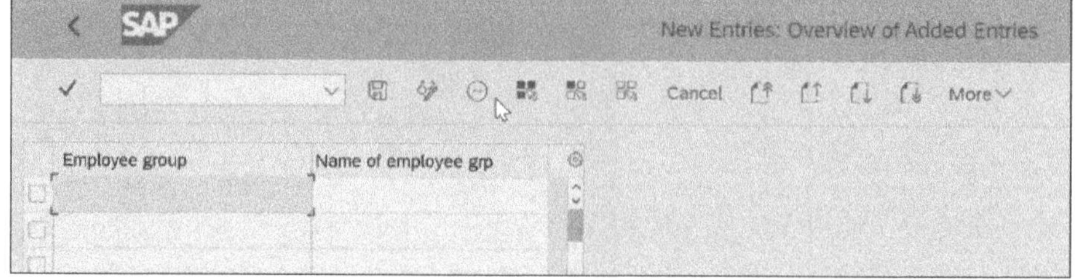

Figure 5.10 Adding a New Employee Group

3. Saving the entry will result in the screen shown in Figure 5.11.

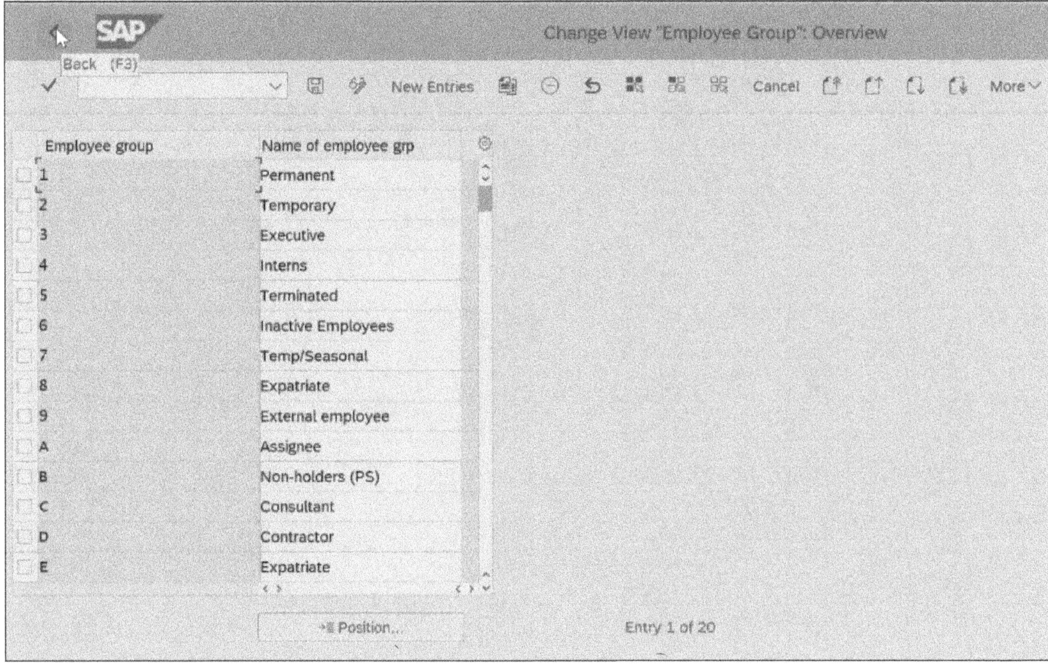

Figure 5.11 Employee Group List

Similarly, employee subgroups can be added for each employee group. First, add the new employee subgroups, by following these steps:

1. Bring up the configuration screens in Transaction SPRO by following the menu path **Enterprise Structure • Definition • Human Resources Management • Employee Subgroup.**
2. Now, click on the **New Entries** button to add a new employee subgroup, as shown in Figure 5.12.

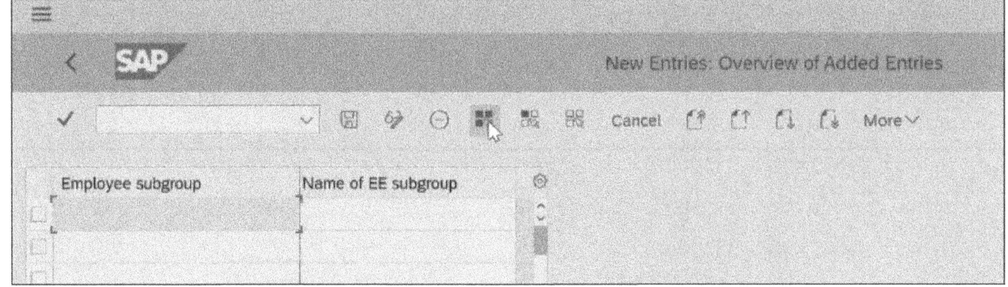

Figure 5.12 Employee Subgroups: New Entry

3. Saving the entry will result in the screen shown in Figure 5.13.

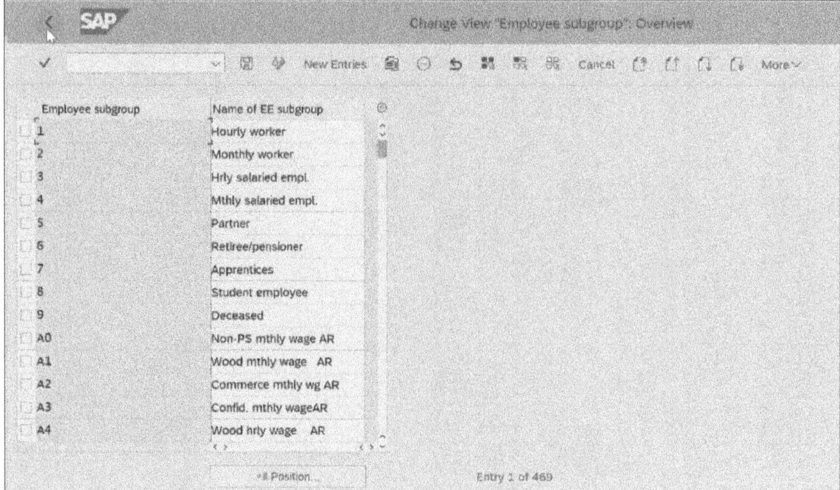

Figure 5.13 Employee Subgroup List

Now, employee groups and subgroups must be linked to each other by following these steps:

1. Bring up the configuration screens in Transaction SPRO by following the menu path **Enterprise Structure • Assignment • Human Resources Management • Assign Employee Subgroup to Employee Group**.
2. Now, click on the **New Entries** button to connect an employee group with a subgroup. Further, select the applicable countries, as shown in Figure 5.14. You'll perform this step for each group and subgroup pairing.

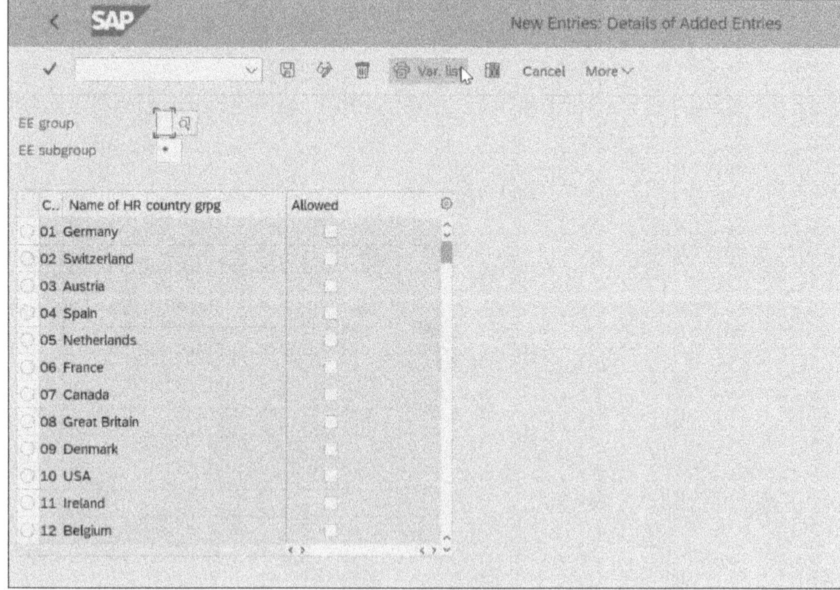

Figure 5.14 Assigning an Employee Subgroup to an Employee Group

3. Saving the entry will result in the screen shown in Figure 5.15.

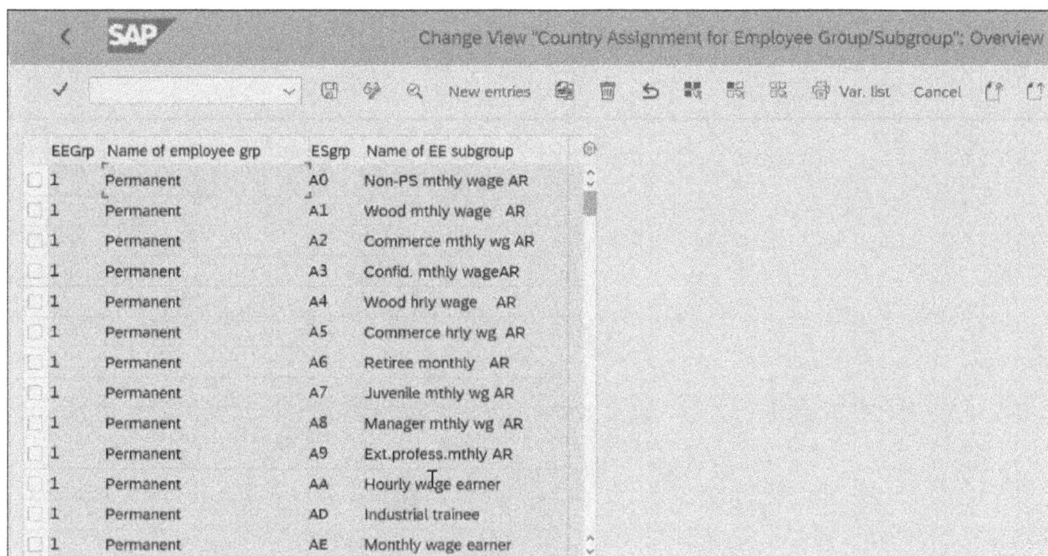

EEGrp	Name of employee grp	ESgrp	Name of EE subgroup
1	Permanent	A0	Non-PS mthly wage AR
1	Permanent	A1	Wood mthly wage AR
1	Permanent	A2	Commerce mthly wg AR
1	Permanent	A3	Confid. mthly wageAR
1	Permanent	A4	Wood hrly wage AR
1	Permanent	A5	Commerce hrly wg AR
1	Permanent	A6	Retiree monthly AR
1	Permanent	A7	Juvenile mthly wg AR
1	Permanent	A8	Manager mthly wg AR
1	Permanent	A9	Ext.profess.mthly AR
1	Permanent	AA	Hourly wage earner
1	Permanent	AD	Industrial trainee
1	Permanent	AE	Monthly wage earner

Figure 5.15 Assigned Employee Subgroups

5.2.2 Basic Settings

In basic settings, four steps should be performed, which we'll walk you through in this section: evaluating country data, defining number ranges, sharing data with infotypes, and setting up bank details.

> **Evaluating Country Data**
>
> The first basic setting requires you to evaluate the country- or region-specific data that is required. You'll configure the required data by matching country data with data that has been enabled in Employee Central, as discussed further in Chapter 7, Section 7.2.

Defining a Number Range

Next on the list is defining a number range. This standard feature of SAP allows for the assignment of a personnel number to a new hire record. However, employees are all created in Employee Central, and thus, you typically don't need to assign a new number. You can choose to create a number range that is different from Employee Central. In this case, you'll have to maintain a mapping, as discussed further in Chapter 8, Section 8.1.

To identify a personnel number range for Employee Central Payroll, bring up the configuration screens in Transaction SPRO by following the menu path **Personnel Management • Personnel Administration • Basic Settings • Maintain Number Range Intervals for Personnel Numbers**.

Now, define the number range that should be used in Employee Central Payroll by clicking on the **Intervals** button with the pencil icon, as shown in Figure 5.16.

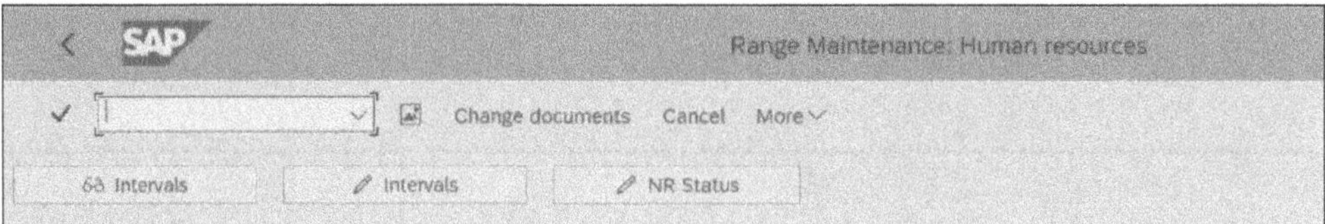

Figure 5.16 Defining a Number Range

Now, set up your number range. Saving the entry will result in the screen shown in Figure 5.17.

No	From No.	To Number	NR Status	Ext
01	00000001	19999999	0	✓
02	90000001	99999999	0	
03	20000000	89999999	0	

Figure 5.17 Personnel Number Ranges in the System

Rarely do companies bother using an additional number range and instead simply adopt the numbers that are generated in Employee Central.

To use the numbers generated by Employee Central, a BAdI must be implemented, as described in more detail in Chapter 7. The name of this BAdI is HRSFEC_B_CE_EXT_PERNR_MAP. The method used is IF_HRSFEC_B_CE_EXT_PERNR_MAP~MAP_EXT_PERNR. Some coding is required to create this mapping, which is beyond the scope of this book. The implementation of this BAdI must be performed in Transaction SE18 (BAdI Builder).

The most commonly used field from Employee Central is Person_ID_External. This value exists only once per employee and thus is not suitable for multiple employment (when an employee serves multiple roles that are paid differently). A change in company code will result in a new personnel number in SAP. The Employment_ID field must be populated in Employee Central for this BAdI to work.

The number range interval must also be marked as external by selecting the **Ext** checkbox shown in Figure 5.17.

A default range must be selected for each country being configured via feature NUMKR, as shown in Figure 5.18. *Features* are items that can be configured based on parameters and can be updated via Transaction PE03—they provide a hierarchical value map. Enter the word "NUMKR" in the **Feature** field and click on **Change**. When the default tree structure appears, edit the values as required, as shown in Figure 5.18. The **Otherwise** category is invoked if the country in question is not covered.

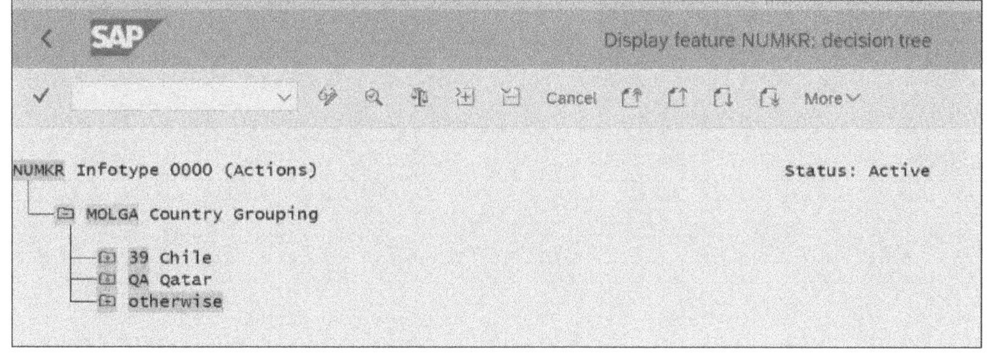

Figure 5.18 Feature NUMKR for Default Number Ranges

Data Sharing with Infotypes

The next setting is known as *data sharing*. SAP's new infotype framework requires organizations to group infotypes to replicate them accurately from Employee Central to Employee Central Payroll. This grouping of infotypes involves two steps:

1. Setting the PC_CURE switch.
2. Setting the grouping reason for the relevant infotypes.

The PC_CURE switch is activated by bringing up the configuration screens in Transaction SPRO by following the menu path **Personnel Management • Personnel Administration • Customizing Procedures • Infotypes • Infotypes in Detached Infotype Framework • Use Decoupled and Not Decoupled IT Framework.**

You can also access the PC_CURE switch in Transaction SM30 by entering the table name "T77SO" in the **Table/View** field, as shown in Figure 5.19, and then clicking on **Maintain.**

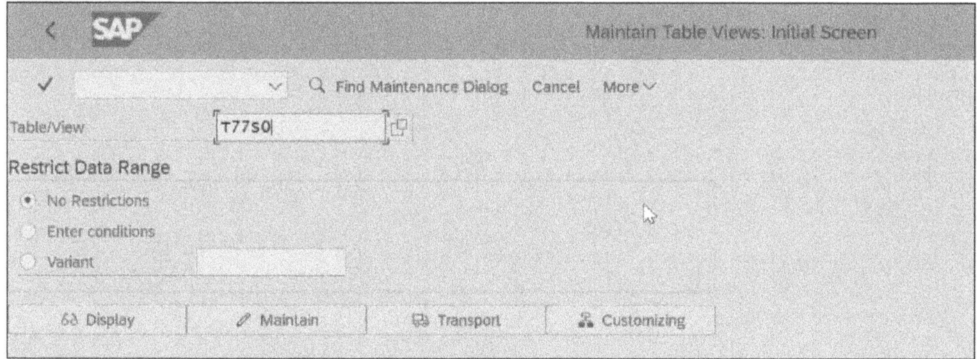

Figure 5.19 Maintaining Tables Using Transaction SM30

On the next screen, toggle the switch by scrolling down to the **CCURE** group, locating the **PC_UI** entry, and entering "X" in **Sem.abbr.** field, as shown in Figure 5.20.

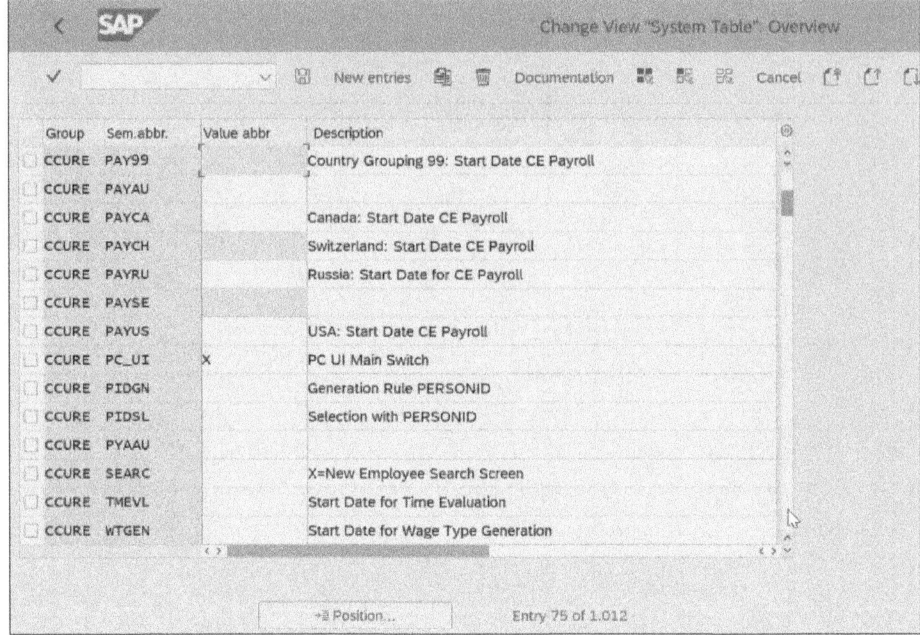

Group	Sem.abbr.	Value abbr	Description
CCURE	PAY99		Country Grouping 99: Start Date CE Payroll
CCURE	PAYAU		
CCURE	PAYCA		Canada: Start Date CE Payroll
CCURE	PAYCH		Switzerland: Start Date CE Payroll
CCURE	PAYRU		Russia: Start Date for CE Payroll
CCURE	PAYSE		
CCURE	PAYUS		USA: Start Date CE Payroll
CCURE	PC_UI	X	PC UI Main Switch
CCURE	PIDGN		Generation Rule PERSONID
CCURE	PIDSL		Selection with PERSONID
CCURE	PYAAU		
CCURE	SEARC		X=New Employee Search Screen
CCURE	TMEVL		Start Date for Time Evaluation
CCURE	WTGEN		Start Date for Wage Type Generation

Figure 5.20 Settings for Data Sharing

Next, bring up the configuration screens for data sharing in Transaction SPRO by following the menu path **Personnel Management • Personnel Administration • Customizing Procedures • Settings for Concurrent Employment • Copy Infotypes.**

Set the **GrpRsn** (grouping reason) field to **XNON** for infotypes that require replication for this country (for instance, infotypes 0002, 0006, etc.). The resulting screen is shown in Figure 5.21.

Infotype/Subtype Assignment - Grouping Reason

IType	Type	IT Text	GrpRsn	+
0002		Personal data	XNON	Personnel Assignment
0006		Addresses	XNON	Personnel Assignment
0006	QB	Addresses	XNON	Personnel Assignment
0009	0	Bank Details	XNON	Personnel Assignment
0021		Family Member/Dependents	XNON	Personnel Assignment
0021	AU01	Family Member/Dependents	XNON	Personnel Assignment
0033	SE1C	Statistics	SEJU	SE Legal Person
0036		Social Insurance CH	XMOL	Country Grouping
0038		Fiscal Data CH	XMOL	Country Grouping
0048	A	Residence Status	XMOL	Country Grouping
0048	AU01	Residence Status	AUBN	Canada Employer - Registration Numb
0048	B	Residence Status	XMOL	Country Grouping

Figure 5.21 Grouping Reason for Infotypes

Now, run report RPUFIXDS to correct your current master data. When an employee moves from one country to another, their personnel number must be changed. The two personnel numbers are automatically linked using infotype 0709. However, on the Transaction PA20 and Transaction PA30 screens, which display and maintain master data in Employee Central Payroll, respectively, that the two personnel numbers represent the same physical person may not be obvious. The **Personnel Assignment** switch resolves this issue. By turning this switch on, the Personnel ID (a single number for the same physical person) is visible, in addition to the personnel number used in Transaction PA20 and Transaction PA30.

To activate this setting, follow these steps:

- Enable the switch in the Customizing activity **Activate Concurrent Employment for Personnel Administration** and follow the menu path **Personnel Management • Personnel Administration • Customizing Procedures • Settings for Concurrent Employment.**
- Change default settings in the following Customizing activities:
  - **Description of Personnel Assignments**: Choose the infotype fields with which you would like to describe the employment.
  - **Personnel Assignment Overview and Personnel Assignment Details**: Choose the layout for these fields.

Bank Details

Bank details are required to send payroll for each employee to the bank. The company usually sends the money to one bank which then distributes the funds based on bank account numbers provided by each employee. This process requires configuration of the disbursing bank. To complete the bank detail configuration, follow these steps:

1. Set up the required payment methods in Transaction SPRO by following the menu path **Personnel Management • Personnel Administration • Personal Data • Bank Details • Define Payment Method.**
2. Define the payment methods required for this country (e.g., bank transfers and checks, as shown in Figure 5.22). The necessary company codes and paying company codes should already be set up by finance, as described in Section 5.1. The specific payment method can be chosen from a standard list for a company code by clicking on **Pmnt methods in company code** button and for a country by clicking on the **Pmnt methods in country** button.
3. Set up house banks and account numbers for these payment methods in Transaction PE03 by activating feature DTAKT for each country, as shown in Figure 5.23.

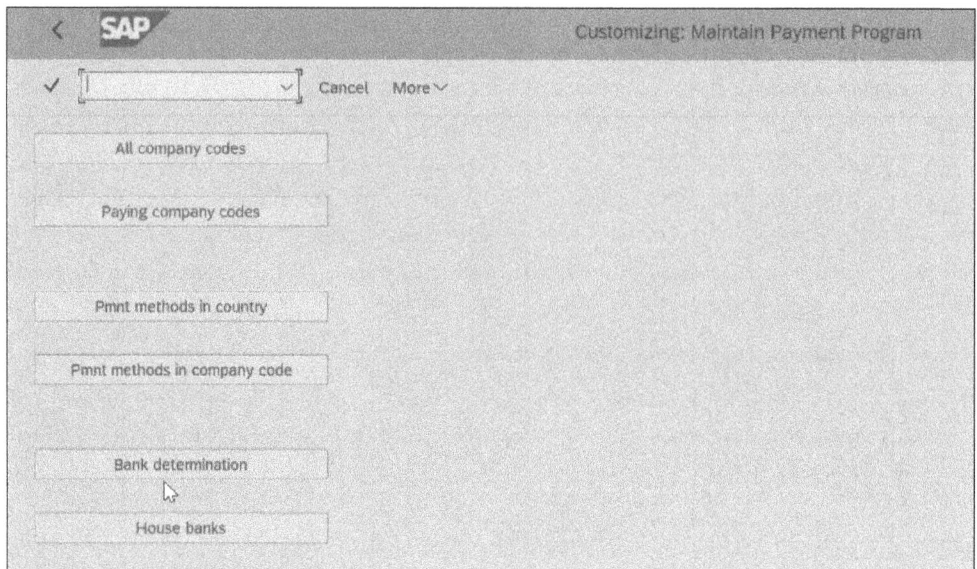

Figure 5.22 Payment Methods

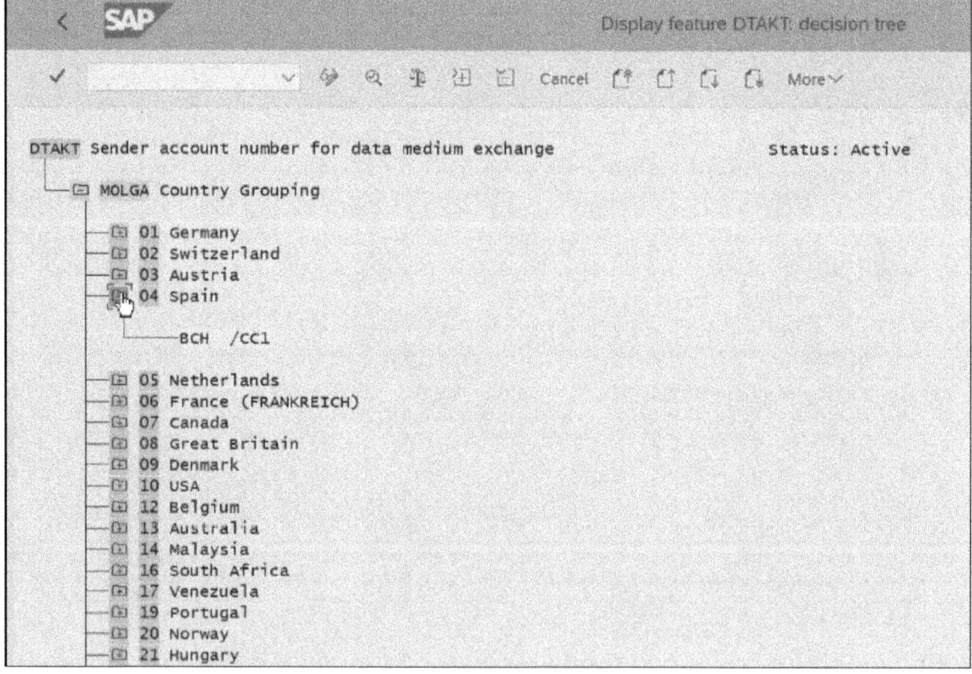

Figure 5.23 Feature DTAKT

4. Next, activate International Bank Account Numbers (IBANs) in Transaction SPRO by following the menu path **Personnel Management • Personnel & Organization • Integration • Integration with Accounting • Activate IBAN.**

5. Maintain the following values, as shown in Figure 5.24:
   - **Group**: Enter "ADMIN."
   - **Sem. abbr.**: Enter "IBAN."
   - **Value abbr.**: Enter "S."

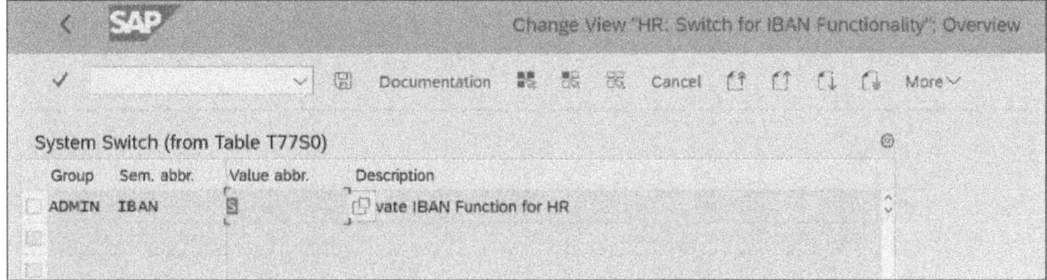

Figure 5.24 IBAN Settings

5.2.3 Payroll Organizational Data

A *payroll area* is a unit for collecting employees processed together for payroll. Create a payroll area in Transaction SPRO by following the menu path **Personnel Management • Personnel Administration • Personal Data • Organizational Assignment • Create Payroll Area.**

A payroll area depends on its date settings to deliver a *payroll calendar*. The **Pay Group** field in Employee Central is the equivalent field for a payroll area. A payroll calendar can be set up in Employee Central Payroll, exported using the standard report PAOCF_EC_PAY_CALENDAR_INFO, and then imported into Employee Central using the **Import Foundation Data** function.

Each payroll area should have a control record. Before running payroll, a new *control record* must be created for each payroll area. Control records keep track of the last retroactive date for all employees that belong to that payroll area, as well as the last payroll processed.

To create a control record, bring up the configuration screens in Transaction SPRO by following the menu path **Personnel Management • Personnel Administration • Organizational Data • Organizational Assignment • Create Control Record.** A control record is created by selecting a previously created payroll area and clicking the **Create** button.

Enabling multiple payroll administrators is important for dividing up the administrative work when many employees must be managed. Once administrators are created, security in Employee Central Payroll can be woven around these administrator codes, which will also be available for assignments in the Payroll Control Center, which we'll cover in detail in Chapter 10.

Create an administrator by following the menu path **Personnel Management • Personnel Administration • Organizational Data • Organizational Assignment • Define Administrator.**

Click on **New Entries**, select an administrator group from the **Group Administrator From** dropdown list, enter a name in the **Administrator Name** field, and enter an SAP user ID in the **SAP Name** field.

A group administrator can also be defined in Transaction PE03. Enter the word "PINCH" in the **Feature** field and then click **Change**. Add a subtree with the **Personnel Area** field. Enter the value "*****" into the **Personnel Area** field to set the group name the same as personnel area.

5.2.4 Basic Pay

The basic pay of an employee is a vital element for calculating payroll. In an Employee Central system, basic pay is stored in the **Compensation** portlet. In Employee Central Payroll, this information is available in infotype 0008. In Employee Central Payroll, basic pay data involves several components, as shown in Figure 5.25.

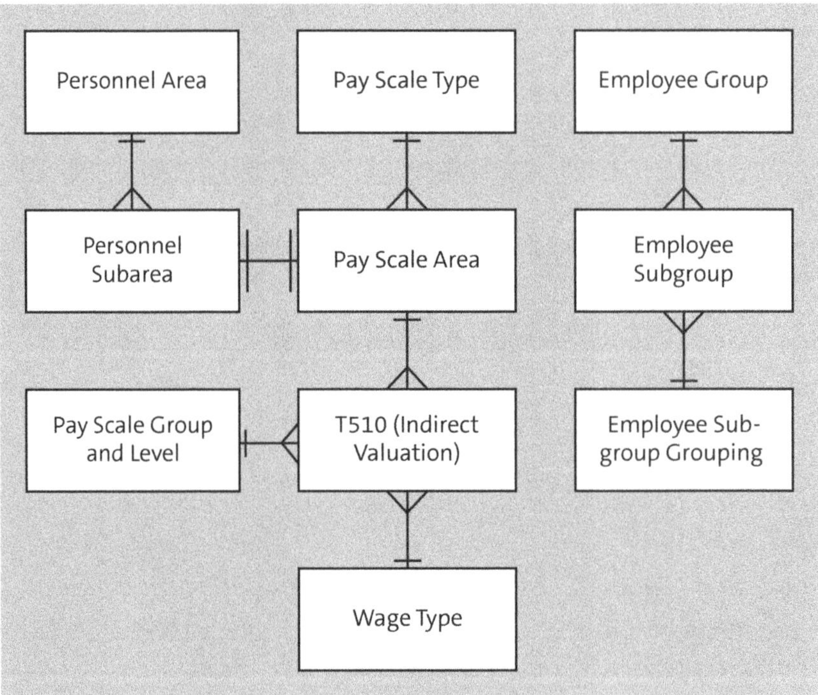

Figure 5.25 Entity Relationship Diagram for Basic Pay

Basic pay (such as an annual salary or an hourly rate) is typically only one component, but several components may be required depending on the employment contract. These components are called *pay* components in Employee Central and wage types in Employee Central Payroll. (Refer to Chapter 1 for a detailed discussion on wage types and to Chapter 7 for mapping pay components to wage types.) For now, we'll only cover the prerequisite configurations for basic pay.

First, define the employee subgroup for pay calculation rules by following the menu path **Personnel Management • Personnel Administration • Payroll Data • Basic Pay • Define Employee Subgroup Grouping for PCR and Collective Assignment Provisions.** These employee subgroups will be used in the valuation of basic pay both for displaying compensation in infotype 0008 and for calculating pay within the payroll engine. Assign a subgroup grouping for each combination of employee group and subgroup.

Next, set the pay scale type and area by following the menu path **Personnel Management • Personnel Administration • Payroll Data • Basic Pay • Check Pay Scale Type.**

You must link the pay scale type and area to the personnel area and subarea, which are location specific. Create this link by following the menu path **Personnel Management • Personnel Administration • Payroll Data • Basic Pay • Check Assignment of Pay Scale Structure to Enterprise Structure.** Choose the personnel area to which the pay scale belongs. Now, set up the group and level by following the menu path **Personnel Management • Personnel Administration • Payroll Data • Basic Pay • Revise Pay Scale Groups and Levels.** This information goes into table T510 and sets up the conditions for *indirect valuation*, which is a technique where a code (in this case, group and level) represents a person's salary, and the actual value of the pay rate is derived from a table entry instead of being stored in the individual employee's record.

If this scenario is not required, you'll still need to set this table up with blanks by first choosing **Pay Scale Type and Area** and then clicking on **New Entries**, as shown in Figure 5.26. Enter the employee subgroup grouping in the **Grpg** field. Enter a name of up to 8 characters for the **PS group** and up to 2 characters for **Lv** and then save your settings.

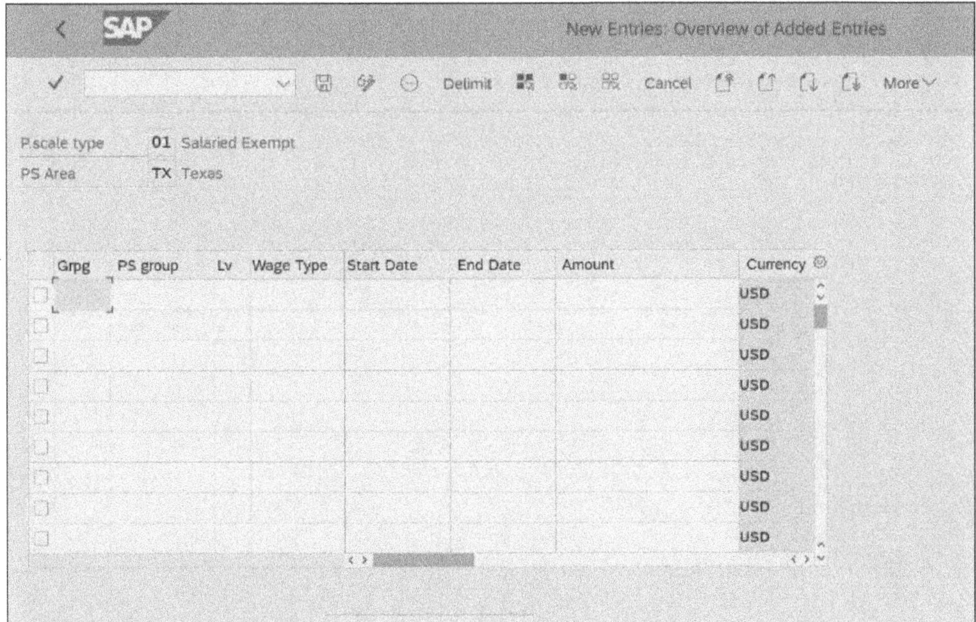

Figure 5.26 Table T510 Entries for Group and Level

5.2.5 Recurring Payments and Deductions

Recurring payments and deductions are controlled by infotype 0014. These values can be directly entered into the Employee Central Payroll system or replicated from Employee Central. This infotype has the structure shown in Table 5.2.

Field Name	Field Description	Field Description Long	Data Type	Field Size	Required
PERNR	Personnel Number	Enter the personnel number in this field.	NUMC	8	Yes
INFTY	Infotype		CHAR	4	No
SUBTY	Subtype		CHAR	4	No
OBJPS	Object ID		CHAR	2	No
SPRPS	Lock indicator		CHAR	1	No
ENDDA	End Date	The date when the record becomes inactive.	DATS	8	Yes
BEGDA	Start date	The date when the record becomes active.	DATS	8	Yes
SEQNR	Infotype record no.		NUMC	3	
AEDTM	Changed On		DATS	8	
UNAME	Changed by		CHAR	12	
HISTO	Historical record		CHAR	1	
ITXEX	Text exists		CHAR	1	
REFEX	Reference fields exist (cost assign.)		CHAR	1	
ORDEX	Conf. fields exist		CHAR	1	
ITBLD			CHAR	2	
PREAS	Reason for Change		CHAR	2	
FLAG1	Reserved Field/ Unused Field		CHAR	1	

Table 5.2 Fields for Infotype 0014

Field Name	Field Description	Field Description Long	Data Type	Field Size	Required
FLAG2	Reserved Field/ Unused Field		CHAR	1	
FLAG3	Reserved Field/ Unused Field		CHAR	1	
FLAG4	Reserved Field/ Unused Field		CHAR	1	
RESE1	Reserved Field/ Unused Field of Length 2		CHAR	2	
RESE2	Reserved Field/ Unused Field of Length 2		CHAR	2	
GRPVL	Grouping Value for Personnel Assignments		CHAR	4	
LGART	Wage Type	Enter the wage type/pay component.	CHAR	4	Yes
OPKEN	Operation indicator		CHAR	1	
BETRG	Amount	Enter the amount for the wage type.	CURR	13	Yes
WAERS	Currency	Enter the currency key (e.g., USD).	CUKY	5	Yes
ANZHL	Number/unit	Enter an amount or number.	DEC	7	
ZEINH	Unit of time/ meas.	This field contains the unit that is the basis for the corresponding calculation (e.g., days/ weeks).	CHAR	3	
INDBW	Ind. valuation		CHAR	1	

Table 5.2 Fields for Infotype 0014 (Cont.)

Field Name	Field Description	Field Description Long	Data Type	Field Size	Required
ZDATE	1st payment date	This date defines the payroll period in which the recurring payment or deduction should be made the first time.	DATS	8	
ZFPER	1st payment period	You can enter the first period in which the recurring payment or deduction is to be made.	NUMC	2	
ZANZL	Interval/Unit	The interval between recurring payments and deductions (number for determining further payment dates).	DEC	3	
ZEINZ	Next payment	Time unit for determining next payment.	CHAR	3	
ZUORD	Assignment Number	Can be used for the storage of text specific to this deduction.	CHAR	20	
UWDAT	Bank Transfer	Date of bank transfer.	DATS	8	
MODEL	Payment model	The payment model defines a time pattern in which deductions are withdrawn and payments made.	CHAR	4	

Table 5.2 Fields for Infotype 0014 (Cont.)

You can configure recurring payments and deductions by following the menu path **Personnel Management • Personnel Administration • Payroll Data • Recurring Payments**

and Deductions. On this screen, the most important activity is to define which wage types are recurring payments or deductions. Detailed discussion on wage types can be found in Chapter 2.

The other activity is to determine payment models (the last field shown in Table 5.2). *Payment models* determine how payments are made with respect to the pay period and can provide a number of useful features for complying with some common business rules. For example, you can specify that the payment or deduction is made weekly, monthly, only in the month of January, or even only in the third pay period of the month in a weekly processing cycle.

To configure payment models, follow the menu path **Personnel Management • Personnel Administration • Payroll Data • Recurring Payments and Deductions • Payment Models • Define Payment Models** and then click on **Define Periodicities.** Then, click on **New Entries** and, as shown in Figure 5.27, maintain the following information:

- **Payment model**: Enter a 4-character name.
- **Name**: Enter a description for the model.
- **Date Modifier** and **Period Parameter** fields: The values that define your payroll area.
- **CumType**: Choose the frequency of deduction required.

When you're done maintaining these settings, click **Save**.

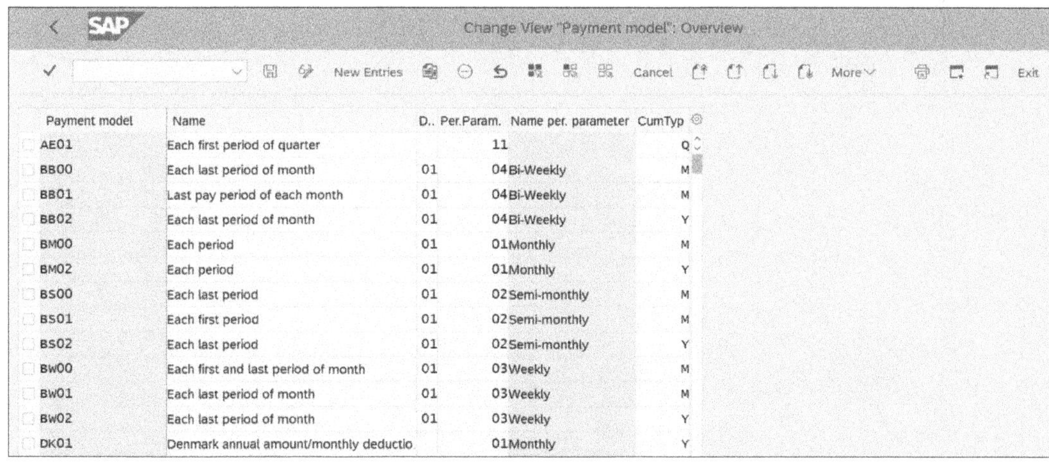

Payment model	Name	D..	Per.Param.	Name per. parameter	CumTyp
AE01	Each first period of quarter		11		Q
BB00	Each last period of month	01	04	Bi-Weekly	M
BB01	Last pay period of each month	01	04	Bi-Weekly	M
BB02	Each last period of month	01	04	Bi-Weekly	Y
BM00	Each period	01	01	Monthly	M
BM02	Each period	01	01	Monthly	Y
BS00	Each last period	01	02	Semi-monthly	M
BS01	Each first period	01	02	Semi-monthly	M
BS02	Each last period	01	02	Semi-monthly	Y
BW00	Each first and last period of month	01	03	Weekly	M
BW01	Each last period of month	01	03	Weekly	M
BW02	Each last period of month	01	03	Weekly	Y
DK01	Denmark annual amount/monthly deductio		01	Monthly	Y

Figure 5.27 Defining Periodicities in a Payment Model

Now, click on **Define Deductions** and enter a validity period in the **From** and **To** fields. Depending on the frequency, select the checkboxes you require, as shown in Figure 5.28, and then save these settings.

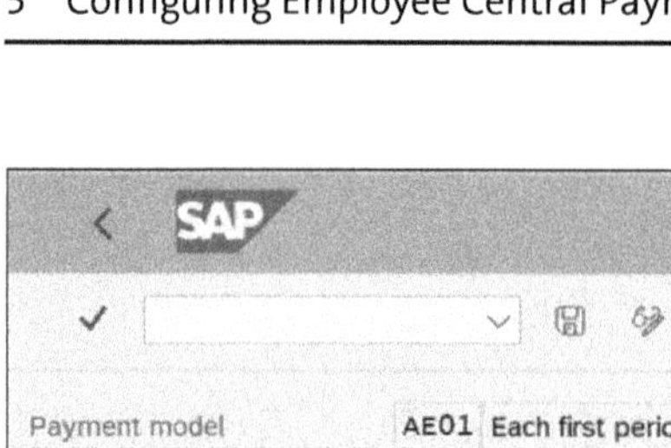

Figure 5.28 Defining Deduction Periods in a Payment Model

5.2.6 Additional Payments

An *additional payment* is defined as a one-time payment or deduction and is controlled by infotype 0015. These values can be directly entered into the Employee Central Payroll system or replicated from Employee Central. We recommend storing deductions directly in the payroll system since this data is not useful in Employee Central. (Its main use is for the computation of net payroll.) Infotype 0015 has the structure shown in Table 5.3.

Field Name	Field Description	Field Description Long	Data Type	Field Size	Required
PERNR	Personnel Number	Enter the personnel number in this field.	NUMC	8	Yes
INFTY	Infotype		CHAR	4	No
SUBTY	Subtype		CHAR	4	No
OBJPS	Object ID		CHAR	2	No
SPRPS	Lock indicator		CHAR	1	No

Table 5.3 Fields in Infotype 0015

Field Name	Field Description	Field Description Long	Data Type	Field Size	Required
ENDDA	End Date	The date when the record becomes inactive.	DATS	8	Yes
BEGDA	Start date	The date when the record becomes active.	DATS	8	Yes
SEQNR	Infotype record number		NUMC	3	No
AEDTM	Changed On		DATS	8	No
UNAME	Changed by		CHAR	12	No
HISTO	Historical record		CHAR	1	No
ITXEX	Text exists		CHAR	1	No
REFEX	Reference fields exist (cost assign.)		CHAR	1	No
ORDEX	Conf. fields exist		CHAR	1	No
ITBLD			CHAR	2	No
PREAS	Reason for Change		CHAR	2	No
FLAG1	Reserved Field/ Unused Field		CHAR	1	No
FLAG2	Reserved Field/ Unused Field		CHAR	1	No
FLAG3	Reserved Field/ Unused Field		CHAR	1	No
FLAG4	Reserved Field/ Unused Field		CHAR	1	No
RESE1	Reserved Field/ Unused Field of Length 2		CHAR	2	No
RESE2	Reserved Field/ Unused Field of Length 2		CHAR	2	No

Table 5.3 Fields in Infotype 0015 (Cont.)

Field Name	Field Description	Field Description Long	Data Type	Field Size	Required
GRPVL	Grouping Value for Personnel Assignments		CHAR	4	No
LGART	Wage Type	Enter the wage type/pay component.	CHAR	4	Yes
OPKEN	Operation indicator		CHAR	1	No
BETRG	Amount	Enter the amount for the wage type.	CURR	13	Yes
WAERS	Currency	Enter the currency key (e.g., USD).	CUKY	5	Yes
ANZHL	Number/unit	Enter an amount or number.	DEC	7	No
ZEINH	Unit of time/meas.	This field contains the unit that is the basis for the corresponding calculation (e.g., days/weeks).	CHAR	3	No
INDBW	Ind. valuation		CHAR	1	No
ZUORD	Assignment Number	Can be used for the storage of text specific to this deduction.	CHAR	20	No
ESTDT	Date of Origin		DATS	8	No
PABRJ	Payroll Year		NUMC	4	No
PABRP	Payroll Period		NUMC	2	No
UWDAT	Bank Transfer	Date of bank transfer.	DATS	8	No
ITFTT	Processing Type	C for calculation, S for simulation, etc.	CHAR	2	No

Table 5.3 Fields in Infotype 0015 (Cont.)

For additional payments, you must also complete some additional configuration steps in Transaction SPRO. Follow the menu path **Personnel Management • Personnel Administration • Payroll Data • Additional Payments.** The most important configuration step is setting up wage types, a process similar to setting up recurring payments. No payment model is required since this payment is a one-time event.

5.2.7 Communications

In an Employee Central Payroll system, the communication infotype (infotype 0105) represents all elements that are used to communicate with the employee. The configuration for this infotype centers around determining its subtypes such as:

- Phone number
- Email
- User ID in the SAP system

The setup screen is accessed by following the menu path **Personnel Management • Personnel Administration • Communication • Create Communication Type.** Click on **New Entries.** Then, enter a 2-character subtype, enter a short description, and choose a time constraint. Time constraints can be 1, 2, or 3, depending on whether gaps are allowed in the applicable timeframe of the record. For example, time constraint 1 means that there can be no gap between records for the same employee. If the first record ends on January 1, 2021, the second record must start on January 2, 2021.

An important subtype is delivered as standard—called ECUS, this subtype stores the user ID of the employee in the Employee Central system. The replication engine uses this information to connect the Employee Central record with the Employee Central Payroll record.

5.2.8 Actions

Actions record an event of significance for an employee, such as hire or termination. Adding actions is a prerequisite for mapping events from Employee Central to the Employee Central Payroll system. Actions and action reasons map directly with events and event reasons from Employee Central, as shown in Figure 5.29.

You can configure actions by following the menu path **Personnel Management • Personnel Administration • Customizing Procedures • Actions.** Actions and action reasons must be created with the objective of mapping them with their corresponding Employee Central values.

For actions, click on **Setup Personnel Actions.** Copy a preexisting action by selecting it and clicking on the **Copy** button or create a new one by clicking the **New Entries** button, as shown in Figure 5.30.

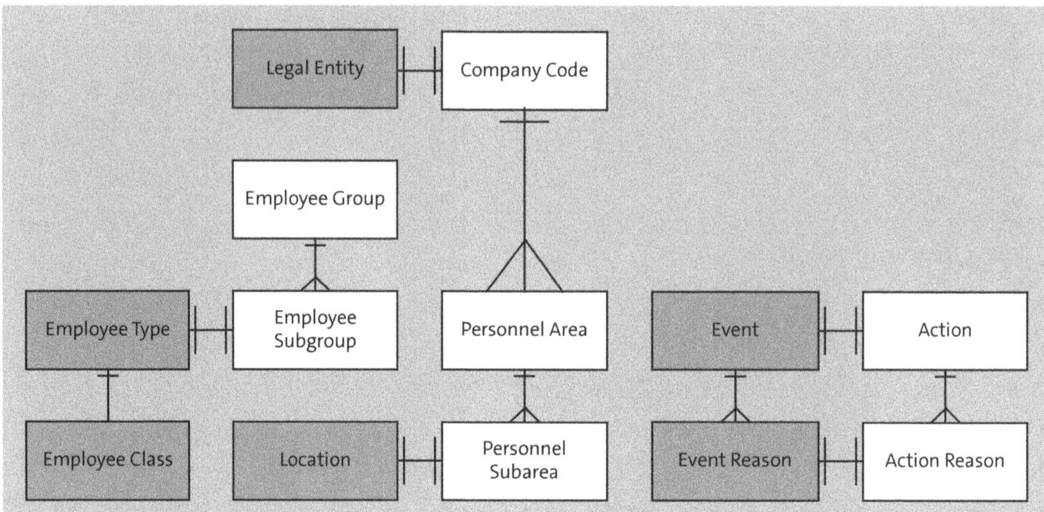

Figure 5.29 Relationship between Events and Event Reasons

Change View "Personnel Action Types": Overview

New Entries Cancel More

Actio...	Name of Action Type	FC	Cus.-...	Empl...	Spec...	Check	PA	EG	ES	IG	D...	U...	U...	C...
01	Hiring	1		3	1		✓	✓	✓	10		✓	✓	
02	Organizational reassignment	0					✓	✓	✓	82		✓	✓	
05	Country reassignment	0					✓					✓	✓	✓
06	Transfer (EDP) - active	0		3	1		✓	✓	✓	12		✓	✓	
07	Transfer (EDP) - retiree	1		2	0		✓	✓	✓	12		✓	✓	
08	Additional pers. assignment	1		3	1		✓	✓	✓	10		✓	✓	
10	Leaving	0		0	0	MSN20	✓	✓	✓	84	1	✓	✓	
12	Reentry into company	0		3	1	MSN21	✓	✓	✓	86		✓	✓	
14	Early retirement/retirement	0		2	0	MSN32	✓	✓	✓	85	1	✓	✓	
15	Semiretirement	0		3	1					83		✓	✓	
16	Change in pay	0								81		✓	✓	
17	Higher grouping	0								81		✓	✓	
18	Lower grouping	0								81		✓	✓	

Figure 5.30 Action Setting

You must maintain the following fields:

- **Action Type**: Enter a 2-character identifier.
- **Name of Action Type**: Enter a description for the action.
- **FC** (function character): Enter "1" for first hiring action, otherwise it should be set to 0.
- **IG** (infotype group number): Enter a 2-digit number to identify the group.
- **PA**: Select this checkbox to indicate if the personnel area value can be maintained from the **Personnel Action** screen.

- **EG**: Select this checkbox if the employee group value can be maintained from the **Personnel Action** screen.
- **ES**: Select this checkbox if the employee subgroup value can be maintained from the **Personnel Action** screen.

Once you've maintained these fields, save these settings and then click on **Create Reasons for Personnel Actions**, which will open the screen shown in Figure 5.31.

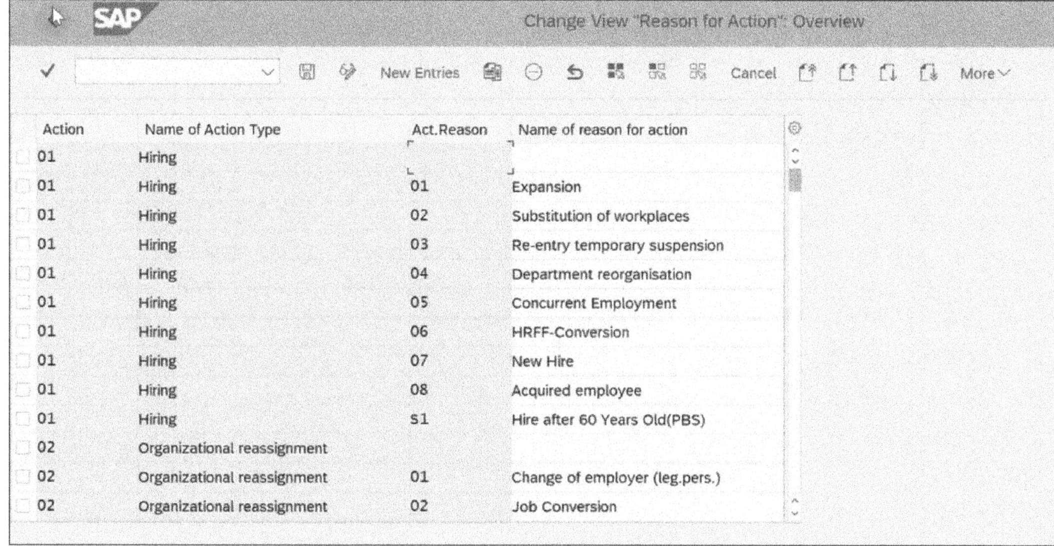

Change View "Reason for Action": Overview

Action	Name of Action Type	Act.Reason	Name of reason for action
01	Hiring		
01	Hiring	01	Expansion
01	Hiring	02	Substitution of workplaces
01	Hiring	03	Re-entry temporary suspension
01	Hiring	04	Department reorganisation
01	Hiring	05	Concurrent Employment
01	Hiring	06	HRFF-Conversion
01	Hiring	07	New Hire
01	Hiring	08	Acquired employee
01	Hiring	S1	Hire after 60 Years Old(PBS)
02	Organizational reassignment		
02	Organizational reassignment	01	Change of employer (leg.pers.)
02	Organizational reassignment	02	Job Conversion

Figure 5.31 Setting Up Action Reasons

Click on **New Entries**. Identify the action we just created in the **Action** field. Enter a 2-digit reason code in the **Act Reason** field. Then, enter a description in the **Name of Reason for Action** field and save.

You can define infogroups (groups of infotypes) specifically for infotypes that won't be replicated. You can also create customer-specific statuses if required. Reasons must be related to actions using the same configuration path for defining actions. A best practice is to copy the standard actions and reasons provided by SAP to your own namespace so that you can still retain the actions predelivered by SAP.

5.2.9 Disabling the Transport Connection for Personnel Planning

A *transport* is a mechanism for moving configuration changes from one SAP system to another (for example, from development to quality assurance). Most configuration items are automatically saved as transports.

Since no major organizational structure exists in an Employee Central Payroll system, the organizational management system in Employee Central Payroll must accordingly having the minimum information required for Employee Central Payroll to work.

Thus, one required task is to turn off transports. By default, SAP will create a transport automatically for any change in an organizational object. This setting can be turned off by following the menu path **Personnel Management • Organizational Management • Transport • Set Up Transport Connection.** On this screen, find the TRSP group with semantic abbreviation CORR. Set the switch value to **X** (no transport).

5.2.10 Integration between Personnel and Organization Data

As mentioned earlier, not much use exists for the sophisticated organizational management system found in the Employee Central Payroll environment. You've learned how to turn transports off, but another default feature of organizational management must also be turned off.

This feature tries to automatically update organizational elements, like positions, with changes recorded in infotype 0001. This behavior is undesirable because the position and infotype 0001 are independently controlled through replication (see Chapter 7). Therefore, to turn this feature off, follow the menu path **Personnel Management • Organizational Management • Integration • Integration with Personnel Administration • Set Up Integration with Personnel Administration.** On this screen, follow these steps:

1. Open **Participate in Integration.**
2. Switch off using PLOGI by removing the **X.**
3. Double click on **Basic Settings.**
4. Switch off ORGA in group PLOGI by removing the **X** from **Value abbr.** field.
5. Clear the group PPABT by deleting the information in it.

The resulting table is shown in Figure 5.32.

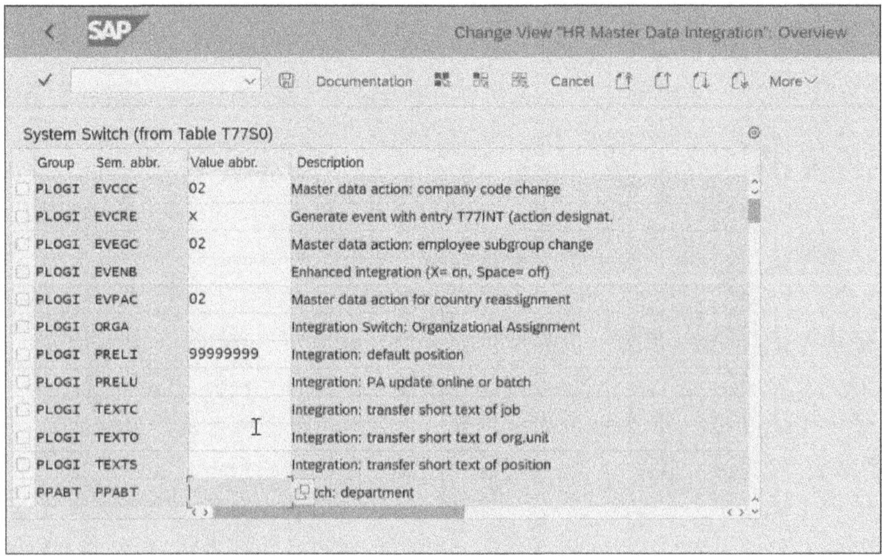

Figure 5.32 Turning Off the ORGA and PPABT Switches

5.3 Setting Up Pay Statements in Employee Central Payroll

Pay statements required the development of an SAP form in a technology called *HR Forms*. This technology enables the use of logic through BAdIs and incorporates security so that your employees can only view their own pay statements (and not the pay statements of others).

5.3.1 HR Forms

A pay statement is developed in Transaction HRFORMS, which is a comprehensive form-building environment. In this transaction, you can pull data from payroll tables and render a report in PDF format. An HR Forms object is built by a developer, which is beyond the scope of this book. In this section, we'll describe the steps you must take after an HR Forms object has been developed.

You'll maintain the settings found by following the menu path **Personnel Management • Employee Self Service • Service Specific Settings • Benefits and Payment • Salary Statement**.

The feature HRFOR must be configured to point to this form in Transaction PE03. You can set up the feature tree by choosing report category C and then adding the country code for your country in the **MOLGA** node. Then, under that country node, you would add a value for the name of the HR Form object that should be used.

If you want traditional SAPscript pay statements, change the value of feature HRFOR to $CEDT$ in Transaction HRFORMS.

5.3.2 Business Add-Ins for Pay Statements

Table 5.4 explains the BAdIs required for various types of pay statements.

Pay Statement Type	Business Add-In
Direct pay statement, where you can click on the required period and navigate directly to the desired pay statement	HRESS_PAYSLIP_BADI_DEF provides information of pay periods. HRSFEC_WAGE_TYPE_BADI_DEF defines the wage types the pay statement service provides. Both have default implementations.
Arrow-based pay statement (also known as the legacy pay statement), where you use arrow buttons to move between chronologically sequenced pay statements	HR_EC_PAYSLIP

Table 5.4 BAdIs for Pay Statements

In some cases, your payroll department may want to restrict the viewing of pay statements until the pay date. In this way, the payroll department can make emergency changes when required, without the embarrassment of needing to offer complicated explanations to employees about last-minute changes. This capability is achieved with the BAdI `IF_HRESS_PAYSLIP_BADI~PROVIDE_FILTERED_RGDIR`. No default implementation exists for this capability, which must be implemented in Transaction SE19.

For this task, you'll require a developer to write the necessary ABAP code and to implement the following two steps:

- Identifying the period from table `CT_RGDIR` and comparing that value with current date.
- If today's date is before the payment date, then deleting the entry from table `CT_RGDIR`.

Pay statements offer many more capabilities, which we'll cover in Chapter 11.

5.3.3 Additional Pay Statement Security

Another capability is available to make pay statements even more secure. Using this feature, the system will throw an error if the URL for viewing the statement is tampered with or used more than once.

The switch is activated by following the menu path **Personnel Management • Personnel Administration • Customizing Procedures • Infotypes • Infotypes in Detached Infotype Framework • Use Decoupled and Not Decoupled IT Framework.**

Alternatively, you can open Transaction SM30 and enter "T77SO" in the **Table/View** field. Then, follow these steps:

1. Position the switch to **Group SFEC**.
2. Find semantic abbreviation SPYSL.
3. Enter "X" into the **Value Abbreviation** field.
4. Save these settings.

Now, employees in the Employee Central system can only view their own pay statements. This visibility is automatically restricted by the role assigned to an employee.

5.4 Setting Up Concurrent Employment

Concurrent employment is a capability, available in Employee Central Payroll, for when an employee works multiple jobs that pay different rates, potentially for different departments, at the same time. A good example is a professor teaching in multiple colleges of the same university and drawing a different salary from each one simultaneously.

Concurrent employment is different from a *global assignment*, where is a person employed in one country but travels to another for long periods of time to work there, thus drawing a combination of home-country and host-country salaries.

In terms of personnel administration, concurrent employment has the following implications:

- Personnel numbers belonging to the same central person are linked.
- Infotypes belonging to the same central person are shared.

In Employee Central Payroll, SAP supports the following capabilities:

- Combined tax, garnishment, and social insurance calculation for several employment IDs (PERNRs) belonging to the same person
- Enhanced legal reporting for concurrent employment

Note

Enabling concurrent employment is an irreversible change, meaning that you cannot go back and turn it off once the capability is activated.

Turning on concurrent employment requires certain switches be activated in table T77S0, as listed in Table 5.5.

Switch Group	Value	Mandatory/Optional	Description
SFEC	CCMM	Mandatory	Enables employees to display data for all their own personnel numbers. With this switch, you run the User Creation report only for the first personnel number of an employee.
CCURE	PAUIX	Mandatory	Enables linked personnel numbers (PERNRs) in Transaction PA20/PA30 using the drop-down list.
CCURE	MAINS	This step is mandatory *only* if concurrent employment is used in your selected country-/region-specific payroll. Otherwise, this switch is optional.	Enables the display of infotype 0709 (Person ID) in Transaction PA20/30. This infotype is filled in during the replication process.

Table 5.5 Enabling Concurrent Employment

Switch Group	Value	Mandatory/Optional	Description
CCURE	PAY*	This step is mandatory *only* if concurrent employment is used in your selected country-/region-specific payroll. Otherwise, this switch is optional.	Where the asterisk (*) stands for the corresponding country/region grouping, for example, PAYAU for Australia (Start Date CE Payroll).

Table 5.5 Enabling Concurrent Employment (Cont.)

5.5 Security in Employee Central Payroll

Due to the sensitive nature of compensation and employee data, securing your payroll system is an important task. In this section, we'll walk through the two steps to ensure security in Employee Central Payroll: defining user roles and creating/maintaining users by assigning roles to them.

5.5.1 Defining Roles

Authorizations in Employee Central Payroll are driven by roles, which are collections of authorization settings, as shown in Figure 5.33.

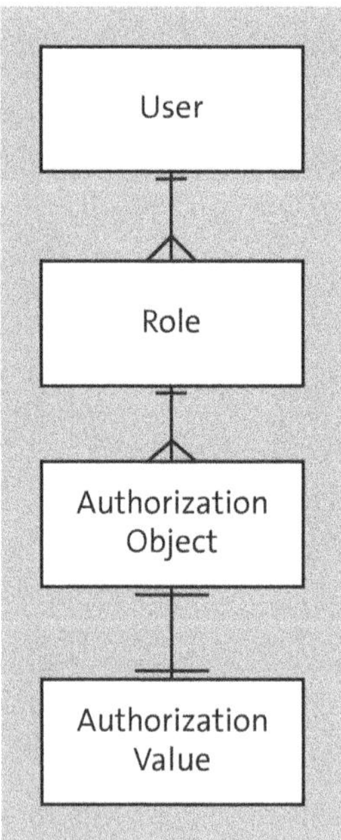

Figure 5.33 Entity Relationship Diagram for Authorizations

An authorization object defines a specific attribute of the authorization. For example, S-TCODE is an authorization object that has an attribute called a *transaction code*, and this code is set to the transaction that the employee should have access to. An infotype is a common attribute for human resources (HR) authorization objects that specify which infotypes the user has access to.

A user is assigned multiple roles depending on the function the user is supposed to perform. Each role is a collection of authorization objects with their value settings, as shown in Figure 5.33. SAP provides several roles for Employee Central Payroll that deliver the functionality required for managing HR data, running payroll, and viewing pay statements.

For any given role to be used, it needs to be copied and assigned. An important best practice is not to use the standard roles directly and instead copy predelivered roles to create new ones.

Figure 5.34 shows which steps of the support process require which roles. Table 5.6 lists standard, predelivered roles and their corresponding process steps. In the names of roles, xx represents ISO country code.

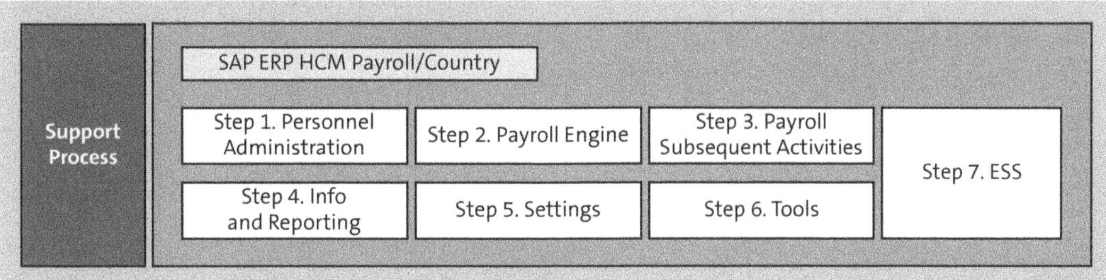

Figure 5.34 Process Steps that Require Specific Roles

Role	Description	Process Step
SAP_CLOUD_HCM_PA-HR-MANAGER	Monitoring master data administration	1, 4
SAP_CLOUD_HCM_PA-US-MANAGER	Additional transactions for US managers using Employee Central Payroll	1,4
SAP_CLOUD_HCM_PA_HR-ADMINISTRA* x	Processing personnel master data, executing personnel actions, and evaluating personnel master data	1,4
SAP_HR_PA_SPECIALIST	Administrating and performing system tasks; logging and monitoring changes to master data	1, 4, 5, 6
SAP_HR_PY_PAYROLL-SPEC	International expert in the content and technical side of payroll	2, 3, 4, 5, 6

Table 5.6 Standard Delivered Roles

Role	Description	Process Step
SAP_HR_PY_xx_PAYROLL-SPEC	National expert in the content and technical side of payroll	2, 3, 4, 5, 6
SAP_HR_PY_xx_PAYROLL-ADM	National payroll processing	2,6
SAP_HR_PY_xx_PAYROLL-MANAGER	National payroll manager	2, 3, 4
SAP_HR_PY_xx_PS_PAYROLL-ADM	National payroll processing for public services	2, 6
SAP_CLOUD_EMPLOYEE_ESS_PAYSLIP*	The Employee Self-Service (ESS) role is assigned automatically to each employee. See the Maintaining Roles to be Assigned to Users section. For integration option Pay Statement (direct), in addition to this role, the SAP_CLOUD_ESS_OAUTH role is required.	7
SAP_CLOUD_MANAGER_ESS_PAYSLIP* SAP_ESS Manager Payslip	The ESS role is assigned to managers or HR administrators that must have access to pay statements of employees in their personnel area. For integration option Pay Statement (direct), in addition to this role, role SAP_CLOUD_ADMIN_OAUTH is required.	7
SAP_CLOUD_ESS_PAYMNT_SUM_AU*	SAP_ESS user for payment summaries in Australia.	7
SAP_CLOUD_ADMIN_OAUTH	This role is required for administrators to access payroll system information and confirm infotype existence, (for example, whether a tax infotype set up for an employee) as well as access pay statements (direct) on behalf of an employee. For pay statements (direct), this role must be assigned to administrators in addition to role SAP_CLOUD_MANAGER_ESS_PAYSLIP.	5,6
SAP_CLOUD_ESS_OAUTH	This role is required for employees to access pay statements (direct). This role must be assigned to employees in addition to the role SAP_CLOUD_EMPLOYEE_ESS_PAYSLIP.	5,6

Table 5.6 Standard Delivered Roles (Cont.)

The roles marked with an asterisk (*) are ESS roles that require the creation of a general purpose role that should be assigned to every employee. To do this, add the authorization object S_BDS_DS to an authorization role with basic authorizations. Fill in the attributes with the following values:

- **ACTVT**: Activity: 03
- **CLASSNAME**: Business document service: Class name: DEVC_STXD_BITMAP
- **CLASSTYPE** (name of the application class type): OT

Roles can be accessed via Transaction PFCG. As shown in Figure 5.35, to copy the role, click the copy icon (to the left of the pencil icon).

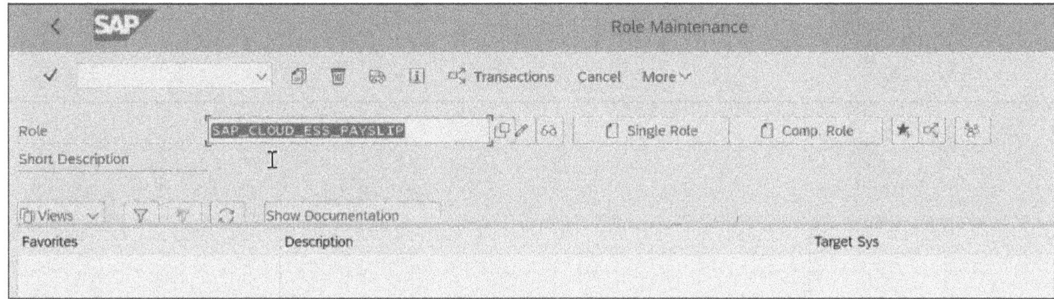

Figure 5.35 Transaction PFCG

Click on the **Authorization** tab to view the authorization objects, as shown in Figure 5.36.

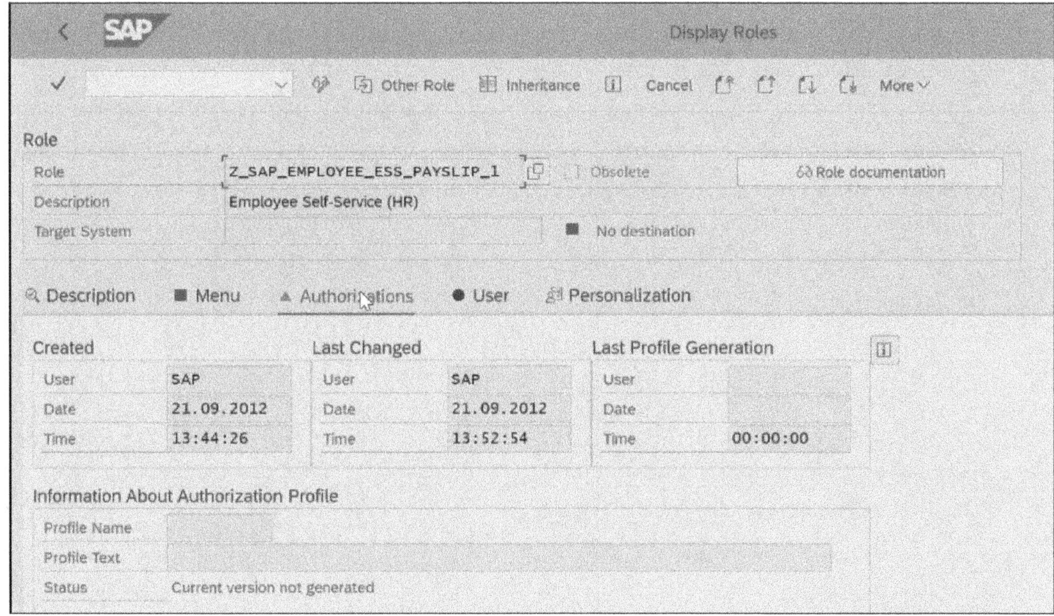

Figure 5.36 Authorization Tab in Transaction PFCG

Click on **Display Authorization Data** to see the authorization settings, as shown in Figure 5.37.

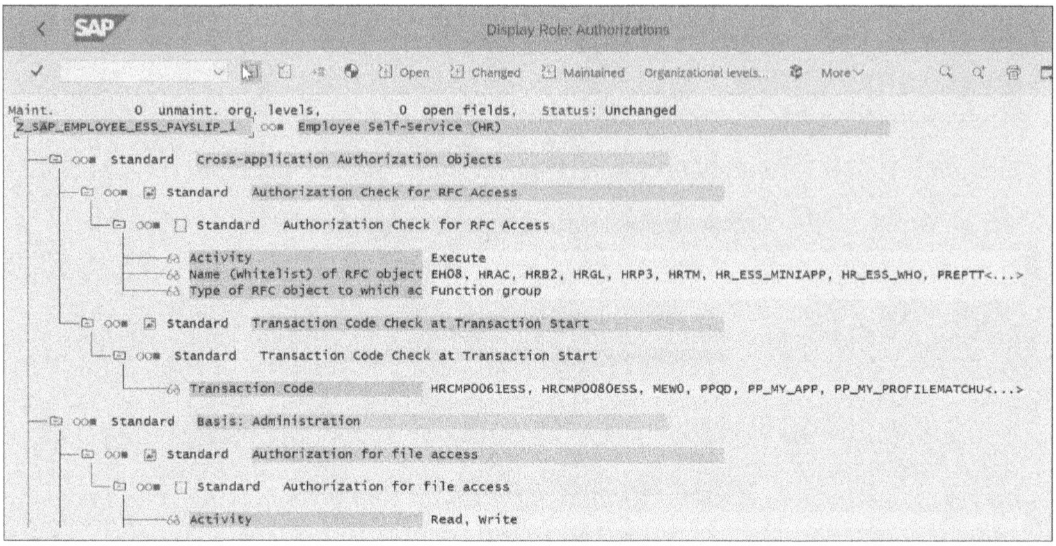

Figure 5.37 Authorization Objects and Their Attributes

5.5.2 Creating Users

The presence of a user in Employee Central does not automatically generate a corresponding user in Employee Central Payroll, where users must either be manually created or created through a report. Once created, a user must be assigned to a role. This task can also be automated by setting up an authorization group, to which the required roles are attached.

SAP provides a standard authorization group for general employee called ESSUSER. When the User Creation report is run, not only is the user created; the user is also assigned to all roles in the specified user group. To execute this report, you'll need permission to read infotypes 0000, 0002, and 0105 and permission to write to infotype 0105, subtype ECUS. In addition, the following permissions are needed:

- **User creation**
  - Authorization object: S_USER_GRP
  - Authorization field: ACTVT = 01
  - Authorization field - CLASS: A user group for user maintenance (for example, ESSUSER)
- **Role assignment**
  - Authorization object: S_USER_AGR
  - Authorization field: ACTVT = 02
  - Authorization field - Role Name: The (composite) role to be assigned to the users

The report, called report HRSFEC_ESS_USER_UPDATE, creates users with the user name and alias set to the employee's user ID in the Employee Central system (Employee Central user ID). This value is the same as the value of subtype ECUS in infotype 0105.

To execute, run Transaction SE38 and enter "HRSFEC_ESS_USER_UPDATE" in the **Program** field, as shown in Figure 5.38.

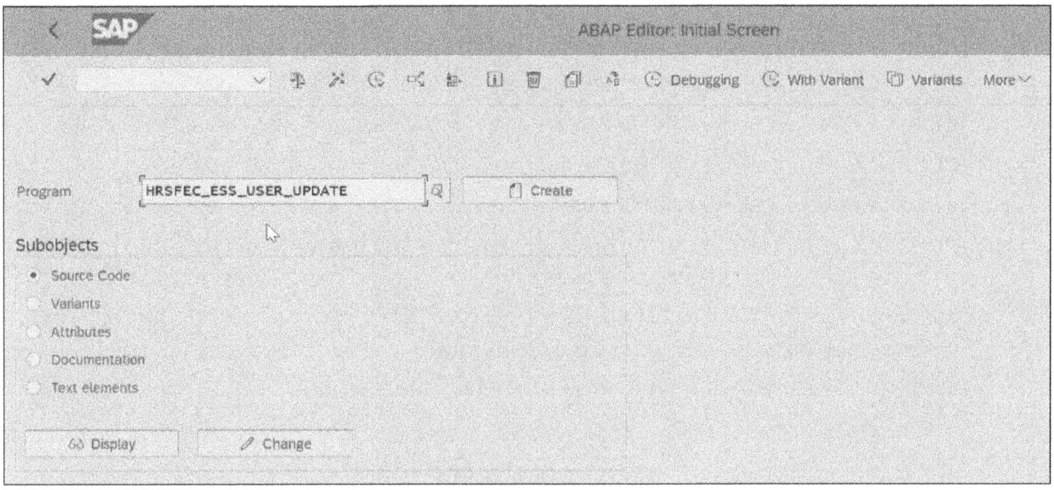

Figure 5.38 Employee Central Payroll: User Creation Report

Click on the **Execute** button, and the screen shown in Figure 5.39 will appear.

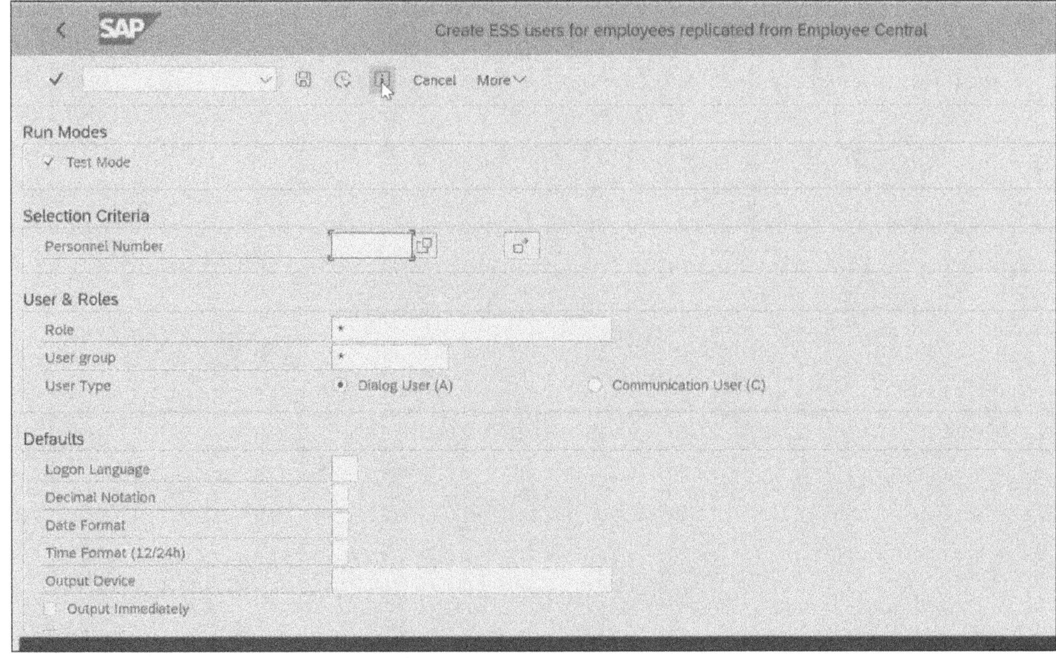

Figure 5.39 Parameters for User Creation

The report can be run in test mode by selecting **Test Mode**. In this case, the system offers a confirmation report but does not update the roles or create new users.

By leaving the **Personnel Number** field blank, you can run this report for all employees. In this case, the report will scan all records and create user IDs for the records that lack one.

Note that this report creates users and assigns roles only once—once the system discovers a user with a name that already exists, that record is skipped. In the **Roles** section, you can assign a single role or a collection of roles in the **User Group** field. The **Default** section is useful for setting up default settings like the date and currency for the country being configured.

These users are created without passwords since authentication is performed through the identity manager from the Employee Central environment. The identity manager automatically authenticates the identity based on credentials entered in Employee Central so that a password does not have to be entered again. These autogenerated users are not meant for logging on directly into the Employee Central Payroll system (without going through the Employee Central system first).

The length of a user ID is limited to 12 characters. Therefore, user IDs created in Employee Central must also be limited to 12 characters.

5.6 Summary

In this chapter, we described the settings necessary to lay the foundation for an Employee Central Payroll system. Now, your payroll system is ready to receive data, and you can set up users with the appropriate permissions to enable them to view and edit their own information. The tasks described in this chapter also prepared the system to begin integration activities from Employee Central to Employee Central Payroll and to prepare the environment for the Payroll Control Center to take over the actual processing of payroll. We'll discuss these topics in the next four chapters, starting with integration.

Chapter 6
Integration

Welcome to the world of collaboration! A payroll system is not fully effective unless it can communicate with other systems for vital pieces of data. After enabling Employee Central Payroll in Chapter 5, integration will play a major role in the rest of the configuration process.

6

This chapter introduces the integration concept in Employee Central Payroll. We'll provide a high-level overview of the different systems Employee Central Payroll needs to interact with to deliver payroll capabilities and discuss various strategies for connecting them successfully.

You must fulfill the following three prerequisites before integration with an Employee Central system can work:

- Employee Central Payroll has been turned on in provisioning (Chapter 3).
- Employee Central has been configured (Chapter 4).
- Employee Central Payroll has been configured (Chapter 5).

Let's start off this chapter by looking at the integration technologies that can be used to integrate with Employee Central Payroll. We'll then jump into the types of data involved in integration: time data, benefits data, and finance data.

6.1 Integration Technologies

The integration between Employee Central Payroll and other systems can use multiple technologies. The traditional method of integrating with other systems was to use files. In the modern world, the following have technologies have evolved:

- Web services
- Application programming interfaces (APIs)
- SAP Integration Suite

We'll discuss each of these technologies in more detail in this section. However, we'll start by covering file-based integration because this approach is still in use and sometimes is the only possible way to integrate with a partner system. Employee Central Payroll integration can only be achieved if the corresponding technology exists in the partner system.

6.1.1 File-Based Integration

Employee Central provides two different methods for exporting files, which we'll look at in the following sections:

- Employee Export
- Integration Center

Exporting Employee Data

To download employee data, launch the **Export Users** page by typing in "Employee Export" in the search bar or by going to the Admin Center and clicking **Update User Information • Employee Export**, as shown in Figure 6.1.

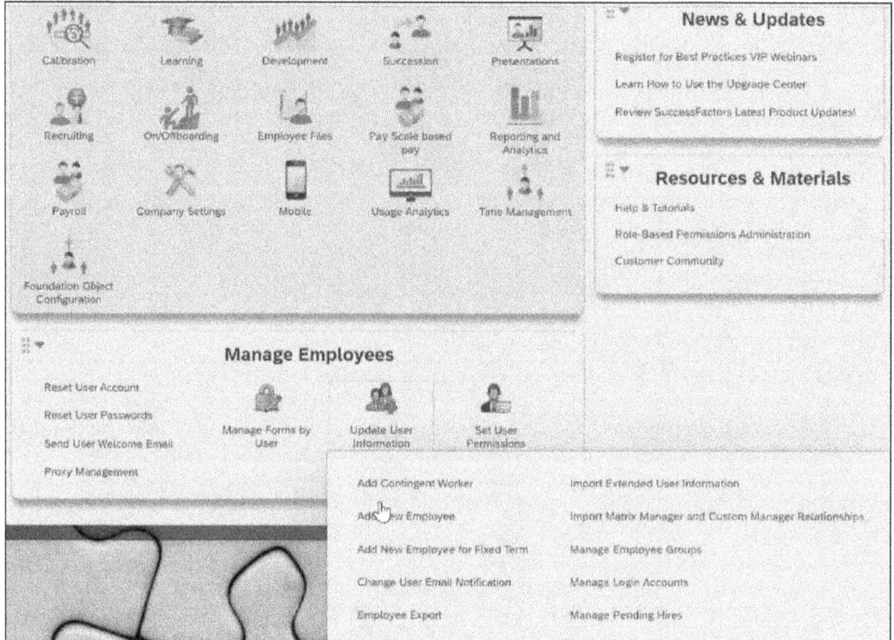

Figure 6.1 Launching Employee Export

You'll be presented with the input screen shown in Figure 6.2.

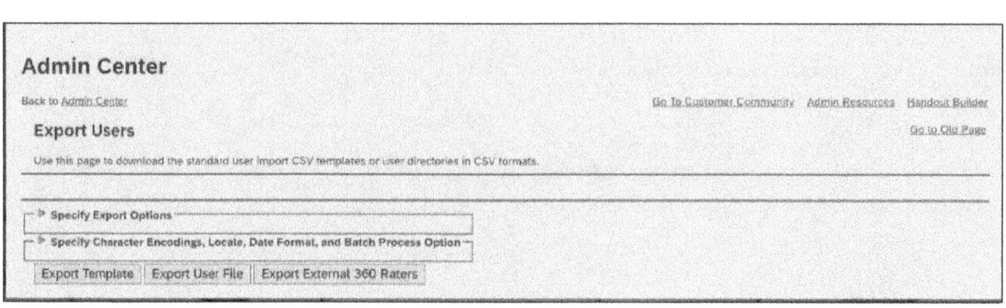

Figure 6.2 Exporting Employee Data from Employee Central

This straightforward screen offers no flexibility. The format in which you'll download your employee data will be specified by SAP, and you cannot change it. By clicking the **Export Template** button, you'll download a template for the format your data will be in, as shown in Figure 6.3.

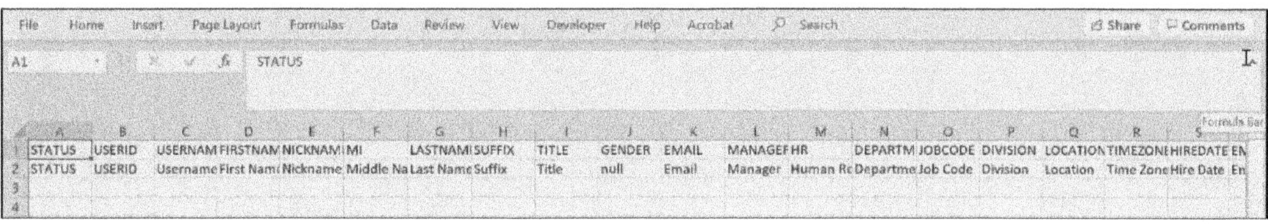

Figure 6.3 Downloaded Employee Export Format

Integration Center

The Integration Center provides a lot more flexibility in how you can download information. You can access the Integration Center from the Admin Center, and you can build customized file extracts using any data that is accessible through the catalog of SAP SuccessFactors OData APIs. For example, you can use the Integration Center to immediately run a job or schedule a job to export a provider-specific OData object to Secure File Transfer Protocol (SFTP) based on defined filter conditions and applying a provider-defined output order and file format. The **Configure Fields** button (shown later in Figure 6.10) provides the ability to choose fields. Destination settings can be maintained so that the file can be delivered to a secure location automatically.

You can use this technique to distribute data selectively in customized formats. You would start with launching the Integration Center. To open the Integration Center, from the Admin Center, click on **Company Settings** and then select **Integration Center**, as shown in Figure 6.4.

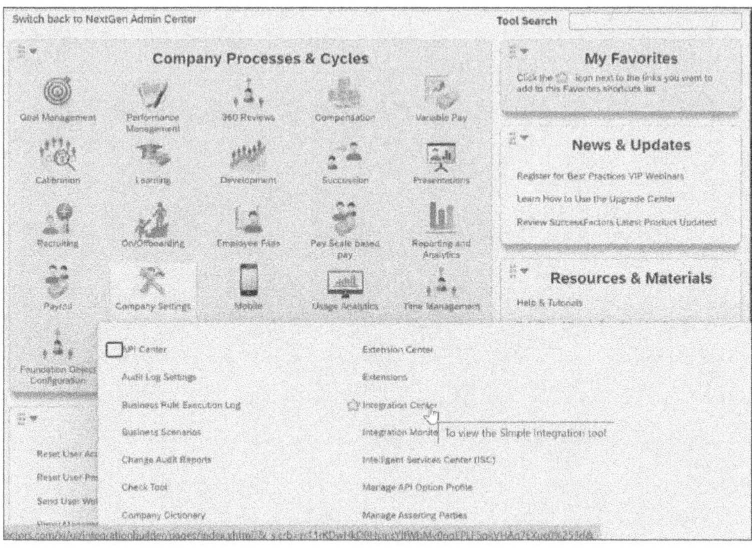

Figure 6.4 Launching the Integration Center

Once launched, you'll see the Integration Center workbench screen, as shown in Figure 6.5.

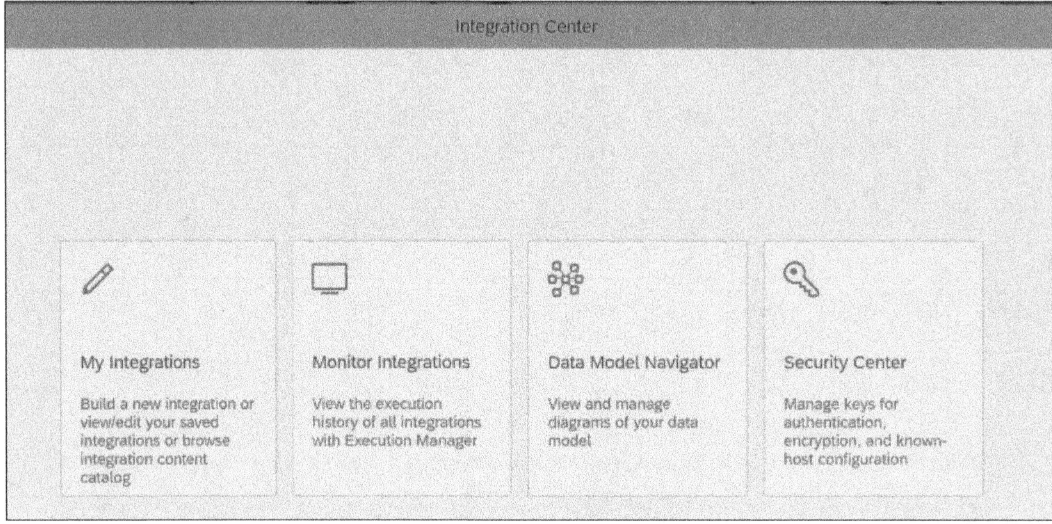

Figure 6.5 Integration Center Main Screen

The Integration Center helps you build, run, schedule, and monitor simple integrations. This tool simplifies most common export/import use cases and allows you to build integrations quickly and easily through a guided workflow. Predefined templates are available, and you can also create your own templates. Let's walk through a quick example of creating a file with user data and scheduling this job to run every day:

1. After launching the Integration Center, click on **My Integrations**. You'll see a screen listing all existing integrations, as shown in Figure 6.6.

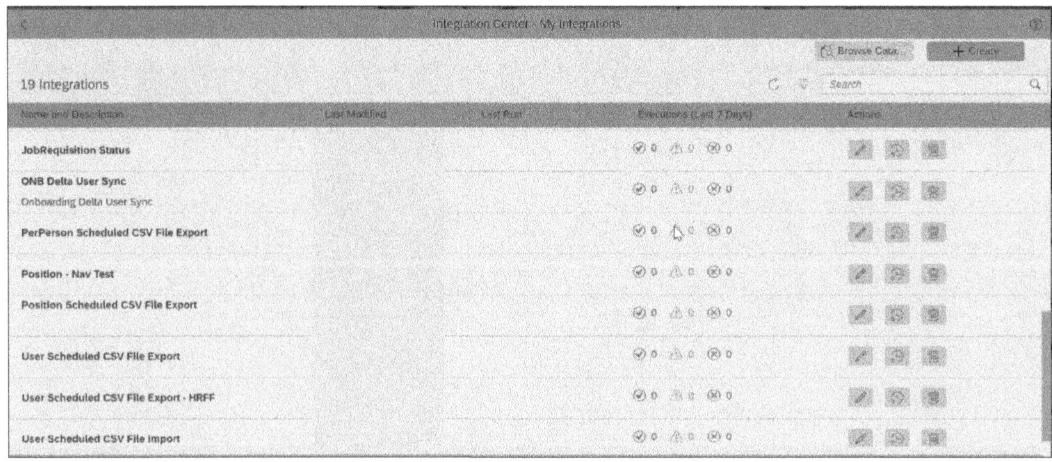

Figure 6.6 My Integrations

2. Click on the **Create** button, which prompts you to choose an integration type, as shown in Figure 6.7.

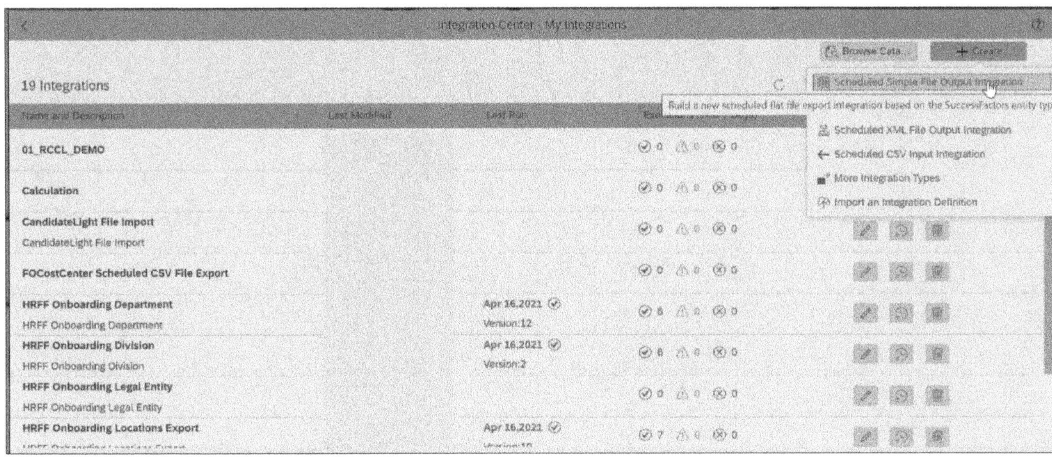

Figure 6.7 Creating a New Integration

3. For a quick demonstration, we'll choose a simple integration type option: **Scheduled Simple File Output Integration**. The system then prompts you to choose an entity, which you can search for by name. In this case, choose **User**, as shown in Figure 6.8.

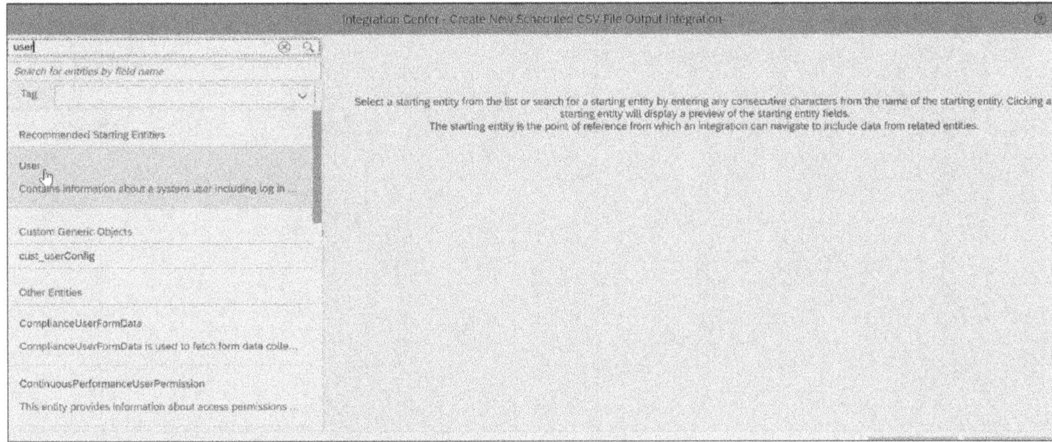

Figure 6.8 Choosing a User Entity for Integration

4. You can now select the fields you need by selecting the checkboxes shown in Figure 6.9. Once done with your selections, click on **Select**.

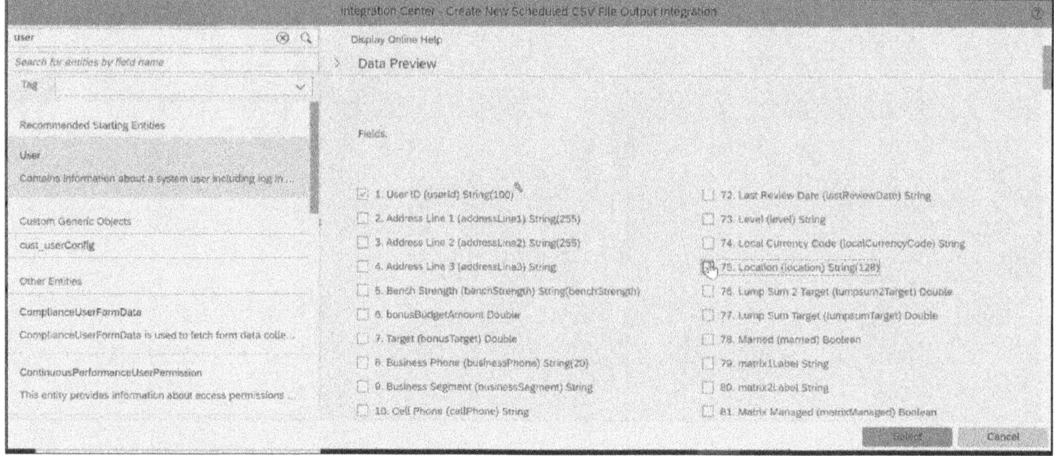

Figure 6.9 Selecting Fields

You'll now see a comprehensive option screen, shown in Figure 6.10, from which you can perform the following actions:

- Further adjust fields
- Set filter criteria
- Set a destination
- Schedule jobs
- Review and run the integration

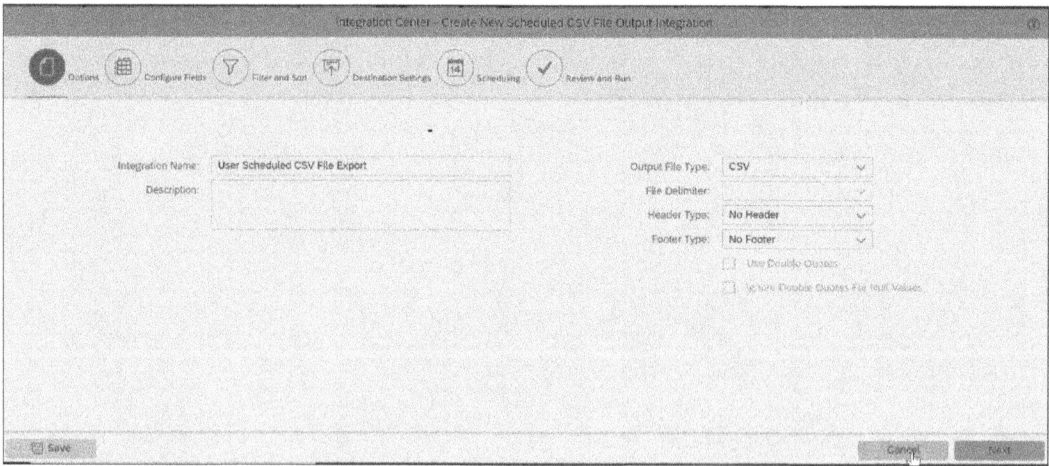

Figure 6.10 Completing the Integration Setting

In the Employee Central Payroll environment, creating file-based exports or imports is possible through the development of custom reports using the ABAP programming language.

A standard file upload utility is available in SAP called the Legacy System Migration Workbench. This tool allows you to import flat files, in restricted formats, directly into infotypes. As the name suggests, the Legacy System Migration Workbench is not popular since many restrictions exist on how it can be used.

6.1.2 Web Services

A *web service* is a piece of software that can be used over the internet to connect one system to another system using standard protocols. This resource is available to other systems and allows them to connect and exchange information with the first system. Generally, no programming is necessary to use a web service since the functionality is already available within the service. Specific web services usually come with a set of instructions on how to configure them to perform the necessary integration.

Table 6.1 lists some example web services in Employee Central Payroll that you can use to connect with Employee Central, including the information that they deal with in each system.

Web Service in Employee Central Payroll	Employee View in Employee Central
HRPAO_PAOM_MASTERDATA	Country-/region-specific master data
PAYSLIP	Pay statement
PAYMNT_SUM_AU	Payment summary form in Australia
HRESS_A_PAYINFO	Superannuation in Australia
HRESS_A_REP_IN_FORM16	Form 16 in India

Table 6.1 Web Services to Connect Employee Central Payroll with Employee Central

6.1.3 Application Programming Interfaces

An API is a programming tool that retrieves or updates the data within an application, depending on the use case. The system that is being connected to provides data item specifications in great detail, in the form of a *data dictionary*. For example, Figure 6.11 shows the OData API data dictionary for employee data. You can access this data dictionary by typing in "OData API Data Dictionary" in the search bar. Select **Employee Central (EC)** from the **Tag** dropdown menu and **LegalEntityUSA** from the **Name** dropdown menu.

Employee Central Payroll uses point-to-point replication to connect with Employee Central. In fact, the point-to-point replication program in Employee Central Payroll uses the API definition found in Employee Central Payroll to connect and retrieve the information required. The great news is that you don't have to learn the API definitions—these definitions are already programmed in the replication program. We'll review the point-to-point replication program in great detail in Chapter 8.

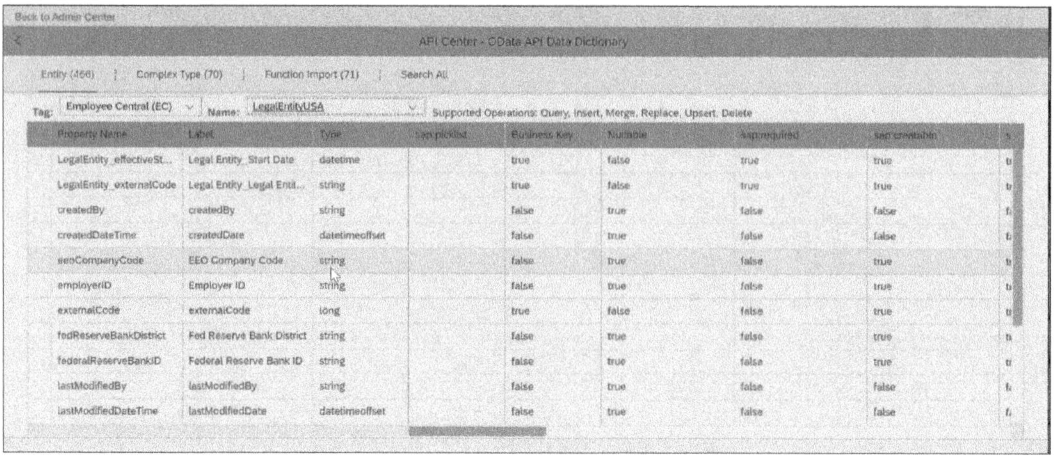

Figure 6.11 API Dictionary Entries for the LegalEntityUSA Object

Let's quickly add one point about this API: SAP currently uses the SOAP API, not the OData API, for the point-to-point replication program in Employee Central Payroll. The SOAP API (also known as SFAPI) has been deprecated except for the CompoundEmployee object, which is used solely for this replication. You can bring up the SOAP API in Employee Central in a similar manner as explained for OData API, by entering "SFAPI Data Dictionary" into the search bar. The elements of the CompoundEmployee object are shown in Figure 6.12.

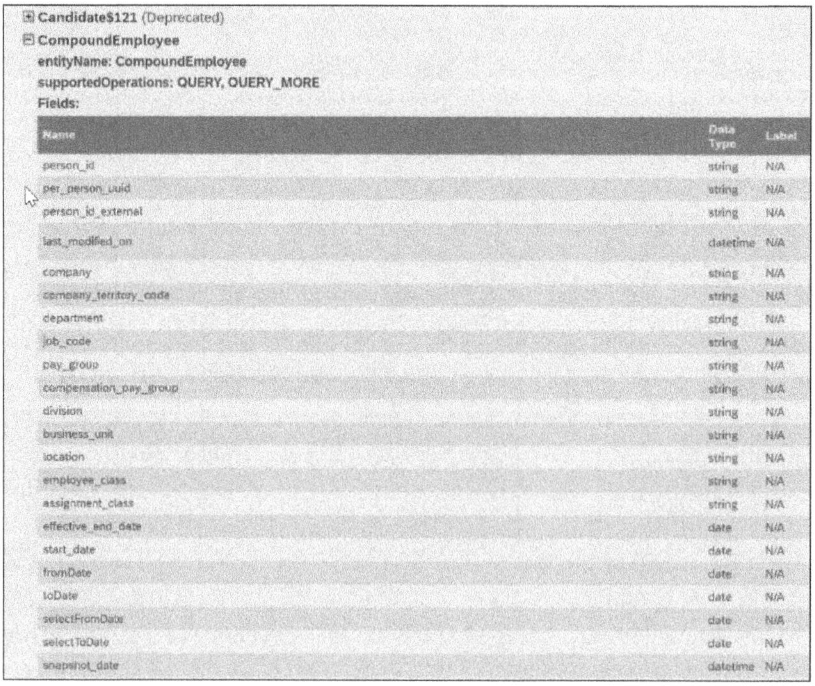

Figure 6.12 SFAPI Data Dictionary for the CompoundEmployee Object

6.1.4 SAP Integration Suite

SAP Integration Suite can serve as a specialized layer you can use to integrate a pair of systems. SAP Integration Suite is an enterprise integration platform-as-a-service (eiPaaS) that allows you to connect applications. While APIs and web services are features within the systems that need to be connected, SAP Integration Suite is a separate environment that can integrate two systems.

You can integrate SAP and non-SAP, cloud, and on-premise applications and process messages in real time. SAP Integration Suite comes with several integration packs that make it possible to connect known pairs of systems. We'll see some of these pairs in the next few pages.

Note that this environment does not require any major programming skills. The tools for connecting a sender with a receiver and for transforming their values are graphical. Of course, uncommon requirements that are not delivered by the integration pack may need some coding.

Since SAP Integration Suite is its own layer of application, this tool comes with the following capabilities:

- Error reporting
- Auditing
- Encryption
- Scheduling

A typical Employee Central Payroll integration landscape will have several systems it must connect with, as shown in Figure 6.13. Employee Central Payroll receives data about employees (i.e., name and address) from the Employee Central system. Hourly employees must log their time to record the hours they need to get paid. This data is collected in the time system. The employer often offers health and other benefits from which the employee can choose. This data is available from the benefit system. Financial data such as cost centers comes from the finance system. Timely direct deposit payments require a connection with the bank's system.

Employee Central Payroll will require the following information:

- Employee data (to know whom to pay and how)
- Time data (to know how much to pay)
- Benefit deduction data (to calculate net pay)

Once the pay is calculated, Employee Central Payroll sends the net pay information to the bank for payment and sends details about the gross pay and net pay to the finance system for accounting.

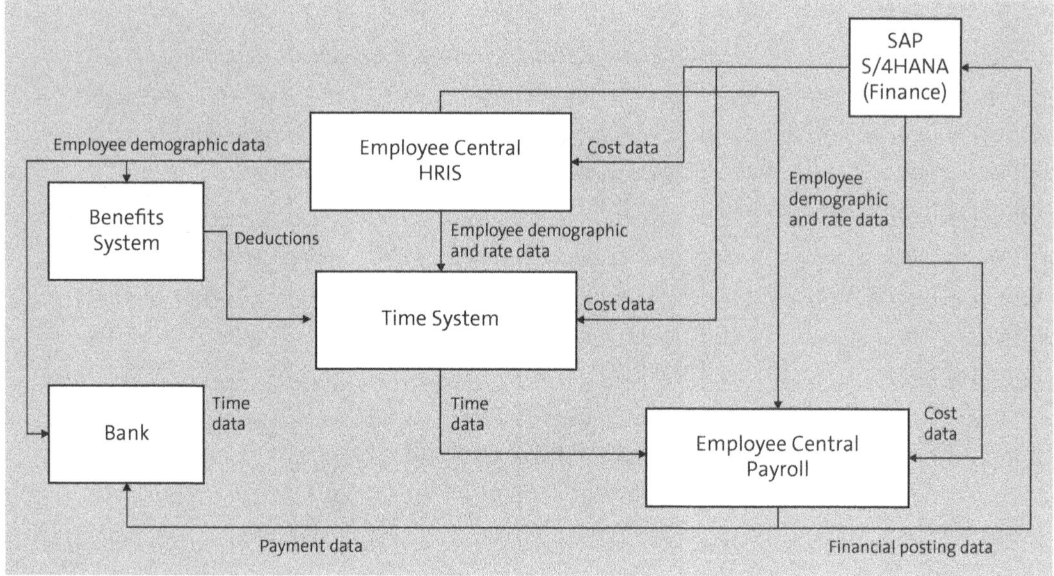

Figure 6.13 Typical Payroll System Landscape

6.2 Time Data

Time data consists of the following four types of data:

- Card punches that capture start and end times, generally produced from a time-clocking system
- Pay codes that specify how a payment must be classified
- Hours to represent the quantity of work done
- Personnel identifiers that represent for whom pay must be calculated

To use time data, two types of integration are required, as shown in Figure 6.13:

- Employee demographic data from the human resources information system (HRIS) system (Employee Central) to the time system
- Time data from the time system to the payroll system

Employee Central Payroll does not include a time capture system. The typical approaches to time systems include the following:

- Having employees enter times in the employee time sheet system, which is part of Employee Central. Employee Central has limited features for evaluating time. For complex union rules and determining overtime and premium time using multiple personnel- and work-related factors, Employee Central may fall short.
- Using a third-party system. Many third-party systems are available for use, but two systems—WorkForce Software's Time and Attendance system and Kronos Workforce Dimensions Timekeeping—are SAP partners and therefore have standard

connections to both the HRIS and Employee Central Payroll systems. Both are cloud-based systems with powerful features, not only time tracking, but also scheduling and regulatory compliance.

Figure 6.14 shows a view of a standard integration pack included with SAP Integration Suite for connecting Employee Central with WorkForce Software.

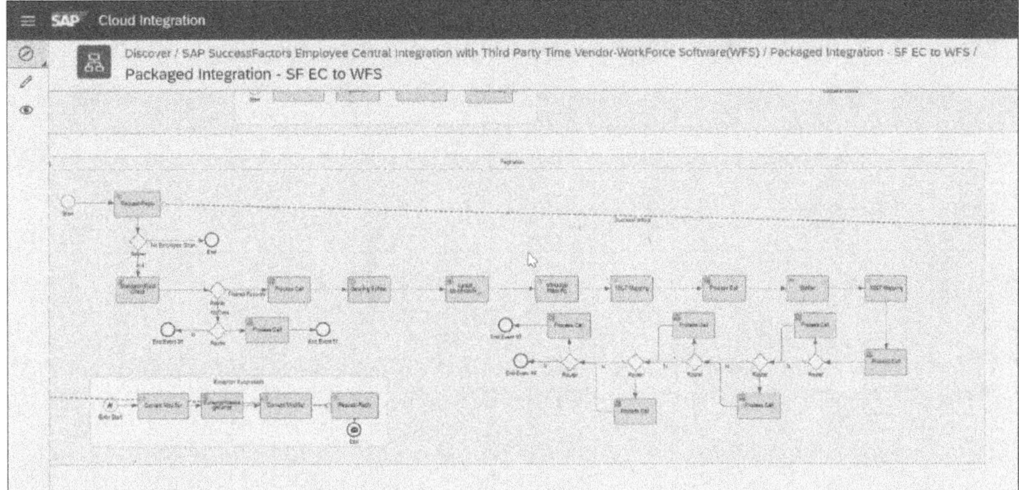

Figure 6.14 Integration from Employee Central to WorkForce Software Time and Attendance System

Figure 6.15 shows a similar view of the standard integration pack included with SAP Integration Suite for connecting Employee Central Payroll with WorkForce Software.

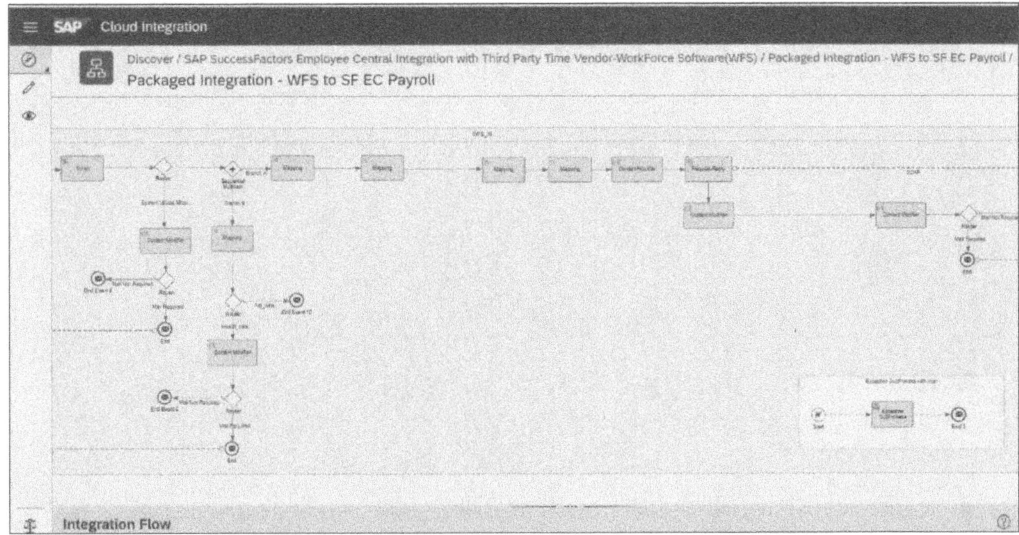

Figure 6.15 Integration from WorkForce Software Time and Attendance System to Employee Central Payroll

Figure 6.16 shows a view of the standard integration pack included with SAP Integration Suite for connecting Employee Central with Kronos.

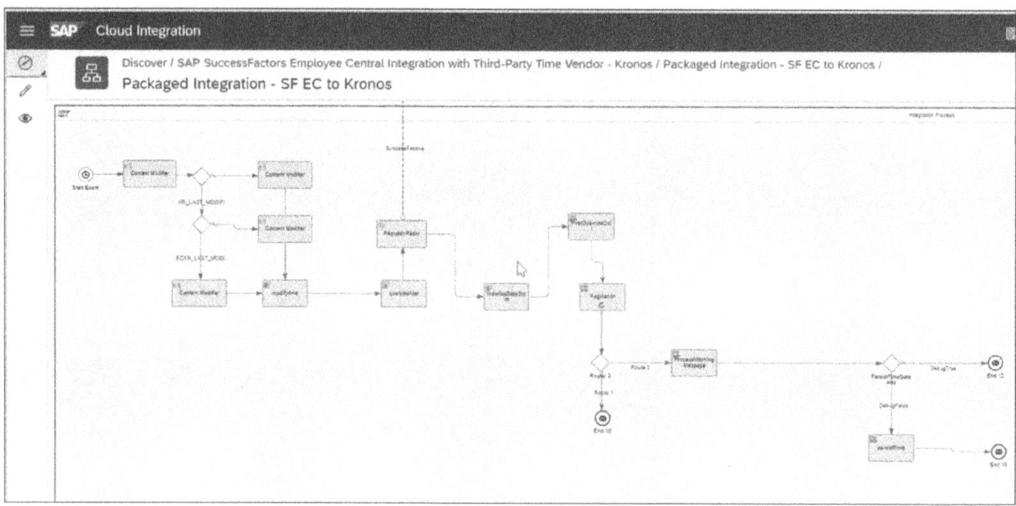

Figure 6.16 Integration from Employee Central to Kronos Workforce Central

Figure 6.17 shows a similar view of the standard integration pack included with SAP Integration Suite for connecting Employee Central Payroll with Kronos.

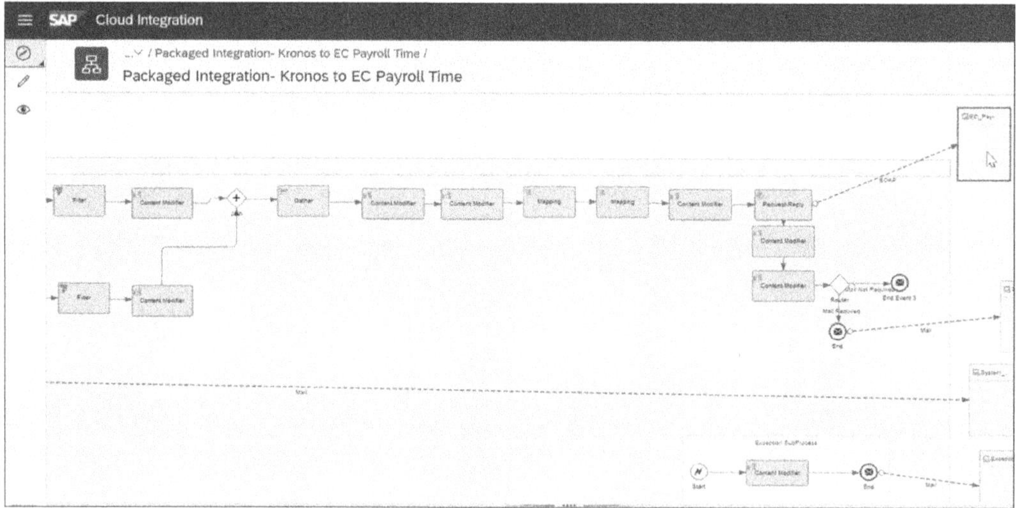

Figure 6.17 Integration from Kronos to Employee Central Payroll

6.3 Benefits Data

Benefits are varied and are typically offered differently by country. Some countries like the US have businesses that offer benefit services in the form of health, life, or other insurance products. Employees can use these services to subscribe to the benefits that interest them offered by the company.

Benefits data generally requires two types of basic integrations:

- Demographic employee data to send to the benefit provider to help identify the employee. This data is used by the benefit provider to provide a personalized portal of opportunities to the employee.
- Once the employee has confirmed the services to subscribe to, you'll need deduction data to send to payroll.

Recently, this scenario has driven a new business model: the *aggregator*. One such provider is Benefitfocus. These businesses offer enrollment services only and no products of their own. The advantage is that the employing company won't need to worry about sending files to various providers because the enrollment services company takes care of those tasks. A standard SAP integration package is available for both our scenarios (sending employee data from Employee Central to an aggregator and receiving deductions from an aggregator to Employee Central Payroll). This integration package is available in SAP Cloud Integration, as shown in Figure 6.18.

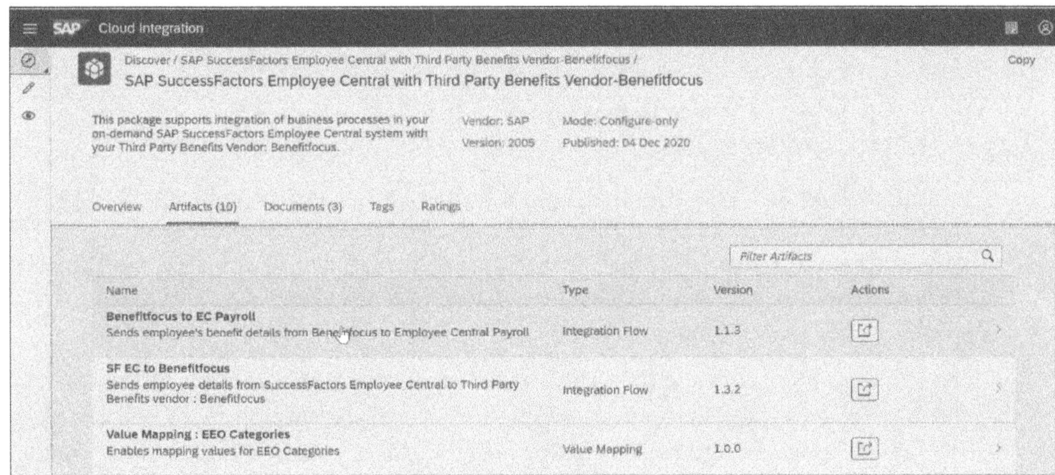

Figure 6.18 Employee Data and Benefits Details Integration

6.4 Finance Data

The *cost center* is an essential financial element that is required in both the HRIS and Employee Central Payroll systems. The cost center is sent automatically from the SAP S/4HANA Finance system to Employee Central through a new feature, called the Finance Master Data Service for Cost Center (SAP_COM_0559) communication scenario in SAP S/4HANA, which you must activate for this feature to work. On the Employee Central end, you must activate the cost center replication capability from SAP Master

Data Integration to Employee Central on the **Business Scenarios** page in the Admin Center. This capability became available from 2018 onwards.

If your system is SAP ERP, standard integration is available using SAP Cloud Integration, as shown in Figure 6.19.

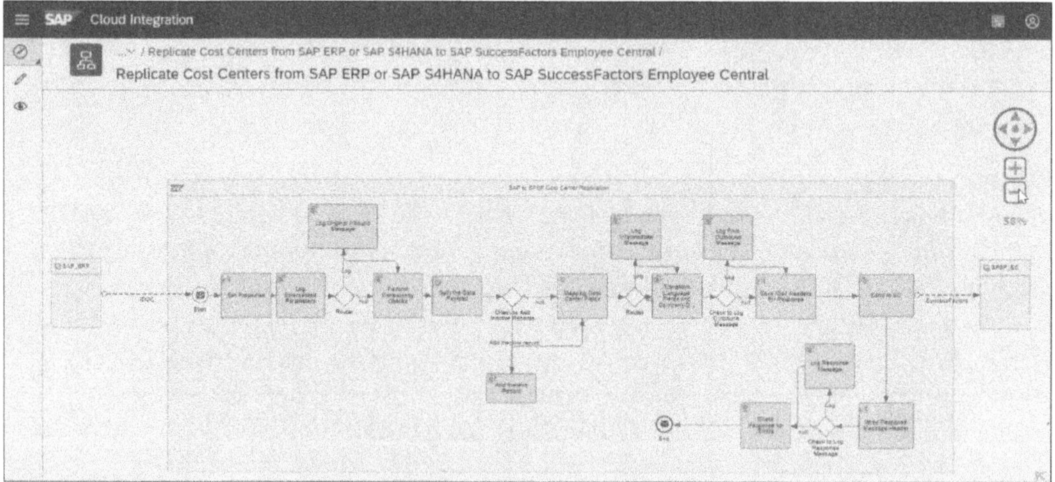

Figure 6.19 Replication of Cost Center Information from SAP ERP to Employee Central

On the Employee Central side, message type COSMAS is dedicated to the sending and receiving of cost center data.

After payroll has been processed and paid, the data must return to the finance system. The finance system is different from the Employee Central Payroll system but is typically another SAP system. SAP provides standard integration capabilities using an IDoc service to connect your Employee Central Payroll system with your finance system for posting to the general ledger.

The IDoc can be configured in both systems using standard messages in Transaction SPRO by following the menu path **BASIS • IDoc Interface / Application Link Enabling (ALE) • Modelling and Implementing Business Processes • Maintain Distribution Model and Distribute Views.** Alternatively, you can use Transaction SALE.

The receiver system must have the IDoc receiver function turned on. Once this function is activated, the receiver system should automatically receive messages.

In configuration, you'll activate the correct methods, set up remote function call (RFC) destinations, and generate partner profiles using the distribution model shown in Figure 6.20. This book does not cover this topic in detail—you'll need the help of a technical expert with knowledge of both the payroll and finance systems to complete this task. Our discussion on Application Link Enabling (ALE) can only be introductory.

To access instructions on configuring the accounting IDoc, follow the menu path **BASIS • IDoc Interface / Application Link Enabling (ALE) • Modelling and Implementing Business Processes • Configure Predefined ALE Business Processes • Human Resources • HR<> AC.**

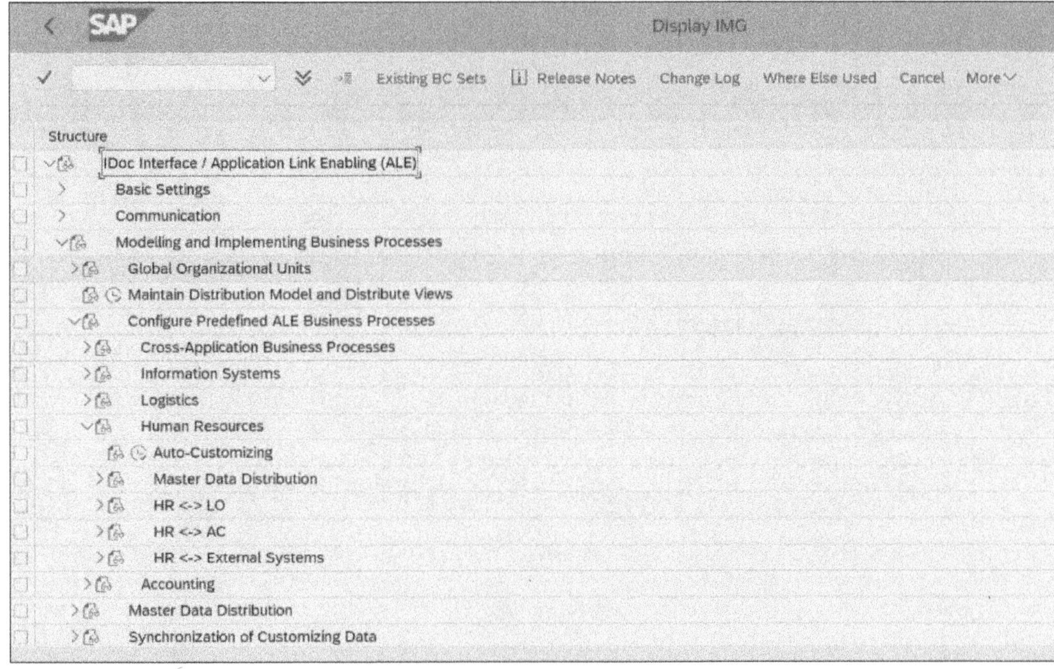

Figure 6.20 Distribution Model and ALE Configuration Screen

The actual configuration activity must be performed in the **Maintain Distribute Model and Distribute Views** setting by editing or creating a distribution model using BAPIs, as shown in Figure 6.21.

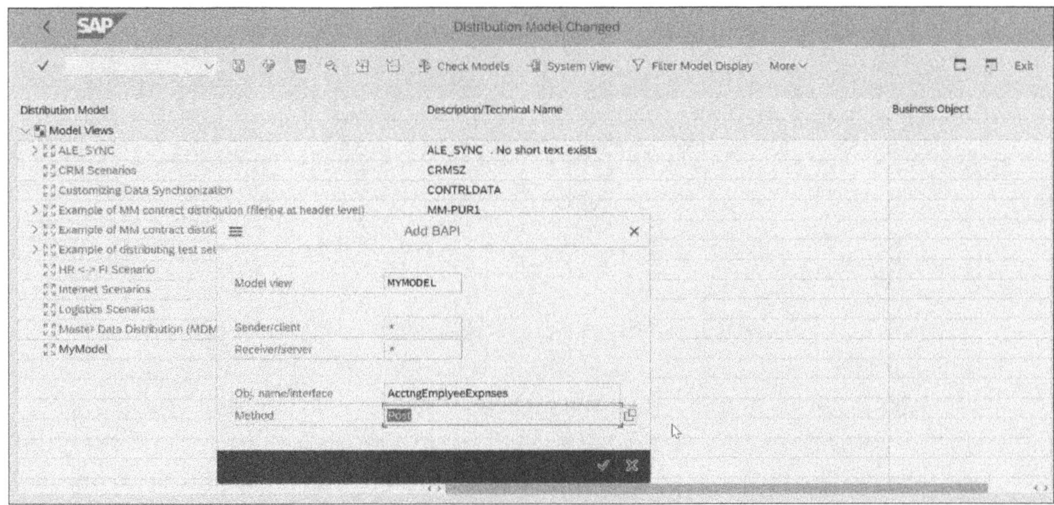

Figure 6.21 Adding a BAPI for the Distribution Model

Enter the Employee Central Payroll system as the sender system and enter the finance system as the receiver for the objects listed in Table 6.2.

Object Method	Message Type
`AcctngEmployeeExpenses.Post`	ACC_EMPLOYEE_EXP
`AcctngEmployeeRcvbles.Post`	ACC_EMPLOYEE_REC
`AcctngEmployeePaybles.Post`	ACC_EMPLOYEE_PAY

Table 6.2 Message Types for Payroll Posting

Enter the finance system as the sender system and enter the Employee Central Payroll system as the receiver for the object listed in Table 6.3.

Object Name	Message Type	Description
PayrollAccDocument	Display_Acc	Display Accounting Document

Table 6.3 Message Type for Document Display

Follow the menu path **BASIS • IDoc Interface / Application Link Enabling (ALE) • Basic Settings • Activate Event Receiver Linkage for IDOC Inbound** and ensure that this setting is turned on in the finance system.

Other Human Resources Information Systems

Integrating with HRIS (Employee Central) is explained in detail in Chapter 8. If you have a different HRIS system, then integration with payroll will require custom development. This effort could be reduced by using SAP Integration Suite. For example, in Section 6.3, we discussed the integration of employee demographic data, which should be exported from the HRIS system, with benefits systems.

6.5 Summary

In this chapter, we described the strategies necessary to integrate other systems with an Employee Central Payroll system. This topic is important to understand to deliver the payroll solution holistically. We discussed the different techniques available at a high level, with just enough detail to get you started. Explaining these strategies in depth is beyond the scope of this book since each technology we discussed in this chapter is extensive. However, this chapter lays the foundation for integration with HRIS systems (Employee Central), which we'll cover in much more detail in the next few chapters.

Chapter 7
Master Data Integration

Making employee data available is a key first step to determine pay for the workforce. Earnings, deduction, security, and operations all depend on this data to be timely and accurate. However, to avoid keying in this data manually—in the spirit of efficiency and elimination of redundant operational effort—we want this data to come directly from the human resources (HR) system.

In the previous chapter, you learned about the various types of integrations possible with Employee Central Payroll. Now, we'll examine the specific interface with Employee Central, an important discussion. In this chapter and the next, you'll learn everything you need to be proficient in connecting your Employee Central Payroll system to your Employee Central environment. The replication process is administered by a configurable web service commonly known as a *point-to-point replication*. This service is at the center of the exchange between the Employee Central Payroll and Employee Central systems, as shown in Figure 7.1.

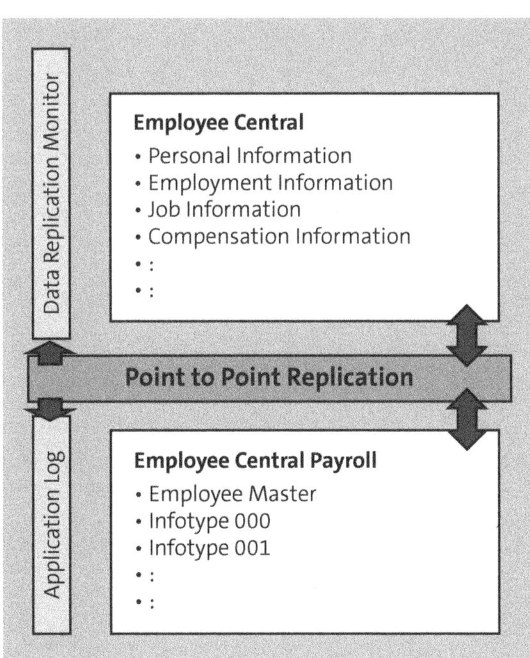

Figure 7.1 HR and Payroll Information Exchange

This entire chapter is devoted to discussing the workings of a single program responsible for this replication: RP_HRSFEC_PTP_EE_REPLICATION.

Two aspects drive the actual process of replication:

- What is requested from the Employee Central Payroll system
- What is flagged as replicable material by the Employee Central system

We'll discuss various aspects of these two operations, including how to control them, in the rest of the chapter. We'll discuss how to be specific about the scope of replication, how to set up connectivity, how to launch replication from both systems, and to how investigate missing configuration once the setup is ready.

7.1 Infotype Filtering

Infotype filtering is the step where you'll determine what is required by the Employee Central Payroll system. This task is not trivial. You must carefully review all the fields and field groups that may be required to drive payroll computation. At an infotype level, a global set is available, and local infotypes required by specific country payrolls can be added easily.

Once the list is finalized, configure the list in the Employee Central Payroll system by following the menu path **Personnel Management • Integration Settings for SuccessFactors Employee Central Payroll • Infotype Filter • Filter Infotypes.** Infotypes will be available for selection for each country, as shown in Figure 7.2. Our example shows the infotypes available for country grouping 06 (France), but other infotypes are available to you. If you need to add other infotypes, click on **New Entries**; choose the country grouping, infotype, and subtype; and then save.

Note that, if you leave this configuration blank, no infotype filtering occurs at all. However, if you add infotypes for countries, only those infotypes for those countries will be considered for replication. SAP recommends that the system is populated with all countries in scope.

Through this setting, you can record your decisions on all infotypes that must be replicated, country by country. You only need to set up the countries for which you'll run payroll. If your company operates in a country where payroll is outsourced, that country doesn't need be included in the infotype filter; the employee master data required by the payroll outsourcer of that country should derive from the Employee Central system.

We'll discuss these infotypes in a lot more detail in the next chapter.

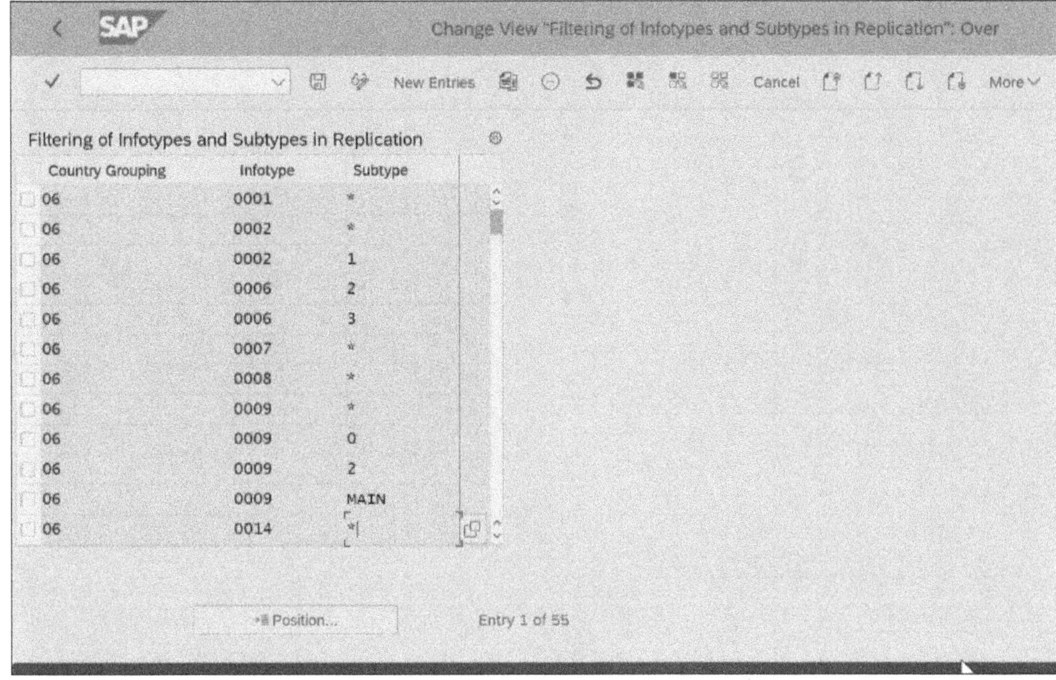

Figure 7.2 Infotype Filtering

7.2 Defining Customer-Specific Activities

The basic task of configuring replication involves several activities that will require planning, especially for dealing with values that are different for each customer based on country, industry, or locality.

Defining these activities involves four categories of tasks:

- Assigning code values from Employee Central to Employee Central Payroll
- Dealing with organizational elements
- Designing and mapping wage types
- Turning on extensibility

You can configure all these details in Transaction SPRO by following the menu path **Personnel Management • Integration Settings for SuccessFactors Employee Central Payroll.** However, before you can perform any of these tasks, you must set up the query by selecting from the available list. The following queries (portlets) can be used in the query:

- `accompanying_dependent`
- `address_information`

- alternative_cost_distribution
- compensation_information
- paycompensation_non_recurring
- paycompensation_recurring
- deduction_non_recurring
- deduction_recurring
- dependent_information
- email_information
- employment_information
- global_assignment_information
- ItDeclaration
- job_information
- job_relation
- national_id_card
- payment_information
- person
- personal_documents_information
- personal_information
- person_relation
- phone_information

7.2.1 Assigning Code Values

Settings enable the mapping of code values, field by field, for the standard SAP fields that are always required for running payroll.

To begin the process of assigning code values, bring up Customizing in Transaction SPRO and follow the menu path **Personnel Management • Integration Settings for SuccessFactors Employee Central Payroll • Assignment of Code Values • For Point-to-Point Replication • Display Code Value Mapping as Delivered by SAP.** The default code value mapping delivered by SAP, shown in Figure 7.3, will be displayed.

In the next step, you'll specify the following properties for these fields/portlets:

- Whether they are country specific by selecting the **Country** checkbox
- The mapping mode, in the **Mode** column
- The dates that they are effective, by maintaining the **Start Date** and **End Date** columns

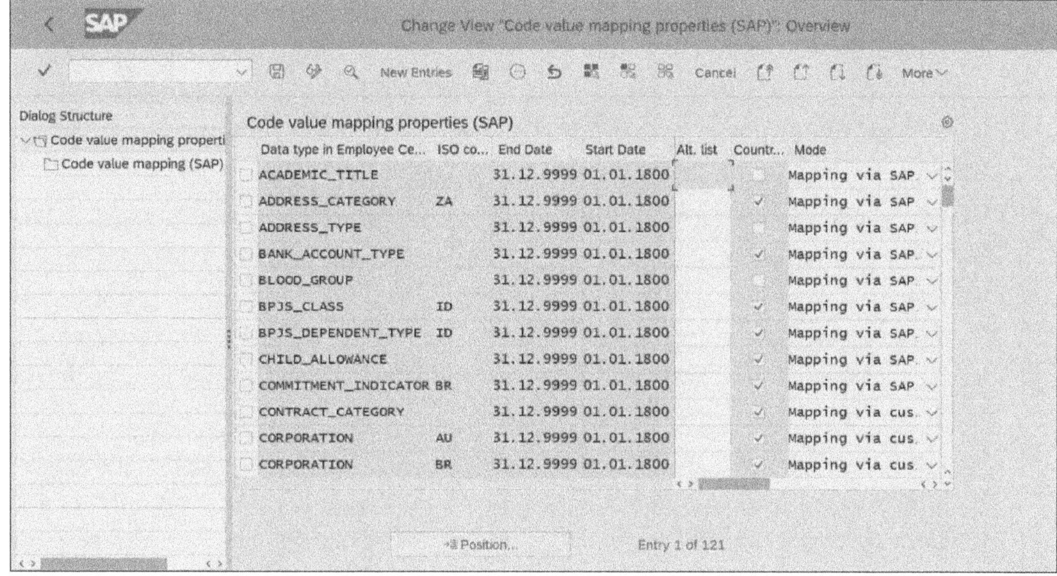

Figure 7.3 Standard Code Mapping

To maintain these properties for code value mapping, follow the menu path **Personnel Management • Integration Settings for SuccessFactors Employee Central Payroll • Assignment of Code Values for Point-to-Point Replication • Maintain Properties for Code Value Mapping**. These properties are revealed, as shown in Figure 7.4.

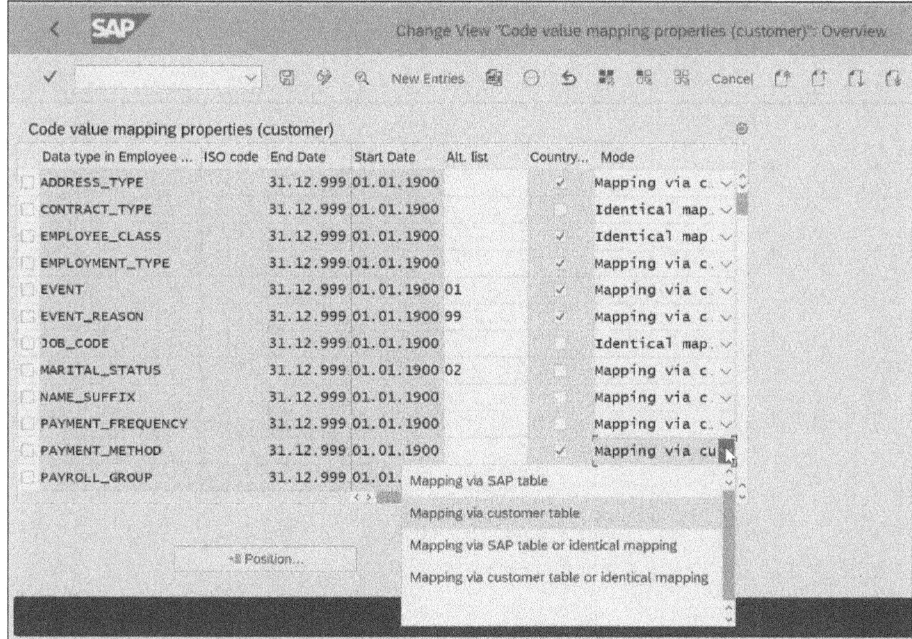

Figure 7.4 Properties of the Mapping

Let's focus on the **Mode** column. When you map code values, five methods are available to choose from:

1. Identical mapping (same as provided by Employee Central)
2. Mapping via an SAP table (provided in table T77SFEC_CVMAPS)
3. Mapping via a customer table (provided in table T77SFEC_CVMAPC)
4. Mapping via an SAP table or identical mapping. If the value is available in a standard table, use that value; otherwise, keep the same value as available in Employee Central.
5. Mapping via a customer table or identical mapping. If the value is available in a customer table, use that value; otherwise, keep the same value as available in Employee Central.

For choices 3 and 5, you'll need to map the values for your organization. In Customizing, follow the menu path **Personnel Management • Integration Settings for SuccessFactors Employee Central Payroll • Assignment of Code Values for Point-to-Point Replication • Maintain Code Value Mapping.** This task can also be performed by maintaining table T77SFEC_CVMAPC through Transaction SM30, as shown in Figure 7.5.

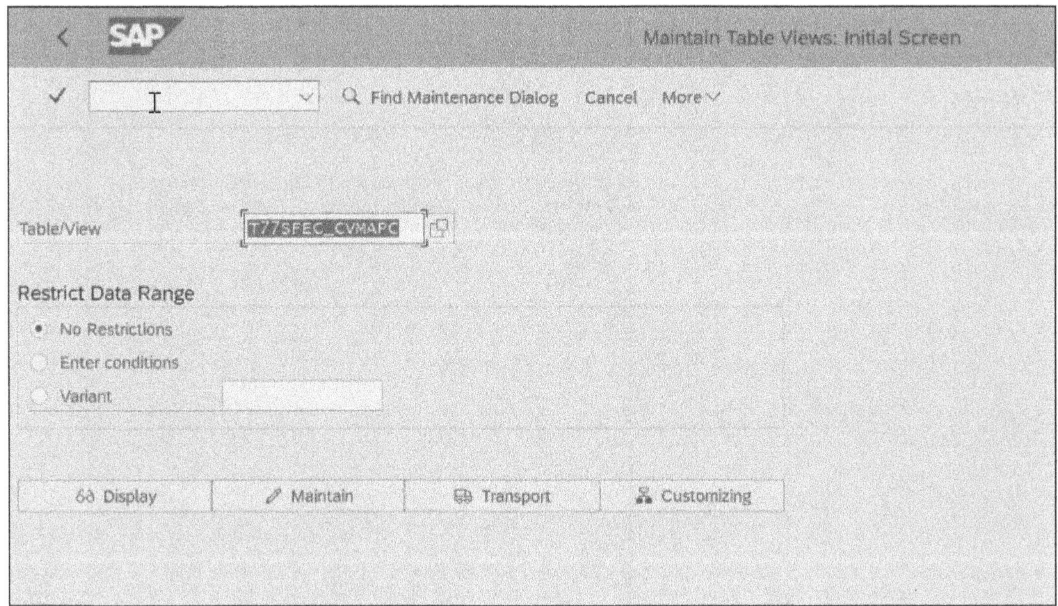

Figure 7.5 Transaction SM30

When you're finished, click on the **Maintain** button to see the code value map, as shown in Figure 7.6.

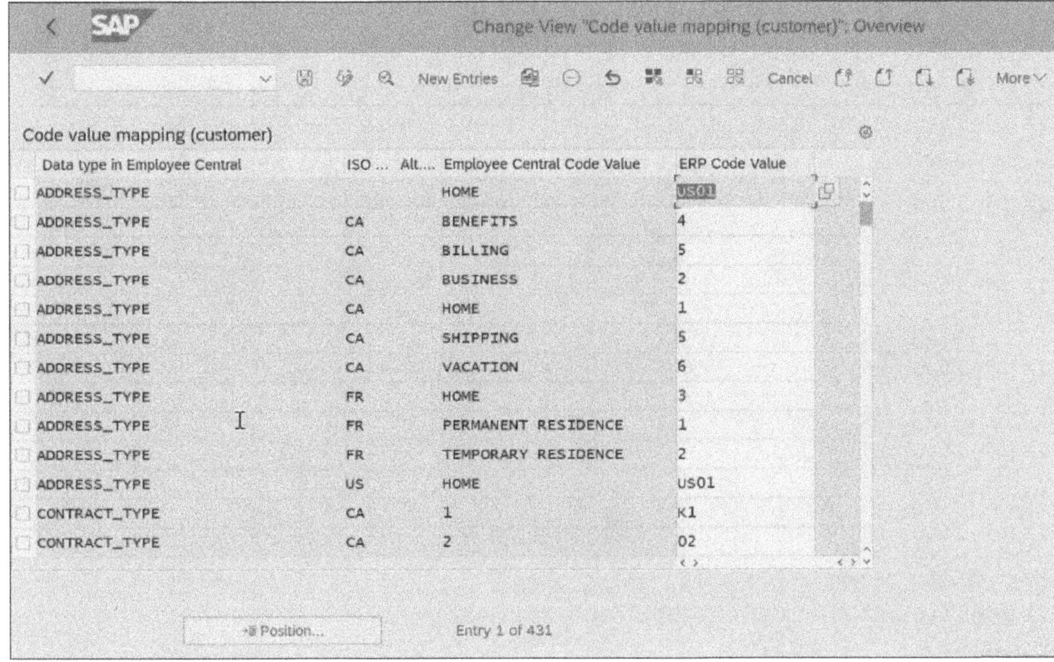

Figure 7.6 Customer Code Value Mapping

An additional step for country-specific items is required: The ISO code for that country must be selected. Follow these steps:

1. Click on **New Entries**.
2. Add the **Employee Central** field in the **Data Type in Employee Central** field. In our example, this field is **address_type**.
3. Select the ISO code for the country from the **ISO** dropdown list.
4. Enter the appropriate code from the Employee Central Payroll system in the **ERP Code Value** field.
5. Save your settings.

Two more data groups require special handling for code value assignment:

- Dates
- National IDs

To map dates, follow the menu path **Personnel Management • Integration Settings for SuccessFactors Employee Central Payroll • Assignment of Code Values • For Point-to-Point Replication • Maintain Date Type Mapping.** You'll map Employee Central date code values to Employee Central Payroll date code values, as shown in Figure 7.7 by following these steps:

1. Click on **New Entries**.
2. Select the country grouping in the **CGrpg** field.

3. Select the date segment and date field name from the corresponding dropdown fields.
4. Enter the Employee Central Payroll date type field in infotype 0041 in the **DT** (date type) field. These date types would have already been set up according to business requirements. Date types can be set up by following the menu path **Personnel Management • Personnel Administration • Contractual and Corporate Agreement • Evaluation Basis • Date Specifications.**
5. Save your settings.

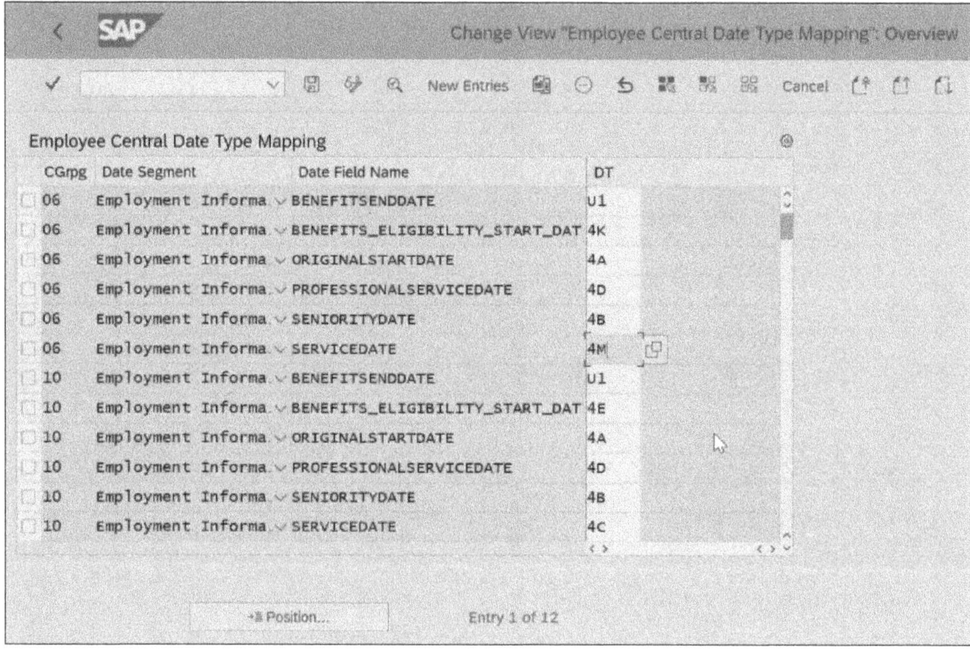

Figure 7.7 Maintaining a Date Mapping

For national IDs, first follow the menu path **Personnel Management • Integration Settings for SuccessFactors Employee Central Payroll • Assignment of Code Values for Point-to-Point Replication • Maintain Countries for IT0185 Replication**. Then, enable the replication of infotype 0185 for the specific countries that you want to implement. Start by clicking on **New Entries**, as shown in Figure 7.8.

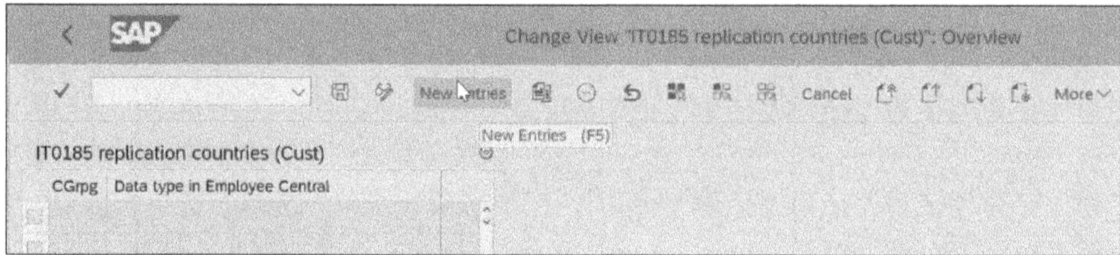

Figure 7.8 Infotype 0185 Countries

In the popup window that appears, enter the required values, as shown in Figure 7.9. Select the country from the **CGrpg** field and save.

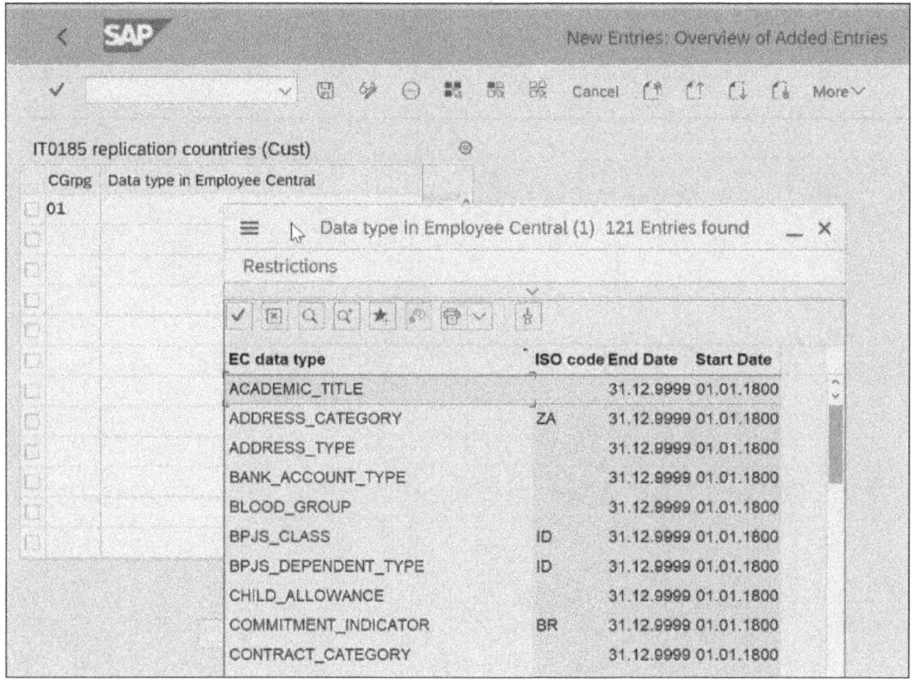

Figure 7.9 Infotype 0185 Replication Countries

Next, follow the menu path **Personnel Management • Integration Settings for SuccessFactors Employee Central Payroll • Assignment of Code Values for Point-to-Point Replication • Maintain Date for Replicating Personal ID from National ID Portlet.**

On this screen, you'll define the start date of the infotype 0185 record, as shown in Figure 7.10, by first clicking on **New Entries.**

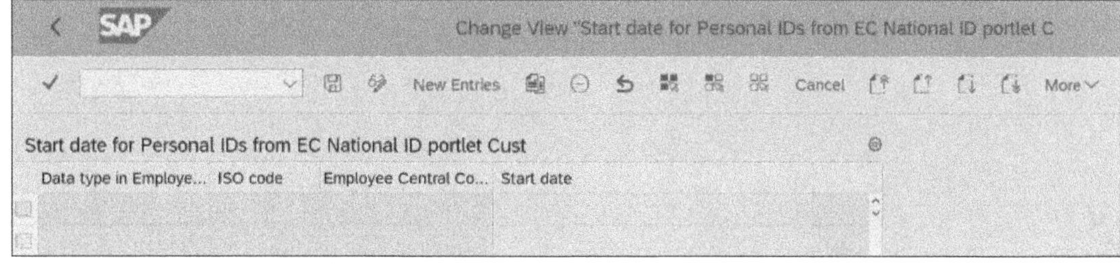

Figure 7.10 Start Date for Infotype 0185

Now, enter the values required, as shown in Figure 7.11, by following these steps:

1. Choose the type of data the **Data Type in Employee Central** field.
2. Choose the ISO code for the country you're configuring in the **ISO Code** field.

3. Choose the Employee Central code in the **Employee Central Code** field.
4. Select either **Hire Date** or **System Date** in the **Start date** column. Since this date is the date for the national ID, its validity before today may not be relevant. In this case, use the **System Date** option. However, if using this type of national ID is a criterion for hire, then choose **Hire Date**.
5. Save these settings.

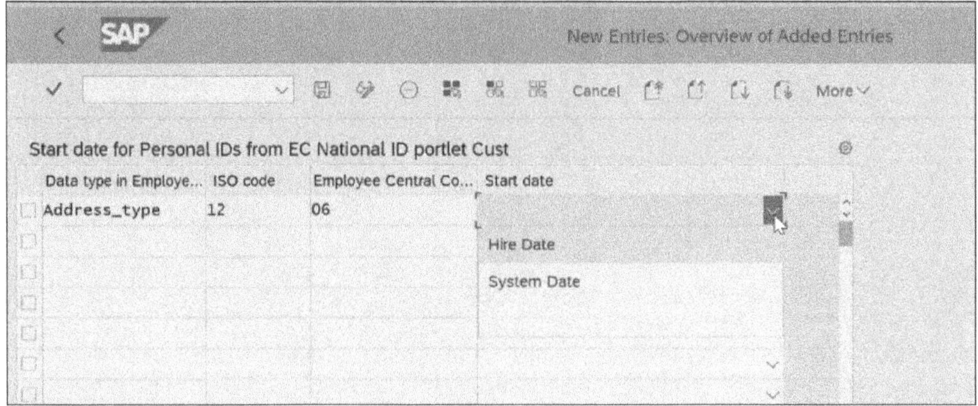

Figure 7.11 Start Date of Fields for a Country

Some countries have certain special methods for infotype 0185, as shown in Table 7.1.

Country	Method for Enabling Infotype 0185	Method for Setting Start Date of Infotype 0185
Saudi Arabia	Default is enabled. Use SAP table HRSFEC_D_IT0185S to change.	Hire date is configured as the start date by default. Use SAP table T77SFEC_DATEMAPS to change.
Hong Kong	Default is enabled. Use SAP table HRSFEC_D_IT0185S to change.	Hire date is configured as the start date by default. Use SAP table T77SFEC_DATEMAPS to change.
Thailand	Default is enabled. Use SAP table HRSFEC_D_IT0185S to change.	No default. Use SAP table T77SFEC_DATEMAPC to set the start date.
China	Default is enabled. Use SAP table HRSFEC_D_IT0185S to change.	Hire date is configured as the start date by default. Use SAP table T77SFEC_DATEMAPS to change.
United Arab Emirates	Default is enabled. Use SAP table HRSFEC_D_IT0185S to change.	Hire date is configured as the start date by default. Use SAP table T77SFEC_DATEMAPS to change.

Table 7.1 Special Methods for Infotype 0185

Country	Method for Enabling Infotype 0185	Method for Setting Start Date of Infotype 0185
Egypt	Default is enabled. Use SAP table HRSFEC_D_IT0185S to change.	Hire date is configured as the start date by default. Use SAP table T77SFEC_DATEMAPS to change.
India	Default is enabled. Use SAP table HRSFEC_D_IT0185S to change.	No default. Use SAP table T77SFEC_DATEMAPC to set the start date.
Oman	Default is enabled. Use SAP table HRSFEC_D_IT0185S to change.	Hire date is configured as the start date by default. Use SAP table T77SFEC_DATEMAPS to change.
Other countries	Default is disabled. Use SAP table HRSFEC_D_IT0185C to change.	No default. Use SAP table T77SFEC_DATEMAPC to set the start date.

Table 7.1 Special Methods for Infotype 0185 (Cont.)

7.2.2 Dealing with Organizational Elements

In terms of organizational elements, three elements must be configured:

- Cost center
- Company code
- Place of work

We'll look at each element in the following sections.

Cost Center

To configure a cost center, you first must assign external keys to internal keys in Customizing by following the menu path **Personnel Management • Integration Settings for SuccessFactors Employee Central Payroll • Assign External Cost Center Keys to Internal Cost Center Keys.**

To map external key values to internal key values, as shown in Figure 7.12, follow these steps:

1. Click **New Entries.**
2. Enter the Employee Central cost center in the **Cost Center ID** field.
3. Enter the company code in the **Company Code** field.
4. Enter the Employee Central Payroll cost center in the **Cost Center** field.
5. Save these settings.

Change View "Cost Centre Key Mapping": Overview

Cost Centre Key Mapping

Cost Centre ID	Company Code	Cost Center
10000	UAEC	10000
12113333	0001	66666666
200000	0001	200000
33000	0001	33000
7001342	0001	799999

Figure 7.12 Cost Center Mapping

Note that Employee Central automatically sends external keys if they are exactly 14 characters or 10 or less characters. If the key is 14 characters long, this key is a combination of 4 digits of the controlling area and 10 digits of the cost center. These keys do not need any assignment. If the external key is between 11 and 13 characters, the first few characters are used to search for a controlling area in Employee Central. If a match exists, the remaining characters are considered as passthrough cost centers and are transmitted to Employee Central Payroll as they appear.

Company Code

To configure a company code, you first must assign external keys to internal keys in Customizing by following the menu path **Personnel Management • Integration Settings for SuccessFactors Employee Central Payroll • Assign External Company Code Keys to Internal Company Code Keys • Value Mapping.**

To map external key values to internal key values, as shown in Figure 7.13, follow these steps:

1. Click **New Entries.**
2. Enter the Employee Central company ID in the **Company ID** field.
3. Enter the Employee Central Payroll company code in the **Company Code** field.
4. Save these settings.

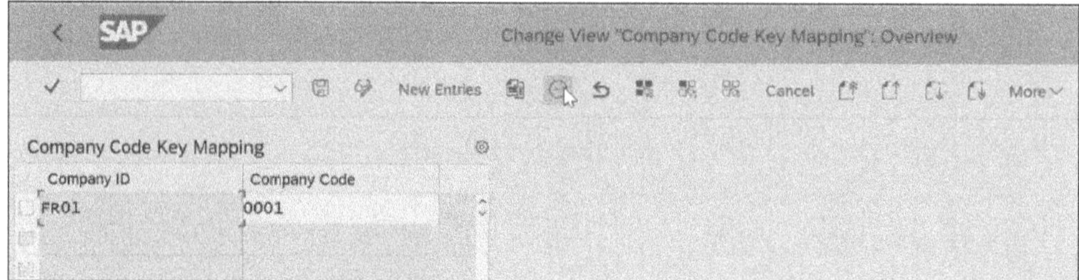

Figure 7.13 Company Code Mapping

Place of Work

To configure a place of work, you first must assign external keys to internal keys in Customizing by following the menu path **Personnel Management • Integration Settings for SuccessFactors Employee Central Payroll • Assign External Place of Work Keys to Internal Place of Work Keys.**

To map external key values to internal key values, as shown in Figure 7.14, follow these steps:

1. Click **New Entries.**
2. Enter the Employee Central field for the work location in the **Permanent Establishment ID** field.
3. Enter the personnel area from Employee Central Payroll in the **Personnel Area** field.
4. Enter the personnel subarea from Employee Central Payroll in the **Personnel Subarea** field.
5. Save these settings.

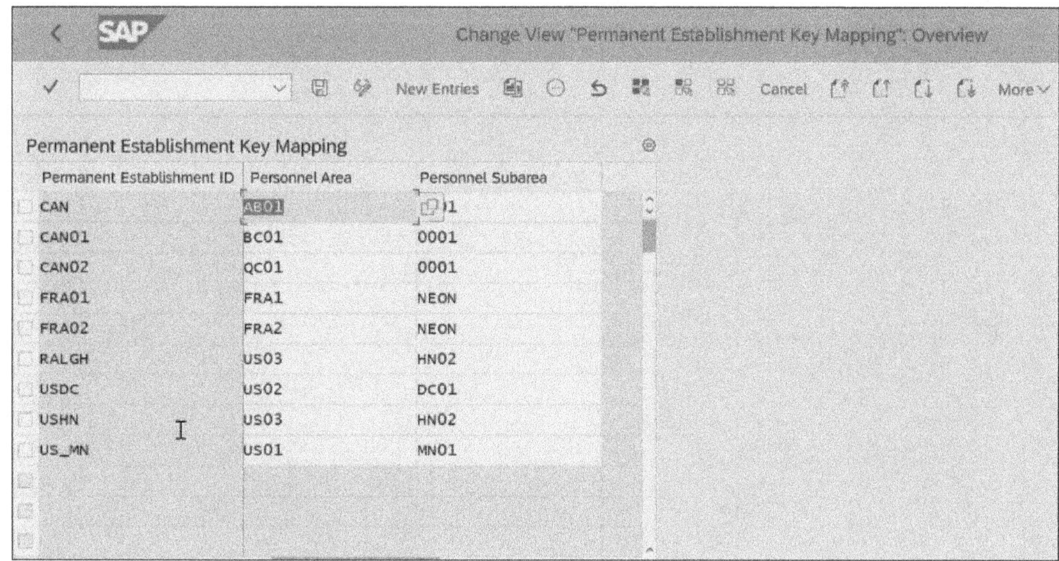

Figure 7.14 Place of Work Mapping

7.2.3 Wage Types

Wage types are required for replication in cases where a value represents an earning. Earnings are typically entered in Employee Central or transferred to Employee Central from SAP SuccessFactors Recruiting when an employee is hired. Deductions are not entered into Employee Central as a general rule—they are entered directly into the Employee Central Payroll system. The name for a wage type in Employee Central is *pay component*.

The following infotypes deal with earnings (wage types):

- Infotype 0008 (Rate of Pay)
- Infotype 0014 (Recurring Earnings)
- Infotype 0015 (One-Time Earnings – On Cycle)
- Infotype 0267 (One-Time Earnings – Off Cycle)

Let's examine wage type assignment with all four of these infotypes and then move on to assigning currencies and checking for accuracy.

Assigning Wage Types for Infotypes 0014 and 0008

For this infotype, you must define the wage type and country in Customizing by following the menu path **Personnel Management • Integration Settings for SuccessFactors Employee Central Payroll • Wage Type Processing • Assign Wage Types to Infotypes.**

To map wage types, as shown in Figure 7.15, follow these steps:

1. Click **New Entries.**
2. Enter the country grouping in the **Country Grouping** field.
3. Enter the wage type from Employee Central Payroll in the **Wage Type** field.
4. Enter the infotype from Employee Central Payroll in the **Info Type** field.
5. Save these settings.

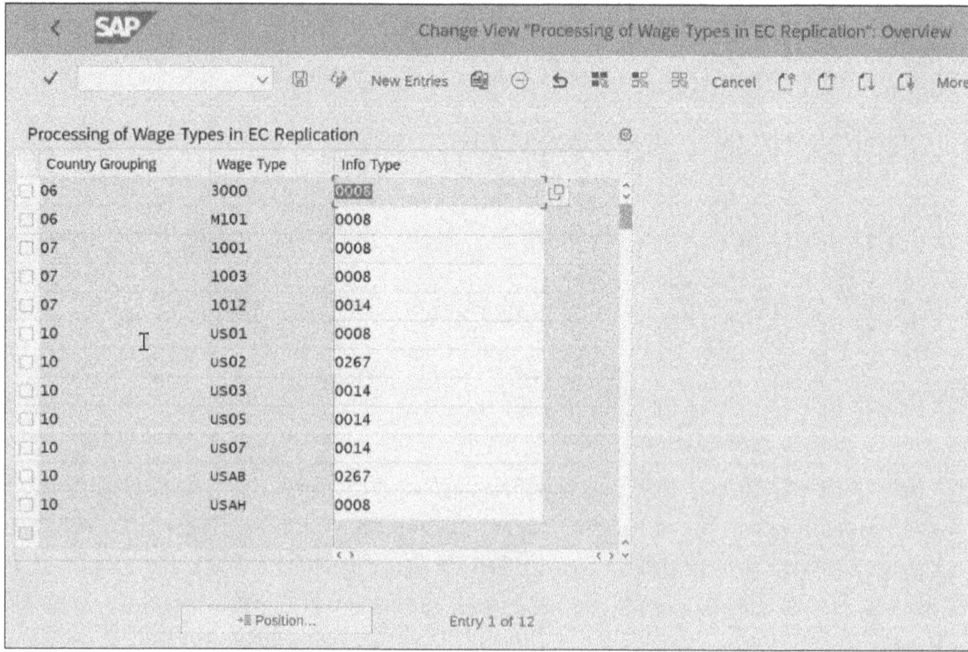

Figure 7.15 Mapping Wage Types

Assigning Wage Types for Infotypes 0015 and 0267

For this infotype, you must define the wage type and country in Customizing by following the menu path **Personnel Management • Integration Settings for SuccessFactors Employee Central Payroll • Wage Type Processing • Assign Non-Recurring Payment Wage Types to Infotypes.**

To map these wage types, as shown in Figure 7.16, follow these steps:

1. Click **New Entries.**
2. Enter the country grouping in the **Country Grouping** field.
3. Enter the wage type from Employee Central Payroll in the **Wage Type** field.
4. Enter the infotype from Employee Central Payroll in the **Infotype** field.
5. Save these settings.

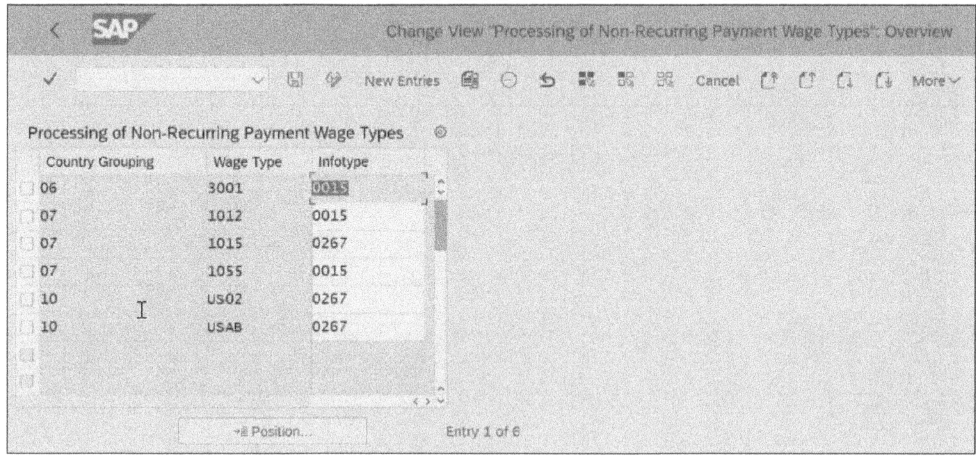

Figure 7.16 Non-Recurring Payment Wage Types

Assigning Currencies

Currencies are provided as standard in all SAP systems. But sometimes in payroll systems (not just Employee Central Payroll!), the number of decimal places required is higher than what is provided as standard. For example, in the US, the standard currency holds 2 decimal places (100 cents = $1). However, you may need to store values with a third or fourth decimal place to accurately record someone's rate of pay. Quite often, the rate of pay is fixed for a week, month, or year, which requires conversion to derive an hourly rate. For example, the hourly rate for a person earning $100,000 USD per year is approximately $48.077 USD per hour based on a 52-week year and 40 hours of work per week.

To store this data, a custom currency may need to be used, since the standard currency USD only stores two decimal places. You can set a custom currency by following the menu path **SAP NetWeaver • General Settings • Currencies • Check Currency Codes.** On this screen, you'll see all the standard available currencies, as shown in Figure 7.17.

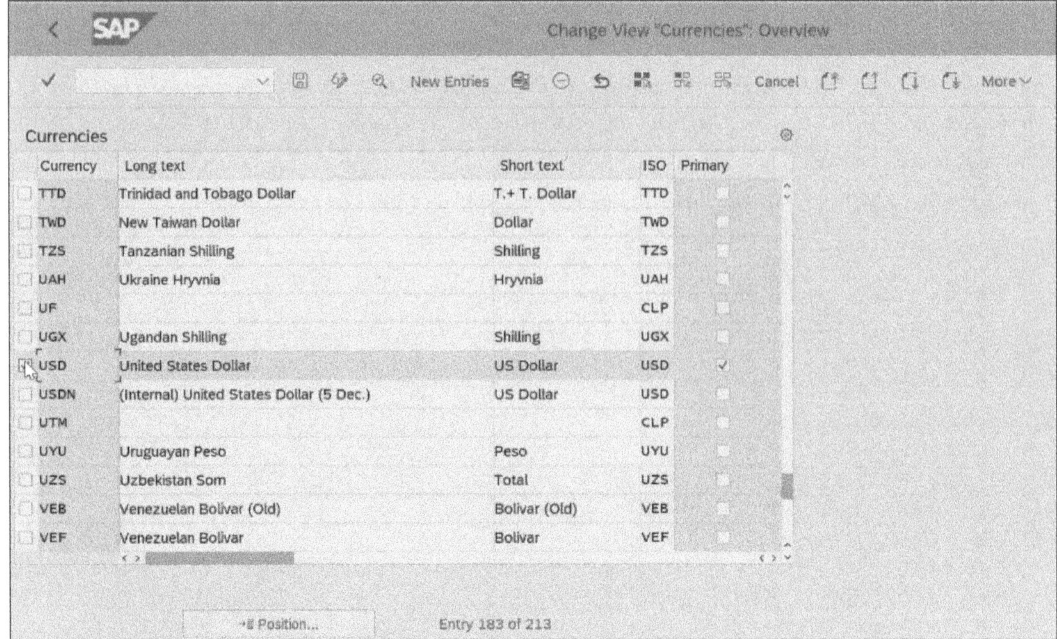

Figure 7.17 Standard Currencies

Click on your currency and then click the **Copy As** button, as shown in Figure 7.18.

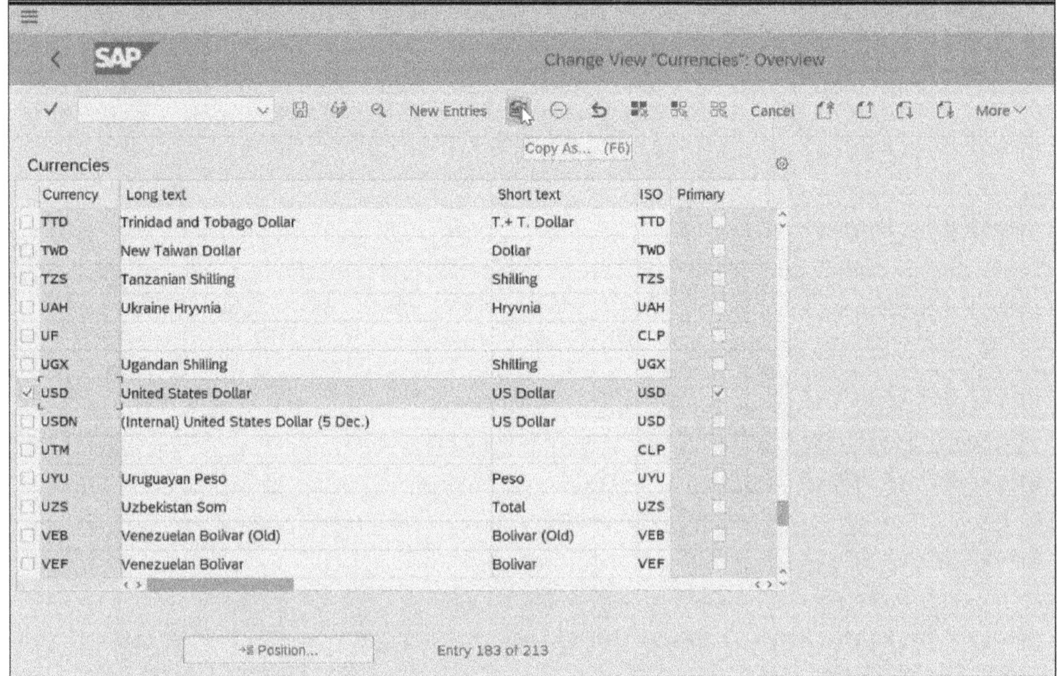

Figure 7.18 Copying a Currency

Change the currency name in the **Currency** field to a different value (something as simple as "USD1" can work), as shown in Figure 7.19, and then save.

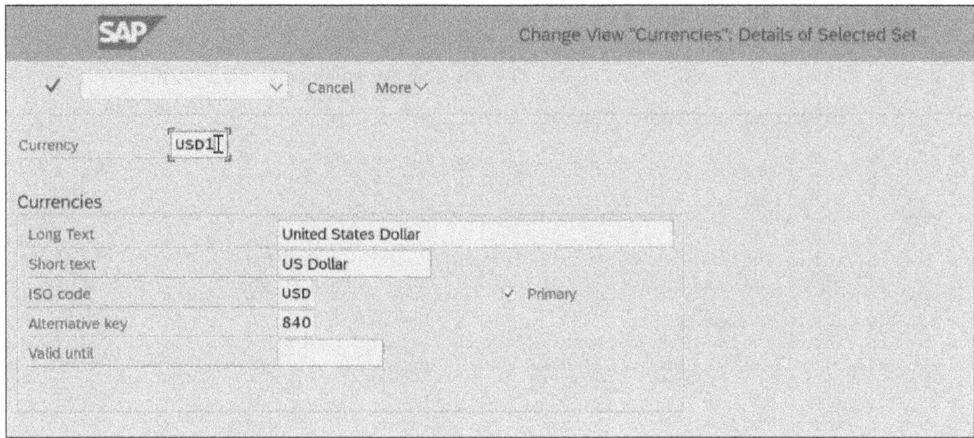

Figure 7.19 Creating a New Currency

Since this step only copied an existing currency and gave it a new name, this currency is still currently set at 2 decimal places. You can increase the number of decimal places by following the menu path **SAP NetWeaver • General Settings • Currencies • Set Decimal Places for Currencies.** Change the number decimal places, as shown in Figure 7.20, by finding the currency we just created and increasing the value in the **Decimals** field.

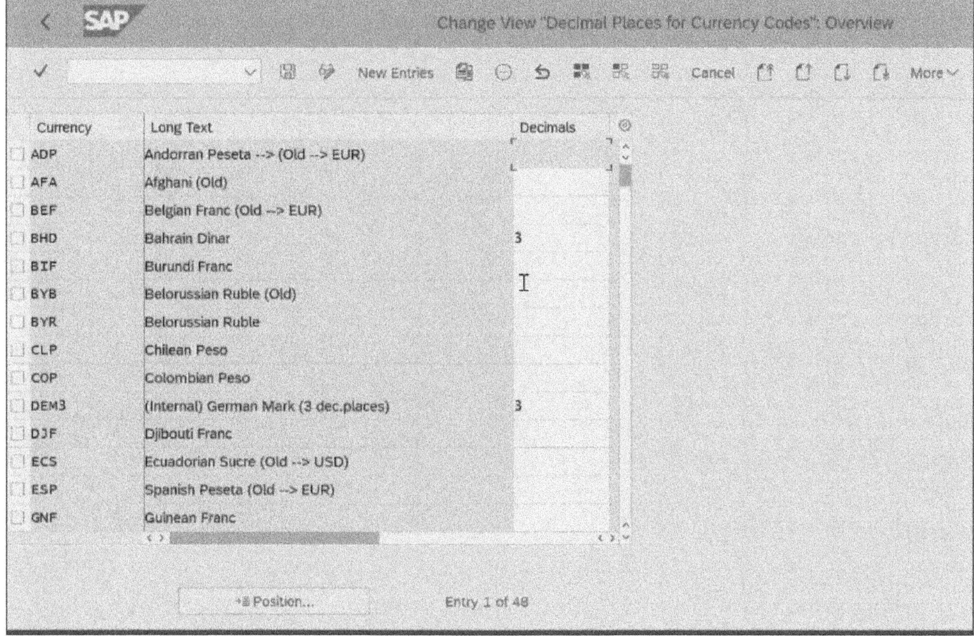

Figure 7.20 Changing the Decimal Places of Currency

Once this step is complete, you'll need to assign this currency (or a standard one) to wage types by following the menu path **Personnel Management • Integration Settings for SuccessFactors Employee Central Payroll • Wage Type Processing • Assign Currencies to Wage Types.**

Map wage types to currencies, as shown in Figure 7.21, by following these steps:

1. Click on **New Entries**.
2. Enter country grouping in the **Country Grouping** field.
3. Choose the wage type in the **Wage Type** dropdown menu.
4. Enter the currency to convert to in the **Currency to Convert To** field. The currency in the Employee Central Payroll system will rarely be different from the currency in the Employee Central system.
5. Enter the currency in Employee Central in the **Replicated Currency** field.
6. Save these settings.

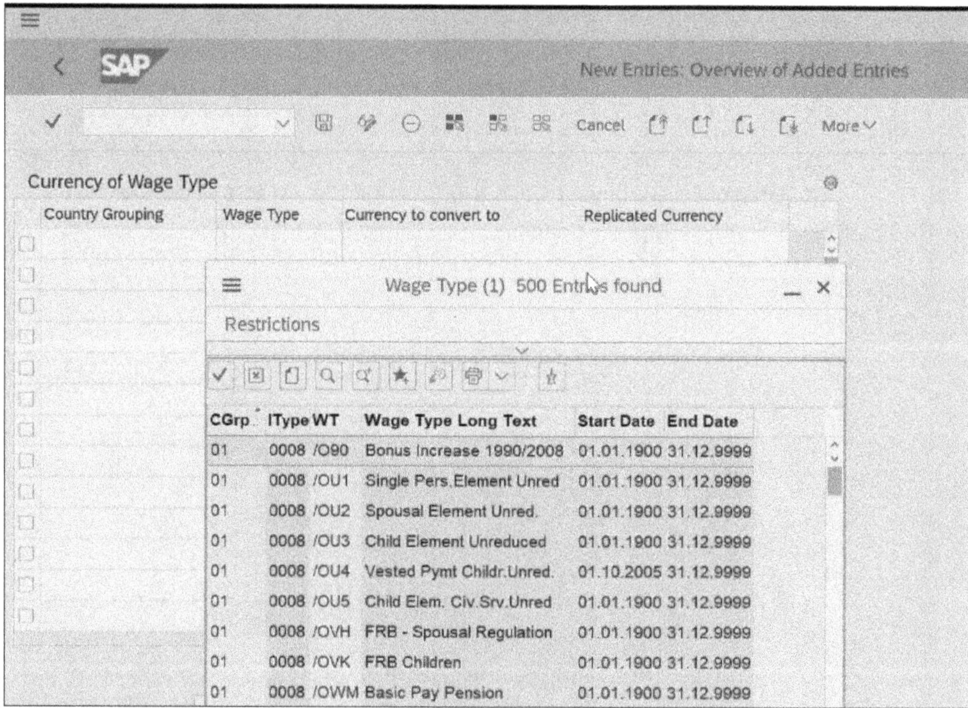

Figure 7.21 Mapping Wage Types to Currencies

Checking for Accuracy

You can check that the configuration is accurate using the standard report RP_HRSFEC_CHECK_PAYC_WAGETYPE. Execute the report using Transaction SE38, as shown in Figure 7.22.

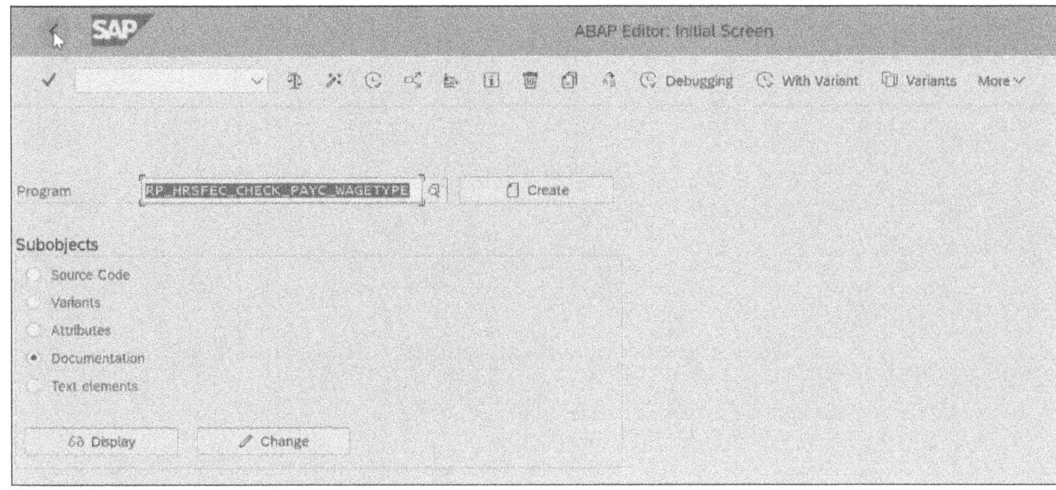

Figure 7.22 Executing the Check Report

You'll now see the selection screen shown in Figure 7.23. Choose the country in the **Country Grouping** field. Use the **Wage Type** field to select the wage types you need to check. If you want to hide entries that do not require any further changes, select the **Hide correct entries** checkbox.

SAP
Pay Component and Wage Type Configuration Check
Cancel More
Select
Import (OData) from Employee Central System HRFFDEMO
Country Grouping
Wage Type to
Layout
Hide correct entries
Layout Variant

Figure 7.23 Selection Screen of a Check Report

Employee Central pay components that have not been mapped are displayed in the list and need further configuration. All records with an error are also displayed in the list. Records belonging to the same pay components and wage types are still displayed for completeness.

You can save your report selection by clicking on the **Save** button and providing a variant name, as shown in Figure 7.24. Now, execute the report by clicking on the **Execute** button.

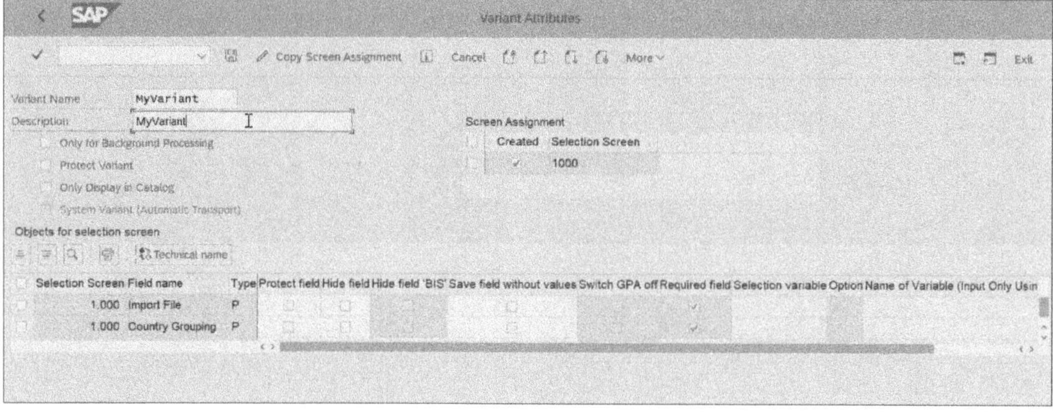

Figure 7.24 Naming a Selection Variant

The report uses color coding to communicate errors in the following way:

- Blue (validity period) means the time period displayed in the **Start Date** and **End Date** fields is relevant for each row separately, including pay component, mapping, and payroll data.
- Green (Employee Central data) displays pay component properties including the **External Code**, **Pay Comp. Text** (pay component text), **Status**, **Pay Comp.** (pay component type), **Currency**, **Freq.** (frequency code), **Is Earning**, and **Is Recurring** fields.
- Yellow (Employee Central Payroll data) displays wage type properties such as the **Wage Type** and **WT** (wage type text) fields. The actual required assignment of wage type to infotype is indicated in the **IT0008**, **IT0014**, **IT0015**, and **IT0267** columns and whether it is permitted is shown in the **IT0008 permitted**, **IT0014 permitted**, **IT0015 permitted**, and **IT0267 permitted** columns, as shown in Figure 7.25.
- Red (error) covers the following errors:
  - Wrong deduction, for instance, the operation indicator for the wage type is wrong and must be corrected.
  - Wage type text is missing.
  - Assignment to infotypes 0008, 0014, 0015, and 0267 is not permitted.
  - Overlapping time validity: Shows whether the start and end date of a pay component or wage type change overlap. Correct the database table entry provided in that column, so the report can process the other checks for this entry.

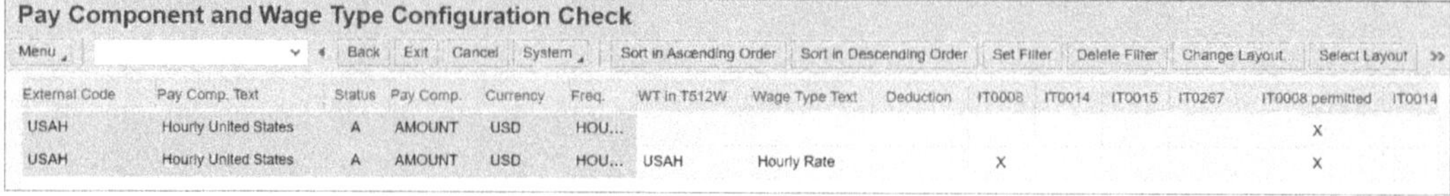

Figure 7.25 Columns of a Wage Type Check Report

If pay component or wage type properties have been added or changed in any of the systems, a row is added to the list. This row shows the new or changed values. You can also download the report to Microsoft Excel by clicking the **Local File...** button, as shown in Figure 7.26.

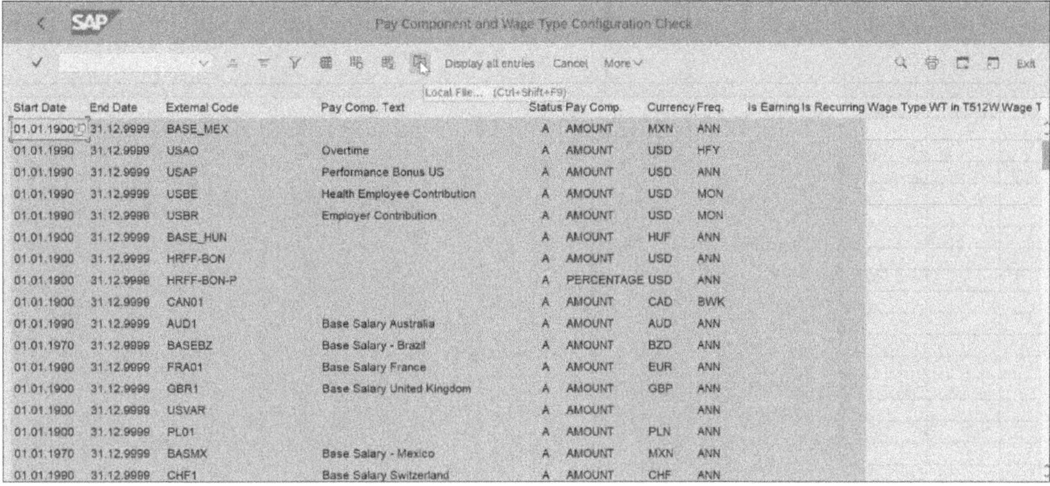

Start Date	End Date	External Code	Pay Comp. Text	Status	Pay Comp.	Currency	Freq.	Is Earning	Is Recurring	Wage Type	WT in T512W	Wage T
01.01.1900	31.12.9999	BASE_MEX		A	AMOUNT	MXN	ANN					
01.01.1990	31.12.9999	USAO	Overtime	A	AMOUNT	USD	HFY					
01.01.1990	31.12.9999	USAP	Performance Bonus US	A	AMOUNT	USD	ANN					
01.01.1990	31.12.9999	USBE	Health Employee Contribution	A	AMOUNT	USD	MON					
01.01.1990	31.12.9999	USBR	Employer Contribution	A	AMOUNT	USD	MON					
01.01.1900	31.12.9999	BASE_HUN		A	AMOUNT	HUF	ANN					
01.01.1900	31.12.9999	HRFF-BON		A	AMOUNT	USD	ANN					
01.01.1900	31.12.9999	HRFF-BON-P		A	PERCENTAGE	USD	ANN					
01.01.1900	31.12.9999	CAN01		A	AMOUNT	CAD	BWK					
01.01.1990	31.12.9999	AUD1	Base Salary Australia	A	AMOUNT	AUD	ANN					
01.01.1970	31.12.9999	BASEBZ	Base Salary - Brazil	A	AMOUNT	BZD	ANN					
01.01.1990	31.12.9999	FRA01	Base Salary France	A	AMOUNT	EUR	ANN					
01.01.1900	31.12.9999	GBR1	Base Salary United Kingdom	A	AMOUNT	GBP	ANN					
01.01.1900	31.12.9999	USVAR		A	AMOUNT		ANN					
01.01.1900	31.12.9999	PL01		A	AMOUNT	PLN	ANN					
01.01.1970	31.12.9999	BASMX	Base Salary - Mexico	A	AMOUNT	MXN	ANN					
01.01.1990	31.12.9999	CHF1	Base Salary Switzerland	A	AMOUNT	CHF	ANN					

Figure 7.26 Running and Saving a Copy of a Check Report

> **Note**
>
> To run this check, you'll need to wait until the connection between the two systems is established, as discussed in more detail in Section 7.6.

7.2.4 Extensibility

Mapping Employee Central custom fields to Employee Central Payroll infotypes is possible through extensibility capabilities, which also allows for changes in the logic of how the system fills in Employee Central Payroll infotypes of the standard set (infotypes 0001, 0014, and 0015). Two types of extensibility are available: basic or custom. In this section, we'll explain both options.

Basic

With basic extensibility, you can map Employee Central custom or standard fields to Employee Central Payroll custom or standard fields. You can extend or overrule the standard mapping between Employee Central employee data fields and infotypes in Employee Central Payroll in various ways.

You can map standard and custom fields from the **Job Information**, **Personal Information**, and **Compensation Information** portlets to the Employee Central Payroll infotypes 0001, 0002, 0007, and 0008 in Employee Central Payroll. Table 7.2 lists further details.

Employee Central Portlet	Employee Central Payroll Infotype	Target Fields in Employee Central Payroll
Job Information	Infotype 0001 (Organizational Assignment) Infotype 0007 (Planned Working Time)	All standard and custom fields
Personal Information	Infotype 0002 (Personal Data)	All standard and custom fields
Job and Compensation Information	Infotype 0008 (Basic Pay)	TRFGR, TRFST, BSGRD, DIVGV, STVOR, ANSAL, ANCUR, and CPIND

Table 7.2 Scope of Basic Extensibility

Let's say you want to map the standard **contract_type** field from Employee Central to the **Work Contract** field of the **Personnel Structure** section (field ANSVH) of infotype 0001 in Employee Central Payroll.

First, make sure that the data type field **contract_type** is available in Employee Central by checking the data definition in the **Manage Business Configuration** screen in Employee Central (as described in Chapter 3, Section 3.2). Then, start the mapping by following the menu path **Personnel Management • Integration Settings for SuccessFactors Employee Central Payroll • Extensibility • Define Extensibility of Field Assignment.** You can also access this screen in Transaction SM30 by entering view V_HRSFEC_EXTMAP. Either way, you'll see the screen shown in Figure 7.27.

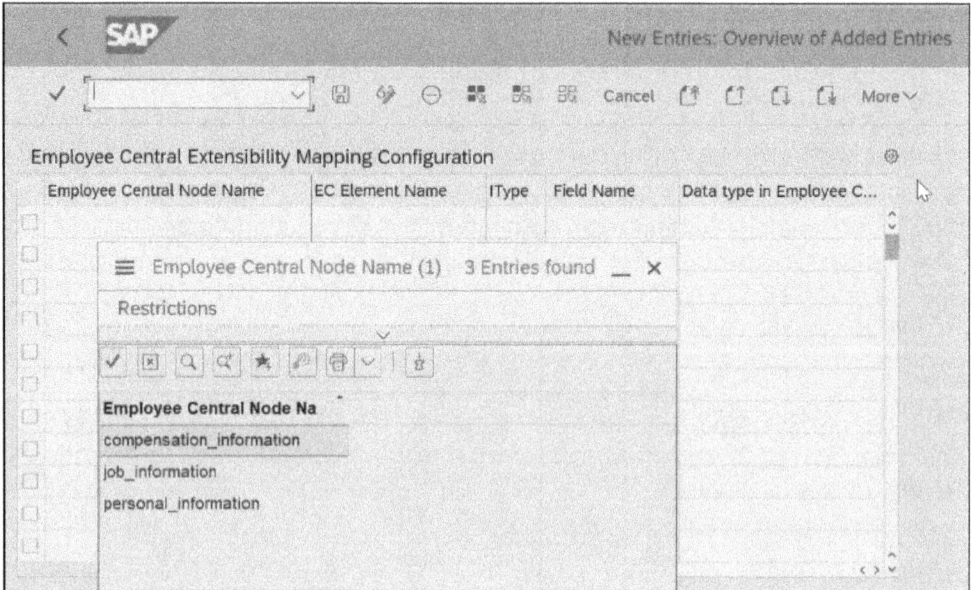

Figure 7.27 Extensibility Field Assignment

On this screen, you'll need to perform the following actions:

1. Press F4 in the **Employee Central Node Name** column and select **job_information**.
2. Select **Employee contract_type** for the **EC Element Name** column.
3. Select **0001** for the **iType** column.
4. Select **CONTRACT_TYPE** for the **Data type in Employee Central** column.

Custom Example

Let's say you want to map the `custom_double20` field from Employee Central to the **Supervisor** field of the **Organizational Plan** section (field `MSTBR`) of infotype 0001in Employee Central Payroll. For this scenario, you would follow these steps:

1. Make sure that the **data type** field `custom_double20` is available in Employee Central.
2. On the same screen shown in Figure 7.27, press F4 in the **EC Element Name** column. You'll get the list of Employee Central elements; select **custom_double20** from them.

Let's look at another, more advanced example. Changing the logic of how infotypes are filled in or using infotypes outside the standard set requires advanced extensibility using Business Add-Ins (BAdIs). Recall that a BAdI is a predelivered enhancement spot in the replication engine where you can place your own code, with some restrictions.

To reveal available BAdIs, follow the menu path **Personnel Management • Integration Settings for SuccessFactors Employee Central Payroll • Extensibility • Business Add Ins for Point To Point Replication.** You'll see the screen shown in Figure 7.28.

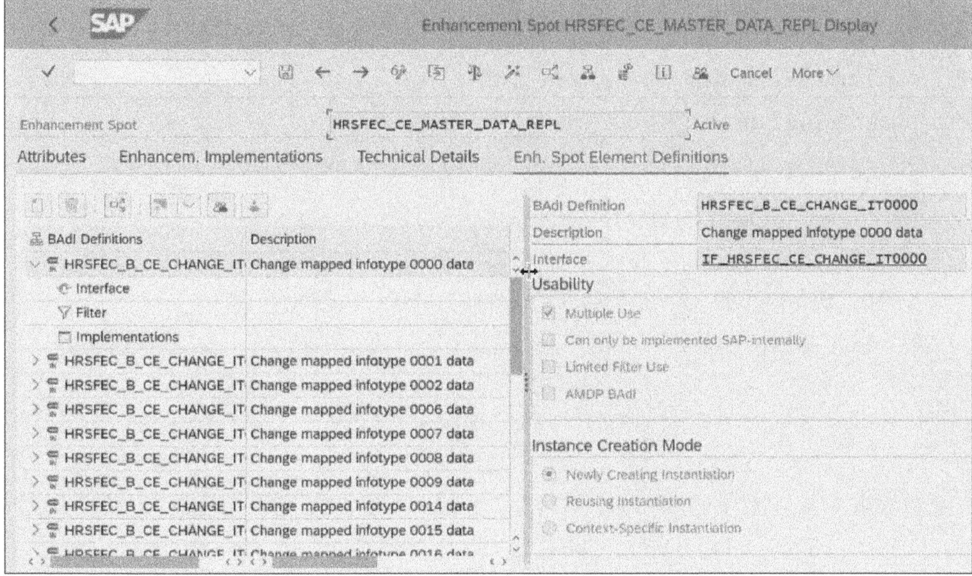

Figure 7.28 Extensibility BAdIs

This activity leads you to the enhancement spot HRSFEC_CE_MASTER_DATA_REPL, where you can find all the BAdIs available for use in replication. Two types of BAdIs are available for enhancing the default logic in specific Employee Central Payroll infotypes:

- BAdIs that are processed after the standard mapping logic but before the fields in the infotype are updated with a value. A BAdI exists for each infotype. Use these to extend or modify the mapping logic for the respective infotypes of the *standard set*, when basic extensibility is not sufficient. You do not need to update the infotype framework in the BAdI.
- BAdIs that are processed after the standard import logic for an employee but before saving the employee record. Use these BAdIs to extend the replication scope to infotypes that are not a part of the standard set. The comparison between old data and mapped data and the update to the infotype framework must be done in the BAdI. We recommend that you use generic update logic, as follows:
  - ITF_CREATE
  - ITF_DELETE
  - ITF_MODIFY
  - ITF_READ
  - ITF_READ_ALL
  - ITF_READ_ALL_IT
  - ITF_READ_FOR_MODE
  - UPDATE_MASS_PNNNN_DATA
  - UPDATE_MASS_PNNNN_DATA_SUBTY

The main operations performed for any update include the following:

- Comparison of old and new values
- Update of the infotype framework
- Deletion of records that did not come through in the replication

Note that every replication sends all the records for that employee from the designated date of system go-live, also known as the full transmission start date (FTSD), discussed in more detail in Section 7.4, which is stored in table PAOCFEC_PN_FTSD.

Note

Using BAdI involves coding, which means you'll need the assistance of a developer.

The standard infotype set for the first type of BAdI is shown in Table 7.3.

Infotype	Description	Employee Central Portlet	BAdI
Infotype 0000	Actions	**Job Information** and **Compensation Information**	HRSFEC_B_CE_CHANGE_IT0000
Infotype 0001	Organizational Assignments	**Job Information** and **Compensation Information**	HRSFEC_B_CE_CHANGE_IT0001
Infotype 0002	Personal Data	**Personal Details** and **National ID**	HRSFEC_B_CE_CHANGE_IT0002
Infotype 0006	Addresses	**Address Information**	HRSFEC_B_CE_CHANGE_IT0006
Infotype 0007	Planned Working Time	**Job Information**	HRSFEC_B_CE_CHANGE_IT0007
Infotype 0008	Basic Pay	**Job Information** and **Compensation Information**	HRSFEC_B_CE_CHANGE_IT0008
Infotype 0009	Bank Details	**Payment Information**	HRSFEC_B_CE_CHANGE_IT0009
Infotype 0014	Recurring Payments/ Deductions	**Job Information** and **Compensation Information (Recurring Payments/Deductions)**	HRSFEC_B_CE_CHANGE_IT0014
Infotype 0015	Additional Payments	**Job Information, Spot Bonus (Additional Payments),** and **One Time Deductions (Additional Deductions)**	HRSFEC_B_CE_CHANGE_IT0015
Infotype 0016	Contract Elements	**Employee** and **Employment Information**	HRSFEC_B_CE_CHANGE_IT0016
Infotype 0021	Family Member/ Dependents	**Person**	HRSFEC_B_CE_CHANGE_IT0021

Table 7.3 BAdIs to Change Infotype Data

Infotype	Description	Employee Central Portlet	BAdI
Infotype 0027	Cost Distribution	**Job Information** and **Cost Distribution**	HRSFEC_B_CE_CHANGE_IT0027
Infotype 0041	Date Specifications	**Job Information** and **Employment Information**	HRSFEC_B_CE_CHANGE_IT0041
Infotype 0105	Email data	**Email Information**	HRSFEC_B_CE_CHANGE_IT0105_EMAI
Infotype 0185	Personal IDs	**Employee** and **Employment Information**	HRSFEC_B_CE_CHANGE_IT0185
Infotype 0267	Additional Off-Cycle Payments	**Job Information**, **Spot Bonus (Additional Payments)**, and **One Time Deductions (Additional Deductions)**	HRSFEC_B_CE_CHANGE_IT0267
Infotype 0709	Person ID	**Employee ID**	HRSFEC_B_CE_CHANGE_IT0709

Table 7.3 BAdIs to Change Infotype Data (Cont.)

The second type of BAdI is useful for extending replication to other infotypes not included in the standard set above. Only two BAdIs fall under this category:

- HRSFEC_B_CE_PROCESS_EMPLOYEE
- HRSFEC_B_CE_PROCESS_ET

In addition to these infotype-focused BAdIs, three other BAdIs provide special capabilities for replication not available with configuration:

- Change Mapping of External Personnel Number (HRSFEC_B_CE_EXT_PERNR_MAP)
- Exclude Certain Subtypes of Infotype Records from Deletion (HRSFEC_B_EXCLUDE_FROM_DELET)
- Decide on Hire or Rehire (HRSFEC_B_CE_DECIDE_HIRE_REHIRE)

Custom Metadata Framework Objects

Sometimes, a business scenario requires the capturing of HR data that is not available in a standard Employee Central system. SAP provides a mechanism for creating a table structure in Employee Central, known as a *custom Metadata Framework (MDF) object*. To bring this object over to Employee Central Payroll, follow these steps:

1. In Employee Central, ensure the custom MDF object has the following values:
   - **API Visibility**: Read
   - **Status**: Active
   - **MDF Version History**: Complete history
2. Still in Employee Central, enter "Manage Data" in the search bar. When the screen shown in Figure 7.29 appears, enter your custom object in the **Object Type** field of the Compound Employee application programming interface (API) Object Types MDF, as shown in Figure 7.30, and save.

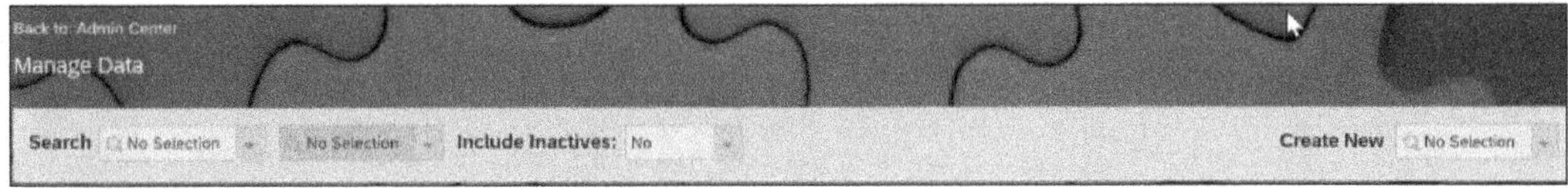

Figure 7.29 Managing Data in Employee Central

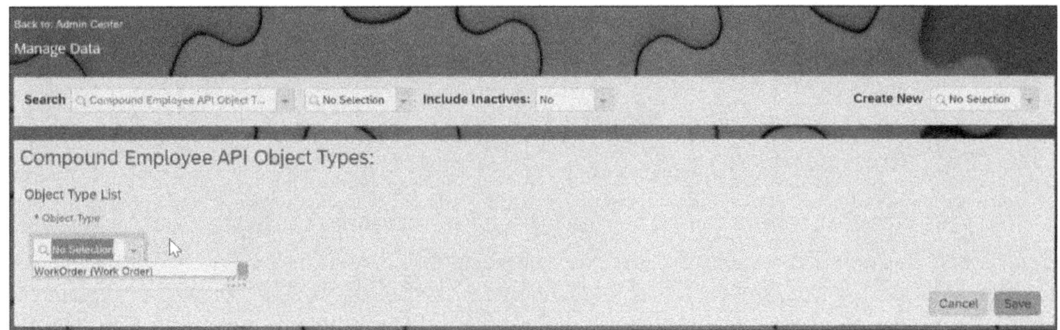

Figure 7.30 Compound Employee API Object Type

3. Let's move to Employee Central Payroll. Each segment of data from Employee Central is brought across to Employee Central Payroll based on the query segment configured by following the menu path **Personnel Management • Integration Settings for SuccessFactors Employee Central Payroll • Configuration of Point-to-Point Replication • Configure Compound Employee API Query**. Select the configuration ID you want to use and click on **Query Segments.** You'll see a screen with the query segments already configured, as shown in Figure 7.31.

Specify the custom object as another segment, as discussed in Section 7.1, as shown in Figure 7.31. After saving, you can now pull in this data into Employee Central Payroll through replication. Remember, you still must decide on the infotype and map the fields using one of the BAdIs mentioned earlier—this mapping does not happen automatically for custom MDF objects.

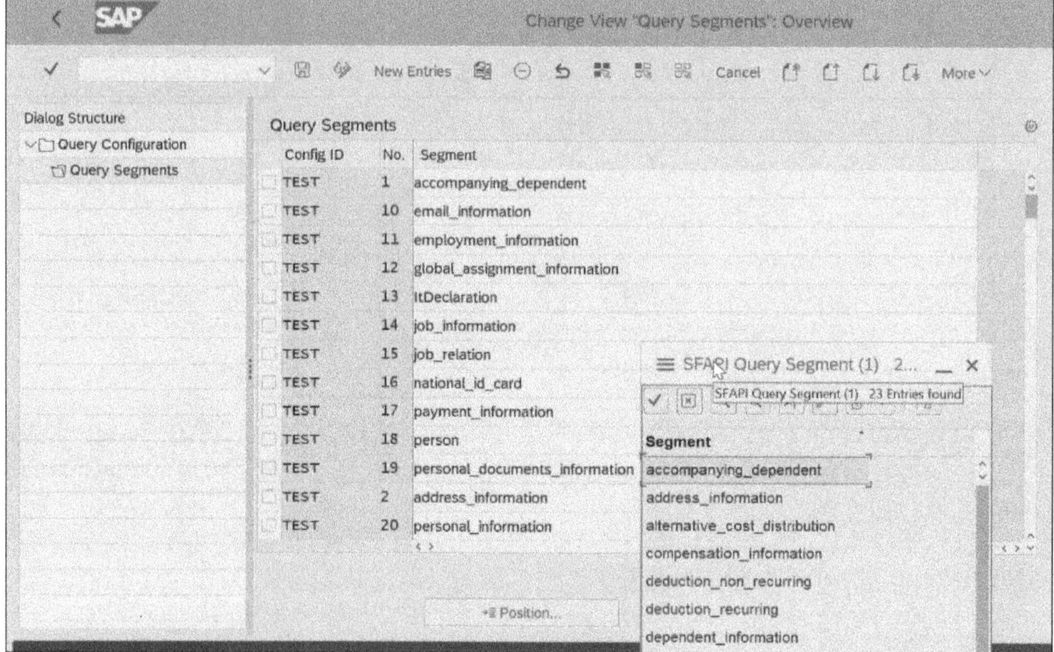

Figure 7.31 Query Segments in Replication Configuration

7.3 Country Filtering

Sometimes, you need to restrict data replication to certain countries. For example, let's say you need to exclude countries where payroll is outsourced due to the relatively small number of employees. You can bring up the configuration screen for country filtering by following the menu path **Personnel Management • Integration Settings for SuccessFactors Employee Central Payroll • Country Filter • Restrict Employee Master Data Replication to Given Countries.**

You'll see the screen shown in Figure 7.32. In the **ISO code** column, select the countries you want to replicate and then save.

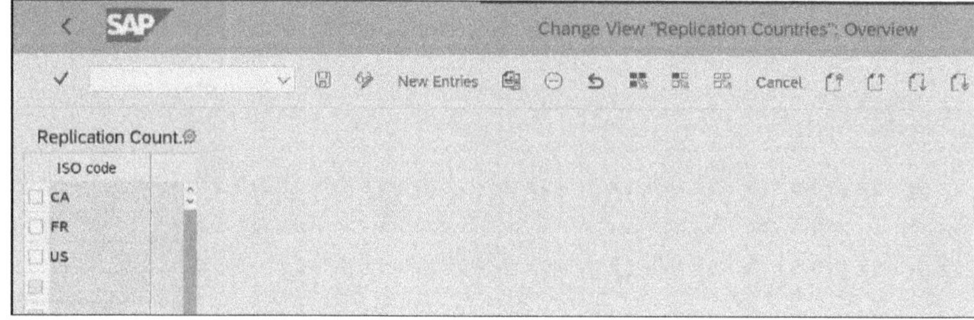

Figure 7.32 Country Filtering

7.4 Migration and Cutover Scenarios

Figure 7.33 shows a typical cutover scenario for an implementation. The data maintenance process passes over from old to new systems in a short period of time called a *cutover*. This time is when the data is migrated into the new system.

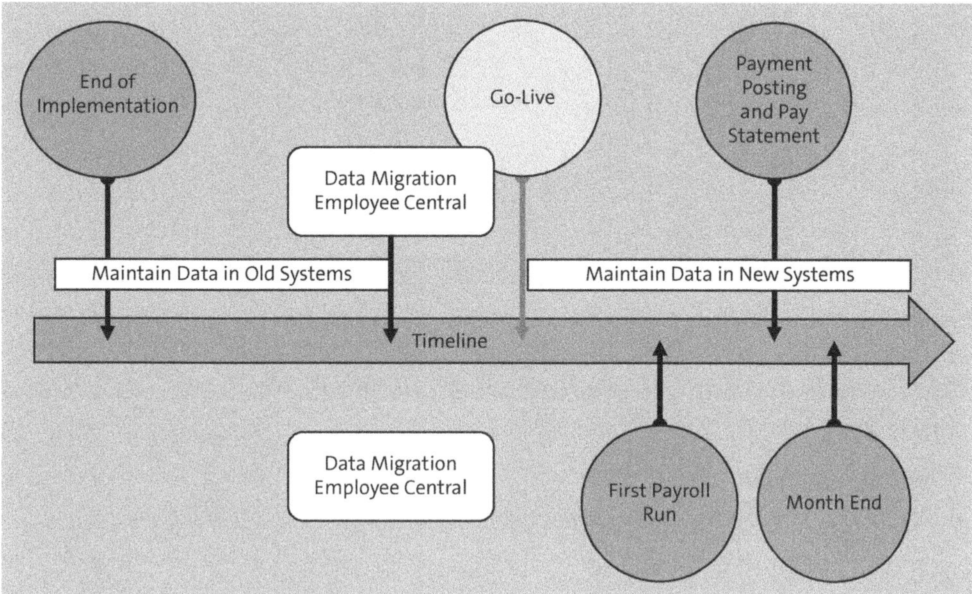

Figure 7.33 Sample Cutover Plan

For a payroll system, the following four tasks are relevant during conversion of data, as shown in Figure 7.34:

1. Extracting employee data from a legacy system (personnel information and employment information including organizational information and compensation information) and loading this data into Employee Central.
2. Loading employee data into Employee Central Payroll using the standard data replication capability.
3. Extracting employee data from a legacy system that is not maintained in Employee Central (tax information, social insurance information) and loading this data into Employee Central Payroll.
4. Extracting payroll results from a legacy system and loading this into Employee Central Payroll.

Note

Note that this first task listed is specific to Employee Central, so we won't cover this topic in detail in this book. The third and fourth tasks will be discussed in Chapter 16.

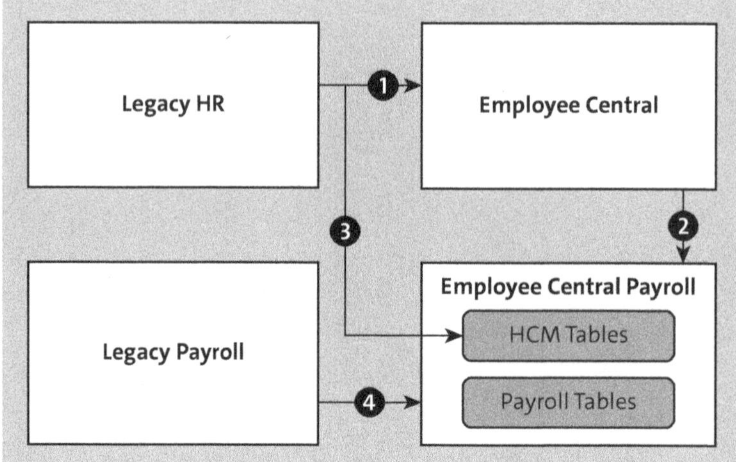

Figure 7.34 Data Migration Model

The second task, loading employee data, is critical to plan, with several questions that you must answer:

- How will the payroll history be loaded?
- How should terminated employees be handled?
- How should employees who are not paid out from the payroll system handled?

At this stage, we need to return to an important concept: the FTSD, which is used to delineate a time period from which the replication starts. This line in the sand is where the system starts looking for changes. In other words, the replication engine does not care about the history of the employee before this date and only looks forward from this date. Typically, this date is the go-live date for the system. Once the FTSD is specified, the following becomes true:

- Only records after this date are extracted from Employee Central.
- Records that end before this date are ignored.
- Replication errors out if a record in Employee Central has a start date before the FTSD and an end date after it.

Therefore, you must be careful about choosing the FTSD date, and you must ensure that the start dates of all records being replicated from Employee Central meet the outlined criteria; otherwise, they will never be successfully replicated.

Before going live with Employee Central Payroll, you must set a start date that matches the FTSD for *all* employees who are in scope for replication in the Employee Central system. This step is achieved through a mass loading process in Employee Central similar to task described earlier.

The most typical implementation starts off with a full load covering all historical data, and then the FTSD can be changed so that subsequent replications can focus on what is modified in the future, not the data in the past.

If you have a long data history, we recommend that you do not transfer all of this data to Employee Central Payroll. Some organizational or compensation structures may be so obsolete that creating those structures will require effort and may lack value for the future. Some companies transfer a history with a "fake" structure to ease the effort.

To set up the FTSD, follow the menu path **Personnel Management • Integration Settings for SuccessFactors Employee Central Payroll • Configuration of Point-to-Point Replication • Configure Compound Employee API Query**. From this screen, only three steps are required: Click **New Entries** (which will open the screen shown in Figure 7.35), set the date, and save these settings.

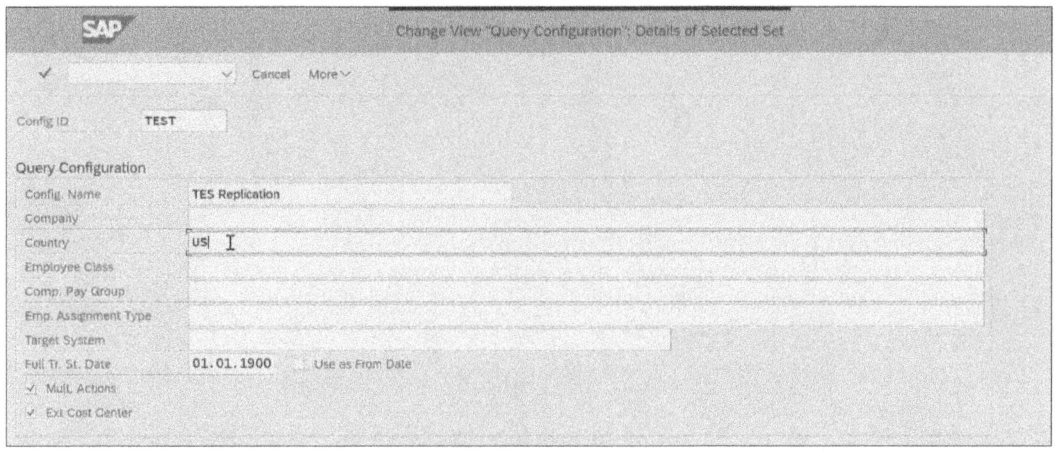

Figure 7.35 FTSD Parameter in Query Configuration

One scenario with terminated employees needs a small digression. If a terminated employee was not loaded during go-live, what happens if this person is rehired? You can exclude these employees from the replication. However, if this employee is rehired, the replication logic recognizes this fact and includes the employee in the replication processes after the rehire date. Unlike active employees, you do not need to split the data records of terminated employees on the cutoff date. This mechanism requires you turn on a switch by following these steps:

1. In Employee Central Payroll, go to Transaction SM30, as shown in Figure 7.36.
2. Enter "V_T77SO" in the **Table/View** field.
3. Click **Maintain**.
4. Click **New Entries**.

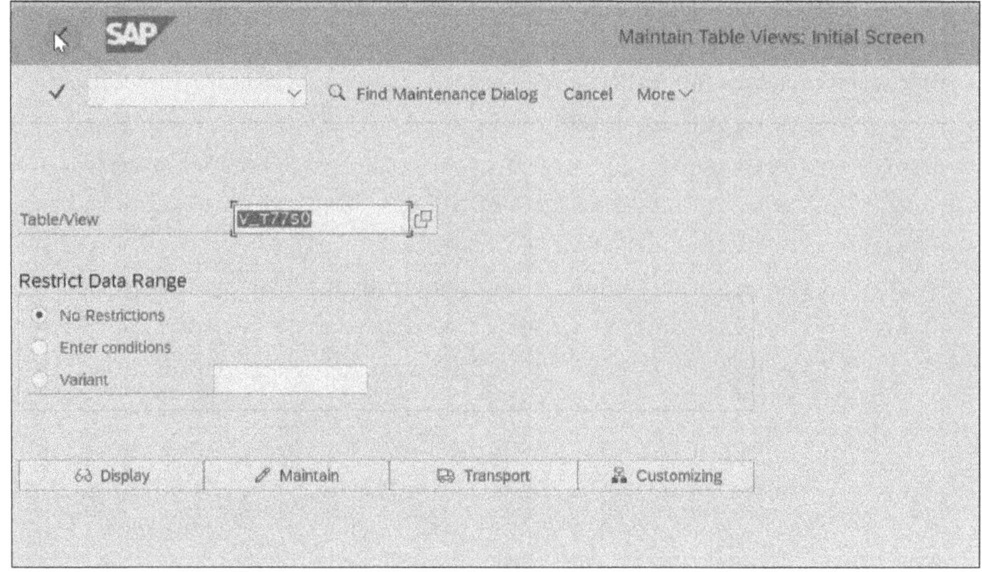

Figure 7.36 Transaction SM30 to Turn On Switch

5. Find **SFEC** under the **Group** field and then semantical abbreviation **EECOH** under the **Sem.abbr.** field, as shown in Figure 7.37.
6. Set the value abbreviation in the **Value Abbr** field to "X."
7. Save these settings.

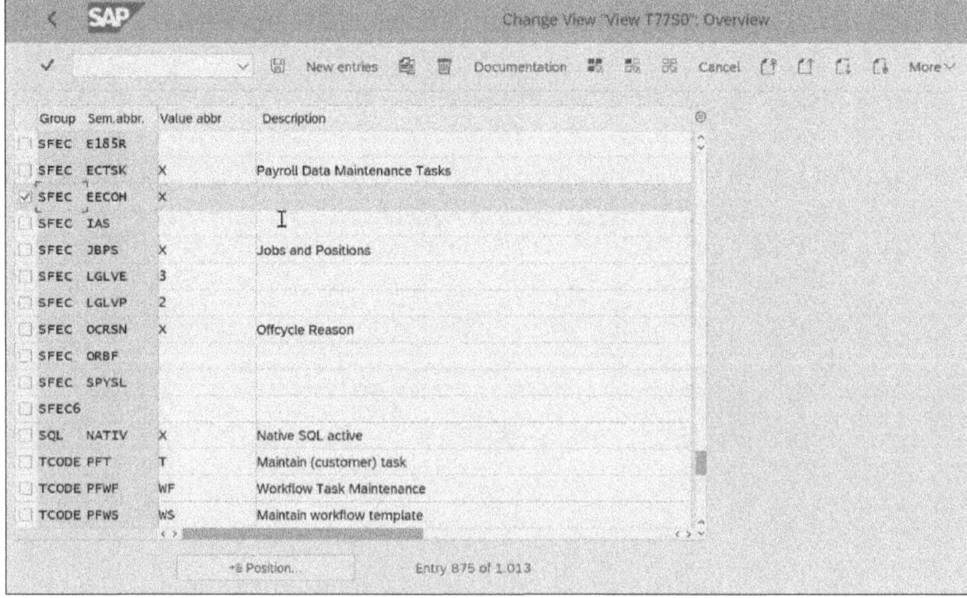

Figure 7.37 Switch Enabling Rehire of Terminated Employees without History

Note

This option is not currently supported if you have turned on multiple actions for the replication in the point-to-point replication configuration. For more information, see Chapter 8, Section 8.2.

In addition, quite a few infotypes are needed by Employee Central Payroll that are never captured in Employee Central. Table 7.4 shows the infotypes required for tax and benefit calculations in the US.

Infotype	Description
0045	Loans
0078	Loan Payments
0161	IRS Limits
0194	Garnishment Document
0195	Garnishment Order
0216	Garnishment Adjustments
0207	Residence Tax Area
0208	Work Tax Area
0209	Unemployment State
0210	Withholding Information W4/W5
0234	Additional Withholding Information
0235	Other Taxes
0506	Tip Indicators

Table 7.4 Infotypes Not Captured in Employee Central

Two methods for loading this data into Employee Central Payroll are popular:

- Using the Legacy System Migration Workbench, a standard method for SAP systems for many years
- Writing a special custom ABAP program to load the data

In both cases, the data is made available in a fixed or CSV file format that closely matches the infotype record. Transferring pay results before go-live is not common because of the complexity involved. But if required, you can transfer pay results before go-live. Usually, instructions for this step vary by country. For the US, you would load three tables and execute a payroll run for conversion. Instructions are available by following the menu path **Payroll (Country) • Transfer Legacy Data • Define Period Table/ Define Wage Type Table/Define Tax/Arrears Table.**

Generally, you would perform the following four steps:

1. Extract the pay results that must be loaded and assign period and wage types.
2. Format the data in two tables T558B (Periods) and table T558C (Wage Types).
3. Load tables T558B and T558C using a custom program or using the configuration path mentioned earlier.
4. Run the payroll schema ULK9.

Even though your company may decide not to migrate payroll results from the past, you may still need to load cumulative balances of the last tax year if the go-live is planned in the middle of the tax year. This requirement stems from tax calculations that rely on information about deductions that have already taken place and calculations to determine limits in deductions—if the limits have been reached, then the deduction must stop.

7.5 Activating User Interface Integration Services

As described in Chapter 6, Section 6.1.2, the integration between Employee Central and Employee Central Payroll depends on web services. We described two services in an Employee Central system: the OData API and SFAPI. In the Employee Central system, equivalent services interact with web services on the HR side.

These services perform many things, but one goal is close to the user—the user interface (UI) integration service. A UI represents the way a human being communicates with Employee Central Payroll. By using these services, an employee can interact with Employee Central Payroll from the Employee Central frontend, enabling easy access through a single system. The service calls Employee Central Payroll behind the scenes and serves up the requested product seamlessly. The following UI services are available to use:

- HRPAO_PAOM_MASTERDATA
  Allows the updating of Employee Central Payroll infotypes from the Employee Central frontend.
- PAYSLIP
  Displays pay statements on the Employee Central frontend by retrieving pay results from Employee Central Payroll.
- PAYMNT_SUM_AU
  Displays the Payment Summary Form for Australia in the Employee Central system by using pay results from the payroll system.
- HRESS_A_PAYINFO
  Displays the Superannuation statement for Australia in the Employee Central system by using pay results from the payroll system.

- `HRESS_A_REP_IN_FORM16`
  Displays Form 16 statement for India in the Employee Central system by using pay results from the payroll system.

Note that these services are all optional, but each must be turned on if required. You can turn on these services in Transaction SICF, as shown in Figure 7.38.

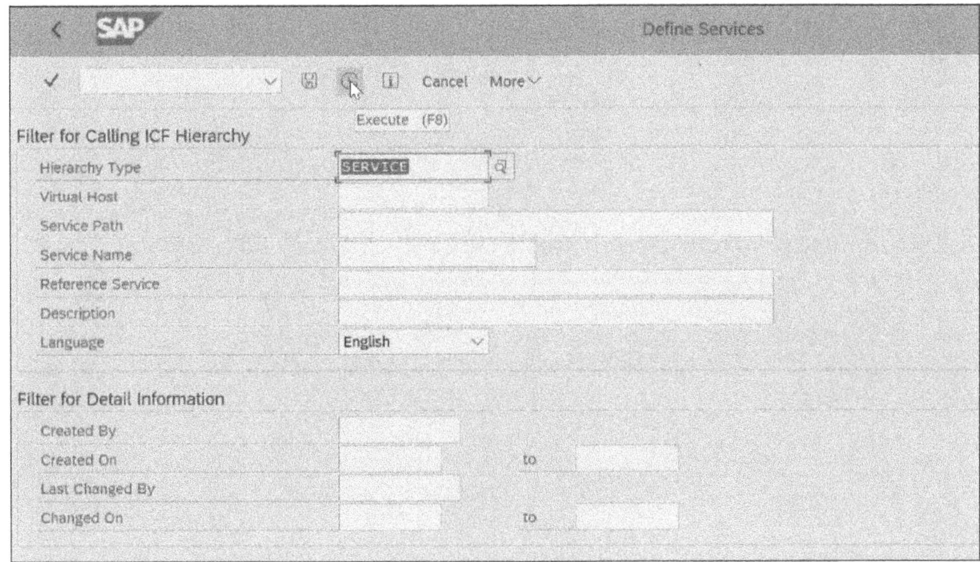

Figure 7.38 Transaction SICF

Click the **Execute** icon at the top of the screen. Then, as shown in Figure 7.39. Right-click on a service and select **Activate Service**. Save.

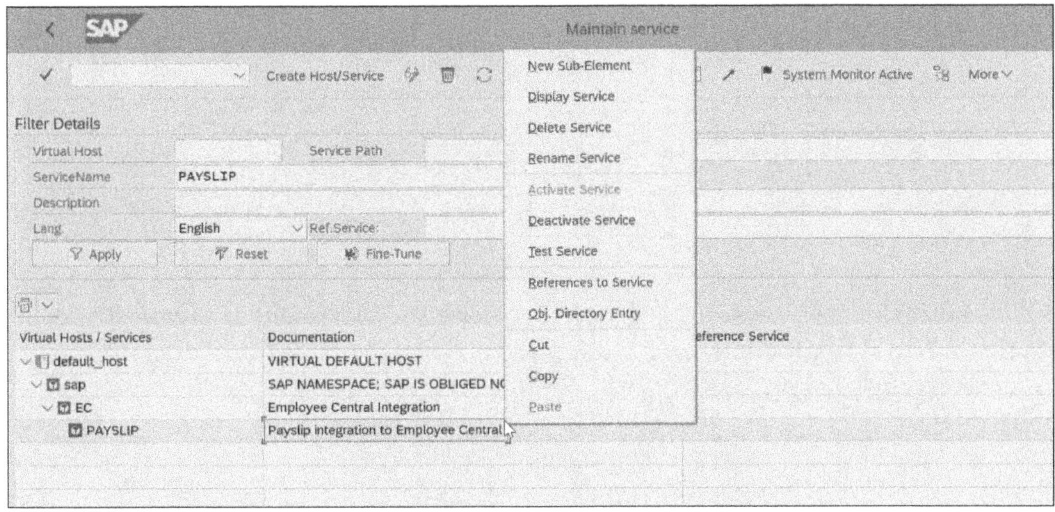

Figure 7.39 Activating Payroll Services in Employee Central Integration

7.6 Setting Up the Connection to Employee Central

The connection to Employee Central can be set up entirely in Transaction HRSFEC_PTP_CONFIG, which uses the program RP_HRSFEC_PTP_CONFIGURATION.

Before executing this program, make sure you've completed the following steps:

- Collect the SFAPI user name and password for the user that will be used to connect. This user needs permission to access Employee Central APIs and OData service communications.
- Create the replication target system.

Creating a replication target can be achieved in two ways. In the first approach, you would follow these steps:

1. Find the logical system name of Employee Central Payroll using Transaction SCC4, as shown in Figure 7.40.

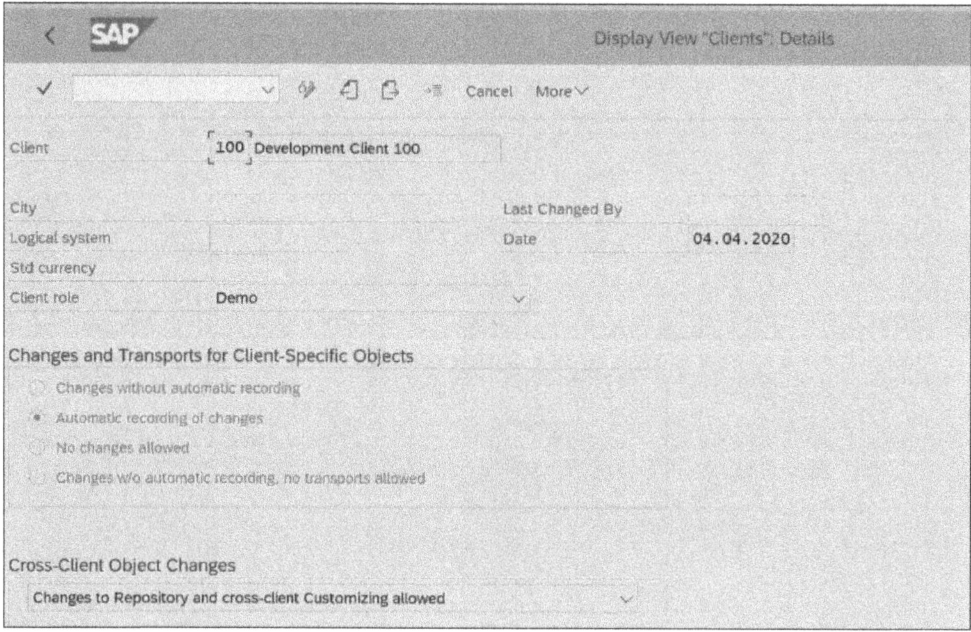

Figure 7.40 Transaction SCC4

2. Log on to the Employee Central system and go to **Admin Center • Manage Data**.
3. Click on the **Create New** button and search for "replication target system" in the first dropdown menu, as shown Figure 7.41.
4. In the **External Code** field, enter the logical system name you found from Transaction SCC4.
5. In the **External Name** field, define an external name according to your preferences.

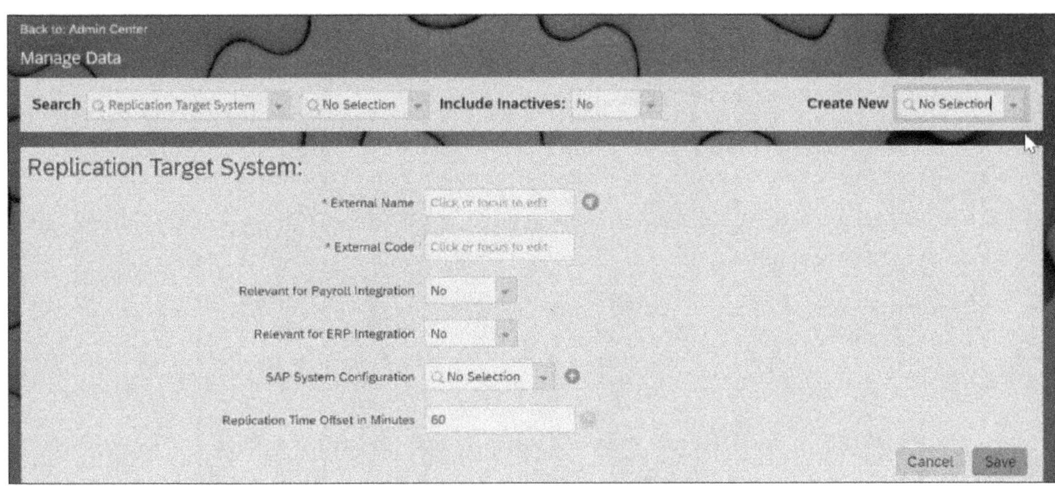

Figure 7.41 Setting Up Replication Target System

6. Choose **Yes** from the **Relevant for Payroll Integration** dropdown list.
7. In the **Replication Time Offset in Minutes** field, enter the delay in minutes you want the resending of data to be postponed by when the payroll area is locked or in correction phase. If you leave this field empty, the default time until resend is 60 minutes.
8. Click **Save**.

The second method of creating a replication target system is much simpler. Simply open Transaction HRSFEC_PTP_CONFIG and select the **Create Repl. Target System** checkbox. However, this method only works if the connectivity between the two systems is working perfectly. To avoid this dependency, defining a replication target system using the first method may be advisable.

Now that you've created the replication target system, let's return to setting up the connection to Employee Central. In Transaction HRSFEC_PTP_CONFIG, follow these steps:

1. Select the **Set Connection Data** radio button.
2. Select the correct API server URL, based on your data center, from the dropdown list.
3. Maintain the **Company**, **User**, and **Password** fields for Employee Central.
4. Click **Execute**.

The program creates everything you need, and you should see the following success messages:

- **Creation of RFC destination ECP_PTP_SOAP_NNN was successful.**
- **Creation of RFC destination ECP_PTP_ODATA_NNN was successful.**
- **Employee Central credentials were saved successfully.**

You can now test your connection via Transaction HRSFEC_PTP_CONFIG, but this time, select the **Ping Employee Central APIs** radio button. Now, click the **Execute** button. Three connections can be tested:

- Employee Master Data API: A logon and logoff to the Compound Employee API are executed.
- Confirmation API: An OData API query is performed on the `EmployeeDataReplicationConfirmation` object.
- Time Data API used for time-off data replication: An OData API query is performed on the `DataReplicationProxy` object.

7.7 Scheduling Master Data Replication from Employee Central to Employee Central Payroll

Once you've finished configuring the replication process and tested the connection, you're now ready to schedule the replication process. Scheduling is a technique for automatically running a transaction in the Employee Central Payroll system at a certain time on a certain day.

To begin scheduling master data replication, enter Transaction HRSFEC_PTP_EE_REPL and click **Execute.** Now, maintain the **Configuration ID** field, as shown in Figure 7.42. For our example, choose the configuration ID that we set up earlier.

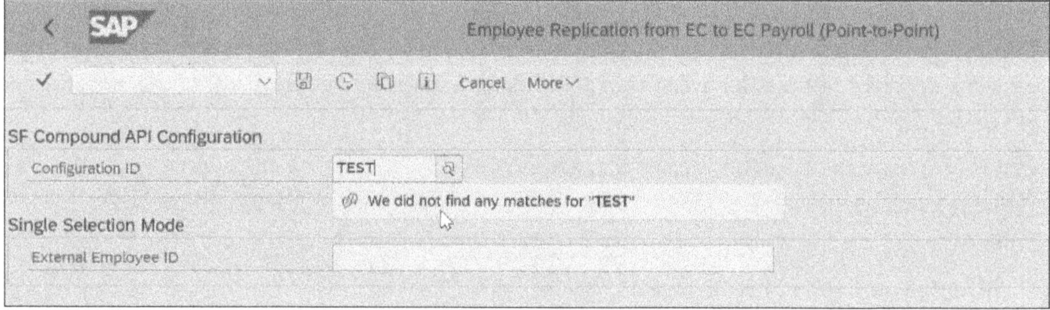

Figure 7.42 Transaction HRSFEC_PTP_EE_REPL

Click on the **Save** icon, and a variant screen appears, as shown in Figure 7.43. Maintain the **Variant Name** and **Description** fields and then click the **Save** icon.

Now, trigger a test run by entering a **Personnel Number** (Person_Id_External) and clicking the **Execute** button. Next, enter "/N HRSFEC_REPL_STATUS." You should see a screen to find the latest replication status, as shown in Figure 7.44.

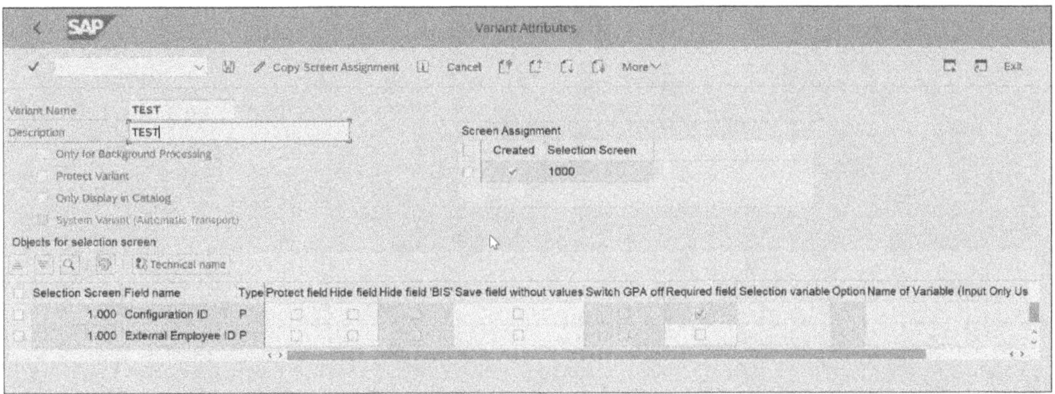

Figure 7.43 Variant for HRSFEC_PTP_EE_REPL

Display Replication Status
Cancel
More
Replication Country
Selection by Employee
External Employee ID
Selection by Personnel Number
Personnel Number
Selection by Status and Time
Replication Status
Successful
Failed
Replication Time
All Until Now
Last Week
Last 24 Hours
Last Hour
Last 15 Minutes

Figure 7.44 Checking the Replication Status

If the replication process was successful, the display will be green or yellow. If you see any red on the screen, a problem occurred. Double-click on the message to reveal the cause of the issue. (This report will be covered in more detail in Chapter 14.) Once the issue is fixed, run the replication again with a test employee and confirm the process is working. Now, run Transaction SM36, maintain the **Job Name** field, and choose **Start condition**, as shown in Figure 7.45.

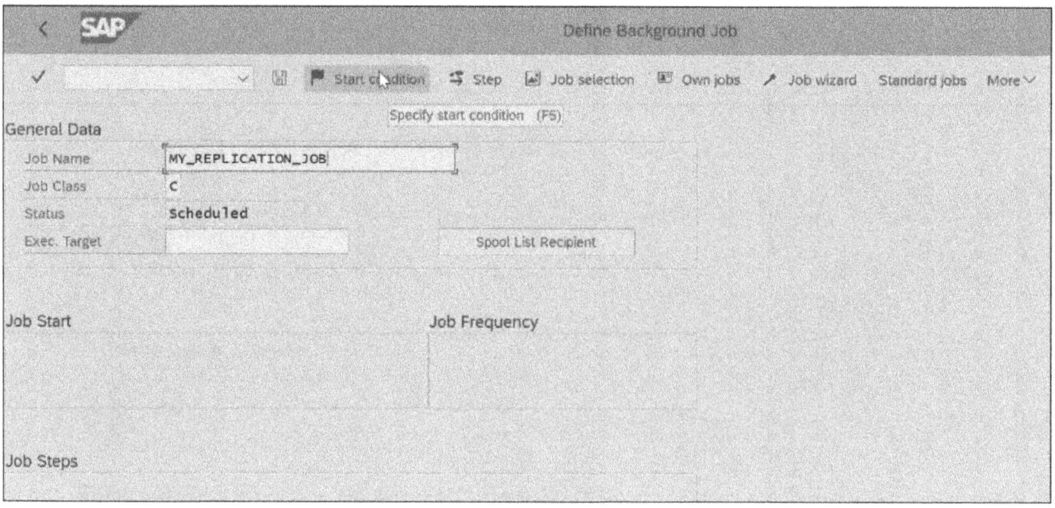

Figure 7.45 Transaction SM36

Click on **Step** and enter the ABAP program name "RP_HRSFEC_PTP_EE_REPLICATION," which is the variant that we saved earlier, and save these settings.

Next, you'll need to define the schedule, as shown in Figure 7.46. Click **Immediate** and select the **Periodic job** checkbox. Select **Period Values** and choose **Other Period.** Enter the timeframe of your choice (e.g., 2 hours) and save.

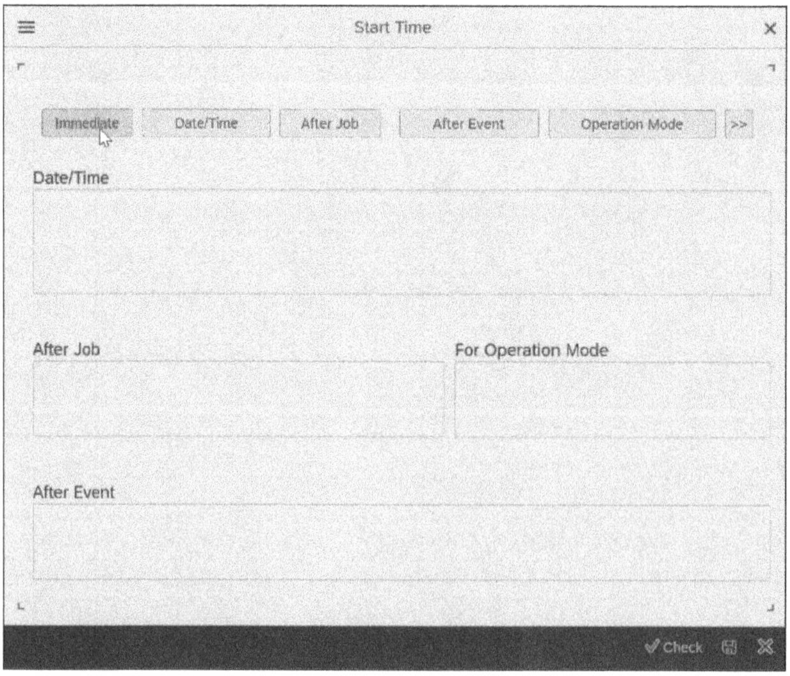

Figure 7.46 Job Timing

Now, the job will execute according to your specified periodicity and pick up all changes in Employee Central data, send these changes over to the Employee Central Payroll system.

7.8 Enabling Push Replication

Sometimes, having a scheduled job to replicate master data is not enough. However frequent the transfer of data may be, this replication is not in real time and therefore is sometimes unsuitable for certain business demands.

In these situations, a nearly real-time replication method can be enabled by utilizing SAP SuccessFactors's Intelligent Services. In this way, the data replicates as soon as critical events like hire, rehire, or termination actions occur.

Before you begin, ensure that Intelligent Services is enabled in provisioning. As a customer, you may not have access to these provisioning screens and instead will need to log a service ticket with SAP Support.

Next, import rules to trigger external events in Employee Central by performing the following steps:

1. Log on to Employee Central and go to the Admin Center.
2. Choose **Import and Export Data** from the menu shown in Figure 7.47.

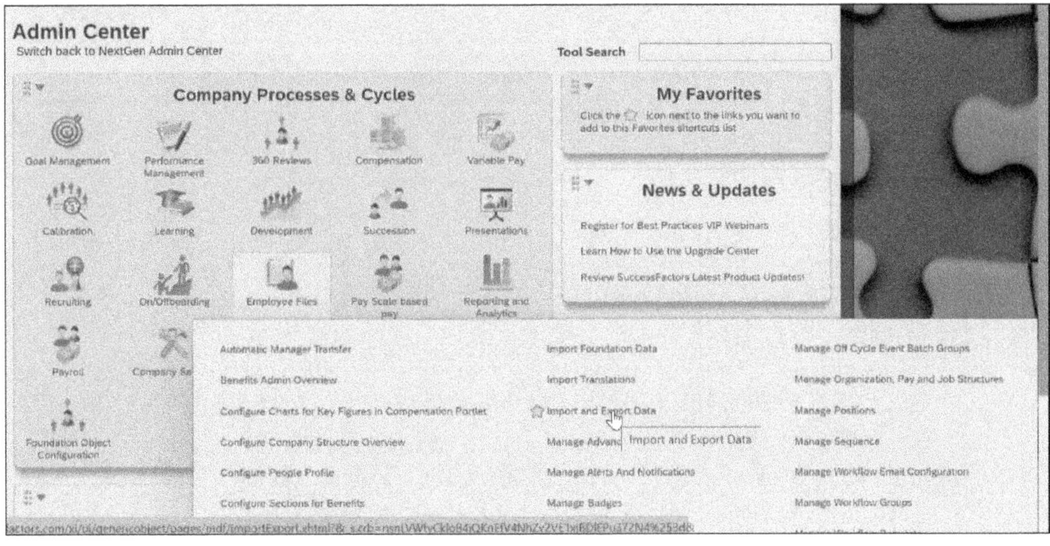

Figure 7.47 Import and Export Data

3. Now, choose **Import Data** from the **Select the action to perform** dropdown list, as shown in Figure 7.48.

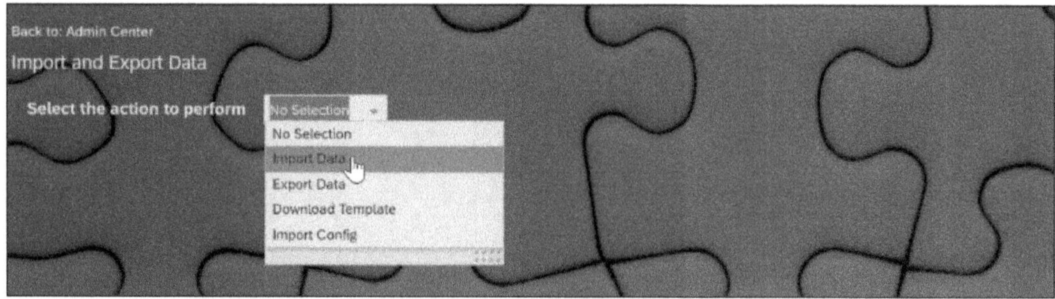

Figure 7.48 Import Data

4. Choose **Success Store.** In the **Package Name** column, select the **Intelligent Service Rules** radio button. Click **Import**, as shown in Figure 7.49.

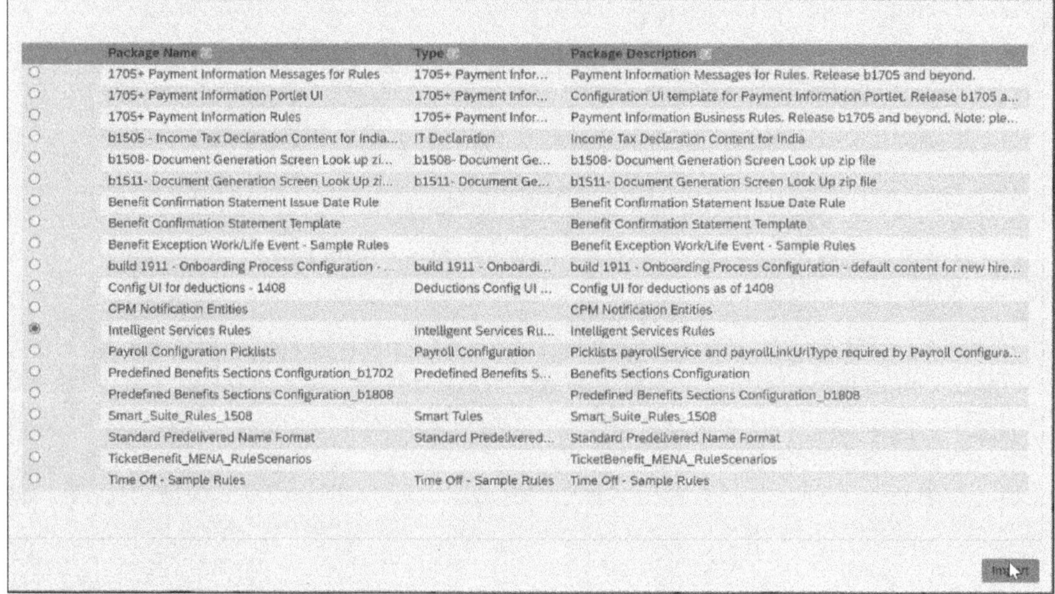

Figure 7.49 Import Package

5. Now, go to the **Manage Business Configuration** screen by typing "Manage Business Configuration" in the search bar.
6. Click **Job Info** and scroll down to **Trigger Rules**, as shown in Figure 7.50. You should see that the following rules have been successfully uploaded:
   - **Event_Hire**
   - **Event_Rehire**
   - **Event_Terminate**

Figure 7.50 Trigger Rules

Assign the rules you just uploaded to this group by adding the following for each rule:

- **Base Object**: Select **Job Information Model**.
- **Event Type**: Select **onPostSave**.

Your next step is to find the access URL, as follows:

1. In the Employee Central Payroll system, execute Transaction SOAMANAGER.
2. Under **Service Administration**, choose **Simplified Web Service Configuration**
3. As shown in Figure 7.51, enter "*masterdatapush*" in the **Search Pattern** field.

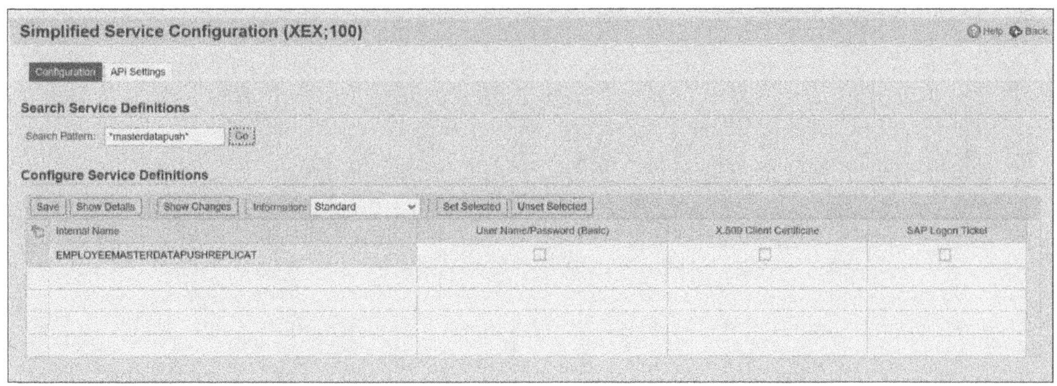

Figure 7.51 SOA Manager

4. Select the relevant service definition, called EMPLOYEEMASTERDATAPUSHREPLICAT.
5. Select the checkbox in the **User Name/Password (Basic)** column and click **Save**.
6. Choose **Show Details**.

7. From the **Configuration Details** section, copy the **Access URL**, as shown in Figure 7.52.

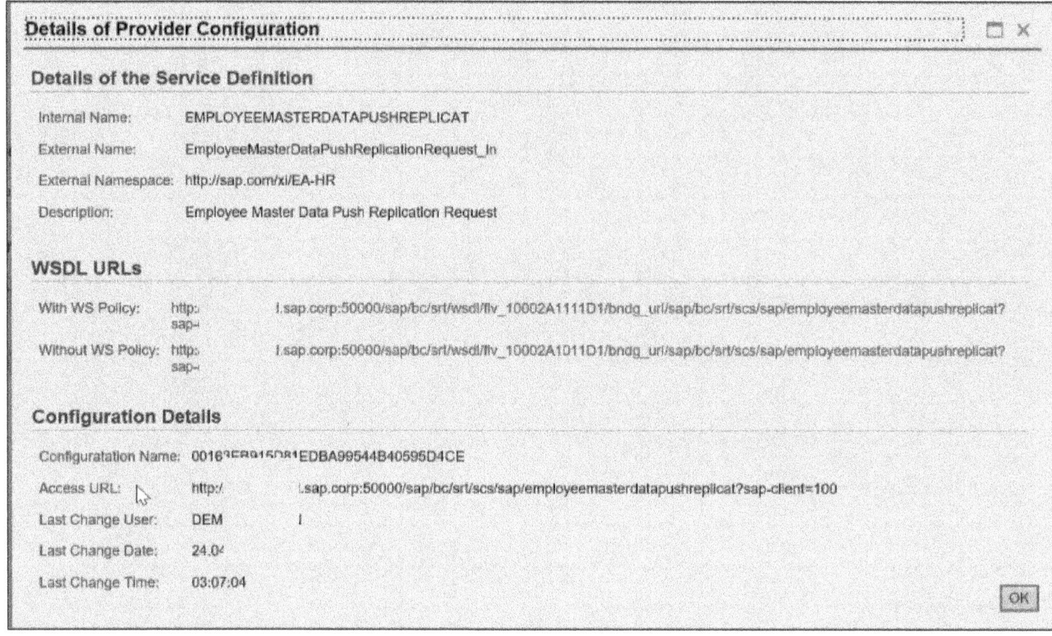

Figure 7.52 Access URL

8. Paste this URL into an open Microsoft Word document so you can use it later in this task.
9. The final task is to set the event notification subscription by entering "Event Notification Subscription" into the search bar, which will result in the screen shown in Figure 7.53.

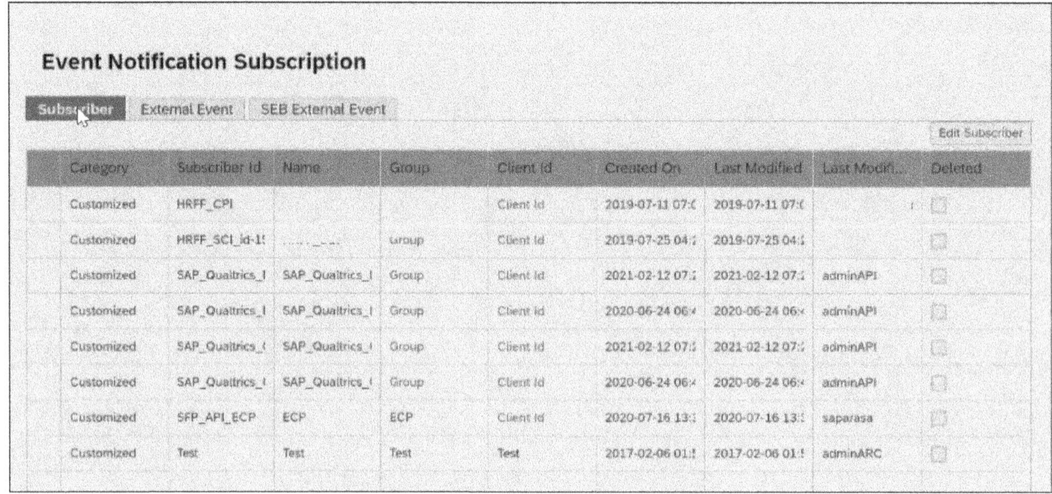

Figure 7.53 Event Notification Subscription

10. Go to the **Subscriber** view and create a subscriber.
11. Then, under the **SEB External Event** tab, click **Add Topic** and maintain the following information:
    - **SEB Event Type**: Select an external event alert type.
    - **Endpoint URL**: Enter the URL you copied in the previous step.
    - **Subscriber**: Enter the subscriber you created from the previous step.
    - **Protocol**: Select **SOAP over HTTP/HTTPS.**
12. Click **Authentication – Basic.**

A quick way to test if the configuration was successful is to follow these steps:

1. In the Employee Central search bar, enter "Event Notification Audit Log."
2. In the **Event Details** column, you can access the log by clicking **View** in the **Event Details** column, as shown in Figure 7.54.

Figure 7.54 Event Notification Audit Log

7.9 Using Mapping Check Tools in Employee Central Payroll

Two tools are available in Employee Central Payroll to detect errors in configuration that might lead to errors or dropouts (records going missing) in data integration. Of course, to run these tools, the connectivity between the two systems must have been set up already.

The first tool, the Pay Component and Wage Type Configuration Check tool, specifically focuses on wage type inconsistencies. (Recall our discussion of wage type synchronization in Section 7.2.3 and how the setup depends on appropriate pay components on the Employee Central system.) This tool will tell you if links are missing between wage types and pay components, highlighting wage types that should be assigned to infotypes 0008, 0014, 0015, and 0267 but are not. It also highlights missing text, identifies if the validity of a pay component fails to match with the validity of a wage type, and identifies pay components or wage types that have been changed or recently added, as discussed earlier in Section 7.2.

The second tool, the Check Inconsistencies in Code Value Mapping and Properties tool, shows discrepancies in the mapping of code values, such as employee classes and action types. To use this tool, run Transaction SE38 and access report RP_HRSFEC_CHECK_EC2ECP_MAPPING. You should see the screen shown in Figure 7.55.

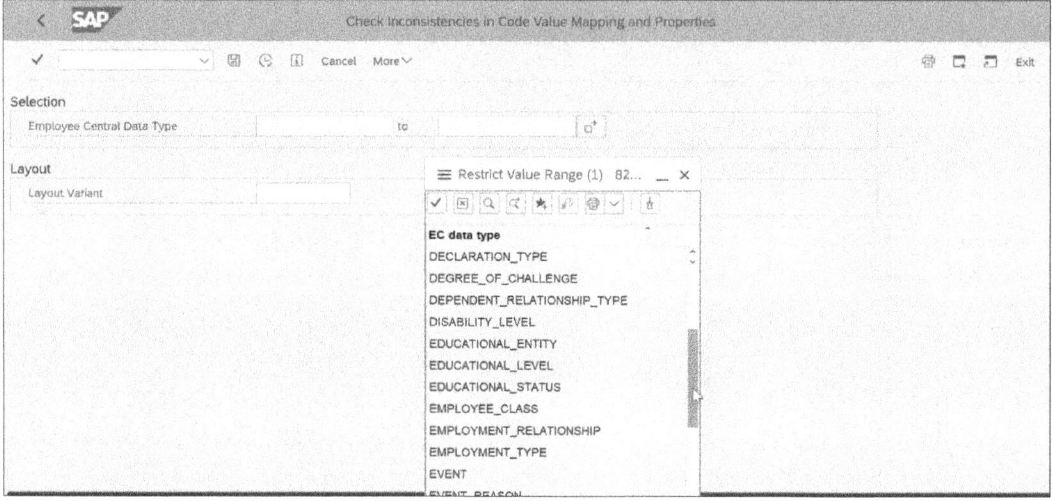

Figure 7.55 Check Tool for Data Type Mapping Inconsistency

Choose the data type you're checking from the **Employee Central Data Type** dropdown menu or leave it blank. Click on the **Execute** button.

You'll then see a report that identifies where an incorrect mapping occurs between the two systems.

7.10 Summary

This chapter explained the integration components related to point-to-point replication. You're now familiar with the basic tools for setting up connections between Employee Central and Employee Central Payroll systems. We also covered the scheduling of replication processing and explored how to simulate nearly real-time transfers through push replication. In the next chapter, we'll examine further intricacies related to the data replicated between the two systems.

Chapter 8
Master Data Replication

Managing the employee data is both a science and an art. While the integration technology takes care of the science, there are many other things that need to be finely tuned to ensure automation, accuracy, and compliance.

In the previous chapter, you learned about the various replication settings in the services available for enabling the integration between Employee Central and Employee Central Payroll systems. Connecting the systems is only half the story, however. We still must ensure the data is received in the right place and in the format required for the business processes linked with the data. In certain cases, you'll make choices made about data elements that may have long-term impacts to how the data can be used. In addition, your company may have adopted (or would like to adopt) certain data standards that will impact how receiving objects must be configured. In this chapter, we'll review the settings for those receiving objects and explore the implications of different options at an infotype level.

These receiving objects are fields that are stored in infotypes. An infotype is a set of related fields that can be assembled together based on their function. Data elements can be broadly classified into two groups:

- Pay
- Non-pay

Pay infotypes directly deal with paying employees. They represent a rate, an amount, a bank account—typical things that affect how much an employee gets paid and how. Accordingly, the affected infotypes control the amount and nature of the payment, as shown in Figure 8.1.

On the other hand, non-pay infotypes record other details about the employee. These details may also affect how or what the employee gets paid, but not directly. The employee's address, position, work schedule, and more are great examples of data elements that fall into this category, as shown in Figure 8.2.

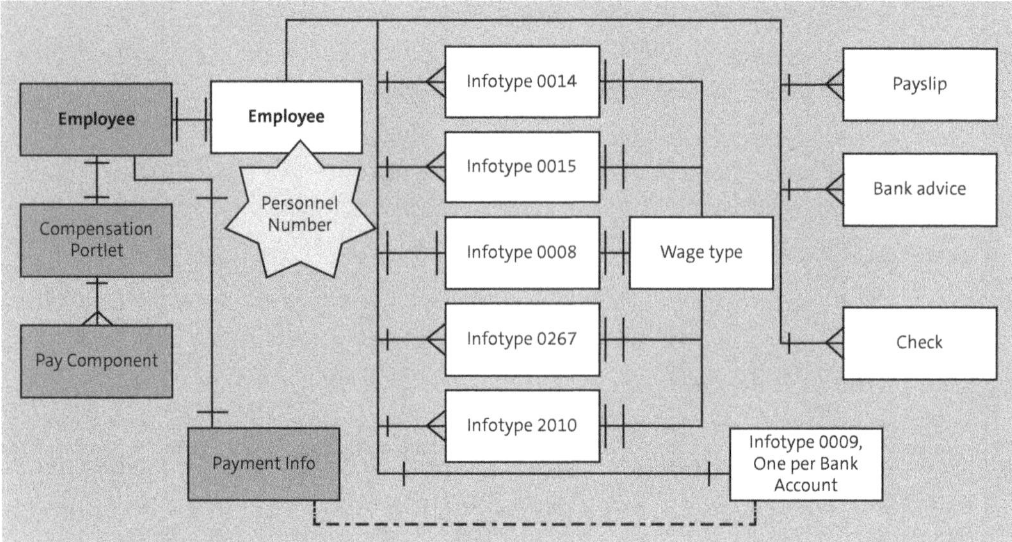

Figure 8.1 Pay Infotypes

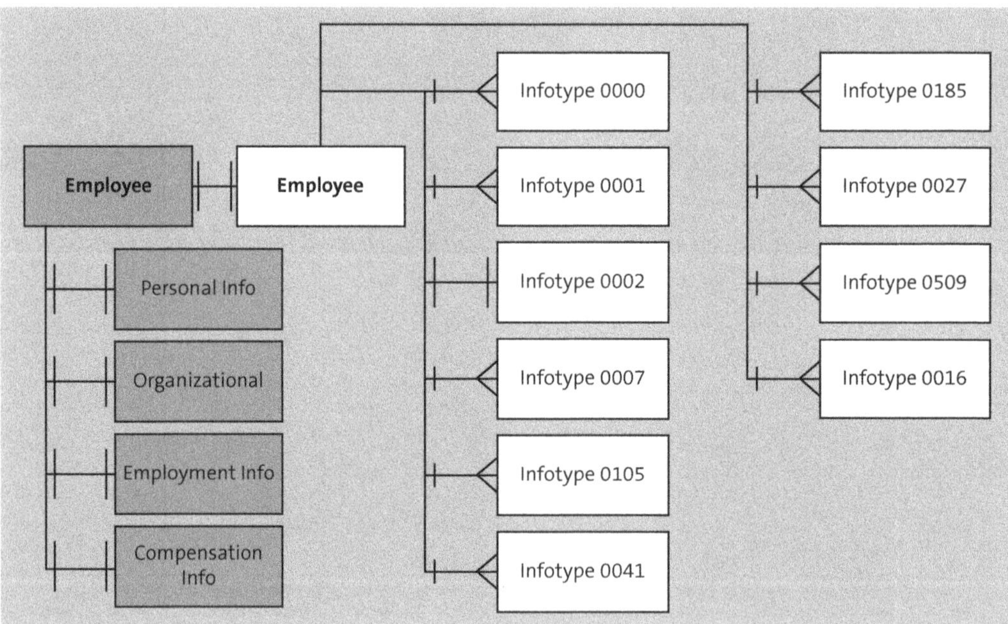

Figure 8.2 Non-Pay Infotypes

This entire chapter is devoted to discussing the infotypes shown in Figure 8.1 and Figure 8.2 and the complex data elements associated with them. Our discussion is organized by Employee Central portlets like the **Job Information**, **Compensation Information**, and **Personal Information** portlets. Each portlet is discussed in detail with a focus on the significance and characteristics of its target infotypes.

Let's begin our journey with a discussion of a key element that is a lynchpin for all these infotypes: the personnel number.

8.1 Maintaining Personnel Numbers

In this step, you determine how the personnel number will be created and maintained during replication. The personnel number is important because it is the single key that holds all the information about an employee in the system. Every infotype in Employee Central Payroll has this number as a key or as part of a key. Determining the approach for populating this data element is a task that requires prior planning and discussion to determine the best approach for your company.

A personnel number in Employee Central Payroll is up to 8 numeric characters. You have the following three options for generating personnel numbers:

- Using a sequentially generated number from the standard SAP number range
- Using a relevant identifier from Employee Central
- Using a custom identifier based on rules

To ensure data integrity, the replication of human resources (HR) master data from Employee Central to the Employee Central Payroll system is always recommended. Thus, if a unique number is created in the Employee Central Payroll system, this value is never passed back to Employee Central.

An employee may have more than one personnel number. For instance, a personnel number can be linked to one and only one company code at a time. This situation may arise in the following scenarios:

- Global employment
- Concurrent employment
- International transfer
- Rehire

In a global employment scenario, the employee retains a job in their home country as an expat but performs a different job in a host country as an inpat. In a concurrent employment scenario, the employee works in multiple locations at different times of the day (or different parts of the week or month), where these locations are under the jurisdiction of different legal entities. In both these situations, the employee is working for multiple entities, and the employee's number needs to be split. When this scenario arises, a strategy must be in place to link these numbers together so that the personnel numbers can be grouped together because they represent the same person.

For an international transfer, the recommendation is to terminate one employment and start another. In this case, the personnel number assigned previously can be retained. But the company may find that having two numbers is more beneficial from an organizational perspective, particularly if the personnel numbers are part of preallocated ranges. Having a meaningful personnel number is not a good practice in general, due to loss of technical flexibility—the meaning attached to a personnel number (such as which country it belongs to) can always be determined through other attributes. An integration nightmare can easily occur if the same number is used when the employee's job conditions change.

An employee leaving an organization for some time and then coming back to work is known as a *rehire*. Typical rehire scenarios maintain the same personnel number.

The default logic for replication is to check the mapping table HRSFEC_D_EEKEYMP, as shown in Figure 8.3. If a PERNR exists in the employee key mapping table, based on the BUKRS (company code) field for all effective Employee Central job information records, the replication uses the existing PERNR for the job information records. When no PERNR exists, a new PERNR is automatically created with the next sequence number.

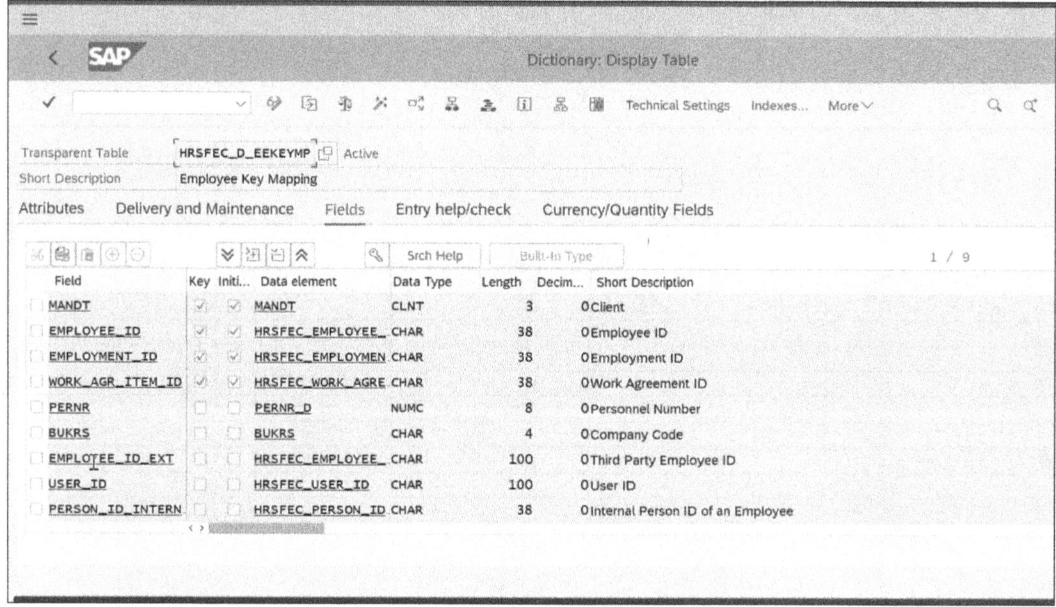

Field	Key	Initi...	Data element	Data Type	Length	Decim...	Short Description
MANDT	✓	✓	MANDT	CLNT	3	0	Client
EMPLOYEE_ID	✓	✓	HRSFEC_EMPLOYEE_	CHAR	38	0	Employee ID
EMPLOYMENT_ID	✓	✓	HRSFEC_EMPLOYMEN	CHAR	38	0	Employment ID
WORK_AGR_ITEM_ID	✓	✓	HRSFEC_WORK_AGRE	CHAR	38	0	Work Agreement ID
PERNR			PERNR_D	NUMC	8	0	Personnel Number
BUKRS			BUKRS	CHAR	4	0	Company Code
EMPLOYEE_ID_EXT			HRSFEC_EMPLOYEE_	CHAR	100	0	Third Party Employee ID
USER_ID			HRSFEC_USER_ID	CHAR	100	0	User ID
PERSON_ID_INTERN			HRSFEC_PERSON_ID	CHAR	38	0	Internal Person ID of an Employee

Figure 8.3 Default Personnel Number Mapping Table

An alternative method is to use the **Assignment ID** field from the Employee Information record in Employee Central, which is always unique for a job assignment and therefore creates the unique PERNR during the replication process—the sequence number is not triggered in this case. You can choose which option the replication process uses. By default, the replication engine will look for the number based on table HRSFEC_D_EEKEYMP.

In this section, we'll review two alternative methods for assigning a personnel number: using an assignment ID and using a business rule. We'll also discuss how to create a business rule to assign this vital number.

8.1.1 Using the Assignment ID

To use the assignment ID feature, you must first enable it from Employee Central by following these steps:

1. Log on to Employee Central.
2. Go to the Admin Center, click on **Employee Files**, and select **Manage Data**, as shown in Figure 8.4.

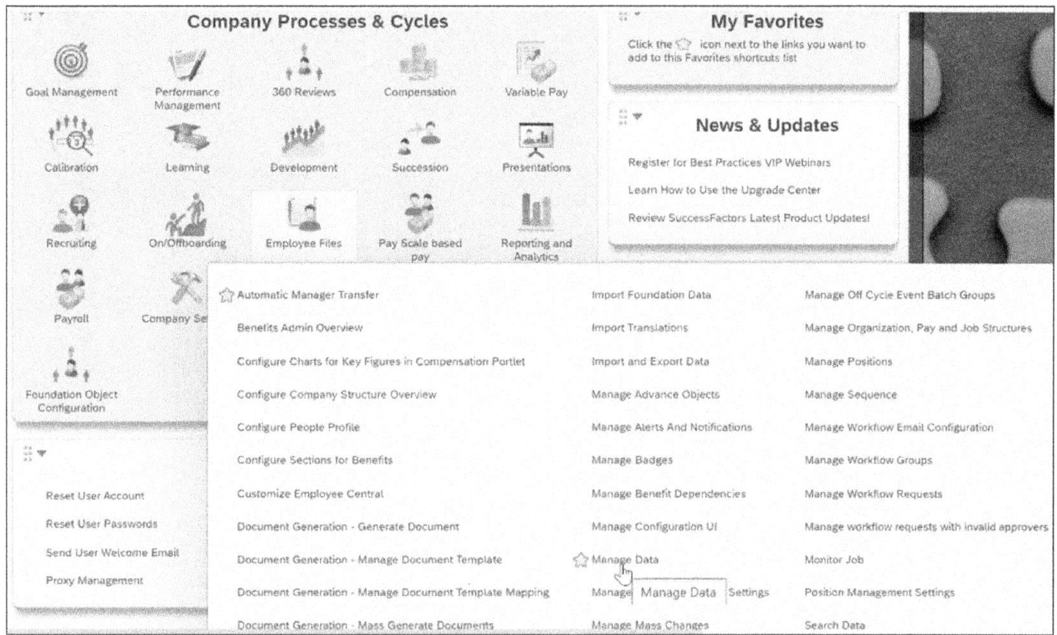

Figure 8.4 Admin Center: Manage Data

3. Click on **Create New** and choose **Sequence**, which opens the screen shown in Figure 8.5.

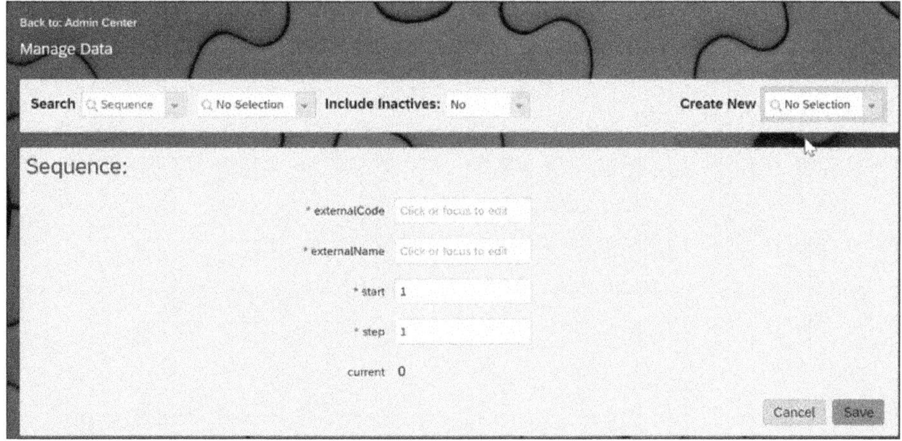

Figure 8.5 Sequence of Employee Numbers in Employee Central

4. Enter a meaningful technical title in the **externalCode** field (e.g., "EXTNUMBER_PERNR_00000000").
5. Enter a meaningful description in the **externalName** field.
6. Enter the beginning number in the **start** field.
7. Enter the increment in the **step** field. Typically, this value is "1" unless you would like to skip some numbers.
8. Enter "0" in the **current** field since numbers have not yet been assigned.
9. Click **Save**.
10. Next, enter "Configure Business Rules" in the search bar. The **Business Rules Admin** view will open, as shown in Figure 8.6.

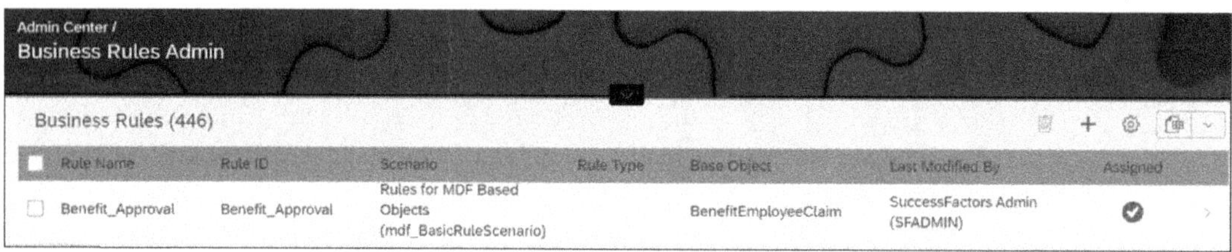

Figure 8.6 Business Rules

11. On the **Business Rules** toolbar, choose **+** (**Add**) to display rule scenario categories.
12. Expand the **Employee Central Core** selection and choose **Generate Assignment ID External**, which then displays a **Generate Assignment ID External** dialog box in the view.
13. Maintain the following fields and then click **Continue**:
    - **Rule Name**: Set to the **externalName** we created earlier.
    - **Rule ID**: Set to the **externalCode** we created earlier.
    - **Start Date**: Specify when the rule is executed.
    - **Description**: Add a meaningful description.
14. Click the **If** bar and check the **Always True**.
15. Under **Then**, select the **Edit** icon.
16. Use the dropdown to select **Set**, then select **Assignment ID** and choose the + to load more options.
17. Select **Assignment ID External** and **Retain** to be **Equal To**.
18. Use the **Text** dropdown to display additional options and select **Format Number()**, as shown in Figure 8.7.

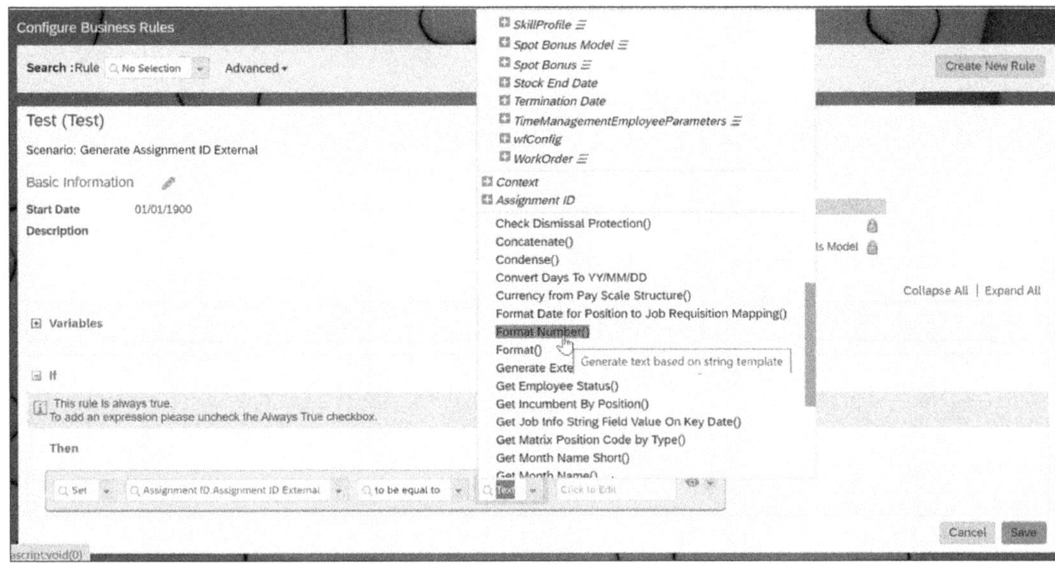

Figure 8.7 Business Rule for External Assignment ID

19. For the **Template**, retain **Text** and enter "%d" in the blank field.
20. For **Number**, use **Get Next Value ()** from the dropdown list.
21. For **Sequence**, retain **Value** and, in the dropdown list, select the sequence name you created in a previous step.
22. Choose **Save** at the bottom of the view.
23. Now, go back to **Business Rules Admin** screen.
24. Locate the business rule we created and choose the **Assigned** column to display the **Assignment of Rule** dialog box.
25. Choose **Assign Rule** and then choose the selection that displays to open a new view, as shown in Figure 8.8. Our example is a test.

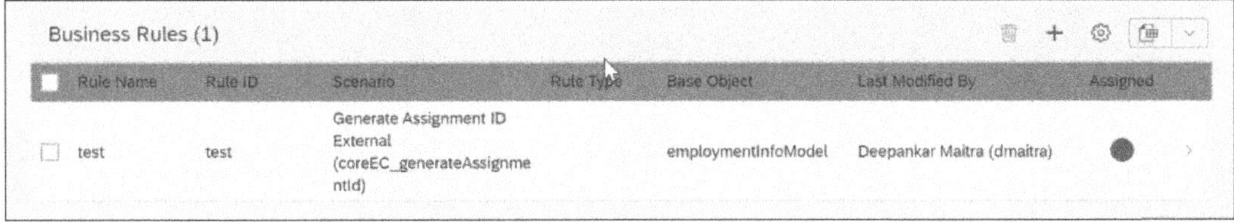

Figure 8.8 Assignment View after Clicking the Assignment Link

26. After clicking the **Assignment** link, select the checkbox for the rule and click on the green circle in the **Assigned** column, which will open the dialog box shown in Figure 8.9.

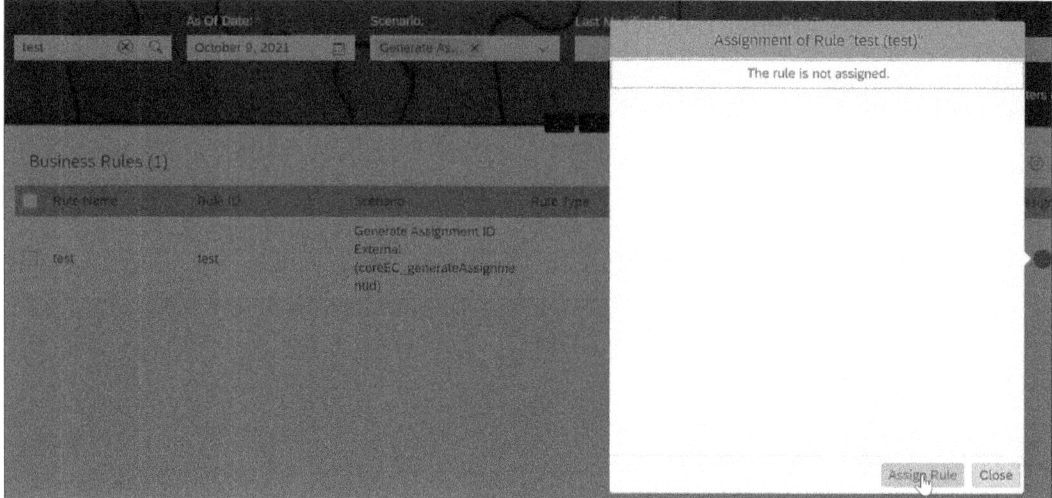

Figure 8.9 Rule Not Assigned

27. Click on the **Assign Rule** button to reveal the screen shown in Figure 8.10. A link to generate the external assignment ID will be displayed. Click on the right arrow icon.

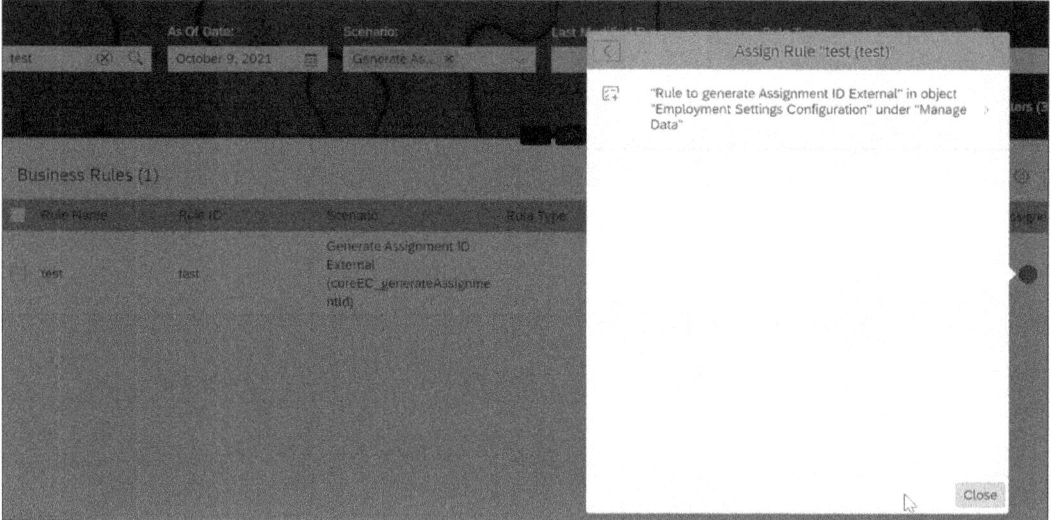

Figure 8.10 Assign Rule Test

28. After clicking this rule to generate the assignment ID, the **Manage Data** screen will automatically open.
29. For **Create New**, select **Employment Settings Configuration** from the dropdown list, and you'll see the screen shown in Figure 8.11.

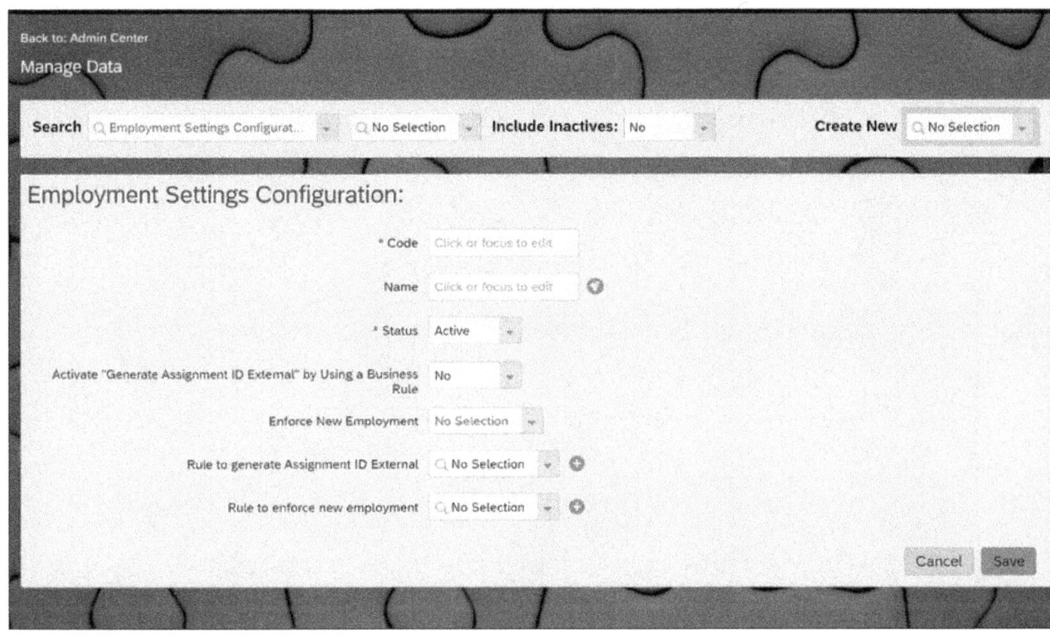

Figure 8.11 Employment Settings

30. Maintain the following fields on the **Employment Settings Configuration** screen:
    - **EntryCode**: EMPLOYMENTSETTINGSCONFIG
    - **Name**: Leave blank
    - **Status**: Active
    - **Activate "Generate Assignment ID External" by Using a Business Rule**: Yes
    - **Enforce New Employment**: On change of Legal Entity
    - **Rule to generate Assignment ID External**: Select the name of the rule we created from the dropdown list

Now, click **Save**. The business rule is activated and displays a green checkmark in the **Assigned** column of the **Business Rules Admin** view. By setting **Enforce New Employment**, you'll be ensuring that a new personnel number is created in Employee Central Payroll whenever the employee moves to a new company code.

Let's now switch to the steps you'll required in Employee Central Payroll. Follow these steps:

1. Log on to Employee Central Payroll.
2. Set feature NUMKR in Employee Central Payroll to **External**, as shown in Figure 8.12, using Transaction PE03. Enter **NUMKR** in the **Feature** field and click on **Change**.

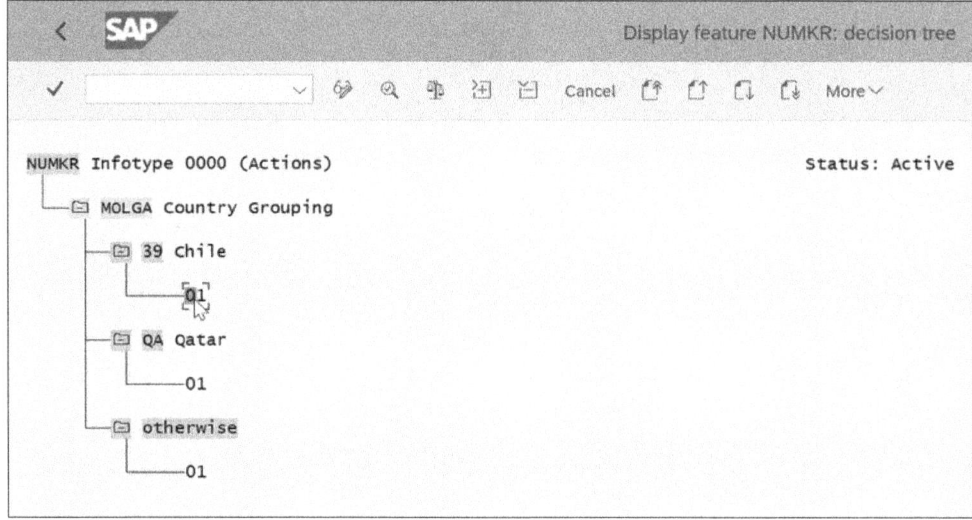

Figure 8.12 NUMKR Feature

3. Still in Employee Central Payroll, bring up the switches we discussed Chapter 7, Section 7.4, by using Transaction SM30 with table T77S0, as shown in Figure 8.13.

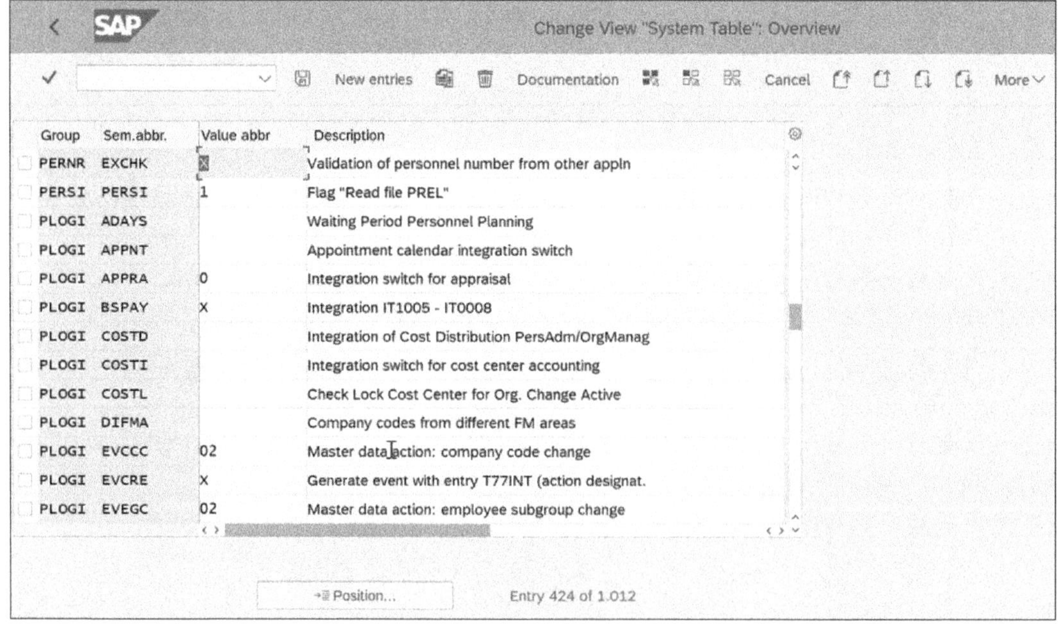

Figure 8.13 T77S0 Setting for Personnel Number

4. Find the group PERNR and set the **Value Abbr.** field to “X.”

5. Save these settings.

8.1.2 Enforcing New Employment Using Business Rules

Some variations of the situation we just discussed may require a new employment event, sometimes with a new number and sometimes not. Table 8.1 lists and describes a few such situations.

Scenario	Requirement	Solution
An employee leaves a company and joins back again in another entity in another country.	Since a new country is involved, a new personnel number is required.	A business rule is created to create a new employment event with a new employee ID in Employee Central, which replicates to Employee Central Payroll and represents the personnel number.
An employee is transferred to another country but then is terminated. Returns to original country and company and is hired back.	Transferring to new country requires a new personnel number, as in the previous scenario. But transferring back should use the same personnel number.	A business rule is created to enforce the same employee ID in Employee Central upon the employee's return, which replicates to Employee Central Payroll and represents the personnel number.
An employee transfers to another paying company in same country.	Since the paying company is different, the personnel number must be different.	A business rule is created to create a new employment with a new employee ID in Employee Central, which replicates to Employee Central Payroll and represents the personnel number.
An employee transfers to another paying company in same country but then comes back to original company.	While transferring to a different company, the employee needed a new personnel number. But upon returning, the employee must have the same personnel number as before.	A business rule is created to enforce the same employee ID in Employee Central upon the employee's return, which replicates to Employee Central Payroll and represents the personnel number.

Table 8.1 Scenarios for Personnel Number Assignment

To create a business rule to enforce new employment, follow these steps:

1. Create an error message from Employee Central by first going to the Admin Center.
2. Click on **Employee Files • Manage Data**.
3. Click on **Create New** field and choose **Message Definition**. You'll see the screen shown in Figure 8.14.

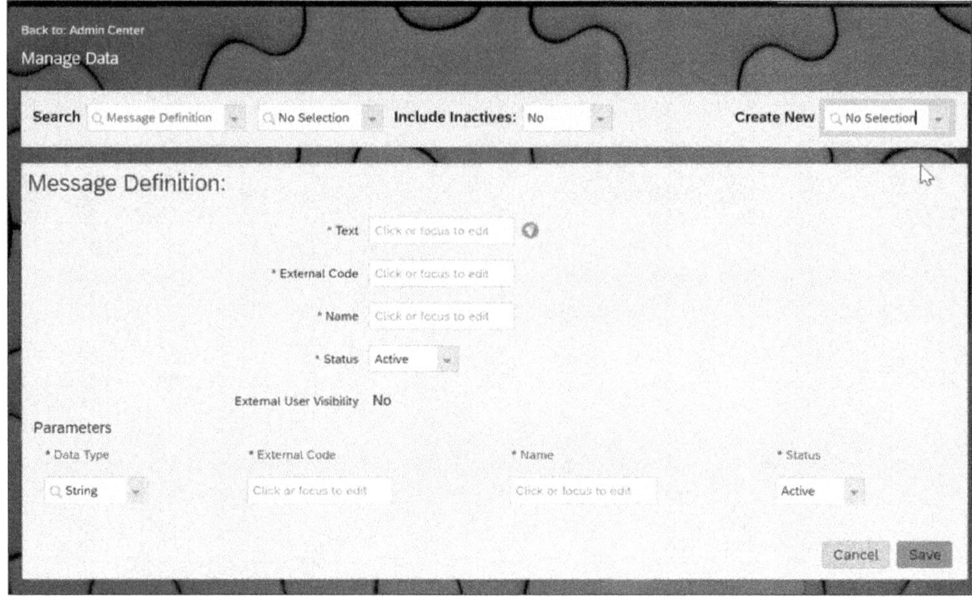

Figure 8.14 Error Message Definition

4. Maintain the following fields:
   - **Text**: Meaningful text message such as "Sorry this number cannot be assigned again."
   - **External Code**: A code that represents the message, for example, "Duplicate_Personnel_Number."
   - **Name**: Same as the **External Code** field.
   - **Status**: Select **Active**.
   - **Parameters**: Leave blank.
5. Click **Save**.
6. Return to the Admin Center and click on **Tools** and then enter "Configure Business Rules" into the search bar. You'll see the screen shown earlier in Figure 8.6.
7. On the **Business Rules** toolbar, choose **+ (Add)** to display the available categories for rule scenarios.
8. Expand the **Employee Central Core** selection, choose **Expand the Employee Central Core** category, and select **Enforce New Employment for Rehire**. The **Enforce New Employment for Rehire** dialog box appears.
9. Maintain the following fields and then click **Continue**:
   - **Rule Name:** Set to a suitable name
   - **Rule ID:** Assigned by the system automatically
   - **Start Date**: When the rule is executed
   - **Description**: A meaningful description

10. Click the **If** bar and set this option according to your needs, as shown in Figure 8.15.

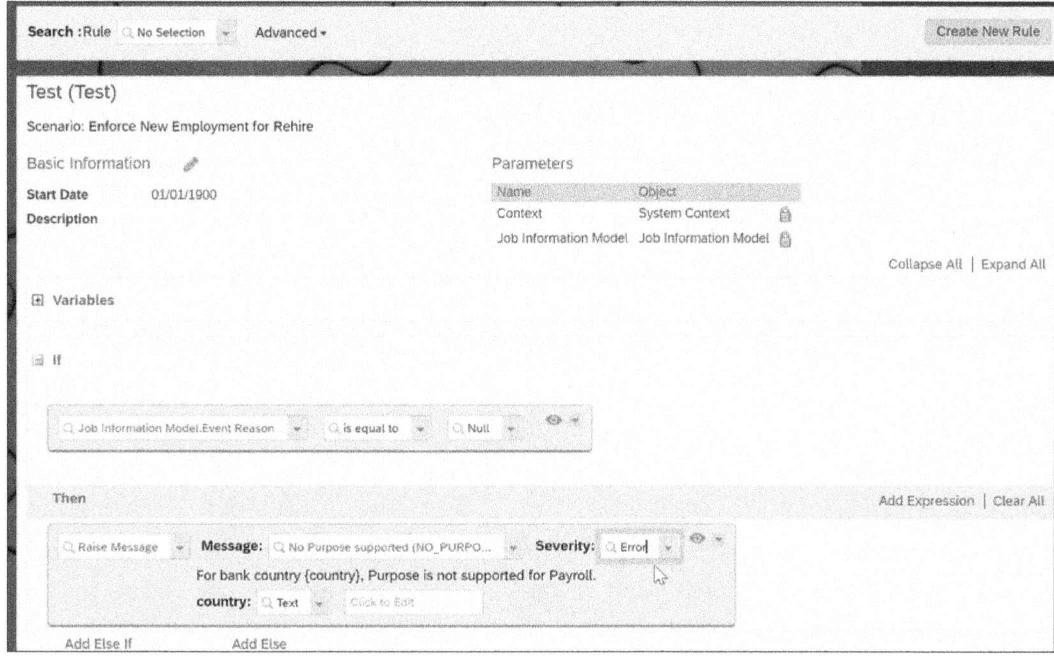

Figure 8.15 Business Rule with Error

11. Click **Save** at the bottom of the view.

On the Employee Central Payroll side, the data model expands to maintain integrity with Employee Central, as shown in Figure 8.16.

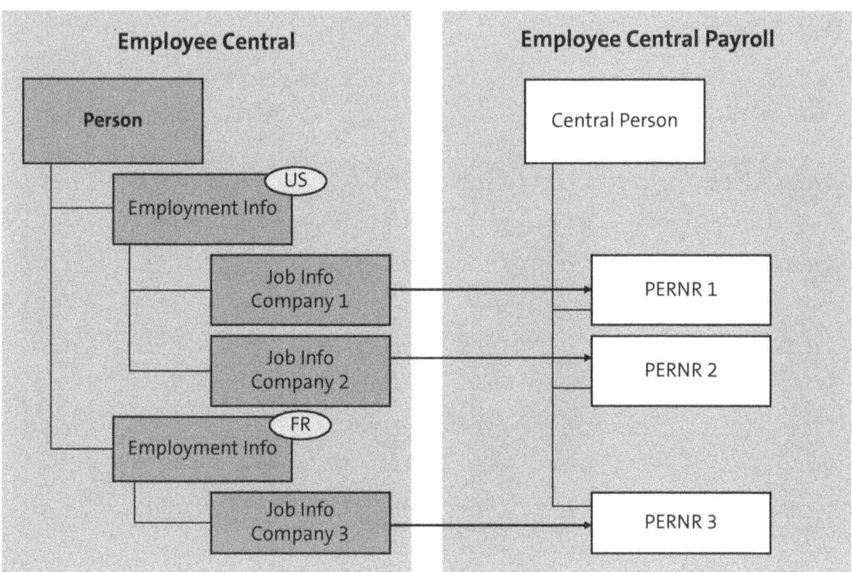

Figure 8.16 Data Model for Employee Central Payroll

Some points to remember about multiple personnel numbers for the same employee include the following:

- Multiple employments can result in multiple personnel numbers in Employee Central Payroll.
- Even the same employment can result in multiple personnel numbers, for instance, when a legal entity changes—in Employee Central Payroll, this feature is standard.
- The central person object keeps personnel numbers grouped together for reporting and other Employee Central Payroll processes.

8.2 Job Information: Infotypes 0000, 0001, 0007, and 0008

Let's now review the fields that are replicated from the **Job Information** portlet of Employee Central into Employee Central Payroll. We'll discuss each infotype and the fields that are derived from the **Job Information** portlet. Infotype 0000 deals with employee events such as being hired, receiving a promotion, and more. A list of these values must be mapped. Similarly—in infotype 0001, some fields such as job, position and organization unit, and location will need closer attention to fully understand how these structures are represented in Employee Central Payroll and to handle the special configuration required for these fields. We'll discuss these fields in detail after reviewing these infotypes as well as review data movements from the **Job Information** portlet in Employee Central.

Five infotypes receive their data from this portlet. We'll cover four now, and the last one—infotype 0016—later in Section 8.11.

8.2.1 Infotype 0000

Infotype 0000 represents the events that take place in an employee's employment cycle that are relevant for payroll. Table 8.2 lists the important fields of this infotype.

Field	Description	Mandatory/Optional
MANDT	Client	Mandatory
PERNR	Personnel number	Mandatory
SUBTY	Subtype	Mandatory
OBJPS	Object Identification	Mandatory
SPRPS	Lock Indicator for HR Master Data Record	Mandatory
ENDDA	End Date	Mandatory
BEGDA	Start Date	Mandatory

Table 8.2 Fields in Infotype 0000

Field	Description	Mandatory/Optional
SEQNR	Number of Infotype Record with Same Key	Mandatory
AEDTM	Changed On	Mandatory
UNAME	Name of Person Who Changed Object	Mandatory
HISTO	Historical Record Flag	Mandatory
ITXEX	Text Exists for Infotype	Mandatory
REFEX	Reference Fields Exist (Primary/Secondary Costs)	Mandatory
ORDEX	Confirmation Fields Exist	Mandatory
ITBLD	Infotype Screen Control	Mandatory
PREAS	Reason for Changing Master Data	Mandatory
RESE2	Reserved Field/Unused Field of Length 2	Mandatory
GRPVL	Grouping Value for Personnel Assignments	Mandatory
MASSN	Action Type	Mandatory
MASSG	Reason for Action	Optional
STAT1	Customer-Specific Status	Optional
STAT2	Employment Status	Mandatory
STAT3	Special Payment Status	Mandatory

Table 8.2 Fields in Infotype 0000 (Cont.)

Of all these fields, only the fields listed in Table 8.3 are inherited from the **Job Information** portlet in Employee Central.

Target Technical Name	Max. Length	Default Mapping Mode	Is Code Value Mapping Country/Region Specific?	Field ID	Field Label
MASSN	2	No mapping	No	event	Personal Event Code
MASSG	2	Customer table	No	event-reason	Event Reason

Table 8.3 Fields in Infotype 0000 Inherited from Employee Central

As listed in Table 8.3, the event field maps to **Action Type** field in Employee Central Payroll and does not change in value. The **Action Reason** field is derived from a mapping table based on business requirements.

8.2.2 Infotype 0001

Infotype 0001 represents the organizational elements that are relevant for payroll. Table 8.4 lists the important fields of this infotype.

Field	Description	Mandatory/Optional
MANDT	Client	Mandatory
PERNR	Personnel number	Mandatory
SUBTY	Subtype	Mandatory
OBJPS	Object Identification	Mandatory
SPRPS	Lock Indicator for HR Master Data Record	Mandatory
ENDDA	End Date	Mandatory
BEGDA	Start Date	Mandatory
SEQNR	Number of Infotype Record with Same Key	Mandatory
AEDTM	Changed On	Mandatory
UNAME	Name of Person Who Changed Object	Mandatory
HISTO	Historical Record Flag	Mandatory
ITXEX	Text Exists for Infotype	Mandatory
REFEX	Reference Fields Exist (Primary/Secondary Costs)	Mandatory
ORDEX	Confirmation Fields Exist	Mandatory
ITBLD	Infotype Screen Control	Mandatory
PREAS	Reason for Changing Master Data	Mandatory
RESE1	Reserved Field/Unused Field of Length 2	Mandatory
RESE2	Reserved Field/Unused Field of Length 2	Mandatory
GRPVL	Grouping Value for Personnel Assignments	Mandatory
BUKRS	Company Code	Mandatory
WERKS	Personnel Area	Mandatory
PERSG	Employee Group	Mandatory
PERSK	Employee Subgroup	Mandatory
VDSK1	Organizational Key	Mandatory

Table 8.4 Fields in Infotype 0001

Field	Description	Mandatory/Optional
GSBER	Business Area	Optional
BTRTL	Personnel Subarea	Optional
JUPER	Legal Person	Optional
ABKRS	Payroll Area	Mandatory
ANSVH	Work Contract	Optional
KOSTL	Cost Center	Mandatory
ORGEH	Organizational Unit	Mandatory
PLANS	Position	Mandatory
STELL	Job	Mandatory
MSTBR	Supervisor Area	Optional
SACHA	Payroll Administrator	Optional
SACHP	Administrator for HR Master Data	Mandatory
SACHZ	Administrator for Time Recording	Optional
SNAME	Employee's Name (sortable by last name and first name)	Mandatory
ENAME	Formatted Name of Employee or Applicant	Mandatory
OTYPE	Object Type	Mandatory
SBMOD	Administrator Group	Mandatory
KOKRS	Controlling Area	Optional

Table 8.4 Fields in Infotype 0001 (Cont.)

Of all these fields, only the fields listed in Table 8.5 are inherited from the **Job Information** portlet.

Target Technical Name	Max. Length	Default Mapping Mode	Is Code Value Mapping Country/ Region Specific?	Field ID	Field Label
WERKS	4	No mapping	No	location	Location
BTRTL	4	No mapping	No	location	Location
BUKRS	4	No mapping	No	company	Company

Table 8.5 Fields in Infotype 0001 Inherited from Employee Central

Target Technical Name	Max. Length	Default Mapping Mode	Is Code Value Mapping Country/ Region Specific?	Field ID	Field Label
KOSTL	10	No mapping	No	cost-center	Cost Center
PERSG	1	Customer mapping	No	employee-class	Employee Class
PERSK	2	Customer mapping	Yes	employment-type	Employment Type
ANSVH	2	Customer mapping	Yes	contract-type	Contract Type
STELL	8	No mapping	No	job-code	Job Classification
PLANS	8	No mapping	No	position	Position
ORGEH	8	No mapping	No	business-unit department division	Business Unit Department Division
SBMOD	4		Filled with country grouping	Not in Employee Central	Not in Employee Central
SACHP	3		Filled from table T526	Not in Employee Central	Not in Employee Central
VDSK1	14		Filled with cost center and personnel area	Not in Employee Central	Not in Employee Central

Table 8.5 Fields in Infotype 0001 Inherited from Employee Central (Cont.)

The following fields are filled in automatically based on certain rules:

- Field SBMOD (Administrator Group) is filled in based on feature PINCH, as shown in Figure 8.17. If feature PINCH is not used or has a default (****) setting, this field is populated with the personnel area determined after translating the **Location** field, as listed in Table 8.5.
- Field SACHP (Administrator for HR Master Data) is populated with the first value in table T526 pertaining to the administrator group determined from field SBMOD.
- Field VDSK1 (Organizational Key) is automatically set to the concatenated value of cost center and personnel area.

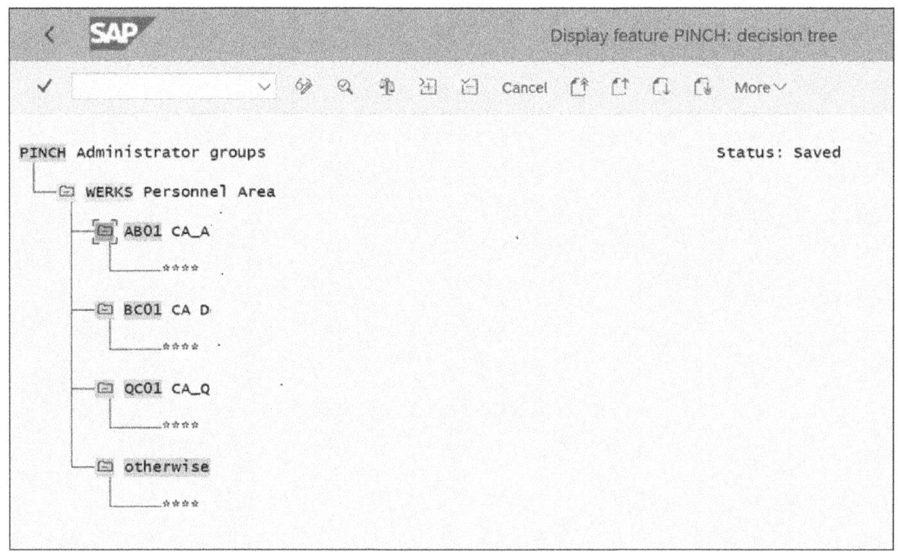

Figure 8.17 PINCH Feature

8.2.3 Infotype 0007

Infotype 0007 represents the work schedules that are relevant for payroll. Note that these schedules may not be the same as the real work schedules that are used in your landscape's time management system, since time rules require granular determination of an employee's planned working time.

The infotype 0007 record has fields that are interrelated, which you'll see in the next few pages. Sometimes, replicating this infotype doesn't make sense, so we'll cover how to turn replication off just for this infotype if required.

Infotype Fields

Table 8.6 lists the important fields of this infotype.

Field	Description	Mandatory/Optional
MANDT	Client	Mandatory
PERNR	Personnel number	Mandatory
SUBTY	Subtype	Mandatory
OBJPS	Object Identification	Mandatory
SPRPS	Lock Indicator for HR Master Data Record	Mandatory
ENDDA	End Date	Mandatory

Table 8.6 Fields in Infotype 0007

Field	Description	Mandatory/Optional
BEGDA	Start Date	Mandatory
SEQNR	Number of Infotype Record with Same Key	Mandatory
AEDTM	Changed On	Mandatory
UNAME	Name of Person Who Changed Object	Mandatory
HISTO	Historical Record Flag	Mandatory
ITXEX	Text Exists for Infotype	Mandatory
REFEX	Reference Fields Exist (Primary/Secondary Costs)	Mandatory
ORDEX	Confirmation Fields Exist	Mandatory
ITBLD	Infotype Screen Control	Mandatory
PREAS	Reason for Changing Master Data	Mandatory
GRPVL	Grouping Value for Personnel Assignments	Mandatory
SCHKZ	Work Schedule Rule	Mandatory
ZTERF	Employee Time Management Status	Mandatory
EMPCT	Employment percentage	Mandatory
MOSTD	Monthly hours	Mandatory
WOSTD	Hours per week	Mandatory
ARBST	Daily Working Hours	Mandatory
WKWDY	Weekly Workdays	Mandatory
JRSTD	Annual working hours	Mandatory
TEILK	Indicator Part-Time Employee	Optional
MINTA	Minimum number of work hours per day	Optional
MAXTA	Maximum number of work hours per day	Optional
MINWO	Minimum weekly working hours	Optional
MAXWO	Maximum number of work hours per week	Optional
MINMO	Minimum number of work hours per month	Optional

Table 8.6 Fields in Infotype 0007 (Cont.)

Of all these fields, only the fields listed in Table 8.7 are inherited from the **Job Information** portlet.

Target Technical Name	Max. Length	Default Mapping Mode	Is Code Value Mapping Country/Region Specific?	Field ID	Field Label
TEILK	1	No Mapping	No	is-fulltime-employee	Fulltime employee indicator
WKWDY	4	No Mapping	No	workingDaysPerWeek	Working days per week
SCHKZ	8	No Mapping	No	workscheduleID	Work Schedule ID

Table 8.7 Fields in Infotype 0007 Inherited from Employee Central

8

Payroll Switch WKWDY

Switch WKWDY can be used to turn off replication if infotype 0007 is being maintained directly in Employee Central Payroll.

Bring up the switches (previously discussed in Chapter 7) by using Transaction SM30 with table T77S0, as shown in Figure 8.18. Find **SFEC** in the **Group** column and **WKWDY** in the **Sem.abbr.** field.

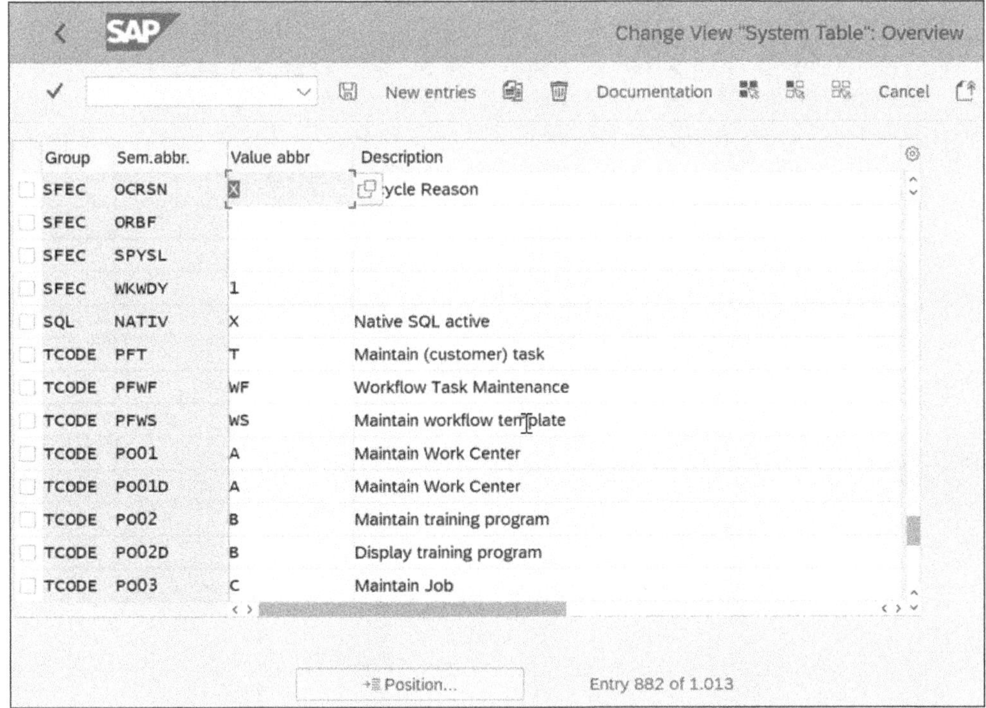

Figure 8.18 Switch for Work Schedule

Set the **Value Abbr** to one of the values listed in Table 8.8, depending on your use case and then save these settings.

Value of Switch WKWDY	Description
Empty	Infotype 0007 is not replicated from Employee Central. A warning message is displayed.
1	The replication stops if the **Working Days Per Week** field is empty for the selected PERNRs. An error message is displayed.
2	Infotype 0007 is replicated if the **Working Days Per Week** field is empty for the selected PERNRs. No error message is displayed.

Table 8.8 Switch WKWDY Values

Working Hours

The following two fields are related to working hours:

- Field ARBST (Daily Working Hours), which is derived from Employee Central standard weekly hours divided by Employee Central working days per week.
- Field EMPCT (Employment Percent), which is derived from the daily working hours and the work schedule rule (field SCHKZ). This field is also used by three other fields, as follows:
  - Weekly Working Hours (field WOSTD), derived from field EMPCT multiplied by its value in table T508A for the work schedule rule
  - Monthly Working Hours (field MOSTD), derived from field EMPCT multiplied by its value in table T508A for the work schedule rule
  - Annual Working Hours (field JRSTD), derived from field EMPCT multiplied by its value in table T508A for the work schedule rule

Whether these fields can be adjusted is based on a setting in feature WRKHR, as shown in Figure 8.19. Use Transaction PE03 to access this screen. Enter "WRKHR" in the **Feature** field and click on **Change**.

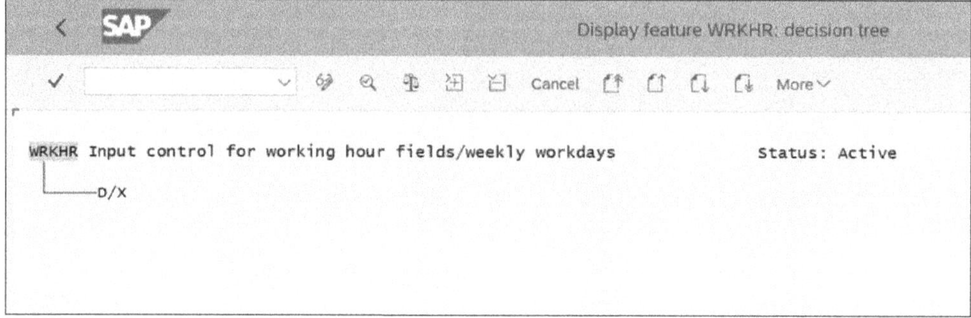

Figure 8.19 Working Hours Feature

The value this feature returns is in the format *<a>*/*<b>*. If *<a>* is D, the **Working Hours per Day** field is enterable, and all the other fields (per week, per month, and per year) are derived. The value *a* can be D (daily), M (Monthly), W (weekly), or Y (yearly/annual). For example, changing this value to M/X will set the **Monthly Working Hours** field as enterable and the daily, weekly, and annual values will be derived from. Changing *a* to Y/X will set the **Annual Working Hours** field as enterable, and the daily, monthly, and weekly values will be derived.

Work Schedule ID

The replication of the work schedule ID is created through a mapping in table T77SFEC_CVMAPC.

We recommend creating a NORM work schedule to represent a generic planned working time for use in prorated payroll for salaried employees, when required.

If the Employee Central value or * (asterisk) is mapped to an empty cell in table T77SFEC_CVMAPC, table T508A is evaluated with parameters based on the **Personnel Area** and **Personnel Subarea** of infotype 0001, as follows:

- If NORM is found in table T508A, infotype 0007 is created with work schedule NORM.
- If NORM is not found in table T508A, then infotype 0007 is created based on the first hit in table T508A after sorting by field SCHKZ.

Working Week

Working week (WWEEK) values can be derived in two ways:

- Through a custom field in Employee Central that is then mapped to infotype 0007 based upon basic extensibility as explained in Chapter 7, Section 7.2
- Using a combination of the Business Add-In (BAdI) HRSFEC_B_CE_CHANGE_IT0007 and feature WWEEK

8.2.4 Infotype 0008

Infotype 0008 represents the compensation data of an employee. This infotype stores an employee's rate of pay that Employee Central Payroll uses to calculate basic pay by multiply the rate by the number of hours worked. This infotype does not store the other earnings besides base pay—additional pay is stored in infotype 0014 or 0015. Table 8.9 lists the important fields of this infotype.

Field	Description	Mandatory/Optional
MANDT	Client	Mandatory
PERNR	Personnel number	Mandatory

Table 8.9 Fields in Infotype 0008

Field	Description	Mandatory/Optional
SUBTY	Subtype	Mandatory
OBJPS	Object Identification	Mandatory
SPRPS	Lock Indicator for HR Master Data Record	Mandatory
ENDDA	End Date	Mandatory
BEGDA	Start Date	Mandatory
SEQNR	Number of Infotype Record with Same Key	Mandatory
AEDTM	Changed On	Mandatory
UNAME	Name of Person Who Changed Object	Mandatory
HISTO	Historical Record Flag	Mandatory
ITXEX	Text Exists for Infotype	Mandatory
REFEX	Reference Fields Exist (Primary/Secondary Costs)	Mandatory
ORDEX	Confirmation Fields Exist	Mandatory
ITBLD	Infotype Screen Control	Mandatory
PREAS	Reason for Changing Master Data	Mandatory
GRPVL	Grouping Value for Personnel Assignments	Mandatory
TRFAR	Pay Scale Type	Mandatory
TRFGB	Pay Scale Area	Mandatory
TRFGR	Pay Scale Group	Mandatory
TRFST	Pay Scale Level	Mandatory
WAERS	Currency Key	Mandatory
BSGRD	Capacity Utilization Level	Mandatory
DIVGV	Working Hours per Payroll Period	Mandatory
ANSAL	Annual salary	Mandatory
FALGR	Case group	Optional
LGA01	Wage Type	Mandatory
BET01	Wage Type Amount for Payments	Mandatory
IND01	Indicator for indirect valuation	Optional
LGA40	Wage Type	Mandatory

Table 8.9 Fields in Infotype 0008 (Cont.)

Field	Description	Mandatory/Optional
BET40	Wage Type Amount for Payments	Mandatory
IND40	Indicator for indirect valuation	Optional
ANCUR	Currency Key for Annual Salary	Mandatory

Table 8.9 Fields in Infotype 0008 (Cont.)

Of all these fields, only the fields listed in Table 8.10 are inherited from the **Job Information** portlet.

Target Technical Name	Max. Length	Default Mapping Mode	Is Code Value Mapping Country/ Region Specific?	Field ID	Field Label
TRFAR	2	Custom Table	Yes	payScaleType	Pay scale type
TRFGB	2	Custom Table	Yes	payScaleArea	Pay scale area
TRFGR	8	Custom Table	Yes	payScaleGroup	Pay scale group
TRFST	2	Custom Table	Yes	payScaleLevel	Pay scale level

Table 8.10 Fields in Infotype 0008 Inherited from Employee Central

8.2.5 Events

Events in the Employee Central Payroll system are mapped to actions in Employee Central. The standard mapping brings in this data without any transformation. Table 8.11 shows the standard map for prominent events. Events and event reasons are available in the **Job Information** portlet.

Employee Central	Employee Central Payroll
H: Hire	01: Hiring
26: Termination	10: Leaving
R: Rehire	12: Re-entry into company
GA: Global assignment	01: Hiring
EGA: End global assignment	10: Leaving

Table 8.11 Event-to-Action Map

Next, let's turn to the concept of artificial actions and payroll events. We'll also explore what happens to the **Action** field when an employee does not show up after being hired.

Artificial Actions

In some situations, an action is automatically created in the Employee Central Payroll system, for instance, in the following cases:

- No hire records exist in the **Job Information** portlet in Employee Central for an employee: A hire action is automatically created in Employee Central Payroll.
- A hire record comes in for an employee that already exists: A termination and rehire action is automatically created to keep the integrity of the data intact since you cannot have a hire record for an employee that already exists.
- A change of location takes place, but the event in Employee Central is not mapped to an action in Employee Central Payroll: An organizational reassignment action is automatically inserted.

An organizational reassignment action can be reprogrammed to something else if desired in Transaction SM30.

Find **SFEC** in the **Group** column with **OCHAT** in the **Sem. Abbr.** column and set the action of your choice in the **Value Abbr.** field, as shown in Figure 8.20.

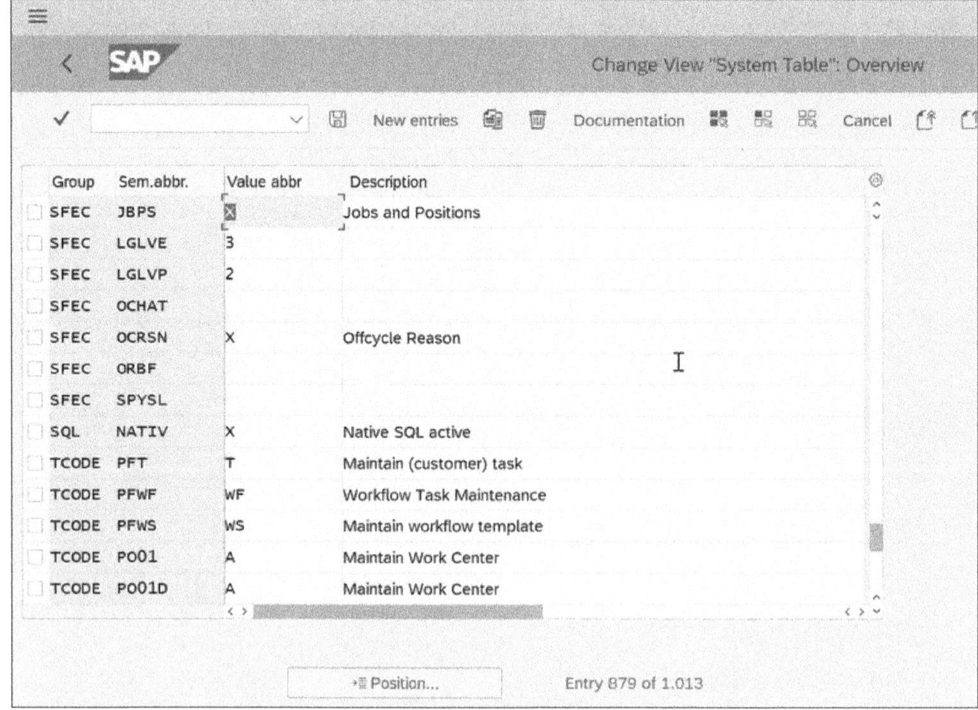

Figure 8.20 Organizational Reassignment Event

Payroll Events

Payroll events are a relatively new feature for tackling two complex scenarios that occur from time to time:

- Two status-changing events taking place on the same day: This case can arise when an employee is changing companies, and changing the personnel number is not desirable. Thus, the person must remain active and not be terminated since this case happens on the same day that the person joins the new company.
- One event with two different statuses desired for payroll purposes: This case is commonly encountered in the event an employee is on leave. The leave of absence is only one event, but the status could be different for payroll depending upon whether the leave is paid or unpaid. Previously, this scenario was difficult to implement because an event could be mapped to only one action. Now, you can map two separate payroll events to this action and identify them in Employee Central with two different event reasons. These statuses map to the respective action—paid or unpaid—in Employee Central Payroll.

To configure payroll events, follow these steps:

1. In Employee Central, go to the Admin Center and go to **Employee Files • Manage Organization, Pay, and Job Structures**, which opens the screen shown in Figure 8.21.

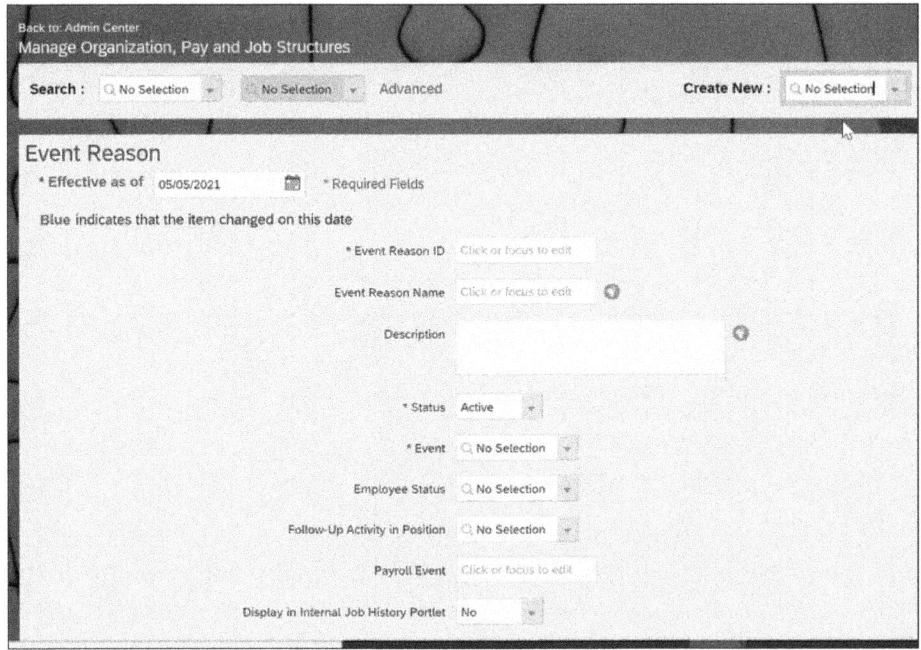

Figure 8.21 Event Reason

2. Select **Create New • Event Reason** and create and save the two new events shown in Table 8.12.

Event	Field	Field Value
Event termination	**Event Reason**	Org assignment rehire (termination)
	Event Reason ID	Org asgt Termination
	Event Reason Name	Org assignment Termination
	Status	Active
	Event	Termination
	Payroll Event	02
	Display in Internal Job History Portlet	Yes
Event rehire	**Event Reason**	Org assignment rehire (org assgt rehire)
	Event Reason ID	Org asgt Rehire
	Event Reason Name	Org assignment rehire
	Status	Active
	Event	Rehire
	Payroll Event	02
	Display in Internal Job History Portlet	Yes

Table 8.12 Setting Event Reason Fields

3. Now, log on to Employee Central Payroll.
4. Use Transaction SM30 to bring up table T529A, as shown in Figure 8.22.
5. Create an organizational reassignment action by clicking on **New Entries** and maintaining the following details:
   - **Action**: Enter "02."
   - **Name**: Enter "Organizational Assignment."
   - **FC** (function character for action): Enter "O."
6. Save these settings.
7. Still in Employee Central Payroll, map the two event reasons we just created to the new Employee Central Payroll action. You can perform this step in Transaction SM30, again for table T77SFEC_CVMAPC, as shown in Figure 8.23. Once completed, save these settings.

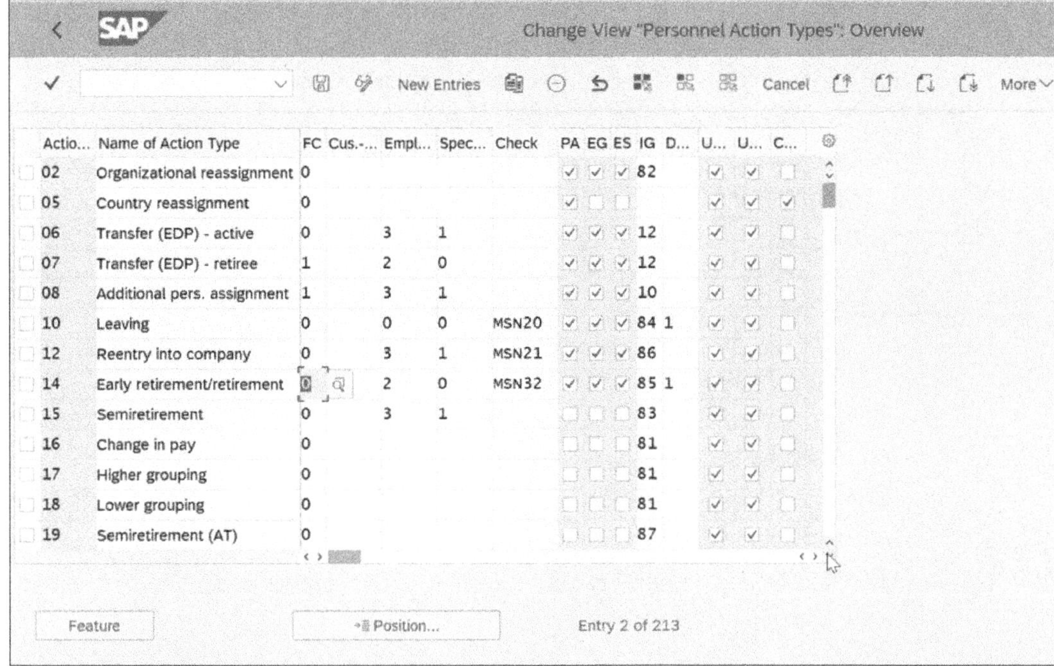

Figure 8.22 Table T529A

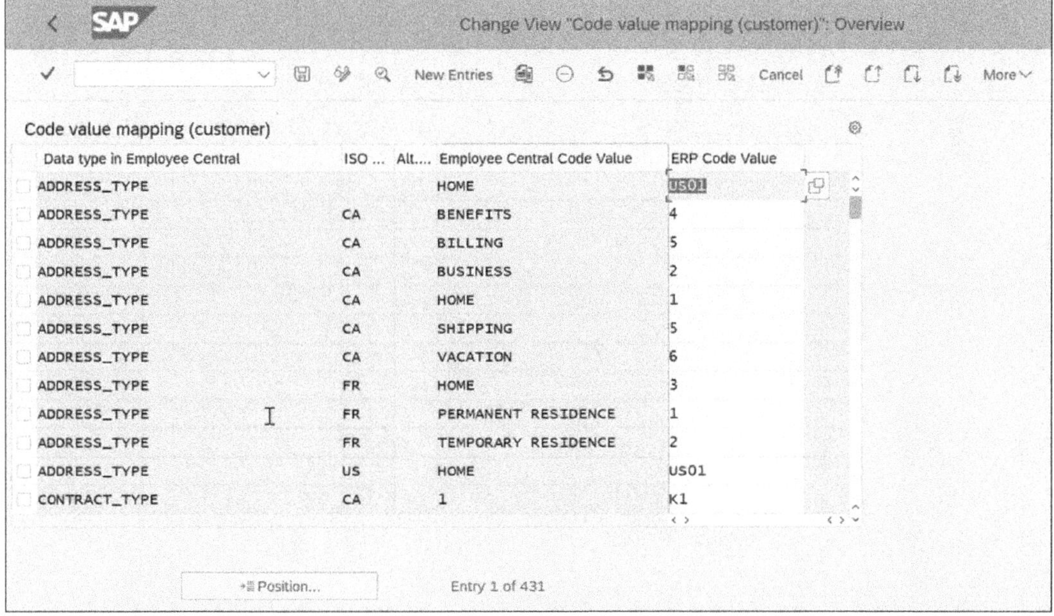

Figure 8.23 Table T77SFEC_CVMAPC

You can use a similar strategy for paid and unpaid leave. Simply define two separate payroll events for the paid and unpaid event reasons using the screen shown earlier in Figure 8.21. Then, using the technique shown in Figure 8.23, map these event reasons to the two separate events with the different statuses as desired.

Multiple Actions on the Same Day

Infotype 0000 is used to record effectively dated primary events (i.e., events that change status like hire and termination) in Employee Central Payroll. However, this infotype does not allow multiple events on the same day. However, Employee Central does allow multiple actions on the same day. To capture these additional events, you'll need to configure infotype 0302 (Additional Actions). SAP designed this infotype to hold values for actions on the same day that cannot be stored in infotype 0000. Follow these steps to configure infotype 0302:

1. Check the settings in table T529A using Transaction SM30 and ensure that the **Update IT 0302** flag has been selected for all actions.
2. Run program RPUEVSUP using Transaction SE38. You'll see the screen shown in Figure 8.24. This step will create infotype 0302 for all employees.

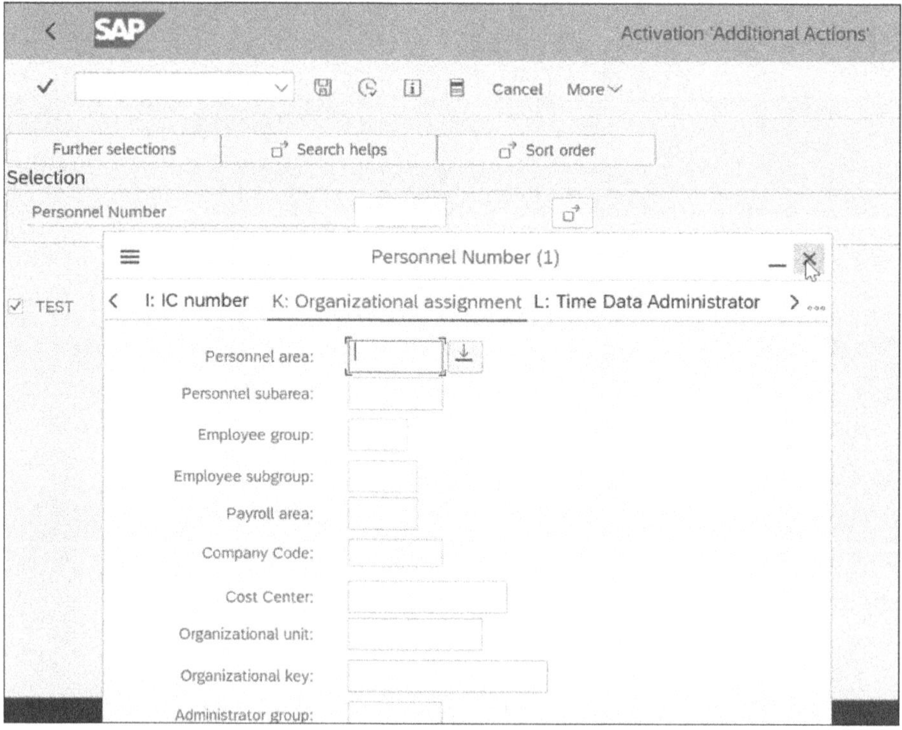

Figure 8.24 Program to Create Additional Actions with Infotype 0302

3. In Transaction SPRO, follow the menu path **Personnel Management • Integration Settings for SuccessFactors Employee Central Payroll • Configuration of Point-to-Point Replication.**
4. Double-click the configuration ID you want to use to open it, as shown in Figure 8.25.
5. Select the **Mult. Actions** checkbox and save these settings.

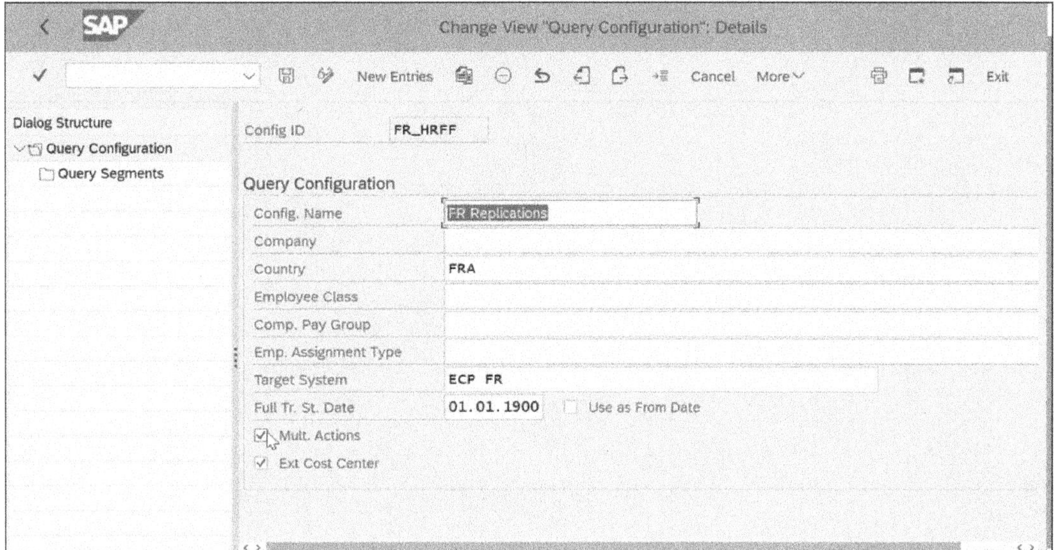

Figure 8.25 Multiple Actions Configuration

No-Shows

A "no-show" is a situation when a prospective employee has been offered a job, has accepted, and is expected to come in to work, but does not show up. The problem this creates for the employer is that the employer has already started processing the employee's record and is ready to receive the employee with an active record in the system. If nothing is done, the system will consider this record to be a valid employee and will process payroll.

Employee Central recently added this event to designate such cases as inactive. To ensure they remain inactive in Employee Central Payroll, this event must be properly mapped by following these steps:

1. Log on to the Employee Central Payroll system.
2. Create an inactive hire action by using Transaction SM30 and table T529A, as shown in Figure 8.22.
3. Set the **Function code** for this action to 1 or 7.
4. Set the **Employment Status** to 0.
5. Using Transaction SM30 to maintain table V_T530 to create a reason for this action.

6. Click on **New Entries** and fill in the **Action** field with the action this reason should be attached to.
7. Then, as shown in Figure 8.26, fill in a 2-character reason code in the **Act.Reason** field and enter a description in the **Name of reason for action** field. Save these settings.

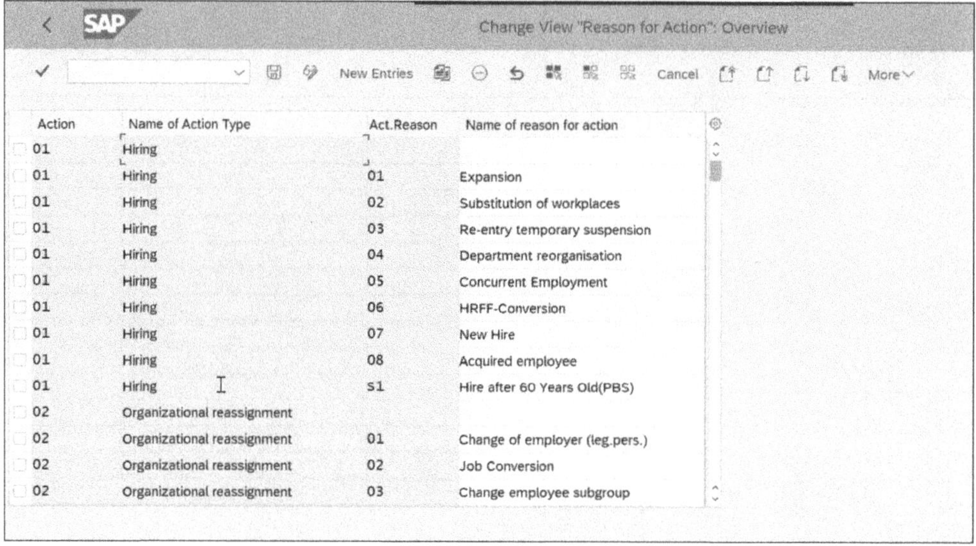

Figure 8.26 Action Reason Setup

8. To bring up the replication configuration screen, in Transaction SPRO, follow the menu path **Personnel Management • Integration Settings for SuccessFactors Employee Central Payroll • Assignment of Code Values • For Point-to-Point Replication • Maintain Properties for Code Value Mapping**.
9. Change the properties of the **EVENT** field by selecting **Mapping via Customer Table** from the **Mapping Mode** dropdown list, as shown in Figure 8.27.
10. Using Transaction SM30, copy all values of the **EVENT** field from table T77SFEC_CVMAPS to T77SFEC_CVMAPC by opening both tables in separate sessions. Use Transaction SM30 and enter "T77SFEC_CVMAPS" in the **Table/View** field. Click on **Display**.
11. Use Transaction SM30 and enter "T77SFEC_CVMAPC" in the **Table/View** field in the second session. Click on **Maintain**. Then, click on **New Entries**.
12. Copy all entries from the first session into the clipboard.
13. Paste the clipboard into the screen of the second session.
14. Change the mapping of the **EC Data Type** field to **Event for NS** for the new action you just created.
15. Save these settings.

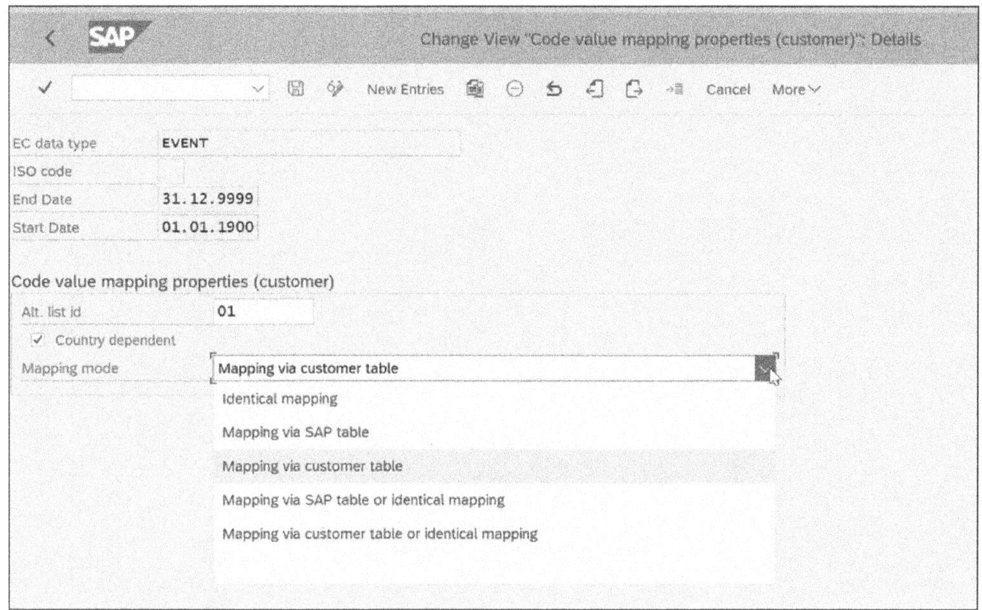

Figure 8.27 Customer Code Value Mapping

Now, you can replicate a "no-show" record and represent an employee in payroll who is inactive on the first day, thus solving a problem that has haunted SAP users for a while.

8.2.6 Jobs, Positions, and Organizational Units

Employee Central Payroll does not need the structural elements that constitute the organizational structure of employees. However, if you need to see this data in Employee Central Payroll, SAP provides a program that you can be run to replicate the data over from Employee Central to Employee Central Payroll.

Note that the organizational elements of job, position, and organizational unit are in Employee Central Payroll *only for viewing*. You cannot use Employee Central Payroll to maintain these elements—which, if required, should be performed in Employee Central. Therefore, some limitations exist. Only a description of these elements is available, with a character limit of 25.

To replicate this structural data, use Transaction RP_HRSFEC_JOB_POS_REPL (Brazil and Hong Kong require report RP_HRSFEC_JOB_POS_REPL_WCS). Also note that the values for these organizational entities do not go to the traditional organizational infotypes stored in the HRP1XXX tables. Jobs are loaded in T513 table; positions, in table T528B; and organizational units, in table T527X.

When you run this transaction, you can make selections based on your needs, as shown in Figure 8.28.

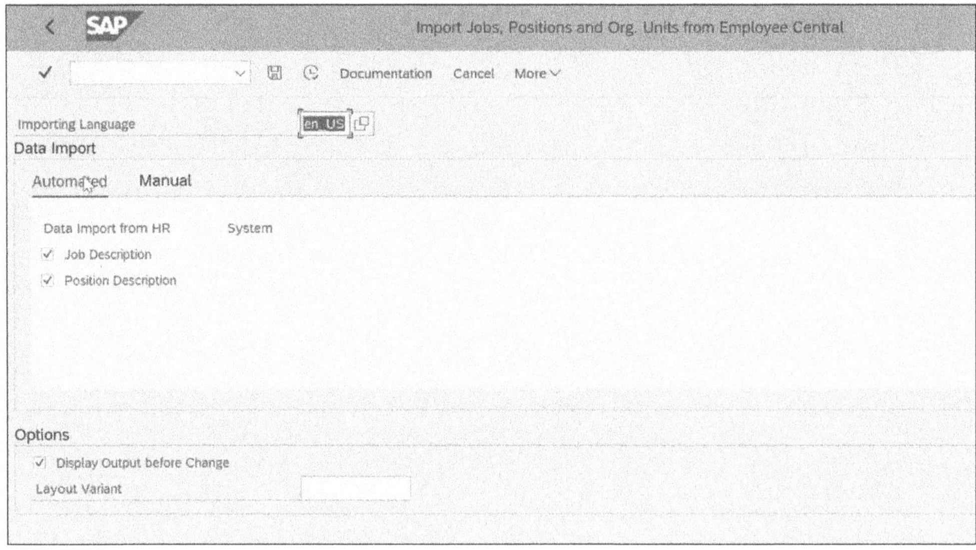

Figure 8.28 Job and Position Replication

8.2.7 Locations

A *location* in Employee Central is a part of a foundation object. This object is mapped to a personnel area and subarea in the Employee Central Payroll system. In Chapter 5, Section 5.2.1, you learned how to create personnel areas and subareas. The mapping between a location and a personnel area/subarea is maintained in table T77SFEC_KMAPPEST.

Note that a personnel area is tightly linked with a company code. Thus, the company code must also exist in Employee Central Payroll before personnel areas can be created. Company codes in Employee Central Payroll map with legal entities in Employee Central, and this mapping is represented in table T77SFEC_KMAPCOMC.

If the company code is less than or equal to 4 characters on the Employee Central side, a mapping to an Employee Central Payroll value is not mandatory as long as a company code with the same name has been created in the Employee Central Payroll system.

8.3 Compensation Information: Infotype 0001

The final piece of information carried into infotype 0001 during replication is the payroll area, as listed in Table 8.13.

Target Technical Name	Max. Length	Default Mapping Mode	Is Code Value Mapping Country/ Region Specific?	Field ID	Field Label
ABKRS	2	Custom table	Yes	pay-group	Pay Group

Table 8.13 Inheriting Payroll Area from Employee Central

This field is especially important for payroll processing and is thus mandatory in Employee Central. The pay-group field is part of the **Compensation Information** portlet.

8.4 Biographical Information: Infotype 0002

The **Biographical Information** portlet represents the personal details of employees for use in identification. All of this information goes into infotype 0002. Table 8.14 lists the important fields of this infotype.

Field	Description	Mandatory/Optional
MANDT	Client	Mandatory
PERNR	Personnel number	Mandatory
SUBTY	Subtype	Mandatory
OBJPS	Object Identification	Mandatory
SPRPS	Lock Indicator for HR Master Data Record	Mandatory
ENDDA	End Date	Mandatory
BEGDA	Start Date	Mandatory
SEQNR	Number of Infotype Record with Same Key	Mandatory
AEDTM	Changed On	Mandatory
UNAME	Name of Person Who Changed Object	Mandatory
HISTO	Historical Record Flag	Mandatory
ITXEX	Text Exists for Infotype	Mandatory
REFEX	Reference Fields Exist (Primary/Secondary Costs)	Mandatory
ORDEX	Confirmation Fields Exist	Mandatory
ITBLD	Infotype Screen Control	Mandatory
PREAS	Reason for Changing Master Data	Mandatory
GRPVL	Grouping Value for Personnel Assignments	Mandatory

Table 8.14 Fields in Infotype 0002

Field	Description	Mandatory/Optional
INITS	Initials	Optional
VORNA	First Name	Mandatory
NACHN	Last Name	Mandatory
NAME2	Name at Birth	Optional
NACH2	Second Name	Optional
CNAME	Complete Name	Optional
TITEL	Title	Optional
TITL2	Second Title	Optional
NAMZU	Other Title	Optional
VORSW	Name Prefix	Optional
VORS2	Second Name Prefix	Optional
RUFNM	Nickname	Optional
MIDNM	Middle Name	Optional
KNZNM	Name Format Indicator for Employee in a List	Optional
ANRED	Form-of-Address Key	Optional
GESCH	Gender Key	Mandatory
GBDAT	Date of Birth	Mandatory
GBLND	Country of Birth	Optional
GBDEP	State	Optional
GBORT	Birthplace	Optional
NATIO	Nationality	Mandatory
NATI2	Second Nationality	Optional
NATI3	Third Nationality	Optional
SPRSL	Communication Language	Mandatory
FAMDT	Valid From Date of Current Marital Status	Optional
ANZKD	Number of Children	Optional
NACON	Name Connection	Optional
PERMO	Modifier for Personnel Identifier	Optional
PERID	Personnel ID Number	Optional

Table 8.14 Fields in Infotype 0002 (Cont.)

Field	Description	Mandatory/Optional
GBPAS	Date of Birth According to Passport	Optional
FNAMK	First name (Katakana)	Optional

Table 8.14 Fields in Infotype 0002 (Cont.)

Of all these fields, only the fields listed in Table 8.15 are inherited from the Employee Central **Biographical Information** portlet.

Target Technical Name	Max. Length	Default Mapping Mode	Is Code Value Mapping Country/Region Specific?	Field ID	Field Label
GBDAT	8	No mapping	No	date-of-birth	Date of Birth
GBLND	3	No mapping	No	country-of-birth	Country of birth
GBORT	40	No mapping	No	place-of-birth	Place of birth
NAME2	40	No mapping	No	birth-name	Birth Name
GBDEP	3	No mapping	No	region-of-birth	Region of Birth

Table 8.15 Fields in Infotype 0002 Inherited from Employee Central

For France and Italy, the **Country of Birth** and the **Region of Birth** are required fields in Employee Central if the relevant employee is a French or Italian national. The **Region of Birth** field is also replicated for Brazil and the United Arab Emirates.

The remaining fields of infotype 0002 come from the **Personal Information** and **National ID Card** portlets, which we'll discuss in the next two sections.

8.5 Personal Information: Infotype 0002

So that country-specific replication can occur smoothly, feature IVWID, shown in Figure 8.29, must be activated. A view indicator is required to maintain the synchronous history of data that sits in more than one infotype. These groupings can be identified by country code, personnel area, employee group, or subgroup. Feature IVWID provides a 2-character code that can be stored in all the infotypes that are related, automatically. This indicator is also stored in infotype 0003 of the employee to be used whenever a change is made to any of the infotypes with related data, so that their start and end dates are synchronized.

To set feature IVWID, use Transaction PE03 and enter "IVWID" in the **Feature** field. Click on **Maintain**. Set up the tree with the fields required to ensure the indicator is correctly derived based on organizational parameters.

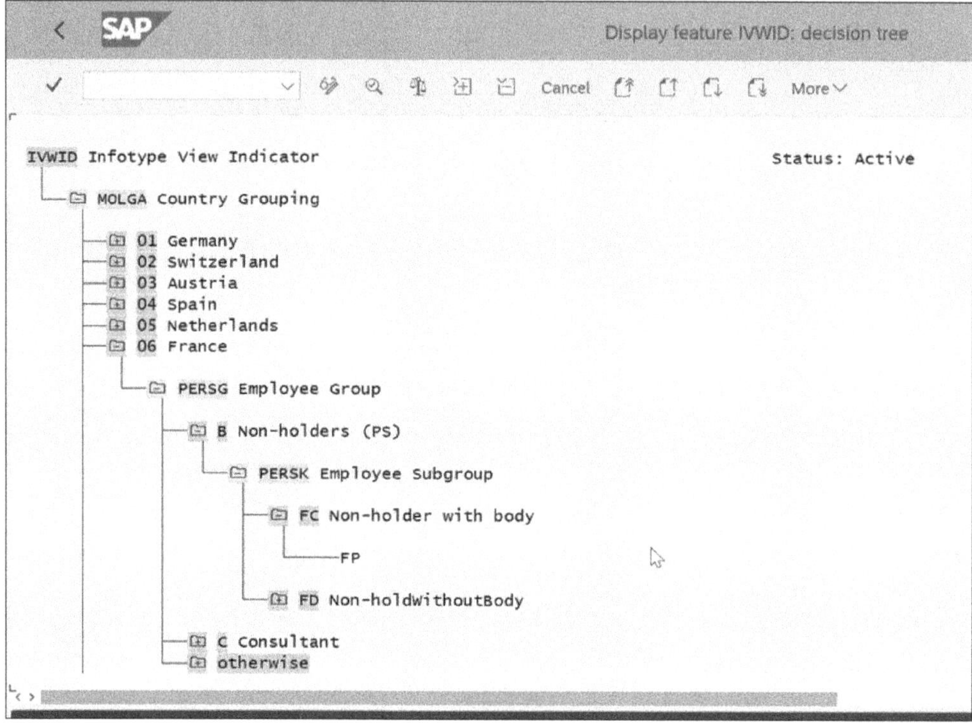

Figure 8.29 Feature IVWID

The fields listed in Table 8.16 are inherited from the **Personal Information** portlet.

Target Technical Name	Max. Length	Default Mapping Mode	Is Code Value Mapping Country/Region Specific?	Field ID	Field Label
FAMST	1	Customer mapping	No	marital-status	Marital status
FAMDT	8	No mapping	No	marital-status-since	Marital status since
SPRSL	1	No mapping	No	native-preferred-lang	Native preferred language

Table 8.16 Fields in Infotype 0002 Inherited from Employee Central

Target Technical Name	Max. Length	Default Mapping Mode	Is Code Value Mapping Country/ Region Specific?	Field ID	Field Label
GESCH	1	SAP standard mapping	No	gender	Gender
NAME2	40	No mapping	No	birth-name	Birth Name
VORNA	40	No mapping	No	firstName	Given name
NACH2	40	No mapping	No	third-name	Third Name
NACHN	40	No mapping	No	lastName	Family name
MIDNM	40	No mapping	No	middleName	Middle name
VORSW	15	No mapping	No	name-prefix	Prefix
VORS2	15	No mapping	No	partner-name-prefix	Partner Name Prefix
NAMZU	15	SAP standard mapping	No	nameSuffix	Suffix
ANRED	1	SAP standard mapping	No	salutation	Salutation
TITLE	15	No mapping	No	title	Title
INITS	10	No mapping	No	initials	Initials
NATIO	3	No mapping	No	nationality	Nationality

Table 8.16 Fields in Infotype 0002 Inherited from Employee Central (Cont.)

8.6 National ID Card: Infotype 0002

The fields listed in Table 8.17 are inherited from the **National ID Card** portlet.

Target Technical Name	Max. Length	Default Mapping Mode	Is Code Value Mapping Country/ Region Specific?	Field ID	Field Label
PERID	20	No mapping	No	national-id	National Id

Table 8.17 Field in Infotype 0002 Inherited from Employee Central

Only national-ids that are flagged as primary in Employee Central are replicated to infotype 0002.

The temporary ID from the national-id block in Employee Central is replicated to Employee Central Payroll when the IS_TEMPORARY field is set as true in Employee Central. The national-id is required for replication for Canada, Czech Republic, Finland, France, Spain, Sweden, the United Kingdom, and the United States. For Brazil, national-ids are not replicated.

Certain country/region versions use additional secondary infotypes for national-id: Austria, Egypt, India, Malaysia, Oman, Mexico, and Singapore (these seven all use infotype 0185), Brazil (infotype 0465), Japan (infotype 0140), and Russia (infotype 0037). These additional infotypes may not be supported yet for replication, except that PAN Card information for India is replicated into infotype 0185. Replication to infotype 0185 is supported for China, Hong Kong, Thailand, the United Arab Emirates, and Saudi Arabia.

8.7 Address Information: Infotype 0006

The **Address Information** portlet represents all the important addresses associated with an employee for use in communication, including mailing checks and benefit letters to the employee. All of this information goes into infotype 0006. Depending on the type of address, this data may be stored in different subtypes.

The labels and layouts for this portlet are country specific. Table 8.18 lists the important fields of this infotype.

Field	Description	Mandatory/Optional
MANDT	Client	Mandatory
PERNR	Personnel number	Mandatory
SUBTY	Subtype	Mandatory
OBJPS	Object Identification	Mandatory
SPRPS	Lock Indicator for HR Master Data Record	Mandatory
ENDDA	End Date	Mandatory
BEGDA	Start Date	Mandatory
SEQNR	Number of Infotype Record with Same Key	Mandatory
AEDTM	Changed On	Mandatory
UNAME	Name of Person Who Changed Object	Mandatory
HISTO	Historical Record Flag	Mandatory
ITXEX	Text Exists for Infotype	Mandatory

Table 8.18 Fields in Infotype 0006

Field	Description	Mandatory/Optional
REFEX	Reference Fields Exist (Primary/Secondary Costs)	Mandatory
ORDEX	Confirmation Fields Exist	Mandatory
ITBLD	Infotype Screen Control	Mandatory
PREAS	Reason for Changing Master Data	Mandatory
GRPVL	Grouping Value for Personnel Assignments	Mandatory
NAME2	Contact Name	Optional
STRAS	Street and House Number	Optional
ORT01	City	Mandatory
ORT02	District	Optional
PSTLZ	Postal Code	Optional
LAND1	Country Key	Mandatory
TELNR	Telephone Number	Optional
ENTKM	Distance in Kilometers	Optional
WKWNG	Company Housing	Optional
BUSRT	Bus Route	Optional
LOCAT	2nd Address Line	Optional
ADR03	Street 2	Optional
ADR04	Street 3	Optional
HSNMR	House Number	Optional
POSTA	Identification of an apartment in a building	Optional
BLDNG	Building (number or code)	Optional
FLOOR	Floor in building	Optional
STRDS	Street Abbreviation	Optional
ENTK2	Distance in Kilometers	Optional
COM01	Communication Type	Optional
NUM01	Communication Number	Optional
COM02	Communication Type	Optional
NUM02	Communication Number	Optional

Table 8.18 Fields in Infotype 0006 (Cont.)

Field	Description	Mandatory/Optional
COM03	Communication Type	Optional
NUM03	Communication Number	Optional
COM04	Communication Type	Optional
NUM04	Communication Number	Optional
COM05	Communication Type	Optional
NUM05	Communication Number	Optional
COM06	Communication Type	Optional
NUM06	Communication Number	Optional
INDRL	Indicator for relationship (specification code)	Optional
COUNC	County Code	Optional
OR2KK	Second address line (Katakana)	Optional
CONKK	Contact Person (Katakana) (Japan)	Optional
OR1KK	First address line (Katakana)	Optional
RAILW	Social Subscription Railway	Optional
STATE	Province	Optional
ANSSA	Address Type	Mandatory
REGIO	Region (State, Province)	Optional
REGIO DESCR	Region (Description)	Optional

Table 8.18 Fields in Infotype 0006 (Cont.)

Of all these fields, only the fields listed in Table 8.19 are inherited from the **Address Information** portlet.

Target Technical Name	Max. Length	Default Mapping Mode	Is Code Value Mapping Country/ Region Specific?	Field ID	Field Label
ANSSA	4	SAP standard mapping	No	Address-type	Address Type
<country specific mapping>	3	SAP standard mapping	Yes	Address1	Country specific

Table 8.19 Fields in Infotype 0006

Target Technical Name	Max. Length	Default Mapping Mode	Is Code Value Mapping Country/ Region Specific?	Field ID	Field Label
<country specific mapping>	Depends on the country	SAP standard mapping	Yes	Address2	Country specific
<country specific mapping>	Depends on the country	SAP standard mapping	Yes	Address3	Country specific
<country specific mapping>	Depends on the country	SAP standard mapping	Yes	Address4 through Address8	Country specific
ORT01	40	No mapping	No	City	City
STATE	3	SAP standard mapping	Yes	County	County
STATE	3	SAP standard mapping	Yes	State	State
STATE	3	SAP standard mapping	Yes	Province	province
PSTLZ	10	SAP standard mapping	Yes	Zip-code	Zip code
LAND1	3	SAP standard mapping	Yes	Country	country

Table 8.19 Fields in Infotype 0006 (Cont.)

The address type has four standard values in Employee Central:

- Home
- Host
- Mailing
- Benefits

While home and host are mapped to infotype 0006, subtype 1 (Permanent Residence), the others do not have a default mapping. Mapping to other types can be created easily by using the mapping table T77SFEC_CVMAPC. If no mapping is found, this type of address is not replicated to Employee Central Payroll. The permanent address is the address type that is typically used for Employee Central Payroll.

As mentioned earlier, the address fields in infotype 0006 are mapped according to the country where the employee is located. SAP has an extensive list of these mappings at

SAP Help Portal (search for "Country-Dependent Mapping of Address Fields (Infotype 0006)"). For example, the mapping for the US is shown in Table 8.20, and the mapping for the UK is shown in Table 8.21.

Target Technical Name	Employee Central Field ID	Employee Central Field Label
STRAS	address1	Address Line 1
LOCAT	address2	Address Line 2
ORT01	city	City
ORT02	county	County
STATE	state	State
PSTLZ	zip-code	ZIP

Table 8.20 US Mapping List for Infotype 0006

Target Technical Name	Employee Central Field ID	Employee Central Field Label
STRAS	address1	Building Number and Street
LOCAT	address2	Address Line 2
ORT02	address3	District
ORT01	city	City
STATE	county	County
PSTLZ	zip-code	Postal Code

Table 8.21 UK Mapping List for Infotype 0006

8.8 Payment Information: Infotype 0009

The payment information of an employee captures various details about how the employee should receive money from your company. Some companies insist on a bank method, in which case this infotype stores a bank routing number, an account number, and the type of account. If your company allows other methods of payment (for example, checks), these options are also available as choices in this infotype.

The human resources information system (HRIS)-based direct deposit/payment information in Employee Central has been replaced by a Metadata Framework (MDF)-based **Payment Information** screen. However, this capability requires that the **Job Information** portlet be set up first.

For bank deposit information to work, bank information must be available. This information can be loaded from downloadable files containing bank names and addresses along with their routing numbers available from banks or other agencies. If you have an SAP ERP Financials or SAP S/4HANA Finance system, this table may have already been populated in that system, and you can simply extract and reload this data into your Employee Central Payroll system since the structure of the data is exactly the same. If not, use Transaction FI01 to load this information manually, as shown in Figure 8.30.

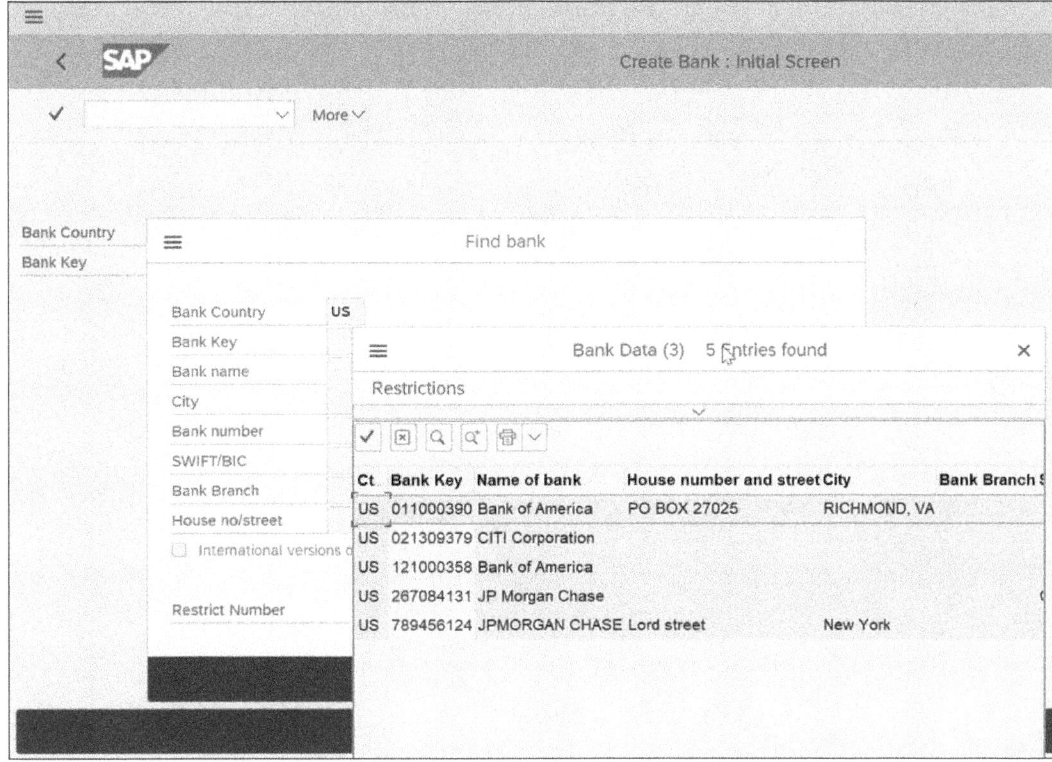

Figure 8.30 Transaction FI01 for Creating Bank Names

Table 8.22 lists the important fields of infotype 0009.

Field	Description	Mandatory/Optional
MANDT	Client	Mandatory
PERNR	Personnel number	Mandatory
SUBTY	Subtype	Mandatory
OBJPS	Object Identification	Mandatory
SPRPS	Lock Indicator for HR Master Data Record	Mandatory
ENDDA	End Date	Mandatory

Table 8.22 Fields in Infotype 0009

Field	Description	Mandatory/Optional
BEGDA	Start Date	Mandatory
SEQNR	Number of Infotype Record with Same Key	Mandatory
AEDTM	Changed On	Mandatory
UNAME	Name of Person Who Changed Object	Mandatory
HISTO	Historical Record Flag	Mandatory
ITXEX	Text Exists for Infotype	Mandatory
REFEX	Reference Fields Exist (Primary/Secondary Costs)	Mandatory
ORDEX	Confirmation Fields Exist	Mandatory
ITBLD	Infotype Screen Control	Mandatory
PREAS	Reason for Changing Master Data	Mandatory
GRPVL	Grouping Value for Personnel Assignments	Mandatory
OPKEN	Operation Indicator for Wage Types	Optional
BETRG	Standard value	Optional
WAERS	Payment Currency	Mandatory
ANZHL	Standard Percentage	Optional
ZEINH	Time/Measurement Unit	Optional
BNKSA	Type of Bank Details Record	Mandatory
ZLSCH	Payment Method	Mandatory
EMFTX	Payee Text	Optional
BKPLZ	Postal Code	Optional
BKORT	City	Optional
BANKS	Bank country key	Mandatory
BANKL	Bank Keys	Mandatory
BANKN	Bank account number	Mandatory
BANKP	Check Digit for Bank No./Account	Optional
BKONT	Bank Control Key	Mandatory
SWIFT	SWIFT/BIC for International Payments	Optional

Table 8.22 Fields in Infotype 0009 (Cont.)

Field	Description	Mandatory/Optional
DTAWS	Instruction key for data medium exchange	Optional
DTAMS	Indicator for Data Medium Exchange	Optional
STCD1	Tax Number 1	Optional
STCD2	Tax Number 2	Optional
PSKTO	Account Number of Bank Account At Post Office	Optional
ESRNR	ISR Subscriber Number	Optional
ESRRE	ISR Reference Number	Optional
ESRPZ	ISR Check Digit	Optional
EMFSL	Payee key for bank transfers	Optional
ZWECK	Purpose of Bank Transfers	Optional
BTTYP	PBS Transfer Type	Optional
PAYTY	Payroll type	Optional
PAYID	Payroll Identifier	Optional
OCRSN	Reason for Off-Cycle Payroll	Optional
BONDT	Off-cycle payroll payment date	Optional
BKREF	Reference specifications for bank details	Optional
STRAS	House number and street	Optional
DEBIT	Automatic Debit Authorization Indicator	Optional
IBAN	IBAN (International Bank Account Number)	Optional
OR1KK	First address line (Katakana)	Optional
RAILW	Social Subscription Railway	Optional
STATE	Province	Optional
ANSSA	Address Type	Mandatory
REGIO	Region (State, Province)	Optional
REGIO DESCR	Region (Description)	Optional

Table 8.22 Fields in Infotype 0009 (Cont.)

Of all these fields, only the fields listed in Table 8.23 are inherited from the **Payment Information** portlet.

Target Technical Name	Max. Length	Default Mapping Mode	Is Code Value Mapping Country/ Region Specific?	Field ID	Field Label
ZLSCH	1	Customer table	No	payment-Method	Payment Method
BANKL	15	No mapping	No	routingNum-ber	Routing Number
BKONT	2	Customer table	Yes	accountType	Account Type
BKONT	2	No mapping	No	checkDigit	Bank Control Key
BANKN	18	No mapping	No	accountNum-ber	Account Number
EMFTX	40	No mapping	No	accoun-tOwner	Owner
BANKS	3	No mapping	No	bankCountry	Bank Country/ Region
WAERS	5	No mapping	No	Currency	Currency
IBAN	34	No mapping	No	IBAN	International Bank Account Number
PSKTO	16	No mapping	No	buildingSoci-etyRollNum-ber	Building Soci-ety Roll Num-ber
ZWECK	40	No mapping	No	Purpose	Purpose

Table 8.23 Fields in Infotype 0009 Inherited from Employee Central

The handling of payment information dates during replication follows these rules:

- If infotype 0009, subtype 1 (Other Bank) is used, the start and end dates are automatically adjusted to the nearest payroll period start and end dates, as shown in Figure 8.31.
- If the changed record has a start date in the current pay period, but the current pay period is closed and the employee has been paid, the date is automatically moved to the start of the next pay period.
- If the changed record has a start date in a prior payroll period that is already closed, the date is automatically moved to the start of the current pay period.

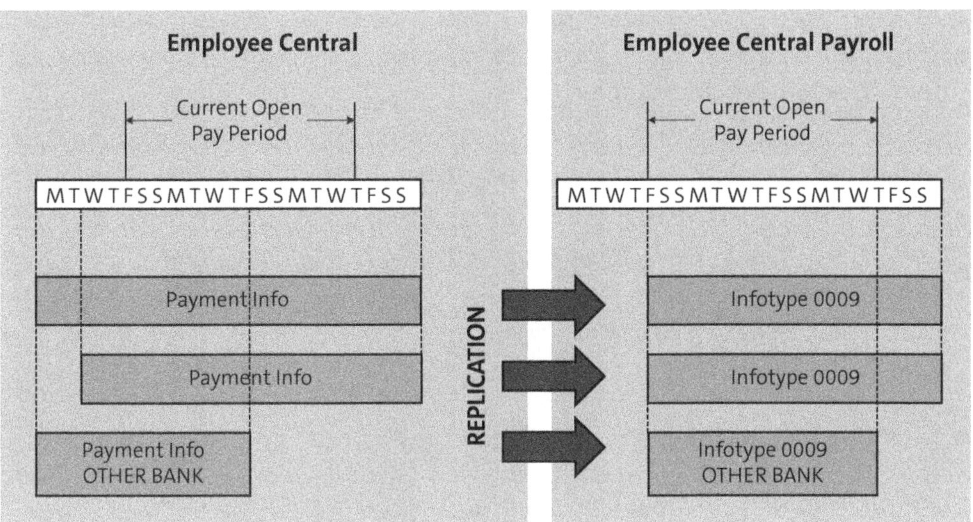

Figure 8.31 Adjusting Start and End Dates for Payment Information

A routing number is a key field for identifying the right bank. The account number identifies the bank account of the employee.

For China, if a bank account number entered in Employee Central is more than 18 digits, the first 18 digits are replicated to field BANKN (Bank Account), and the remaining digits are replicated to field BKREF (Reference Details) of infotype 0009.

For Russia, bank account numbers of 20 digits can be entered into Employee Central. The first 2 digits are replicated into field BKONT (Bank Control Key), and the remaining 18 digits are replicated into field BANKN (Bank Account) of infotype 0009.

Employee Central field accountOwner is a string that is transferred unchanged from Employee Central to Employee Central Payroll, where this information is stored in field EMFTX of infotype 0009. This field is not mandatory. Banking files are created in Employee Central Payroll are based on the formatted name of the employee.

The payment type code specifies three categories of banks in the following way:

- 0 for the main bank, where the amount is deposited
- 1 for the other bank (called "Payroll" in Employee Central as standard)
- 6 for bonus payments

An employee can choose multiple other banks, which will require the specification of dollar amounts or percentages.

For example, if an employee has indicated a value of $50 in the "Other" bank and receives $400 as the salary for that period, $350 will go to the main bank account and $50 to the other bank account.

Option 6 is reserved for bonus payments, and an employee can indicate the payment of a bonus (not regular pay) to a separate bank account. In this case, the "main" and "other" bank types do not play any role.

The **Bank Country** and **Bank Currency** fields are based on the country to which the bank account belongs. Standard payroll processes expect this bank to be in the same country as where the employee works and lives.

An International Bank Account Number (IBAN) is a key that uniquely identifies a bank account across national borders. If provided, this number overrides the routing and account numbers during payroll processing.

8.9 Pay Component Recurring: Infotypes 0008 and 0014

The fields listed in Table 8.24 are inherited from **Pay Component Recurring** portlet. Note that only wage types with time constraints 1 (no time gap) and 2 (time gap but no overlap) are replicated using infotypes 0008 and 0014.

Target Technical Name	Max. Length	Default Mapping Mode	Is Code Value Mapping Country/ Region Specific?	Field ID	Field Label
BEGDA	8	No mapping	No	EffectiveStartDate	Start Date
ENDDA	8	No mapping	No	EffectiveEndDate	End Date
LGA01, LGA02... (for infotype 0008) and LGART (for infotype 0014)	4	Custom mapping	Yes	pay-component	Pay component
WAERS	5	No mapping	No	currency-code	Currency
BET01, BET02... (for infotype 0008) and BETRG (for infotype 0014)	13	No mapping	No	paycompvalue	Amount
ZEINH	3	Standard mapping	No	unit-of-measure	Unit of Measure

Table 8.24 Fields in Infotype 0008 and 0014 Inherited from Pay Component Recurring Fields in Employee Central

Target Technical Name	Max. Length	Default Mapping Mode	Is Code Value Mapping Country/ Region Specific?	Field ID	Field Label
ZEINZ (for infotype 0008) and ZANZL (for infotype 0014)	3	No mapping	No	frequency	Pay Frequency

Table 8.24 Fields in Infotype 0008 and 0014 Inherited from Pay Component Recurring Fields in Employee Central (Cont.)

Since this data contains amounts that will be paid, you must understand the special properties of each field that makes up a line of payment information. Let's walk through these fields briefly:

- **Pay component**
  In Employee Central, this object holds various wage categories for the employee. This object is equivalent to the wage type in Employee Central Payroll (see Chapter 2) that resides in table T512W. The pay component is replicated to both infotypes 0008 and 0014, depending on the configuration.
- **Amount**
  This amount is expressed in the currency used for the company, with the number of decimal places specified, as described in Chapter 7, Section 7.2.3.
- **Unit of measure**
  This field qualifies the rate of pay for the employee and is thus a particularly important field for the determination of periodic rates. The generic object's units of measure must be created with the values listed in Table 8.25.

External Code	Name
ANN	Years
DAY	Day
H87	Pieces
HUR	Hours
MI	Miles
KMT	Kilometers
MIN	Minutes
MON	Months
P1	Percent

Table 8.25 Time Units

- **Frequency**
  The pay frequency is established in Employee Central through the frequency object. This object uses the **Annualization Factor** and **Pay Frequency** fields to translate the rate of pay and to load infotype 0008 into Employee Central Payroll.

  If the **Pay Frequency** field is blank, the system will consider the value of the **Annualization Factor** field for mapping. If the **Pay Frequency** field is not blank, the system will ignore the value of the **Annualization Factor** field. The valid values for the **Pay Frequency** field are 5, 6, 7, 8, 9, 10, 11, and 12.

  For successful replication of infotype 0008, the payroll periodicity can be hourly (stores an hourly rate) or must match the annualization factor of the frequency used. Table 8.26 shows the annualization factor for various frequencies of pay.

Annualization Factor	Pay Frequency	Frequency/ Value	Frequency Code	Compcensation Frequency	Description
0	Blank	1	3	3	Hourly
1	Blank	1	7	7	Yearly
12	Blank	1	6	6	Monthly
24	Blank	99	9	9	Semi-Monthly
26	Blank	2	5	8	Bi-Weekly
4	Blank	3	6	11	Quarterly
52	Blank	1	5	5	Weekly
Any	5	Initial	Initial	5	Weekly
Any	6	Initial	Initial	6	Monthly
Any	7	Initial	Initial	7	Yearly
Any	8	Initial	Initial	8	Bi-weekly
Any	9	Initial	Initial	9	Semi-monthly
Any	10	Initial	Initial	10	Four-weekly
Any	11	Initial	Initial	11	Quarterly
Any	12	Initial	Initial	12	Semi-yearly

Table 8.26 Annualization Factors

- **Deductions**
  Recurring deductions are generally not stored in Employee Central, although this approach is possible using the same objects and replication methods as discussed

earlier. In this case, however, wage types must be deduction wage types instead of earnings.

In a payroll environment, deductions are either calculated based on rules (e.g., taxes) or imported from an external system or agency (e.g., benefits). For this reason, getting this data into Employee Central means pushing this data again to Employee Central Payroll. Thus, most users prefer to maintain deductions directly in Employee Central Payroll using scheduled batch jobs.

8.10 Pay Component Non-Recurring: Infotype 0015

One-time payments like bonuses are often paid directly from the Employee Central Payroll system. Users who disburse one-time payments through Employee Central will generally be calculating bonuses using the SAP SuccessFactors Variable Pay module. A direct integration to Employee Central from this module exists, which reduces manual effort significantly. This data can then be directly replicated to payroll for both off-cycle and on-cycle payroll runs. The fields listed in Table 8.27 are inherited from the **Pay Component Non-Recurring** portlet in Employee Central.

Target Technical Name	Max. Length	Default Mapping Mode	Is Code Value Mapping Country/ Region Specific?	Field ID	Field Label
ANZHL	7	No mapping	No	number of units/ percentage	Number of Units/Value
LGART	4	Customer mapping	Yes	pay-component-code	Pay Component
UWDAT	8	No mapping	No	pay-date	Pay Date
BETRG	13	No mapping	No	paycompvalue	Amount
WAERS	5	No mapping	No	currency-code	Currency
ZEINH	3	Custom	No	unit-of-measure	Unit of Measure
KOSTL	10	No mapping	No	alternative-cost-center	Alternative Cost Center
ZUORD	20	No mapping	No	reference-id	Reference ID

Table 8.27 Fields in Infotype 0015 Inherited from Employee Central

Alternative cost centers are stored in table ASSOB, not in the standard table PA0015.

8.11 Contract Elements: Infotype 0016

Contract elements are used for certain types of employees where maintaining the contract details is important. The fields listed in Table 8.28 are inherited from the **Job Information** portlet in Employee Central and varies in usage from country to country.

Target Technical Name	Max. Length	Default Mapping Mode	Is Code Value Mapping Country/ Region Specific?	Field ID	Field Label
CTTYP	2	Customer mapping	Yes	contract-type	Contract Type
CTEDT	8	No mapping	No	contract-end-date	Contract End Date
EINDT	8	No mapping	No	initial-entry-date	Initial Entry
KONDT	8	No mapping	No	entry-into-group	Entry into Group

Table 8.28 Fields in Infotype 0016 Inherited from Employee Central

If the contract type is fixed term, an end date is mandatory. To ensure the replication process does not fail, you may have to write a business rule to populate this field automatically whenever this contract type is chosen.

8.12 Dependents Information: Infotype 0021

The fields listed in Table 8.29 are inherited from the **Dependents Information** portlet.

Target Technical Name	Max. Length	Default Mapping Mode	Is Code Value Mapping Country/ Region Specific?	Field ID	Field Label
SUBTY/ FAMSA	4	SAP standard	No	relationship-type	Relationship
FANAM	40	No mapping	No	last-name	Last Name
FNAC2	40	No mapping	No	second-last-name	Second Name
FAVOR	40	No mapping	No	first-name	First Name

Table 8.29 Fields in Infotype 0021 Inherited from Employee Central

Target Technical Name	Max. Length	Default Mapping Mode	Is Code Value Mapping Country/ Region Specific?	Field ID	Field Label
FASEX	1	SAP standard mapping	No	gender	Gender
FGBDT	8	No mapping	No	date-of-birth	Date of Birth
FGBOT	40	No mapping	No	place-of-birth	Place of Birth
FGBLD	3	No mapping	No	country-of-birth	Country of Birth
FANAT	3	No mapping	No	nationality	Nationality
FCNAM	80	No mapping	No	display-name	Complete Name

Table 8.29 Fields in Infotype 0021 Inherited from Employee Central (Cont.)

As described in Chapter 7, Section 7.1, the configuration of the infotype filter must include all the infotypes that need replication. Be sure to include infotype 0021. Remember to set it up using country codes if the subtypes differ from country to country. The subtype is a mandatory field, and if the value from Employee Central is not found in the configuration table in payroll, the replication process will abort.

8.13 Cost Distribution: Infotype 0027

The **Cost Distribution** portlet is used to distribute payroll cost using an alternative method. Normally, payroll costs are assigned to the home cost center stored in the infotype record 0001 for the employee. Using this infotype transfers costs to the cost centers identified in this portlet, in the proportion specified in field KPRXX.

The fields listed in Table 8.30 are inherited from the **Cost Distribution** portlet.

Target Technical Name	Max. Length	Default Mapping Mode	Is Code Value Mapping Country/ Region Specific?	Field ID	Field Label
BEGDA	8	No mapping	No	effectiveStartDate	Start Date

Table 8.30 Fields in Infotype 0027 Inherited from Employee Central

Target Technical Name	Max. Length	Default Mapping Mode	Is Code Value Mapping Country/ Region Specific?	Field ID	Field Label
ENDDA	80	No mapping	No	effectiveEndDate	End Date
KST01 through KST12	10	No mapping	No	costCenter	Cost Center
KPR01 through KPR12	5	No mapping	No	percentage	Percentage

Table 8.30 Fields in Infotype 0027 Inherited from Employee Central (Cont.)

The maximum number of cost centers that can be entered at the employee record level is 12. The total must be 100. All the fields are mandatory.

8.14 Dates: Infotype 0041

Dates are important in payroll for different types of processes. For this reason, an entire infotype is dedicated to dates. In infotype 0041, the date has a type and the key date for this event. Typical events are start dates, termination dates, seniority dates, and more.

You can configure these dates in Transaction SPRO by following the menu path **Personnel Management • Personnel Administration • Evaluation Basis • Date Specifications • Define Date Type.** As shown in Figure 8.32, choose the applicable date types for your organization and identify appropriate labels for the date according to your requirements.

The replication of these dates use the following two Employee Central portlets as sources:

- **Job Information**
- **Employment Information**

This infotype itself has a start date and an end date, thus allowing this infotype to have multiple records if necessary. In the **Job Information** portlet, the infotype is effectively dated (i.e., valid for a specific time interval), but not in the **Employment Information** portlet. When a date is replicated from the **Employment Information** portlet, the effective date is determined by the start of the next payroll period.

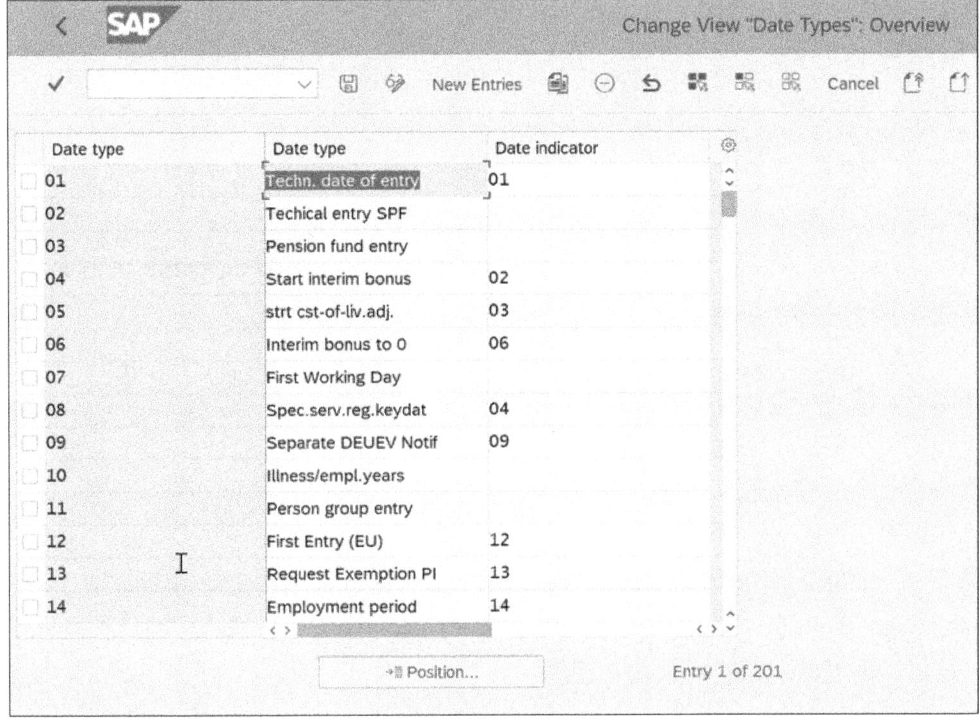

Figure 8.32 Date Types

The following dates are replicated from the **Job Information** portlet:

- `companyEntryDate`
- `departmentEntryDate`
- `jobEntryDate`
- `leaveOfAbsenceReturnDate`
- `hiredate`
- `leaveOfAbsenceStartDate`
- `locationEntryDate`
- `payScaleLevelEntryDate`
- `positionEntryDate`
- `terminationDate`
- `custom_date1-30`

The following dates are replicated from **Employment Information** portlet:

- `benefits_eligibility_start_date`
- `benefitsEndDate`
- `bonusPayExpirationDate`

- endDate
- Termination Date
- firstDateWorked
- lastDateWorked
- originalStartDate
- payrollEndDate
- planned_end_date (global assignment planned end date, pension planned end date)
- professionalServiceDate
- salary_end_date
- seniorityDate
- serviceDate
- startDate (start date, global assignment start date, pension start date)
- StockEndDate
- custom_date1-90

The target fields for these dates can be controlled by following the menu path **Personnel Management • Integration Settings for SuccessFactors Employee Central Payroll • Assignment of Code Values • For Point-to-Point Replication • Maintain Date Type Mapping**. This setting in the replication configuration explained in Chapter 7, Section 7.2.

8.15 Email Address Information: Infotype 0105

The **Email Address Information** portlet is used to record email addresses. This information is only replicated to Employee Central Payroll if the email address(es) are marked as primary. Any email originating from the Employee Central Payroll system will have the sender domain configured to *myXXXX.mail.payroll.ondemand.com* by default. (Note that XXXX is a number that is assigned by SAP for each client.)

The fields listed in Table 8.31 are inherited from the **Email Address Information** portlet.

Target Technical Name	Max. Length	Default Mapping Mode	Is Code Value Mapping Country/Region Specific?	Field ID	Field Label
USRID_LONG	241	No mapping	No	email-address	Email Address
n/a	n/a	n/a	n/a	email-type	ecEmail-Type
n/a	n/a	n/a	n/a	isPrimary	Primary Indicator

Table 8.31 Fields in Infotype 0105 Inherited from Employee Central

8.16 National ID Card: Infotype 0185

Replication of personal IDs in infotype 0185 can be configured in the IMG by following the menu path **Personnel Management • Integration Settings for SuccessFactors Employee Central Payroll • Assignment of Code Values for Point-to-Point Replication • Maintain Countries for IT0185 Replication**, as discussed in Chapter 7, Section 7.2. The fields listed in Table 8.32 are inherited from the **National ID Card** portlet.

Target Technical Name	Max. Length	Default Mapping Mode	Is Code Value Mapping Country/ Region Specific?	Field ID	Field Label
ICTYP	2	No mapping	No	ResidentIdCard	Resident ID Card
ICNUM	30	No mapping	No	national-id	National ID
IDCOT	3	No mapping	No	country	Country/ Region

Table 8.32 Fields in Infotype 0185 Inherited from the National ID Card Portlet

8.17 Work Permit: Infotype 0185

Residence card and work permit information are stored in the same infotype. The fields listed in Table 8.33 are inherited from the **Work Permit** portlet.

Target Technical Name	Max. Length	Default Mapping Mode	Is Code Value Mapping Country/ Region Specific?	Field ID	Field Label
ICTYP	2	No mapping	No	document_type	Document Type
ICNUM	30	No mapping	No	document_number	Document Number
IDCOT	3	No mapping	No	country	Country/ Region
AUTH1	30	No mapping	No	issuing_authority	Issuing Authority
ISSPL	30	No mapping	No	Issue_place	Issue Place
FPDAT	8	No mapping	No	Issue_date	Issue Date
EXPID	8	No mapping	No	expiration _date	Expiration Date

Table 8.33 Fields in Infotype 0185 Inherited from the Work Permit Portlet

8.18 Additional Off-Cycle Payments: Infotype 0267

The format and functionality of infotype 0267 is similar to infotype 15. The only difference is that infotype 0267 will always be paid through off-cycle payroll, while infotype 0015 is paid only through on-cycle payroll. As is true for infotype 0015, infotype 0267 is most often entered directly into the Employee Central Payroll system.

If entry into your Employee Central system is desired, the fields listed in Table 8.34 are inherited from the **Pay Component Non-Recurring** portlet.

Target Technical Name	Max. Length	Default Mapping Mode	Is Code Value Mapping Country/ Region Specific?	Field ID	Field Label
ANZHL	7	No mapping	No	number-of-units/value	Number of Units/ Value
LGART	4	Customer mapping	Yes	pay-component-code	Pay Component
UWDAT	8	No mapping	No	pay-date	Pay Date
BETRG	13	No mapping	No	paycompvalue	Amount
WAERS	5	No mapping	No	currency-code	Currency
ZEINH	3	Custom mapping	No	unit-of-measure	Unit of Measure
KOSTL	10	No mapping	No	alternative-cost-center	Alternative Cost Center
OCRSN	4	n/a	n/a	n/a	n/a
ZUORD	20	No mapping	No	reference-id	Reference ID

Table 8.34 Fields in Infotype 0267 Inherited from Employee Central

Alternative cost centers are stored in table ASSOB, not in the standard SAP table PA0267. The OCRSN field is the off-cycle reason and is essential for running off-cycle payroll.

The following steps are required to configure the derivation of off-cycle reason codes, which do not otherwise exist in Employee Central:

1. To bring up the replication configuration screen, go to Transaction SPRO and follow the menu path **Personnel Management • Integration Settings for SuccessFactors Employee Central Payroll • Wage Type Processing • Assign Off-Cycle Reason to Pay Component.**

2. Assign the relevant wage types, as shown in Figure 8.33. Click on **New Entries**, choose the ISO code for the country in the **ISO code** field, enter the pay component in the **Employee Central Pay Component** field, choose the country grouping in the **CGrpg** field, and choose the off-cycle reason in the **OC reason** field for each pay component. Then, save these settings.

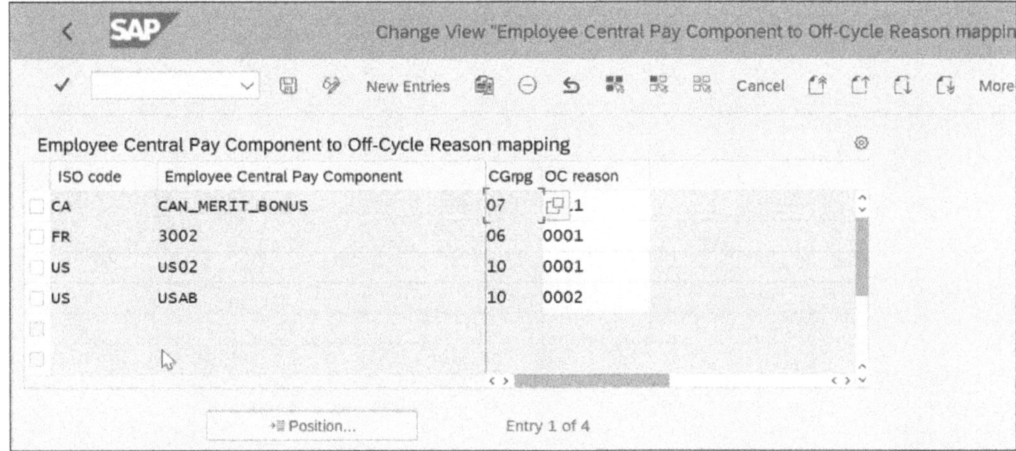

Figure 8.33 Off-Cycle Reason Mapping

3. Go to Transaction SM30 and open view T77SO.
4. Find group SFEC, semantic abbreviation OCRSN, as shown in Figure 8.34.
5. Enter "X" into the **Value abbr** field and save these settings.

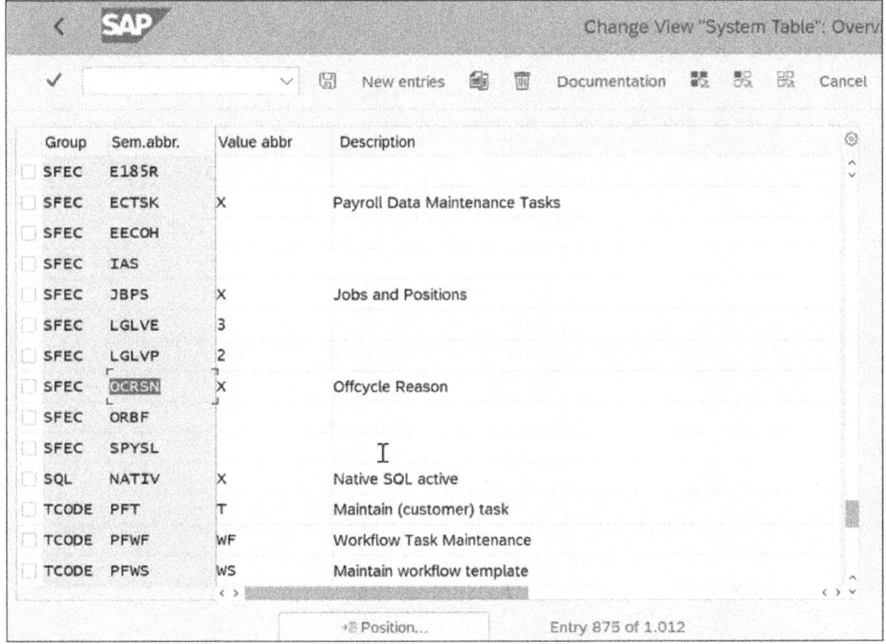

Figure 8.34 Off-Cycle Reason Switch

8.19 Higher Duty or Temporary Assignment: Infotype 0509

The **Higher Duty or Temporary Assignment** portlet is used for the public sector in Australia. This portlet enables pay at a higher rate for specific employment situations in Australian government work environments. When used, the data in these fields replicate to infotype 0509. When replicated, the fields listed in Table 8.35 are inherited from the **Higher Duty or Temporary Assignment** portlet in Employee Central.

Target Technical Name	Max. Length	Field ID	Field Label
EMPCT	5	percentageHDTemp	Higher Duty/Temporary Assignment Percentage
PERTG	5	percentageAllowance	Allowance Percentage
SAIND	1	s17Indicator	Superannuation
CTHFD	1	countingRule	Counting Rule (Half day as full day)
CTAPP	1	countingRule	Counting Rule (Actual
DISHF	1	discardLessThanHalfDay	Discard Less Than Half Day
PLANS	8	hdTempReferencePosition	Higher Duty/Temporary Assignment Position
VPERN	8	hdTempReferenceUserId	Higher Duty/Temporary Assignment User ID
UPHDA and INVAL	1 and 1	allowanceBasis	Allowance Computation Basis (Not Applicable)
DRVAL	1	allowanceBasis	Allowance Computation Basis (Base Salary Only)
INVAL	1	allowanceBasis	Allowance Computation Basis (Base Salary and Allowance)
HAJ01	13	hdTempReferenceSalary	Higher Duty/Temporary Assignment Reference Salary
HLG01	4	hdTempRefSalaryPayComp	Higher Duty/Temporary Assignment Reference Salary Pay Component

Table 8.35 Fields in Infotype 0509 Inherited from the Higher Duty or Temporary Assignment Portlet

Target Technical Name	Max. Length	Field ID	Field Label
NAJ01	13	substantiveReferenceSalary	Nominal Reference Salary
NLG01	4	substantiveRefSalaryPayComp	Nominal Reference Salary Pay Component

Table 8.35 Fields in Infotype 0509 Inherited from the Higher Duty or Temporary Assignment Portlet (Cont.)

8.20 Summary

This chapter examined the replication paradigm in detail, and we provided guidelines for critical fields that you're sure to encounter when enabling Employee Central Payroll. Target data preparation and sound business decisions around the use of critical data elements, such as the PERNR, are important for accuracy, automation, and efficient use by other systems connected in subsequent processes. In the next chapter, we'll dive into the running of payroll and review new features for efficient payroll processing that SAP has incorporated into the Employee Central Payroll application.

Chapter 9
Using the Payroll Control Center

After preparing the payroll system with all the data it needs to calculate pay, you're now ready to navigate to the heart of payroll. Running payroll for a company is one of the most consequential and important jobs for the business and its employees. If payroll runs smoothly, your employees receive money in the bank and can fulfill their financial responsibilities without worry. Often, the impactful work put in by the payroll department goes unnoticed—like a quiet electric train that travels thousands of miles without a skip.

The Payroll Control Center is designed like the cockpit of a plane: Different instruments show you where you are, and alerts warn you of different parts of the plane that could cause trouble. Payroll managers are tasked with running payroll successfully for each period, every period, without fail. To fulfill this mission-critical operation, SAP provides features to help you meet payroll challenges, offering many "instruments" at your fingertips.

This chapter introduces you to the star feature of Employee Central Payroll—the Payroll Control Center. We'll introduce you to the application's key features that support simulations, payroll processing, and post-payroll processing. This chapter sets the stage for configuring the Payroll Control Center, which we'll discuss in the next chapter. Then, the payroll professional can take control of payroll processes. To illustrate how the payroll process works so you can operate the Payroll Control Center efficiently, we'll end this chapter with a practical example.

9.1 Key Payroll Features

The Payroll Control Center offers numerous features to drive efficiency in the payroll process, including the following:

- Validation capabilities
- Key performance indicators (KPIs)
- Alerts
- Solutions for errors
- Collaboration capabilities

In this section, we'll review each feature and explain its role in payroll processing. Before we begin, however, let's look at the regular scheduled and off-cycle processes provided by the Payroll Control Center, to give you the lay of the land.

9.1.1 Scheduled and Off-Cycle Processes

The following process categories are supported by the Payroll Control Center, as shown in Figure 9.1:

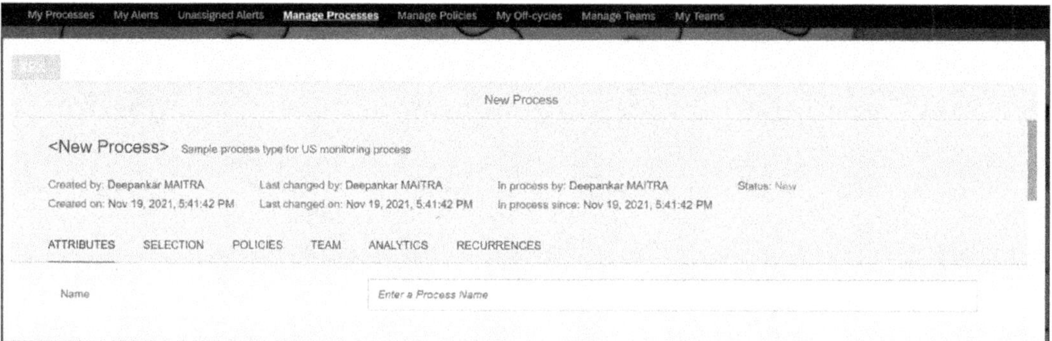

Figure 9.1 Processes in the Payroll Control Center

- **Monitoring processes**
  Monitoring processes are executed before the productive payroll run to detect issues with master data and simulation payroll data as early as possible.
- **Team monitoring processes**
  Team monitoring processes are a special type of monitoring process, an enhanced team functionality for companies with complex human resources (HR) roles and authorizations. The steps covered in these processes include the following:
  - Generating test payroll results for all employees
  - Generating check results for employees who belong to the payroll areas for which the process is executed
  - Monitoring the resolution of the alerts until the start of the productive payroll

  Teams can be set up with one or more team leads and multiple payroll administrators to facilitate team collaboration. Alerts are automatically distributed to teams based on predefined team criteria. A team leader can further assign alerts to individual team members. This type of process has the following features:
  - Segregates alerts belonging to different organizational assignments into teams
  - Refines team collaboration and monitoring
  - Ability to reassign team leads and team members according to their availability in the production system

- **Productive payroll processes**
  Productive payroll processes include all the steps for completing a regular on-cycle payroll run. The steps might include maintaining the control record, identifying and fixing alerts, and executing the payroll schema.
- **Other processes**
  The name of this process is in direct reference to productive payroll: "Other" means "everything else." Typically, can be processes are *before* (e.g., correcting master data) and *after* (e.g., posting to finance) productive payroll.
- **Planned off-cycle productive payroll processes**
  This is a scheduled or preplanned off-cycle payroll for a selection of employees that needs to be run regularly. For example, it might be run to produce the final pay for all terminated employees.
- **Planned off-cycle other processes**
  The name of this process covers all processes other than planned off-cycle productive payroll processes. Typically, these processes can be *before* (e.g., correcting master data) and *after* (e.g., posting to finance) off-cycle productive payroll.
- **Ad hoc off-cycle processes**
  Ad hoc off-cycle processes allows for the execution of payroll for selections of individual employees to quickly produce urgent results required to satisfy immediate business needs. The following types are supported:
  - Payroll adjustment
  - Bonus payment
  - Regular payment
  - Advanced vacation pay

9.1.2 Validations

A payroll policy is a set of validation rules for employee master data and payroll data. Any violation against a payroll policy results in an alert, which can be distributed to administrators for resolution.

SAP provides many types of validations; the standard validation types are listed in Table 9.1.

Number	Validation Rule Example
1	The employment percentage in infotype 0007 does not match the utilization percentage in infotype 0008 (IT7/8).
2	Lived-in state is different from worked-in state.

Table 9.1 Standard Validation Rules

Number	Validation Rule Example
3	An employee receiving zero base payment but also a payment through infotype 0014 or infotype 0015.
4	The timesheets entered in the Cross-Application Timesheet System (CATS) are not fully transferred into time management infotypes.
5	Plant data collection (PDC) errors signify that the time evaluation could not be performed.
6	Employees are missing the main bank details; therefore, a bank transfer for the pay period salary cannot be made.
7	Personnel number locked for payroll.
8	Employees are missing a main address.
9	Unpaid infotype 0011 records exist.
10	This employee has been set up with a time management status of 0 (no time evaluation); thus, time cannot be evaluated for this employee.
11	Active employees in position 99999999 signify that this employee does not have a real position.
12	Employees is missing infotype 0105 records.
13	Position planned compensation does not match infotype 0008 (Basic Pay).
14	Completeness check for employees appointed in the pay period failed—some infotypes are missing.
15	Blocked vendor list may prevent the setup of garnishments.
16	Overview of cash payments that are in error.
17	Gross pay over a preset amount.
18	Retroactive pay over a preset amount.
19	Positive deductions within a range.
20	Variance in net pay over a preset amount.
21	Variance in net pay over a preset percentage.
22	Overtime payment over a certain amount.
23	Employees with overtime exceeding percentage of total pay.

Table 9.1 Standard Validation Rules (Cont.)

Number	Validation Rule Example
24	Employees with large retroactive calculations.
25	Employees with net pay greater than a specific amount.
26	Separated employees with net pay.
27	Negative gross pay.
28	Active, positive time recording employees without pay.
29	Employees with positive deductions.
30	Employees with tax refunds.
31	Employee with deductions greater than gross earnings.
32	Employees with claims.
33	Employees with messages in the payroll log.
34	Employees with tax allowances exceeding threshold.
35	Employees with the **No Tax** flag active.
36	Employees with deductions that have exceeded the annual deduction limit.
37	Employees with a missing cost center.
38	Employees in withdrawn or leave of absence status with an active garnishment.
39	Employees in an off-cycle process whose net pay is greater than a threshold.
40	Employees who were rejected in a post-simulation run.
41	Actions changed or entered in pay period.
42	Wage type in a range over a threshold.

Table 9.1 Standard Validation Rules (Cont.)

These rules are described further in Chapter 10, Section 10.3.3. For example, rule 31 determines if any employee has deductions higher than their gross earnings, which would result in no net pay. Your organization's payroll policy could be satisfied by a selection of rules that analyze thresholds, missing values, and statuses.

9.1.3 Key Performance Indicators

A KPI is a process-context-enabled analytics chart. An example is shown in Figure 9.2.

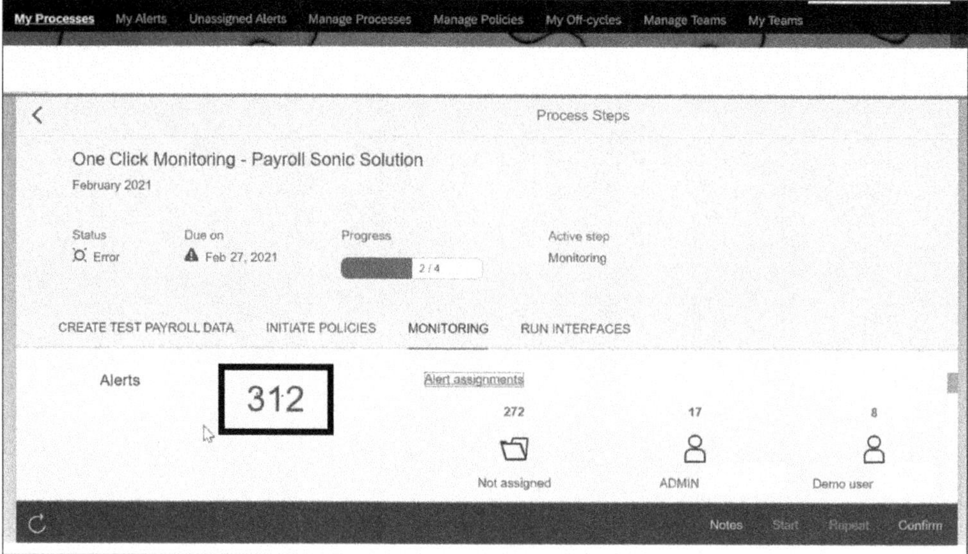

Figure 9.2 An Example KPI

In Employee Central Payroll, a KPI is a key piece of statistical information with a hyperlink that provides details about the indicator once clicked. For example, the **Alerts** KPI in Figure 9.2 shows that 312 errors were produced by the run. If you click on the KPI, details about those errors will be revealed, along with potential solutions.

Similarly, you could specify that the number of employees who earn more than $5000 net pay per pay period should be represented by a KPI. Let's say that the number in a certain month is 5 employees. This number will be displayed with a hyperlink in the audit trail. If you click on the hyperlink, you'll see the names and personnel numbers of the people who fulfilled this condition. You can then determine if these are truly errors or if legitimate reasons exist for an employee to be paid that amount.

SAP provides some standard KPIs, listed in Table 9.2.

Number	KPI Name
1	Bar chart comparison of the total payroll between the last period and the current period
2	Delta chart comparison of the total payroll between the last period and the current period
3	Display the number of employees with organizational changes between the last period and the current period
4	Display an increase or decrease in the growth of employees (as a percentage) between the last period and the current period

Table 9.2 Standard KPIs

Number	KPI Name
5	Display the Harvey Ball chart for the number of employees with a retroactive payroll run between the last period and the current period
6	Show the number of active employees and compares this number between the last period and the current period

Table 9.2 Standard KPIs (Cont.)

9.1.4 Alerts

My Alerts is a page in the Payroll Control Center of Employee Central that helps payroll administrators in Employee Central manage all the alerts that have been assigned to them, as shown in Figure 9.3.

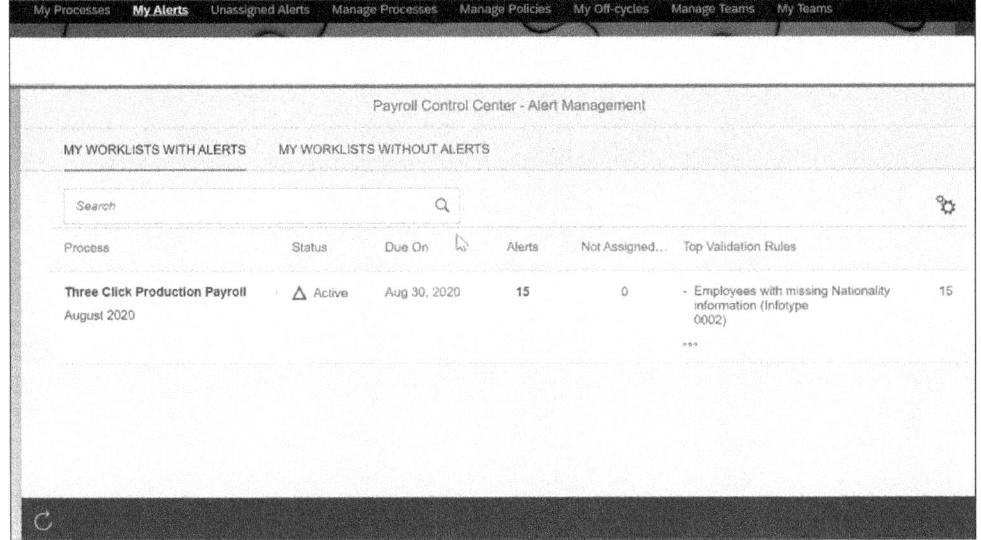

Figure 9.3 My Alerts

On the **My Alerts** page, you can perform the following activities:

- Review solution proposals that might offer a payroll administrator direct entry to correct employee master data (see the example in Section 9.1.5)
- Validate the data after applying a solution by using the recheck function
- Forward an alert that has been assigned to another payroll administrator
- Access a list of unassigned alerts and assign them
- Provide key value indicators for an alert (for example, an employee's wage has increased by 200% as compared to the previous payroll period)
- Read additional help documentation on how to resolve an error

Alerts that are not yet assigned to a payroll processor are grouped together under **Unassigned Alerts**. This section provides a quick overview for payroll administrators to see master data alerts and payroll result alerts that lack a plan to be fixed.

9.1.5 Solutions for Errors

The **Solutions** tab is a great feature of the Payroll Control Center that provides immediate help for the payroll team to deal with errors, without checking with an expert or remembering transaction codes. Instead, the system offers a solution shortcut: a link that you can click with instructions on how to resolve an issue.

In our example shown in Figure 9.4, the Employee Central system has a **Validations PCC** error and suggests the solution **Add missing infotype 9 subtypes 0 data** and further instructing the user to navigate to Transaction PA30.

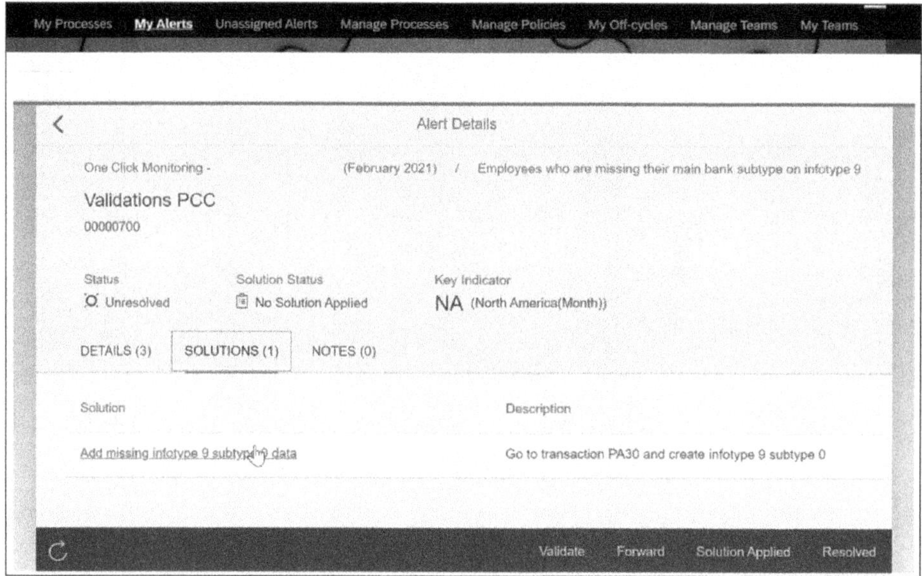

Figure 9.4 Solutions

The solutions feature does require some initial setup, as we'll discuss in Chapter 10, Section 10.3.4. During implementation, each solution must be carefully planned, and a link must be provided for each error scenario that may occur. But this setup is a one-time effort and well worth pursuing since it drives efficiency and accuracy.

9.1.6 Collaboration

The **Manage Teams** page in the Payroll Control Center of Employee Central helps the payroll team collaborate with each other, as shown in Figure 9.5.

On the **Manage Teams** page, you can set up teams of payroll administrators for handling the alerts that arise from team monitoring processes. You can even develop criteria based on employee master data to automatically assign certain types of issues to certain team members based upon their availability and expertise.

The team processes all alerts and finds solutions. The team lead (controllable through authorizations) can perform the following actions:

- Activate and deactivate team members
- Assign alerts to team members
- Assign alerts to another team
- Monitor the progress of alert processing

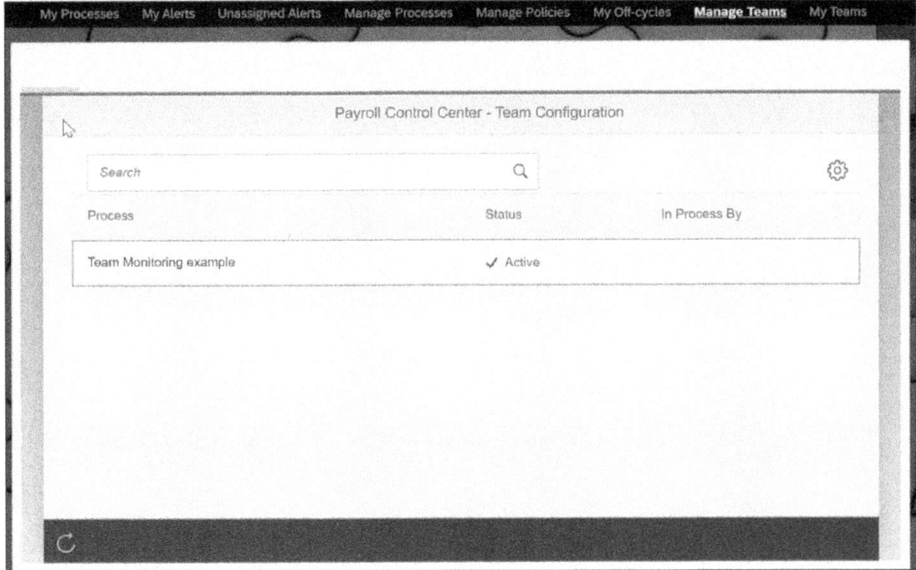

Figure 9.5 Manage Teams

9.2 Payroll Simulation

Simulating a payroll means running payroll in test mode. In a simulation, all the calculations normally done by SAP for live payroll are performed, without saving the results to the live tables.

In previous versions of SAP ERP HCM Payroll, simulation results were only visible during the execution run. Now, you can save simulation results for reporting, an essential capability for the Payroll Control Center because it uses this technique to store results that are then displayed to users through the alert system we described earlier. A simulation is preceded by master data checks that can impact payroll. For example, infotype 0009 (Payment Method) is a mandatory requirement for any employee to be processed in payroll. A simulation will throw an error about the missing infotype 0009 if absent.

An extra layer of validation is provided by the Payroll Control Center through the maintenance of policies. This capability is different from simulation. But some of the same checks on the master data are performed to ensure that errors are caught early and fixed before the actual payroll is run. Simulation and policies are different. Policies do not run the payroll in test mode; only simulation does.

Running a simulation payroll required selection screens, variants, and report export steps in the traditional SAP payroll engine. You can still run a simulation in this older approach (see Chapter 2), but by using the Payroll Control Center, a simulation payroll can be run with a single click on the **My Processes** page, as shown in Figure 9.6. However, this step requires configuration first, which is discussed in Chapter 10, Section 10.3.11.

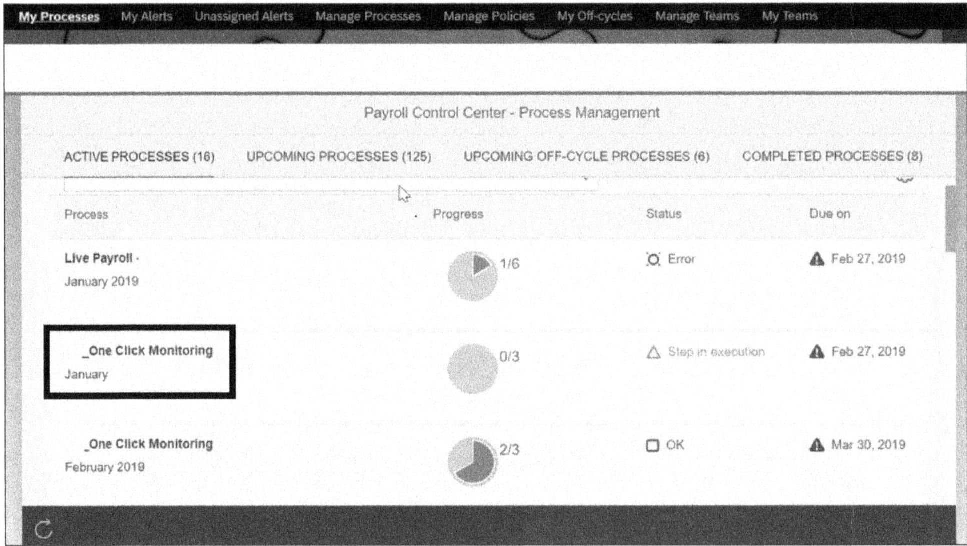

Figure 9.6 Running Payroll Simulation in Payroll Control Center

9.3 Payroll Processing

Payroll processing is the actual calculation of pay and culminates with saving results in live tables that are then available for reporting and subsequent processes. Two distinct methods exist for running payroll:

- An *on-cycle payroll* is the periodic process for all employees, but this payroll can also be run in batches that divided employees into groups for speed or convenience. Sometimes, dividing employees into groups is essential to meet the requirements of different pay periods or check dates (dates of check issuance) that are required by different parts of the organization.

- An *off-cycle payroll* is typically used for a single employee or a small group of employees. This payroll can be used for a regular requirement, such as processing the final pay for all employees terminated during the day (a legal requirement in some parts of the world). Alternatively, this payroll could relate to a processing request for a one-off need for an employee. For example, perhaps an employee did not submit their timesheet on time and has requested immediate payment after submitting their previous week's timesheets.

In the following sections, we'll first look at the steps for performing regular payroll processing and then discuss the reports you can use to confirm that payroll has been processed correctly.

9.3.1 Performing Payroll Processing

The following steps are typically part of a productive (on-cycle) payroll process:

1. Open payroll (release the payroll control record for the payroll process)
2. Productive payroll execution
3. Generation of check results
4. Monitoring
5. Exit payroll

The following steps are typically part of an ad hoc (off-cycle) payroll process:

1. Create posting documents
2. Release posting documents
3. Transfer posting documents
4. Create pay slips
5. Print pay slips
6. Create a pre-Data Medium Exchange (DME) file for transferring money to the designated employee bank accounts
7. Create a DME file

Note that more steps are required for an ad hoc off-cycle payroll than for a regular on-cycle payroll. The reason for this difference is that an on-cycle payroll run is immediately followed by post-payroll processing. The most critical step in this series is the generation of payroll results through the execution of the schema that pertains to the country in which the payroll is running.

Payroll cannot be completed without review, even if no errors have arisen. A payroll manager must look at the results of the payroll as a whole and confirm its accuracy before the payroll continues with the subsequent steps of producing payment files and sending the required payment information to the bank. These activities are usually performed using reports in the backend, which can also be run as part of the payroll process in the frontend as a separate process thread. In the example shown in Figure 9.7, the entire process, with all the steps mentioned earlier, are configured as steps within the same process called *live payroll*. A payroll manager simply must click the **Live Payroll** link. This workflow is again a great improvement from the traditional SAP ERP HCM Payroll, where either each step was manually executed through selection screens and variants and an improvement on the traditional process model using Transaction PUST. None of those steps are required in this context. However, configuration will be required, which is the subject of Chapter 10.

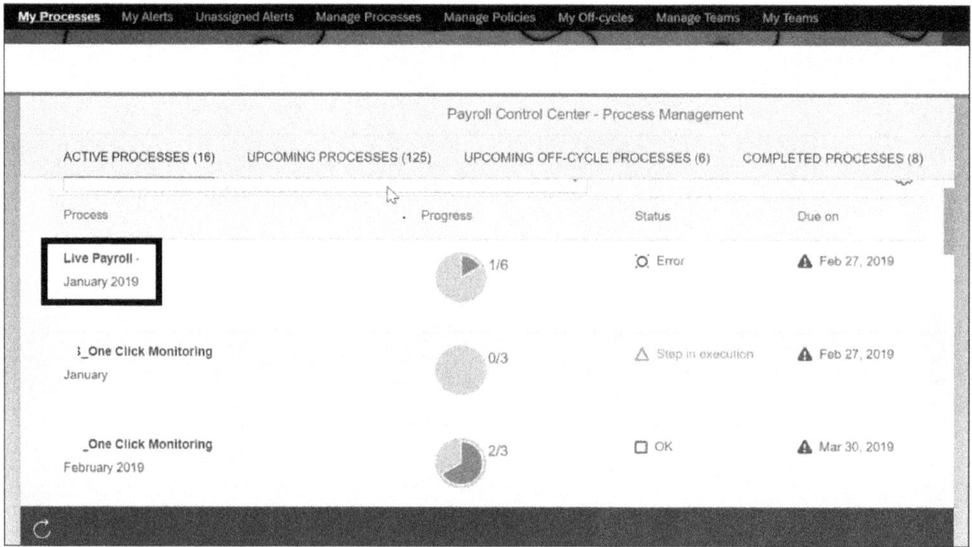

Figure 9.7 Live Payroll Run

9.3.2 Reports to Confirm Periodic Payroll

Three main reports can be used to confirm a periodic payroll:

- Payroll Journal report
- Payroll Reconciliation report
- Pay Results report (for an employee)

You can have these reports run automatically in the processing, along with a manual step to confirm. The idea is that the payroll manager will look at these reports and, if satisfied, use the manual step to go forward with subsequent payment processes. If not satisfied, the process can be repeated as many times as required.

Payroll Journal

The Payroll Journal report is executed with Transaction PC00_MXX_CLJN, where XX stands for the 2-digit country grouping (field MOLGA) in SAP, as shown in Figure 9.8 for the USA (MOLGA = 10).

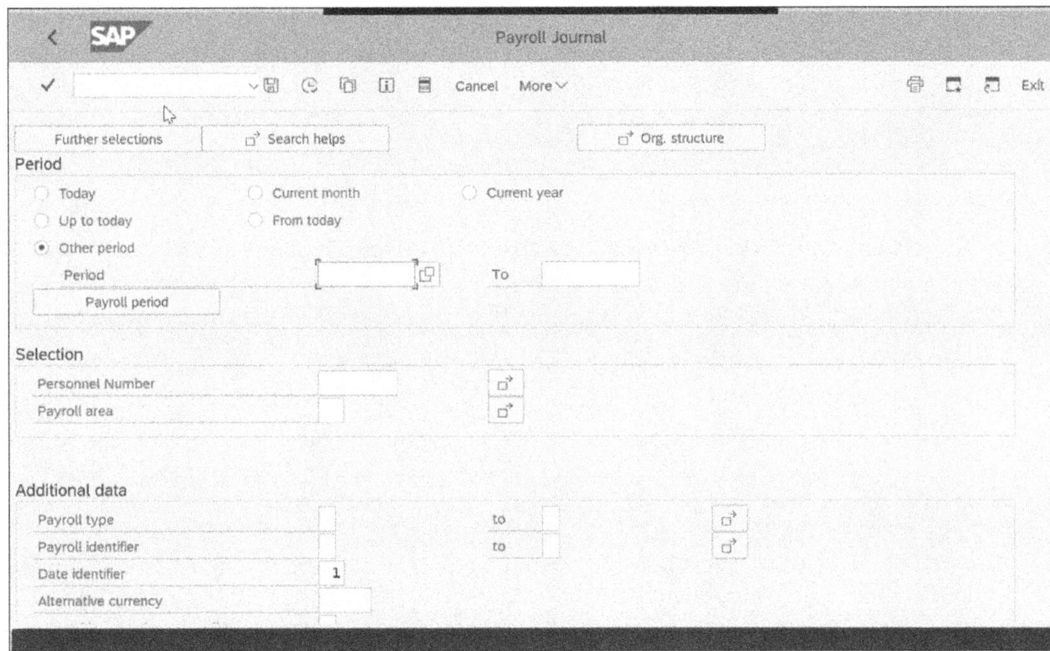

Figure 9.8 Payroll Journal Report

Once executed, the report produces a totals sheet for easy review. In our example shown in Figure 9.9, this report provides the totals of the payroll processed organized by wage type at various organizational levels (**Company Code**, **Cost Center**, **Employee Group**, **Personnel Area**, etc.). These wage types are classified into separate sections, such as **Earnings**, **Deductions**, etc.

```
Payroll Journal
Cancel  More  Exit

Company Code:  UAE   Cost Center:  *    PPBegin:  24.02.2021  PPEnd:  28.02.2021  Check Date:  01.03.2021  Page:        2
Org Unit:  00000000  Personnel Area:   US03  Personnel Subarea:   HN02  Payroll Area:   *                  Period:  02/2021
            8.904,00          8.904,00  MedcarEE          129,11          185,66
                                                        2.120,12

Company Code:  UAE   Cost Center:  *    PPBegin:  24.02.2021  PPEnd:  28.02.2021  Check Date:  01.03.2021  Page:        3
Org Unit:  00000000  Personnel Area:   US03  Personnel Subarea:   *     Payroll Area:   *                  Period:  02/2021
*****************************************************************************************************************************
Sum of the structure
*****************************************************************************************************************************
Earnings:       Current    Year to Date  Taxes:          Current    Year to Date  Dedn:          Current      YTD   Net current
-----------------------------------------------------------------------------------------------------------------------------

Reg time        7.004,00          7.004,00  W/H    EE      1.438,96        2.203,75                              Paymen    6.783,88
CarAllow        1.900,00          1.900,00  OASDI EE         552,05          793,85                                        6.783,88
                8.904,00          8.904,00  MedcarEE         129,11          185,66
                                                          2.120,12

Company Code:  UAE   Cost Center:  *    PPBegin:  24.02.2021  PPEnd:  28.02.2021  Check Date:  01.03.2021  Page:        4
Org Unit:  00000000  Personnel Area:   *     Personnel Subarea:   *     Payroll Area:   *                  Period:  02/2021
*****************************************************************************************************************************
Sum of the structure
*****************************************************************************************************************************
Earnings:       Current    Year to Date  Taxes:          Current    Year to Date  Dedn:          Current      YTD   Net current
-----------------------------------------------------------------------------------------------------------------------------
```

Figure 9.9 Organizational Totals in the Payroll Journal Report

9

A payroll manager should generally have an idea of the amount of money that should go through payroll, along with totals of deductions and specific categories of wages. With this understanding, a payroll manager should be able to quickly confirm the amounts are correct at a summary level.

Payroll Reconciliation Report

Sometimes, this review reveals a total that isn't quite right, such as an overly bloated Federal Insurance Contributions Act (FICA) tax or cumulative bonus amount in a non-bonus period. If this problem arises, a payroll manager can run the Payroll Reconciliation report, which can be used to compare multiple year-to-date (YTD) amounts across multiple payroll periods.

To execute this report, start Transaction PC00_MXX_REC, where XX stands for the 2-digit country grouping (field MOLGA) in SAP, as shown in Figure 9.10 for USA (MOLGA = 10).

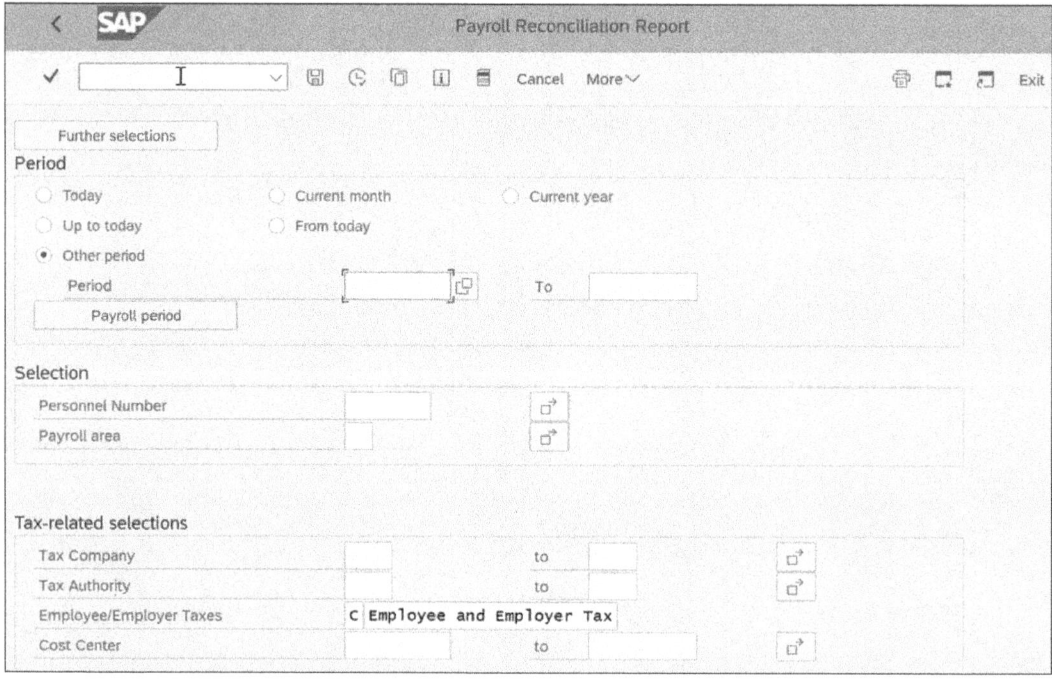

Figure 9.10 Payroll Reconciliation Report

Once executed, the report produces a YTD amount for each selected wage type, across chosen periods, for easy review. In our example shown in Figure 9.11, the fields that are visible are configurable on the report selection screen. You can see the YTD values for the various wage types of an employee, along with month-to-date and quarter-to-date figures if required. These periodic totals can be extremely helpful for reconciling taxes when submitting files to the federal government or a state/local tax jurisdiction.

Payroll Reconciliation Report

Selection Period 02.2021 (01.01.2021 - 15.01.2021)
Run Date/Time 09.05.2021 / 16:01:06
Wage Type Selections Wage Type

Tx.	Auth	PersNo	Payroll Results	WT	Wage Type Text	Current Pe	Current Pe	Current Pe	Current Pe	M-T-D Numb	M-T-D Amou	Q-T-D Numb	Q-T-D Amou	Y-T-D Numb	Y-T-D
U9E		9889	Normal Period Results	/101	Total gross	02.21-02.21	17.01.2021		2.000,00						2.0
U9E		9889	Normal Period Results	/102	401(k) Wages	02.21-02.21	17.01.2021		2.000,00						2.0
U9E		9889	Normal Period Results	/110	Net payments/Deductions	02.21-02.21	17.01.2021		834,02-						8
U9E		9889	Normal Period Results	/11	Base wage for BSI	02.21-02.21	17.01.2021		2.000,00						2.0
U9E	FED	9889	Normal Period Results	/301	TG Withholding Tax	02.21-02.21	17.01.2021		2.000,00						2.0
U9E	FED	9889	Normal Period Results	/303	TG EE Social Security Tax	02.21-02.21	17.01.2021		2.000,00						2.0
U9E	FED	9889	Normal Period Results	/304	TG ER Social Security Tax	02.21-02.21	17.01.2021		2.000,00						2.0
U9E	FED	9889	Normal Period Results	/305	TG EE Medicare Tax	02.21-02.21	17.01.2021		2.000,00						2.0
U9E	FED	9889	Normal Period Results	/306	TG ER Medicare Tax	02.21-02.21	17.01.2021		2.000,00						2.0
U9E	FED	9889	Normal Period Results	/310	TG ER Unemployment Tax	02.21-02.21	17.01.2021		2.000,00						2.0
U9E	AK	9889	Normal Period Results	/310	TG ER Unemployment Tax	02.21-02.21	17.01.2021		2.000,00						2.0
U9E	AK	9889	Normal Period Results	/320	TG EE Unemployment Tax	02.21-02.21	17.01.2021		2.000,00						2.0

Figure 9.11 Wage Type Level Current and Year-to-Date Amounts

A payroll manager can review the details about the current period in comparison with past results and decide whether the payroll is ok to release.

Pay Results Report

If you still suspect that something isn't quite right, as a payroll manager, you can run a report based on individual employees using Transaction PC_PAYRESULT, as shown in Figure 9.12.

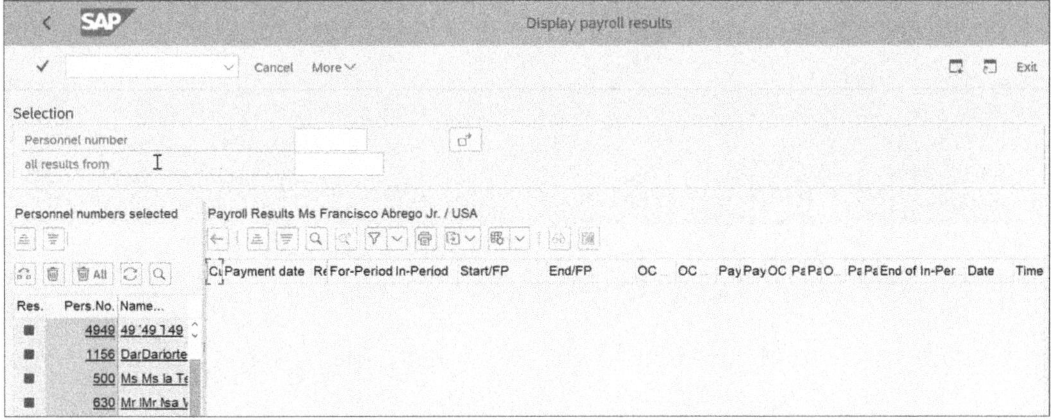

Figure 9.12 Transaction PC_PAYRESULT Report

Each employee's entire payroll calculation is then displayed in a chronologically organized stack in this report. Now, you can double-click on an employee to review. All pay periods processed (including off-cycles) for that chosen employee will be displayed. In our example shown in Figure 9.13, the employee's results appear in a list—one line per pay period. The details for each period can be accessed by double-clicking on a line.

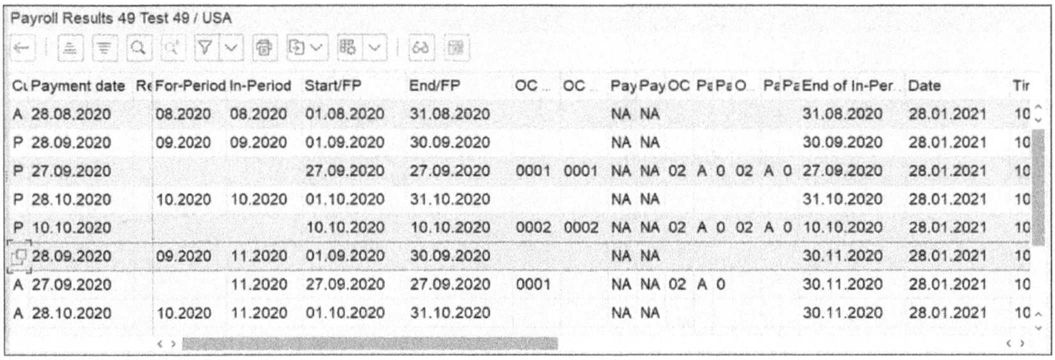

Payroll Results 49 Test 49 / USA

Ci	Payment date	Re	For-Period	In-Period	Start/FP	End/FP	OC ...	OC ...	Pay	Pay	OC	Pa	Pa	O...	Pa	Pa	End of In-Per...	Date	Tir
A	28.08.2020		08.2020	08.2020	01.08.2020	31.08.2020			NA	NA							31.08.2020	28.01.2021	10
P	28.09.2020		09.2020	09.2020	01.09.2020	30.09.2020			NA	NA							30.09.2020	28.01.2021	10
P	27.09.2020				27.09.2020	27.09.2020	0001	0001	NA	NA	02	A	0	02	A	0	27.09.2020	28.01.2021	10
P	28.10.2020		10.2020	10.2020	01.10.2020	31.10.2020			NA	NA							31.10.2020	28.01.2021	10
P	10.10.2020				10.10.2020	10.10.2020	0002	0002	NA	NA	02	A	0	02	A	0	10.10.2020	28.01.2021	10
	28.09.2020		09.2020	11.2020	01.09.2020	30.09.2020			NA	NA							30.11.2020	28.01.2021	10
A	27.09.2020			11.2020	27.09.2020	27.09.2020	0001		NA	NA	02	A	0				30.11.2020	28.01.2021	10
A	28.10.2020		10.2020	11.2020	01.10.2020	31.10.2020			NA	NA							30.11.2020	28.01.2021	10

Figure 9.13 Results for Multiple Pay Periods

In this way, you can drill down into the payroll period of concern, which reveals all the pay results tables, as shown in Figure 9.14. (Pay results tables were described in detail in Chapter 2, Section 2.3.)

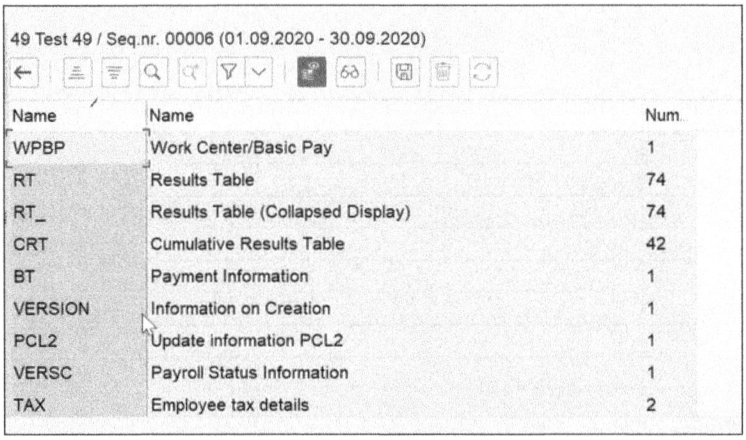

49 Test 49 / Seq.nr. 00006 (01.09.2020 - 30.09.2020)

Name	Name	Num...
WPBP	Work Center/Basic Pay	1
RT	Results Table	74
RT_	Results Table (Collapsed Display)	74
CRT	Cumulative Results Table	42
BT	Payment Information	1
VERSION	Information on Creation	1
PCL2	Update information PCL2	1
VERSC	Payroll Status Information	1
TAX	Employee tax details	2

Figure 9.14 Narrowing Down to One Pay Period

For example, if you double-click on table RT_, details about all wage types in the payroll results will be displayed, as shown in Figure 9.15.

This screen provides payroll managers the ultimate depth of information to investigate an issue at the lowest level possible.

```
Payroll Results
Personnel No.       4949 49 Test 49 - USA
Seq. number         00006 - accounted on 28.01.2021 - current result
For-Period          09.2020 (01.09.2020 - 30.09.2020)
In-Period           11.2020 (Fin.: 30.11.2020)

Table RT - Results Table (Collapsed Display)

A WT   WT Text       APC1C2C3aBKoReBTAwvTvNUnit Amt/Unit No.            Amount

* /101 Total gross                                                    5.400,00
* /102 401(k) Wage                                                    4.400,00
* /110 Net payment                                                       50,00-
* /114 Base wage f                                                    4.400,00
* /301 TG Withhold  01                                                4.400,00
* /301 TG Withhold  02                                                4.400,00
* /303 TG EE Socia  01                                                4.400,00
* /304 TG ER Socia  01                                                4.400,00
* /305 TG EE Medic  01                                                4.400,00
* /306 TG ER Medic  01                                                4.400,00
* /310 TG ER Unemp  01                                                4.400,00
* /310 TG ER Unemp  02                                                4.400,00
* /316 TG Employme  02                                                4.400,00
* /341 TG EE Disab  02                                                4.400,00
```

Figure 9.15 Pay Results from Table RT_

Off-Cycle Reports

Off-cycle payroll results can also be verified through the three reports described earlier in this section. On the selection screen, you'll need to identify the off-cycle payroll by its date and type. In our example Payroll Journal report shown in Figure 9.16, the **Payroll Type** field identifies the type of off-cycle payroll (in our example, A indicates a bonus run). The **Payroll Identifier** field indicates the specific number of the off-cycle payroll run on the same day. Once a date has been entered into the **Period** fields at the top, the **Date Identifier** field identifies the type of date, for instance, a pay date (**2**) versus a period-end date (**3**).

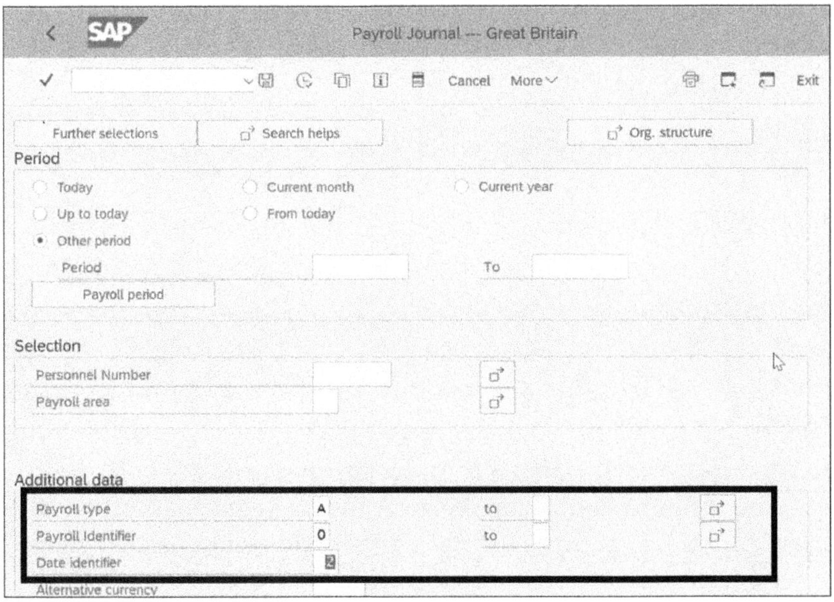

Payroll Journal --- Great Britain

Cancel More Exit

Further selections | Search helps | Org. structure

Period

Today | Current month | Current year

Up to today | From today

Other period

Period | To

Payroll period

Selection

Personnel Number

Payroll area

Additional data

Payroll type: A to

Payroll Identifier: 0 to

Date identifier: 2

Alternative currency

Figure 9.16 Off-Cycle Run for Payroll Journal (UK)

9.4 Post-Payroll Processing

Post-payroll processing covers all the steps required after the payroll is calculated, generally involving the following three tasks:

- Ensuring the money reaches the employee
- Complete the accounting for the money that was paid (typically in SAP ERP Financials or SAP S/4HANA Finance)
- Sending details about deductions to external agencies, such as to medical insurance providers for continuation of service or to government collection agencies for compliance

Notice that these three activities occur outside Employee Central Payroll, perhaps even outside your SAP system. These steps always require some technical support, especially for managing the communication between the SAP system and the various vendors/ agencies. We discussed this topic in Chapter 6 when we reviewed integration. These integrations are time critical and sometimes rely on middleware to automate the transfer of data between systems.

9.5 Practical Example

Let's end this chapter with an example of payroll processing from start to finish. Stephen is a payroll process manager. He looks after payroll for 10,000 employees who are all in a monthly payroll area. In other words, payroll is run once a month. and employees are paid 12 times a year.

Some important dates for Stephen's payroll process, as shown in Figure 9.17, include the following dates:

- Start of payroll period: March 1
- End of payroll period: March 31
- Check date: March 31
- Final payroll process start: March 23

This final date is dependent on many factors. In this instance, the company has decided to pay employees in advance because the payroll process start date is 7 days before the end of the payroll period. Thus, all the data that is required for running payroll must come in before March 24. At least the last 7 days must be made an advance because the data for this period has not come in.

Another important aspect of this timeline is that companies generally don't want to pay too much in advance—doing so creates the risk of overpayment and employee dissatisfaction due to incorrect pay, since the more times pay is made in advance, the greater the chance of inaccuracy.

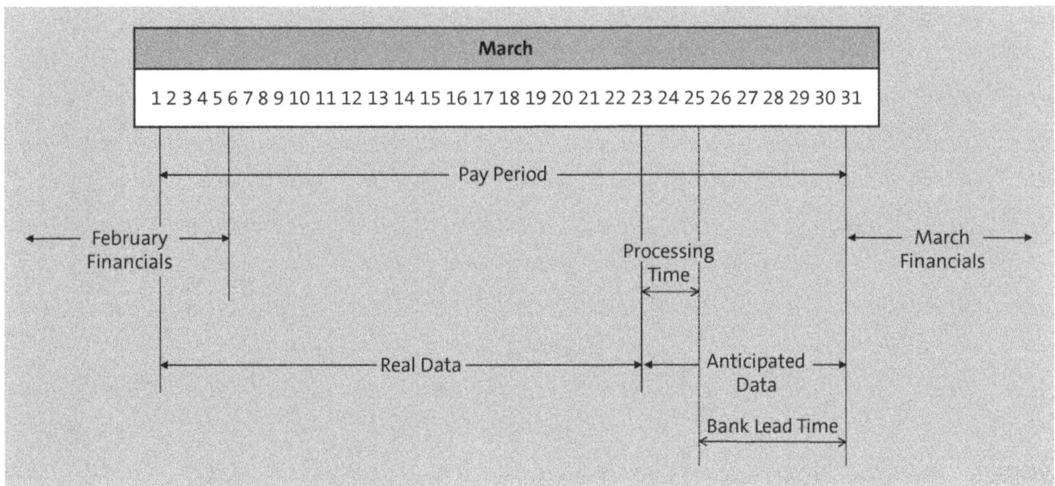

Figure 9.17 Payroll Processing Timeline

Since this calendar involves a tight timeline (due to a tight deadline from the bank, only 2 days are available to process live payroll, between March 23 and 25), and Stephen needs some help. He has a team of subordinate helpers, as shown in Figure 9.18.

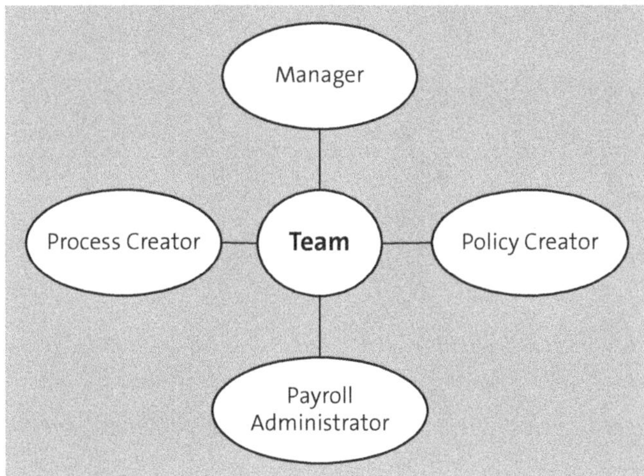

Figure 9.18 Payroll Team

First, March 6 is 1 day after the full close of the previous payroll. On this day, Stephen can open the Payroll Control Center and review the payroll process. In his role as a payroll manager, he is primarily concerned with the following processes: complete processes, upcoming processes, active processes, and upcoming off-cycle processes. These processes should all be populated into his Payroll Control Center cockpit.

Stephen's Payroll Control Center cockpit includes a **Completed Processes** tab, as shown in Figure 9.19, which shows the processes from last month that were completed.

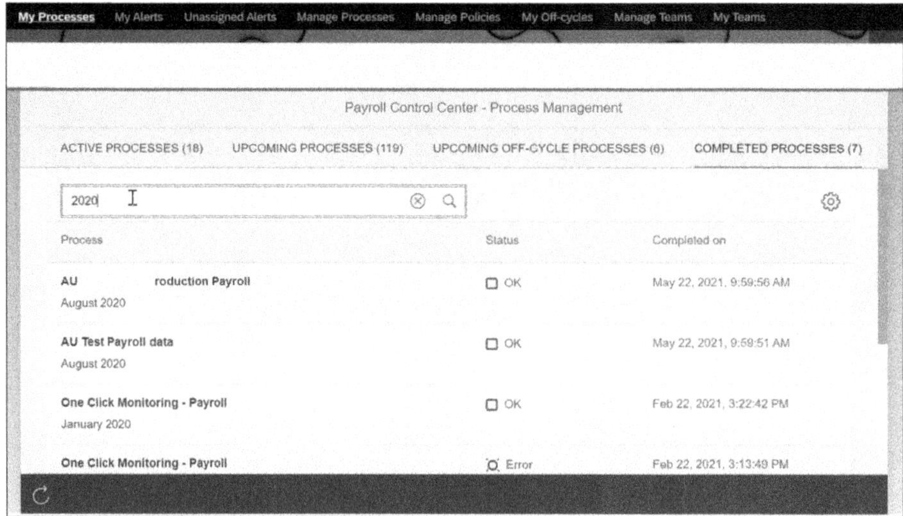

Figure 9.19 Completed Processes Tab

Stephen now opens the **Upcoming Processes** tab, as shown in Figure 9.20, which lists the processes that are scheduled to run.

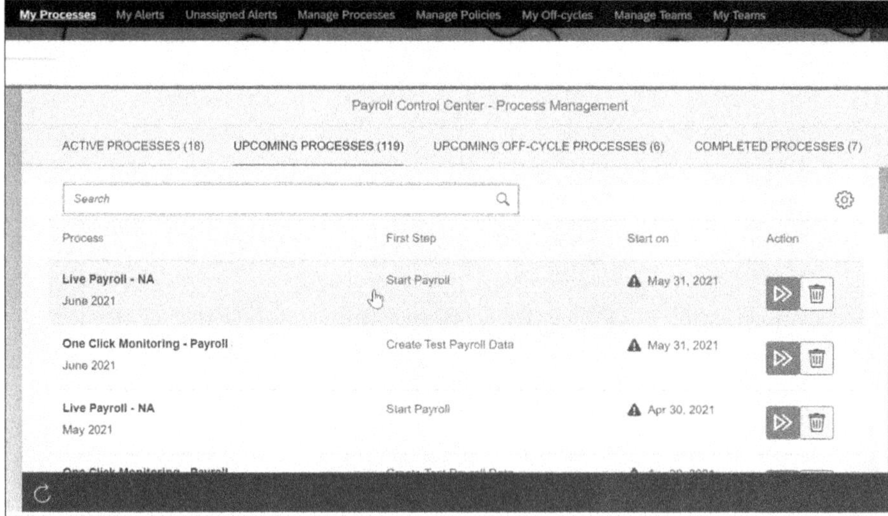

Figure 9.20 Upcoming Processes Tab

Stephen can start the monitoring process by clicking on the **Payroll** tab in Employee Central, which pulls up processes organized by category, as shown in Figure 9.21. Monitoring is a combination of the following steps:

- Generating test payroll results for all employees
- Generating check results for employees who belong to the payroll areas for which the process is executed

- Assigning alerts to payroll administrators and monitoring the resolution of the alerts until the start of the productive payroll

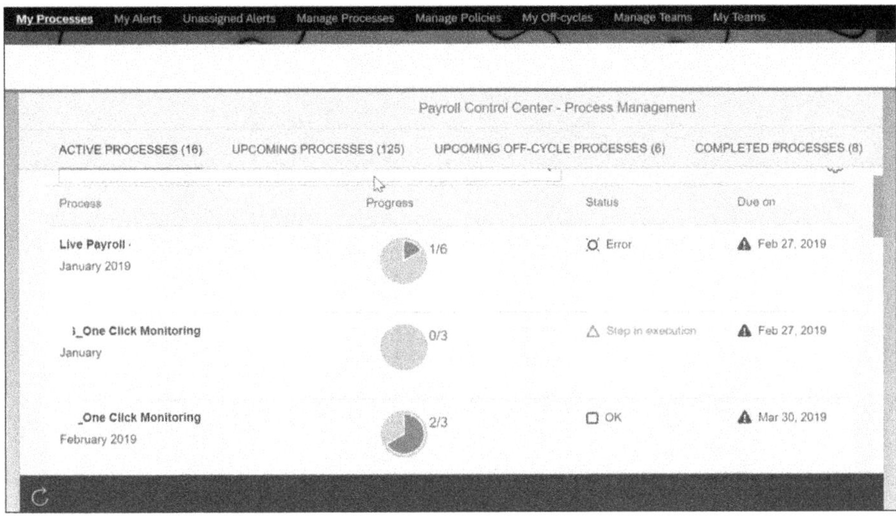

Figure 9.21 Monitoring

On March 8, Stephen goes into the **Active Processes** tab and examines the status of his monitoring process again. Starting the payroll process automatically moves him from the **Upcoming Processes** to the **Active Processes** tab. He finds 312 alerts, as shown in Figure 9.22.

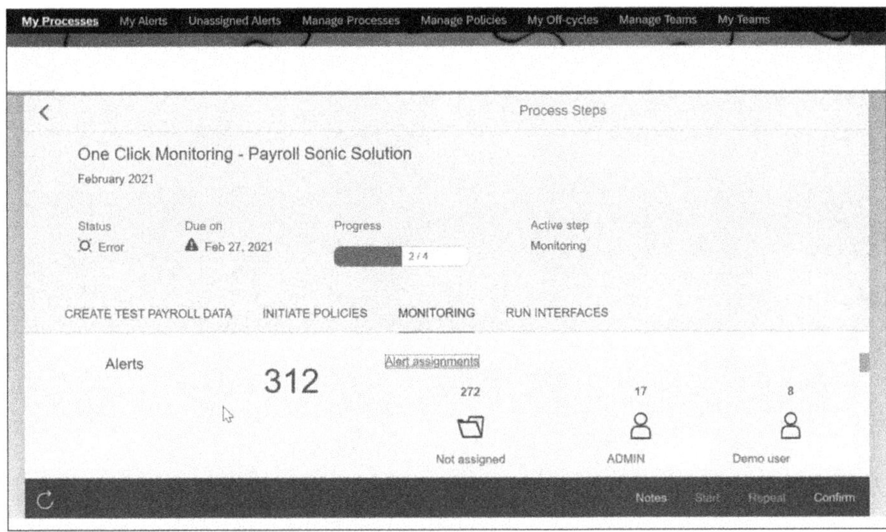

Figure 9.22 Alerts Waiting to Be Fixed

These different conditions reveal potential problems with the payroll run. He clicks on the **Not Assigned** button, which lists alerts that need attention, as shown in Figure 9.23.

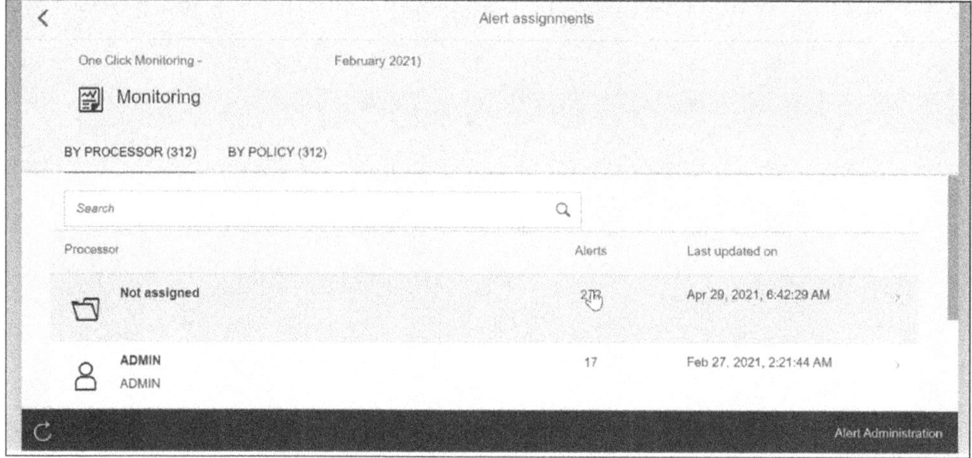

Figure 9.23 Alerts That Need Assignment

He examines the alerts to see what might be wrong by clicking on the **Not Assigned** button. He is now able to see the types of errors, as shown in Figure 9.24.

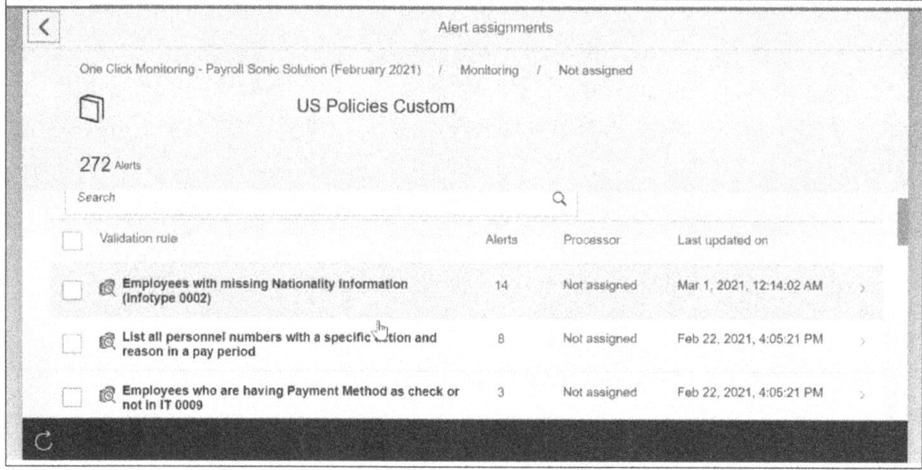

Figure 9.24 Types of Errors

Stephen can now distribute the various alerts to his payroll administrators based on their expertise and availability. He makes these assignments through the assignment function in the **Active processes** window, as shown in Figure 9.25.

On March 10, Roger, one of Stephen's payroll administrators, logs on and finds that Stephen has assigned him 20 alerts. Roger picks them up and starts resolving these issues. He can fix 19 of the 20, but the 20th alert says, "Net pay greater than $10,000." After researching the employee who is receiving the high net pay, Roger discovers that this employee is a new hire who has just been given a sign-on bonus. Roger marks this alert as **Resolved**, as shown in Figure 9.26.

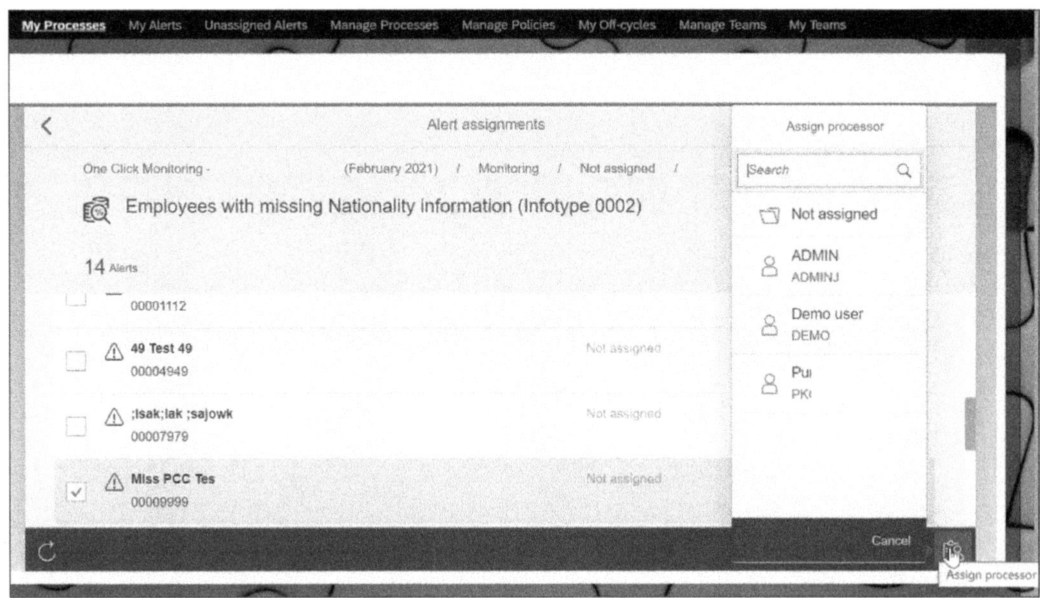

Figure 9.25 Assigning an Alert

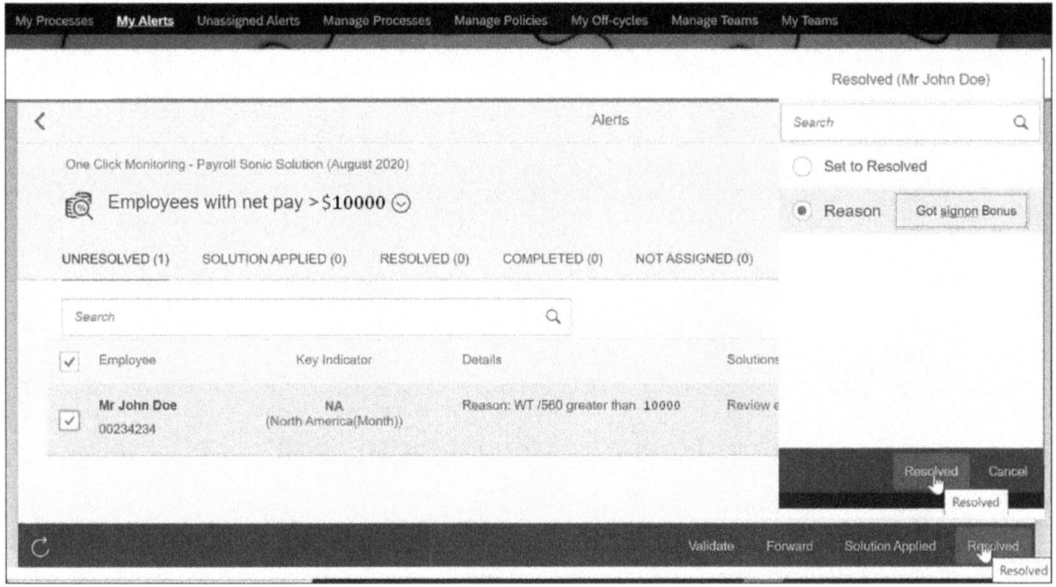

Figure 9.26 Resolving an Alert

On March 11, Stephen goes back to the **Active Processes** window and checks on how the error corrections are progressing. He identifies the monitoring process and clicks on **Repeat**, which kicks off the process again. After some time, he reviews the list and finds the list of alerts has dropped from 500 to only 50. He repeats this process every other day until March 22.

On March 23, Stephen goes to the **Active Processes** window and finds that the process he had started the previous day has completed, and no alerts need to be addressed. Stephen now feels ready to run the productive process, so he kicks off the productive payroll by clicking the **Start Process** button, as shown in Figure 9.27.

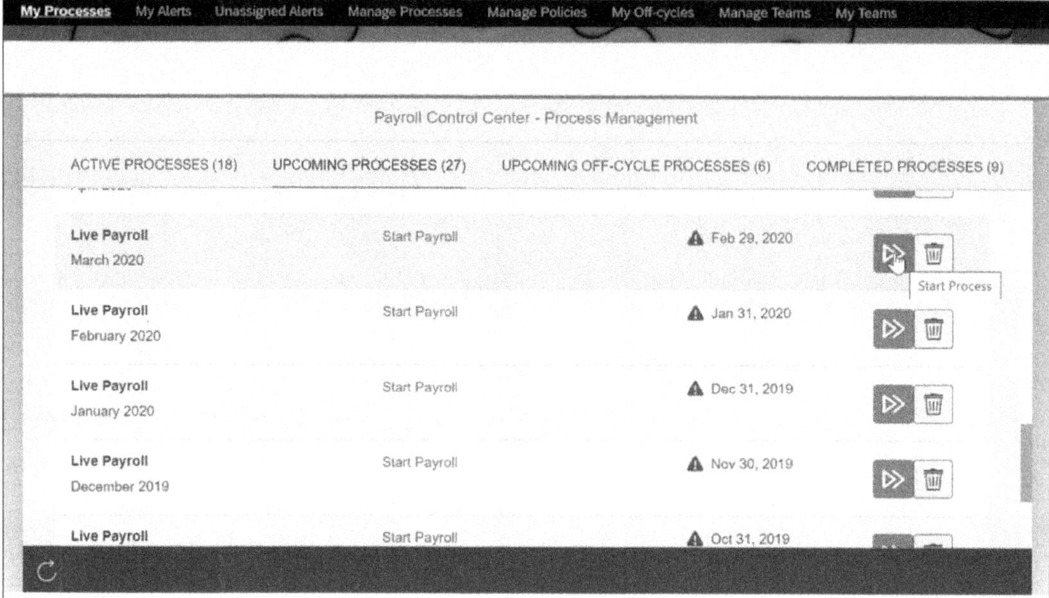

Figure 9.27 Start Button

On March 25, Stephen returns to the **Active Processes** tab and reviews the alerts. He sees none.

Now, Stephen now kicks off the next process in Employee Central Payroll, which performs the following tasks:

- Generating pay slips
- Generating bank files
- Running the financial posting in simulation mode

The bank file is saved in the designated secure folder and then picked up by an automated process that always looks for a file of this type. Once found, this file is moved to the secure server owned by the bank through Secure File Transfer Protocol (SFTP), as is customary for such high-value bank transactions. On March 31, the payment file is duly credited by the bank. The employees can receive their pay.

On April 1, Stephen reviews the alerts from the financial posting simulation he started earlier. He finds some errors that he can fix himself. Now, he can kick off the final process under the **Active Processes** tab. This process performs the live financial postings in the finance system.

This timeline of events gives you an idea of a typical payroll run, just as an example. The method, type, and sequence of operations will be generally similar for all organizations. The details will of course be different, based upon the frequency of payroll and the population of your company. Some of these processes can be automated, through configuration in the Payroll Control Center, although in our example Stephen kicks off all these processes manually.

9.6 Summary

This chapter explored the fundamental principles behind payroll processing, highlighting some key points that require careful attention and efficient management to meet tight timelines. Paying employees in a timely manner is an expectation held by every employee in your organization, including the CEO. You're now familiar with the features that make the Payroll Control Center indispensable for completing payroll efficiently. This solution also provides an audit trail capturing how payroll was calculated so that auditors can confirm that payroll has been completed without any (intentional or unintentional) lapses. We'll now examine how to set up these features so that you can fine-tune these features to enhance your own payroll process.

Chapter 10
Configuring the Payroll Control Center

Timeliness and quality in payroll processing depend not only on the people who are entrusted with this critical process in your organization, but also on the depth and reach of the tools they are using. Configuring the Payroll Control Center ensures that the payroll team has the tools they need to succeed in the race to deliver accurate payroll on time, every time. Configuring this tool is thus complex and maintained in multiple systems.

This chapter describes the details of configuring the star feature of Employee Central Payroll—the Payroll Control Center. There are three areas where configuration will need to be performed:

- Provisioning
- Employee Central
- Employee Central Payroll

Employee Central must be connected successfully to Employee Central Payroll for the configuration to work, a topic we covered in Chapter 6 and Chapter 7 in detail.

We'll begin by discussing the various components of the Payroll Control Center through the use of an entity relationship diagram. Once we understand the concepts behind the items that make up the solution, we'll dive into the roles and applications that drive the use of the solution.

The rest of the chapter will be devoted to the configuration tasks required in Employee Central Payroll and Employee Central. Most tasks take place in Employee Central Payroll, with several new concepts like declustering that we'll cover in this chapter. Configuring Employee Central Payroll is all about the frontend, and this chapter will include a discussion of security and audit trails.

10.1 Concepts

The Payroll Control Center uses several elements as part of its design. Understanding these elements fully is necessary to their configuration. The different components of the Payroll Control Center are shown in Figure 10.1.

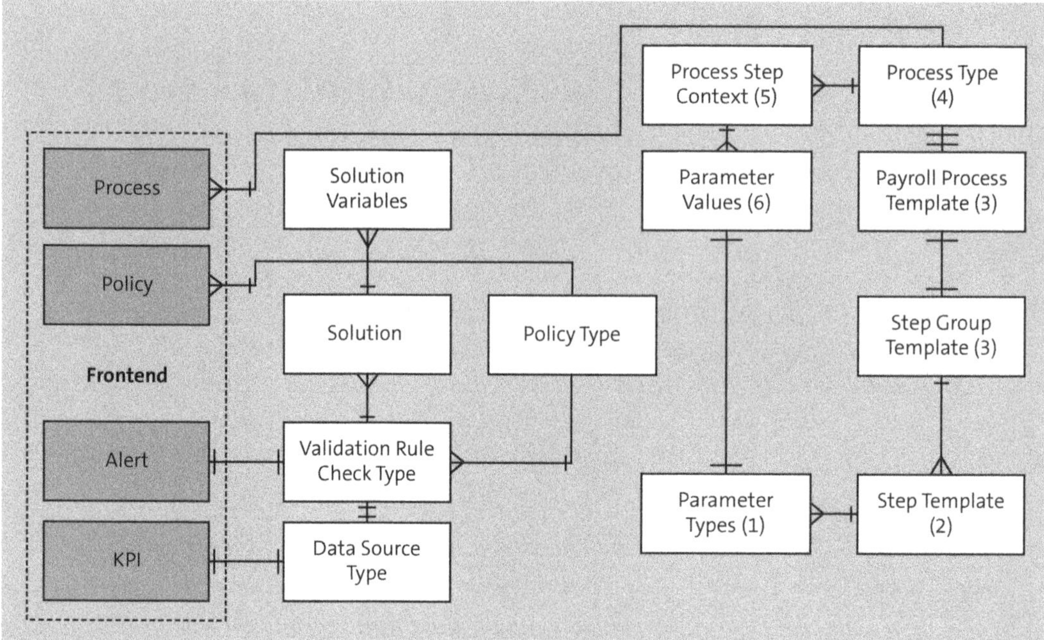

Figure 10.1 Payroll Control Center Entity Relationship Diagram

Table 10.1 lists all these elements.

Term	Definition	Example
Validation rule	This rule consists of logic executed by the system to identify discrepancies that can save time later and prevent incorrect processing.	A validation rule that shows all salaried employees with 20% more pay, compared to the previous payroll period.
Alert	An alert is a message that a validation rule produces.	An alert indicating a missing cost center missing.
Solution	A solution is connected with a validation rule, providing the instructions on how errors detected by the validation rule can be fixed. A solution has many variables that are components of the solution.	For an error with missing cost center, a solution could be to provide a link to navigate directly to the Employee Central portlet where the cost center can be entered. The link is stored in a variable used for multiple solutions.

Table 10.1 Elements of the Payroll Control Center

Term	Definition	Example
Data source type	Data sources are used to model steps of a payroll process. Operations can be performed on the data. The data source type defines the type of data source. A KPI is the type of data that we're particularly interested in since KPIs identify aspects of the payroll or validation check that is performed.	Count of employees in payroll area US1.
Policy	A policy is a set of validation rules grouped according to the needs of the process.	A policy for tax payroll checks as distinct from a policy for master data checks.
Parameter types and values	Parameter types define a step template, providing a value at runtime to execute steps with the process context.	The input is a parameter type, and the report name is the parameter value.
Analytics chart	An analytics chart is a graphic that provides analytical statistics about master data and payroll data that changes over time.	The number of employees that had pay increases in this period compared to the last period.
Analytics	A set of analytics charts grouped according to your needs. Analytics can provide a statistical overview of master data and payroll data changes over time.	Analytics for gross pay.
Step templates	A step template is a single activity within a payroll process.	A program with a predefined variant and relevant values required to execute the program, for example, program RPCALCU0 with variant VAR001.
Payroll process	This process is a sequence of all steps that are required to execute payroll and all related activities (for example, legal reporting) for a specific entity.	A payroll process that includes data validation, running payroll, and pay statements.
Recurrence of payroll process	This information defines a payroll process for a specific entity and period.	Payroll process for October of the current year and payroll area XX.

Table 10.1 Elements of the Payroll Control Center (Cont.)

Term	Definition	Example
Policy type	A policy type groups the relevant technical context for defining a policy (such as selection type and country/region). A policy type is relevant only when policies are created with business function `HCM_LOC_CI_92` (Manage Policies with the Simplified Configuration).	Policy type for monthly payroll for hourly paid employees in the United States.
Process type	Groups the technical context for defining processes later, such as the step context, program, and variant; the process authorization prefix; and the process category. A process type is relevant only when policies are created with business function `HCM_LOC_CI_92` (Manage Policies with the Simplified Configuration).	Process type for productive payroll for employees in the UK.

Table 10.1 Elements of the Payroll Control Center (Cont.)

Configuring the Payroll Control Center has been rather complicated, until recently, when SAP introduced the technique of *simplified configuration*. This chapter will make use of this new technique, as recommended by SAP.

Business function `HCM_LOC_CI_92` must be activated through Employee Central Payroll using Transaction SWF5, as shown in Figure 10.2. Select the **Payroll Control Center Simplified Configuration (Reversible)** checkbox.

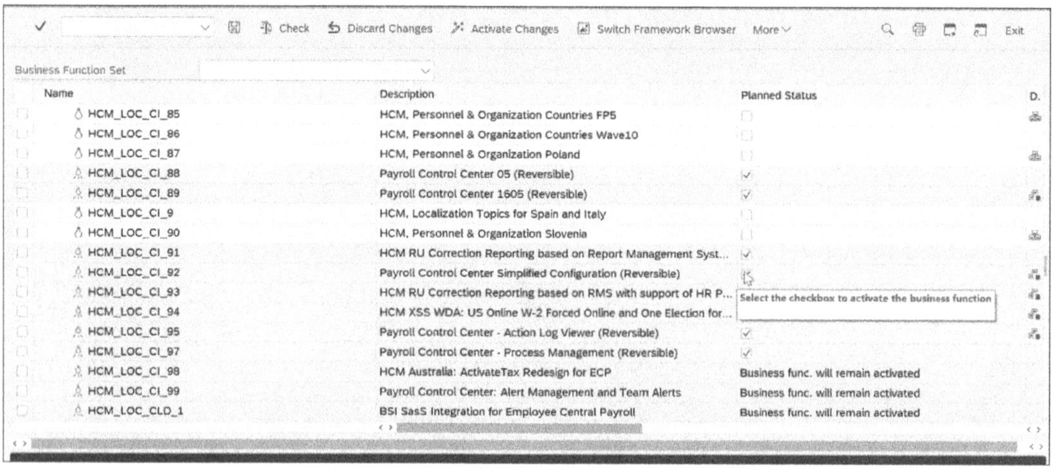

Figure 10.2 Transaction SFW5 to Activate the Simplified Configuration for the Payroll Control Center

Using this method, process types and policy types can be configured in the Configuration Workbench (Transaction SPRO). However, the policies and processes themselves can be set up directly by business users in production without going through configuration. Also keep in mind that business users do not log on to the backend SAP system—the point of Employee Central Payroll is to run payroll directly from Employee Central with a single user interface (UI) for both Employee Central and Employee Central Payroll.

10.2 Roles and Applications

The Payroll Control Center offers a number of applications to help business users complete payroll tasks. Table 10.2 lists the various roles and the applications that each role has available to it.

Role	Task	Application
Policy creator	Creates policies using check types.	Manage Policies
Process creator	Creates processes and assigns policies to processes.	Manage Processes
Payroll process manager	Ensures that the payroll has been processed successfully. In charge of the periodic payroll, the payroll process manager assigns alerts to team members and monitors the entire process.	My Processes My Off-cycles
Team creator	Creates the team to process payroll.	Manage Teams
Team lead	Activates and deactivates the team.	My Teams
Payroll administrator	Resolves issues and reviews unassigned alerts.	My Alerts Unassigned Alerts
Auditor	Reviews action logs for policies and processes	Audit Trail

Table 10.2 Roles and Their Applications

10.3 Configuring the Payroll Control Center in Employee Central Payroll

The Payroll Control Center is configured primarily through the backend where two distinct activities need to be performed in Employee Central Payroll:

- Configuring rules and settings
- Initiating daemon processes

The Payroll Control Center invokes daemon processes according to process logic that you configure. This approach is a bit different from the typical configuration activities in SAP applications. But daemon processes are essential to provide the automated processing capability promised through the Payroll Control Center. Think of this capability as an advanced scheduling system with a "listener" that is always ready and waiting to execute commands issued from the Payroll Control Center.

To accomplish the first activity of configuring rules and settings, SAP has provided Transaction PYC_ CONF_WB to group all the different components together. You can use this transaction as a checklist to ensure you don't miss an important step. The Configuration Workbench also has features for validating your configuration settings to ensure that missing components are detected early.

The Configuration Workbench can also be accessed from Customizing by following the menu path **Human Resources • Payroll • International • Tools • Payroll Control Center • Configuration Workbench.**

To set up the Payroll Control Center, configure the following components in the Configuration Workbench:

- Validation rules
- Analytics charts
- Analytics
- Process step templates
- Policy types (policies are not configurable from the backend using the simplified configuration for the Payroll Control Center.)
- Process types (processes are not configurable from the backend using the simplified configuration for the Payroll Control Center.)

We'll also cover authorizations to control who can do what and describe the technical tasks required to make the settings operational.

The second activity, launching the required processes, is also performed through a single transaction provided by SAP—the admin report available through Transaction PYC_ADMIN_TRANSACTION.

We'll discuss all these activities in detail throughout the rest of this chapter.

10.3.1 Preparing the Backend System and Enabling Declustering to Store Payroll Result

Before we begin performing the configuration activities, we must prepare the Employee Central Payroll system by activating various functions and services and enabling the storage of payroll results in tables, which is referred to as *declustering*.

Software Components and Business Functions

Prior to configuration, you must ensure the following software components are installed and contact SAP Support if necessary:

- SAP_HR and EA_HR 608+ with SP Level 80+
- SAP_UI 751 with SAP_UI51.44+
- SAP_GWFND 740 with SP Level 0010+

You must also ensure the following functions are activated using Transaction SFW5, as shown in Figure 10.2:

- HCM_LOC_CI_50
- HCM_LOC_CI_62
- HCM_LOC_CI_63
- HCM_LOC_CI_68
- HCM_LOC_CI_72
- HCM_LOC_CI_75
- HCM_LOC_CI_76
- HCM_LOC_CI_79
- HCM_LOC_CI_81
- HCM_LOC_CI_88
- HCM_LOC_CI_89
- HCM_LOC_CI_92
- HCM_LOC_CI_95
- HCM_LOC_CI_97
- HCM_LOC_CI_99
- HCM_LOC_CI_100
- HCM_LOC_CI_101
- HCM_LOC_CI_103
- HCM_LOC_CI_107
- HCM_LOC_CI_110

SAPUI5 Applications

Activate the SAPUI5 applications listed in Table 10.3 through Transaction SICF, where you'll right-click each node and choose **Activate**, as shown in Figure 10.3.

SAPUI5 Application	Service Path
Manage Teams	*/sap/bc/ui5_ui5/sap/hrpy_pcc_tm_1*
My Teams	*/sap/bc/ui5_ui5/sap/hrpy_pcc_mc_1*
My Processes	*/sap/bc/ui5_ui5/sap/hrpy_pcc_proc_3*
Alert Management	*/sap/bc/ui5_ui5/sap/hrpy_pcc_errm_2*
Unassigned Alerts	*/sap/bc/ui5_ui5/sap/hrpy_pcc_em_t_2*
Manage Policies	*/sap/bc/ui5_ui5/sap/hrpy_pcc_conf_2*
Manage Processes	*/sap/bc/ui5_ui5/sap/hrpy_pcc_conf_1*
My Off-cycles	*/sap/bc/ui5_ui5/sap/hrpy_pcc_oc_1*
Audit Trail	*/sap/bc/ui5_ui5/sap/hrpy_pcc_al_24*

Table 10.3 SAPUI5 Applications That Must Be Activated

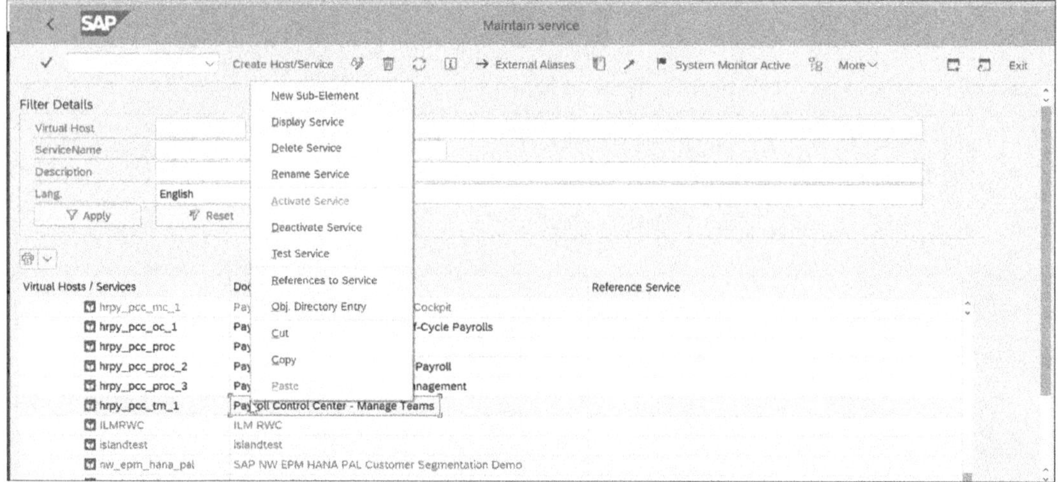

Figure 10.3 Transaction SICF for Turning on Business Functions/Applications in the Payroll Control Center

Activating OData Services

To activate OData services, first navigate to Customizing via Transaction SPRO and follow the menu path **SAP NetWeaver • SAP Gateway • OData Channel • Configuration • Connection Settings • SAP Gateway to SAP System • Manage SAP System Aliases.**

Employee Central Payroll is the destination business system that provides data. If the destination system and client is the current system and client, choose **LOCAL** from the **SAP System Alias** field, as shown in Figure 10.4. (Of course, if you're not using Employee Central Payroll, you would maintain this configuration in the system that will be the source of the payroll data.)

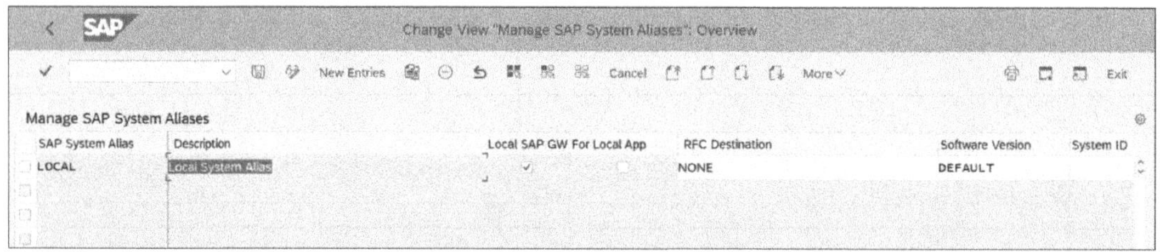

Figure 10.4 OData Gateway Setting

Next, start Transaction /IWFND/MAINT_SERVICE and locate the OData services listed in Table 10.4.

OData Service	Technical Name
My Teams	PYC_TEAM_MAINT_SRV

Table 10.4 OData Services for the Payroll Control Center

OData Service	Technical Name
My Processes	PYC_PROCESS_MANAGER_SRV
My Alerts	PYC_ALERT_MANAGER_SRV
Unassigned Alerts	PYC_TEAM_ALERTS_SRV
Manage Policies/Processes	PYC_CONF_SRV
My Off-cycles	PYC_OFF_CYCLE_SRV
Audit Trail	PYC_CONT_003_SRV

Table 10.4 OData Services for the Payroll Control Center (Cont.)

Add a system alias for each service as **LOCAL** and make each default by selecting the **Default** checkbox, shown in Figure 10.5. Then, activate each service by selecting **Activate** from the **ICF Node** dropdown list.

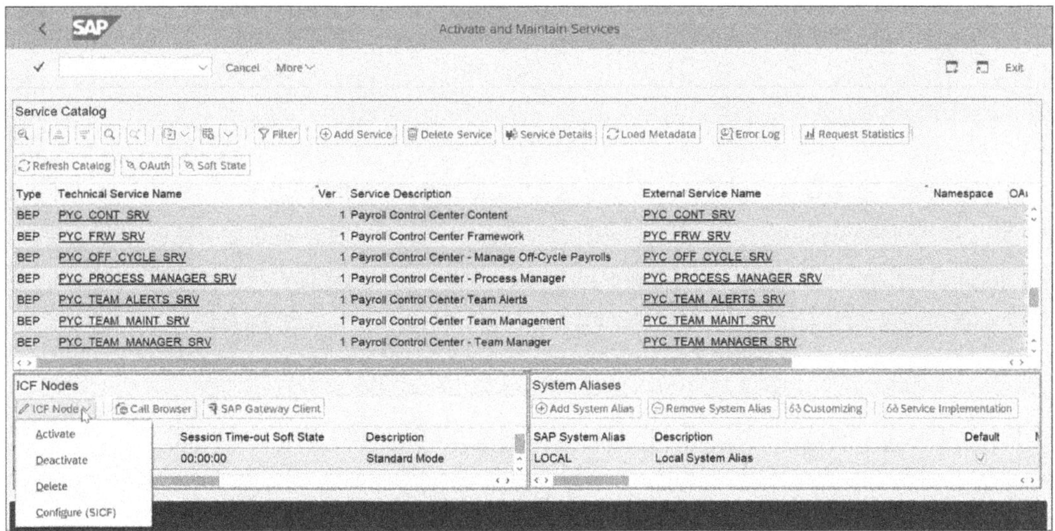

Figure 10.5 Activating OData Services and Setting Aliases

You can use the SAP Gateway client directly for testing. Click on the **SAP Gateway Client** button to open the **Gateway Client** screen. Click on the **Execute** button. A status code of 200 indicates the system is working.

Enabling Declustering to Store Pay Results

Declustering is a prerequisite to enabling the Payroll Control Center. You must switch the storage of payroll results from traditional cluster tables to a set of tables that you can easily query using standard tools like Transaction SE16. To enable declustering, follow these steps:

1. Use Transaction SM30 to maintain table `V_T77DCT_OPTION` on the screen shown in Figure 10.6. Keep the **DB Connection** field empty and, from the **Switch Option** dropdown list, select **Switched on with synchronous declustering**.

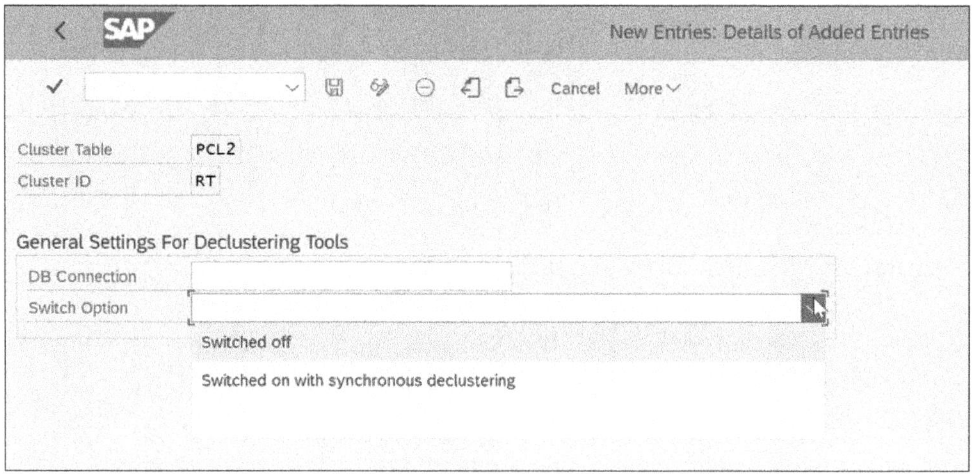

Figure 10.6 Synchronous Declustering Switch

2. Use Transaction SM30 to maintain table `V_T77DCT_REG` on the screen shown in Figure 10.7. Register all necessary tables that are used with validation rules by clicking on **New Entries** and adding the required tables via the dropdown list. Then, save these settings.

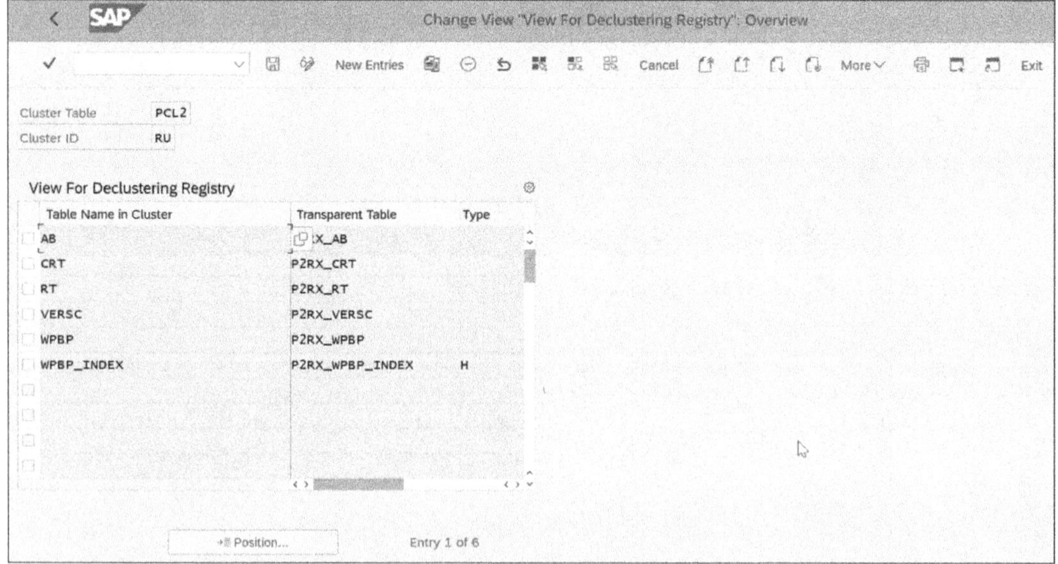

Figure 10.7 Declustering Registry with Denormalized Table Names

3. This step and the next step are required if you intend to use **EVAL_PERIOD** field, which helps you calculate retroactive evaluations. Use Transaction SPRO and define **Settings for Declustering Tools Relid CU** by following the menu path **Payroll • Payroll • International • Declustering Tools.** Now, you'll define general declustering settings for the cluster ID key **CU**. Keep the **DB Connection** field empty and, from the **Switch Option** dropdown list, select **Switched on with synchronous declustering**, as shown earlier in Figure 10.6.
4. Register table EVAL+PERIOD and table T_EVAL_PERIOD by using the same path.
5. If you have previously evaluated payroll results that need to be declustered, you must follow the menu path **Payroll • Payroll • International • Declustering Tools.** Then, you'll choose **Declustering Tools - Initial Load for Payroll Result.** You can also execute this process through Transaction SE38 and report RPCDCT_INITIAL_LOAD, to reach the screen shown in Figure 10.8. Choose the period, personnel numbers, and payroll areas and click on **Execute**.

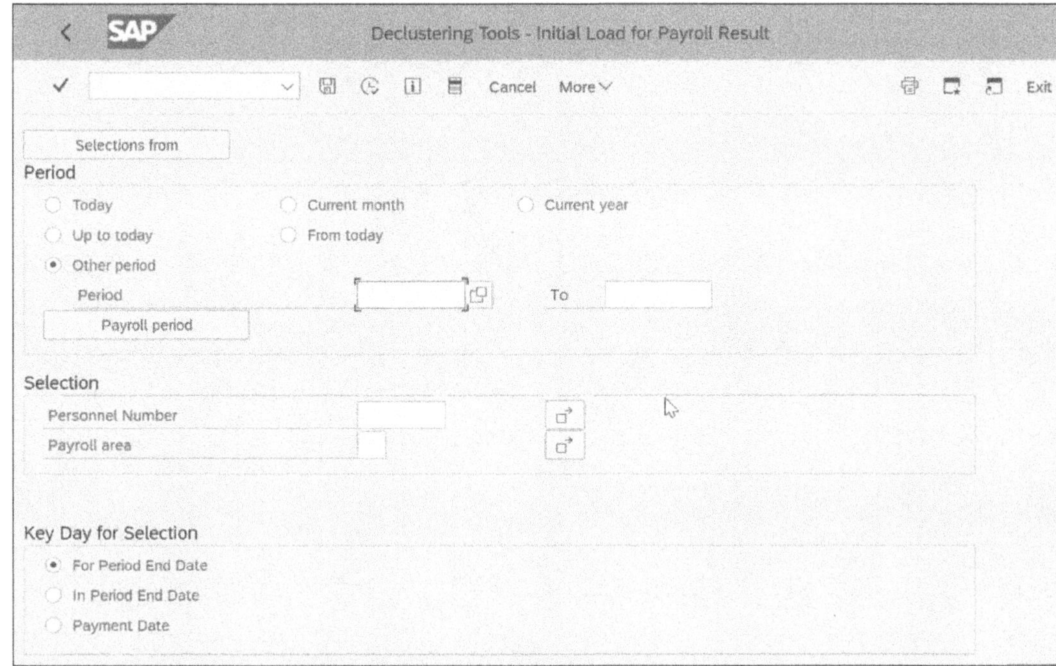

Figure 10.8 Initial Load of Declustered Tables (If Required)

After declustering, the P2RX_XX tables now hold the data for payroll results. For example, the cluster table RT (Results Table) discussed in Chapter 2, Section 2.3.3, can now be found in table P2RX_RT.

Reading the Payroll Results from the Declustered Tables

If you're a former user of SAP ERP HCM Payroll, you might already be familiar with Transaction PC_PAYRESULT. This transaction works the same way in Employee Central Payroll as before and shows the complete results of a payroll processing run, as shown in Figure 10.9.

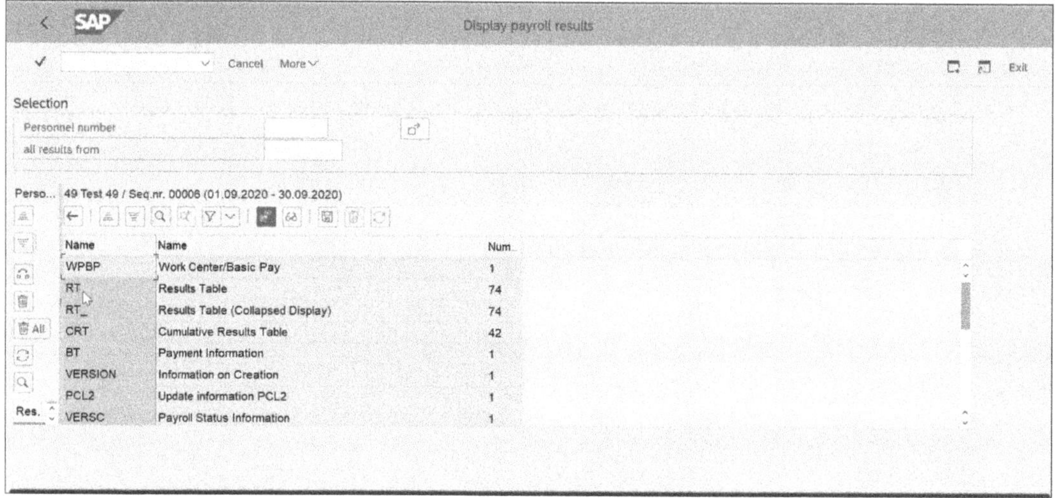

Figure 10.9 Clusters Visible from Transaction PC_PAYRESULT

Run Transaction PC_PAYRESULT and then double-click on the **RT_** cluster to see the pay results, as shown in Figure 10.10.

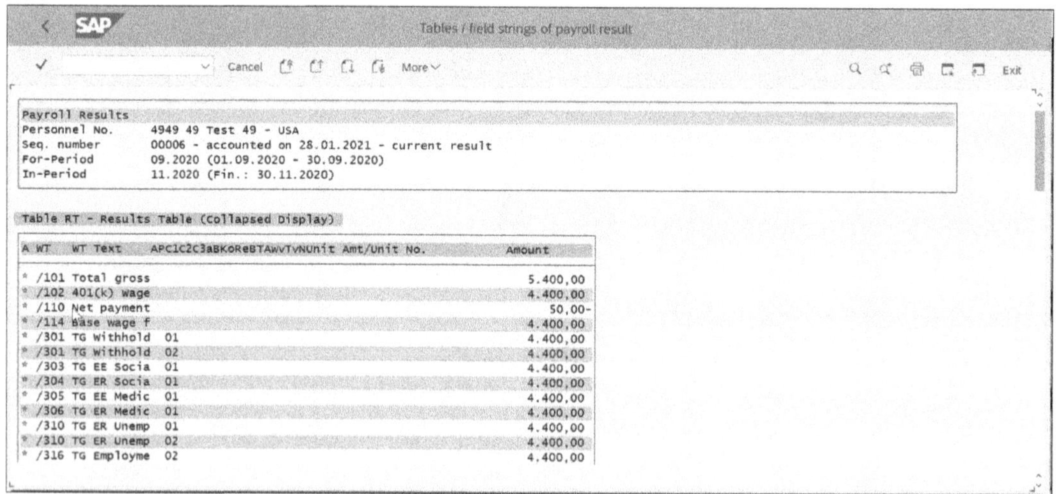

Figure 10.10 Table RT View from Transaction PC_PAYRESULT

Once the declustering is successful, you'll find the table RT in table P2RX_RT, but you'll first need to find the sequence number from table HRPY_RGDIR, as shown in Figure 10.11.

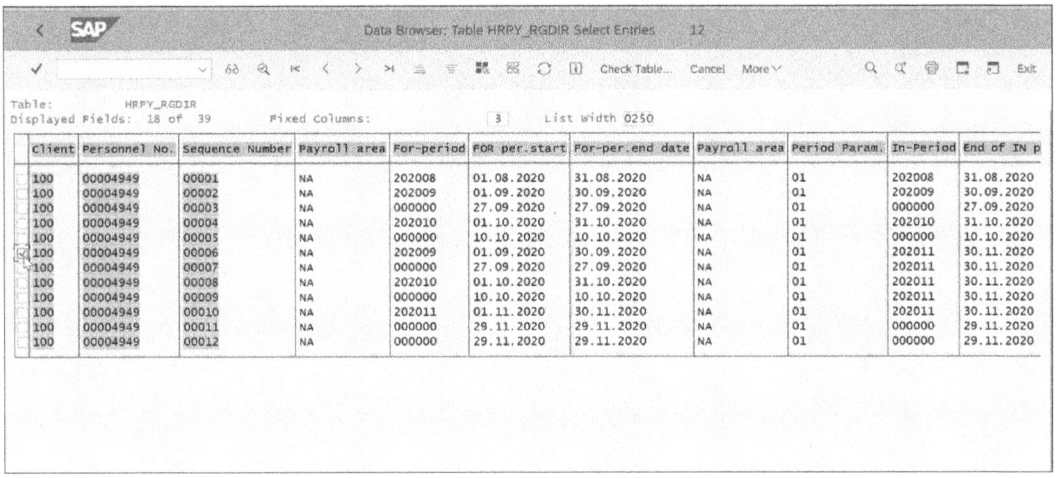

Data Browser: Table HRPY_RGDIR Select Entries 12

Table: HRPY_RGDIR
Displayed Fields: 18 of 39 Fixed Columns: 3 List width 0250

Client	Personnel No.	Sequence Number	Payroll area	For-period	FOR per.start	For-per.end date	Payroll area	Period Param.	In-Period	End of IN p
100	00004949	00001	NA	202008	01.08.2020	31.08.2020	NA	01	202008	31.08.2020
100	00004949	00002	NA	202009	01.09.2020	30.09.2020	NA	01	202009	30.09.2020
100	00004949	00003	NA	000000	27.09.2020	27.09.2020	NA	01	000000	27.09.2020
100	00004949	00004	NA	202010	01.10.2020	31.10.2020	NA	01	202010	31.10.2020
100	00004949	00005	NA	000000	10.10.2020	10.10.2020	NA	01	000000	10.10.2020
100	00004949	00006	NA	202009	01.09.2020	30.09.2020	NA	01	202011	30.11.2020
100	00004949	00007	NA	000000	27.09.2020	27.09.2020	NA	01	202011	30.11.2020
100	00004949	00008	NA	202010	01.10.2020	31.10.2020	NA	01	202011	30.11.2020
100	00004949	00009	NA	000000	10.10.2020	10.10.2020	NA	01	202011	30.11.2020
100	00004949	00010	NA	202011	01.11.2020	30.11.2020	NA	01	202011	30.11.2020
100	00004949	00011	NA	000000	29.11.2020	29.11.2020	NA	01	000000	29.11.2020
100	00004949	00012	NA	000000	29.11.2020	29.11.2020	NA	01	000000	29.11.2020

Figure 10.11 Table HRPY_RGDIR to Find the Sequence Number

This table gives you the sequence number for the payroll run you're looking for. In our case (from the previous PC_PAYRESULT example), we're looking for 09/2020 in the **For-period** field and 11/2020 in the **In-period** field, which corresponds to sequence 6.

Now, you can display the results table data by using the sequence number along with the personnel number in Transaction SE16 and clicking on **Execute**, which will produce the results shown in Figure 10.12.

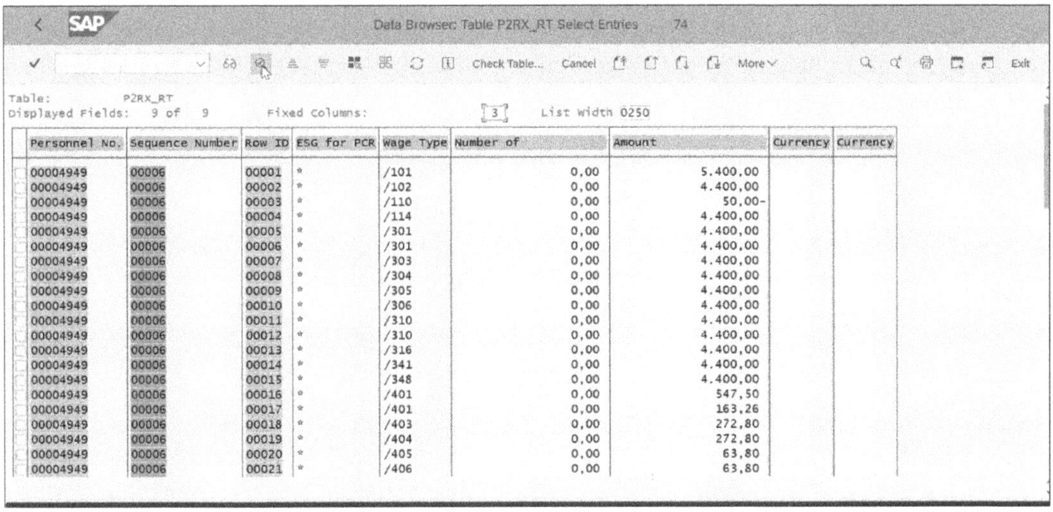

Data Browser: Table P2RX_RT Select Entries 74

Table: P2RX_RT
Displayed Fields: 9 of 9 Fixed Columns: 3 List width 0250

Personnel No.	Sequence Number	Row ID	ESG for PCR	Wage Type	Number of	Amount	Currency	Currency
00004949	00006	00001	*	/101	0,00	5.400,00		
00004949	00006	00002	*	/102	0,00	4.400,00		
00004949	00006	00003	*	/110	0,00	50,00-		
00004949	00006	00004	*	/114	0,00	4.400,00		
00004949	00006	00005	*	/301	0,00	4.400,00		
00004949	00006	00006	*	/301	0,00	4.400,00		
00004949	00006	00007	*	/303	0,00	4.400,00		
00004949	00006	00008	*	/304	0,00	4.400,00		
00004949	00006	00009	*	/305	0,00	4.400,00		
00004949	00006	00010	*	/306	0,00	4.400,00		
00004949	00006	00011	*	/310	0,00	4.400,00		
00004949	00006	00012	*	/310	0,00	4.400,00		
00004949	00006	00013	*	/316	0,00	4.400,00		
00004949	00006	00014	*	/341	0,00	4.400,00		
00004949	00006	00015	*	/348	0,00	4.400,00		
00004949	00006	00016	*	/401	0,00	547,50		
00004949	00006	00017	*	/401	0,00	163,26		
00004949	00006	00018	*	/403	0,00	272,80		
00004949	00006	00019	*	/404	0,00	272,80		
00004949	00006	00020	*	/405	0,00	63,80		
00004949	00006	00021	*	/406	0,00	63,80		

Figure 10.12 Table P2RX_RT for Payroll Results

10.3.2 Using the Configuration Workbench

The Configuration Workbench can be launched using Transaction PYC_CONF_WB, as shown in Figure 10.13. In the rest of this section, we'll review how to create, modify, and delete objects using this screen. You'll also learn how to copy objects and identify where an object might have been used in your rules.

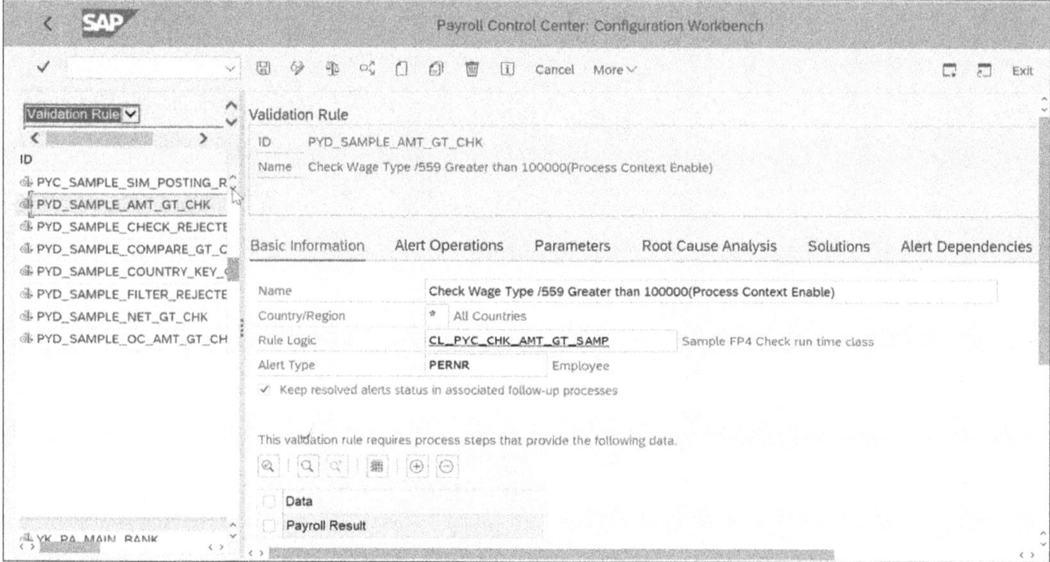

Figure 10.13 Transaction PYC_CONF_WB

Navigation

Navigation in the Configuration Workbench is similar to other SAP programs. Use the **Back** icon, located in the upper-left corner, as shown in Figure 10.13, to return to the previous object.

Saving an object shows a list of messages. From this list, two main actions can be taken, as follows:

- Double-click on a message to reveal its details.
- Click on the **Info** icon to see the cause and recommended fix for the error.

The underlying logic can be viewed for objects driven by rules. Simply double-click the class that appears on the **Basic Information** tab.

Creating an Object

To create an object, switch to edit mode by clicking on the **Display/Change** button and then click the **Create** icon, as shown in Figure 10.14.

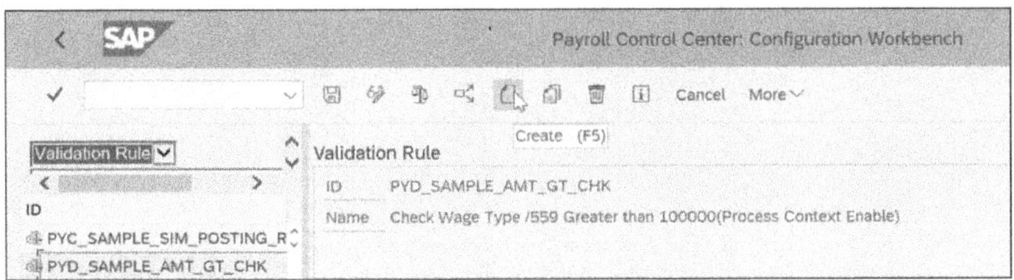

Figure 10.14 Create Button in the Configuration Workbench

To create a new object from scratch (in our case, a validation rule), click the **Create** icon in the application toolbar, enter a unique ID and a name for the new validation rule, and then click **Continue**, as shown in Figure 10.15.

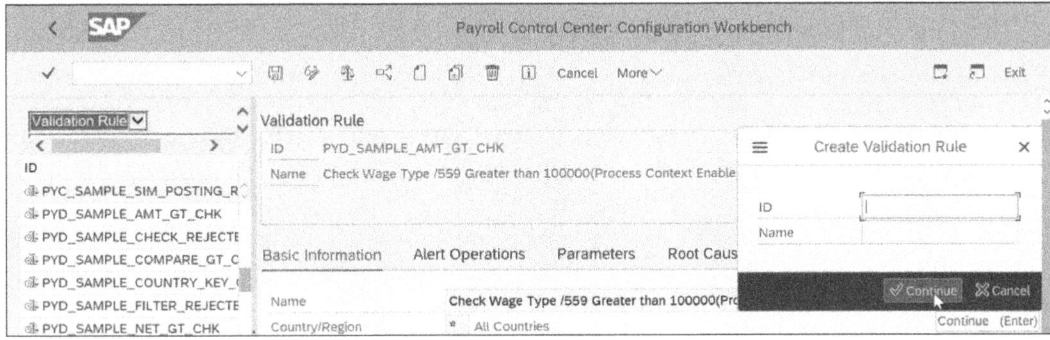

Figure 10.15 Creating a New Validation Rule

Copying an Object

Switch to the edit mode by clicking on the **Display/Change** button. As shown in Figure 10.16, double-click a standard or existing object (here, a validation rule) that you want to copy. Then, click the **Copy** (stack of paper) icon in the application toolbar, to the right of the **Create** icon. In the popup window shown in Figure 10.17, enter a unique ID for the new validation rule in the **To** field and then click **Continue**. All the attributes and values, except for the ID, will be copied to the new validation rule.

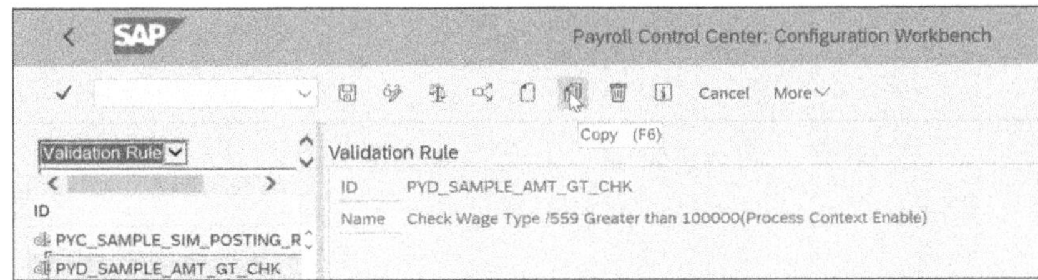

Figure 10.16 Copy Button in the Configuration Workbench

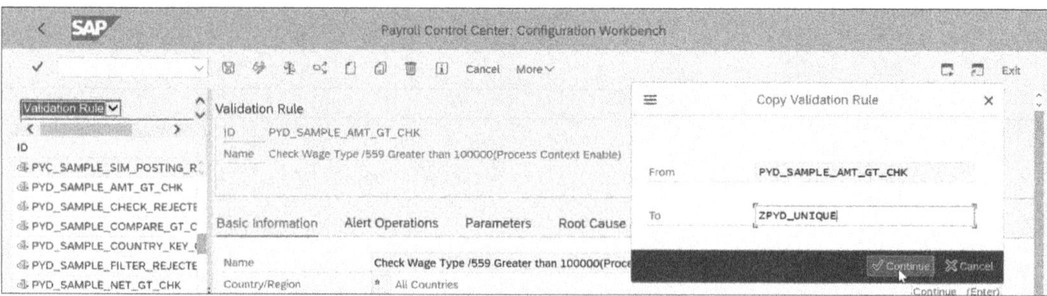

Figure 10.17 Copying an Existing Rule

Saving an Object

Click on the **Save** (floppy disk) icon to save. Since we are working on a configuration, you'll be prompted to create or choose a transport object, as shown in Figure 10.18.

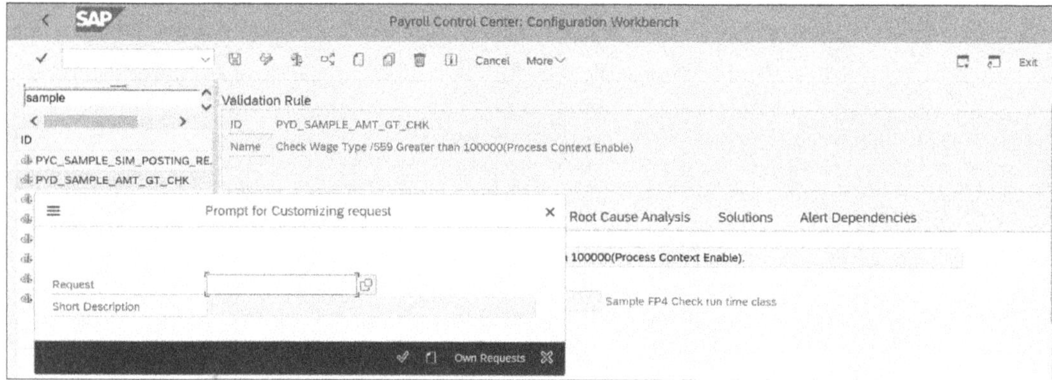

Figure 10.18 Transport Request for Save

Different object types require different logical transport objects, as shown in Table 10.5.

Object Type	Logical Transport Objects
Validation rule	PYVR with validation rule ID
Analytics chart	PYAC with analytics chart ID
Analytics	PYAN with analytics ID
Step template	PYST with step template ID
Policy type	PYTP with policy type ID
Process type	PYTR with process type ID PYPT with generated process template ID

Table 10.5 Logical Transport Object Types for the Payroll Control Center Configuration

Validating Object

Validating the changes made to the configuration helps prevent incorrect settings, which might cause missed actions or incorrect behavior by Payroll Control Center elements. For example, if you're setting up a validation rule and enter a field name in the **Alert Type** field that is not supported or simply spelled incorrectly, clicking the **Validate** icon will display an error prompting you for correction.

The **Validate** icon validates your rules and finds any missing mandatory items, as shown in Figure 10.19. Simply select the object you want to validate from the left and click the **Validate** icon. Figure 10.20 shows the result of clicking the icon.

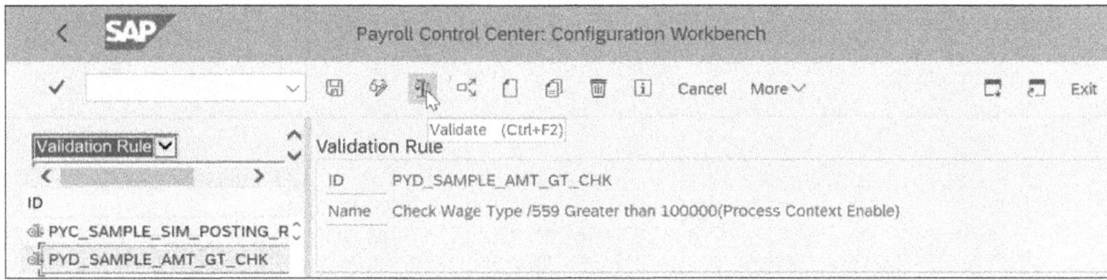

Figure 10.19 Validate Object Icon in the Configuration Workbench

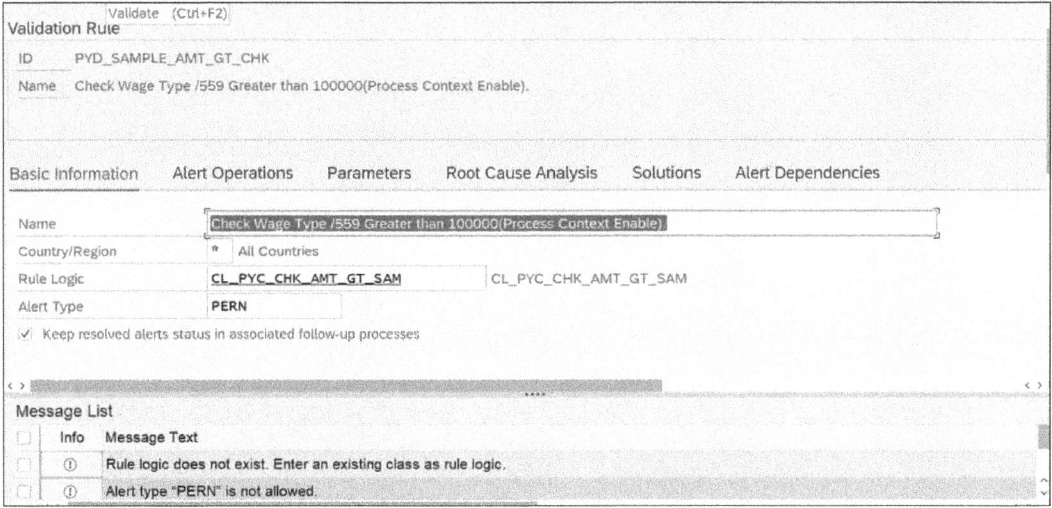

Figure 10.20 Validation Error Messages

Using the Where-Used List

A where-used list is a handy tool for identifying dependent objects that are impacted when you make changes to configuration. For example, if you change a process type, you'll can quickly find all the processes that are based on this process type through its where-used list, so that you can manage changes in the affected processes.

Clicking the **Where-Used List** (box and two arrows) icon in the top left, as shown in Figure 10.21, opens the where-used list to help you identify where the current object is used. Select an object by clicking on it and then click the **Where-Used List** icon. Figure 10.21 shows the related processes that use this process type.

Figure 10.21 Where-Used List Button in Configuration Workbench

This **Where-Used List** icon will display the types of objects listed in Table 10.6.

Object Selected	Objects Covered by the Where-Used List
Validation rule	Policies that the validation rule has been assigned to
Policy	Processes that the policy has been assigned to
Policy type	Process types that the policy type has been assigned to
Analytics chart	Analytics that the analytics chart has been assigned to
Analytics	Processes or process types that the analytics has been assigned to
Step template process	Processes or process types that the step template has been assigned to
Process type	Processes created using this process type

Table 10.6 Types of Objects Displayed by the Where-Used List

Note that, once these objects are created, changing them may not be simple or even possible. This limitation exists because these objects are interrelated. Once they've been used for a live process, further restrictions on what can be changed and how may be active.

Deleting an Object

Deleting an object is possible from the Configuration Workbench in limited circumstances. To delete an object, you must be in edit mode. (Remember, click on the **Display/Change** button to access this mode.) Select the object and click on the **Delete** button, as shown in Figure 10.22.

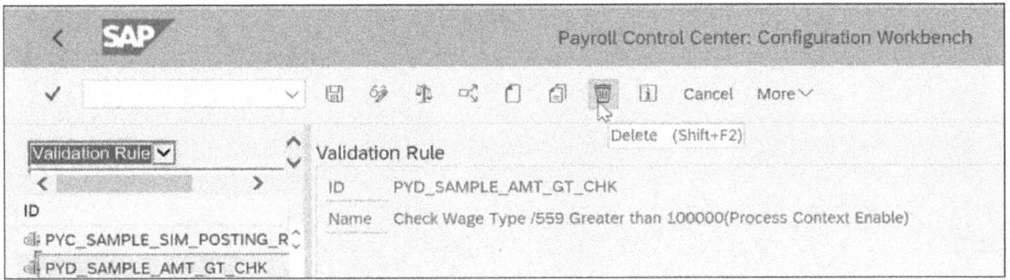

Figure 10.22 Delete Icon in the Configuration Workbench

If the object is being referenced by another object, you cannot delete this object. Click on the **Where-Used List** icon and navigate to the related object by double-clicking it. To proceed with deletion, you must remove the related object first. You must remove all related objects before the system lets you delete your original object.

Remember that the configuration of processes and policies is directly made in the production system and may not be repeated in the production or test systems. Thus, the where-used list may not be identical in all systems.

10.3.3 Creating Validation Rules

A *validation rule* provides a check on the master data in a manner that is especially useful for the payroll department. Going beyond simple checks for values missing in lists and for incorrect data, a validation rule can detect conditions that are syntactically correct but may indicate a manual error early on in the payroll process. For example, a validation rule could identify all employees whose gross pay in an upcoming period differs from their gross pay in the previous pay period by more than 10%. Unless an annual increment period is used in the organization, this magnitude of increase or decrease would not be common, which is why this increase is flagged as an anomaly, for further checks. If the difference is valid (which might be true for employees with an out-of-cycle promotion versus a promotion during the regular merit cycle), the payroll administrator simply needs to flag the increase as "correct" so that another team member doesn't waste time investigating the situation.

Building validation rules requires you set up reason codes and other information to help the root cause analysis after the validation fails. SAP provides standard rule logic and standard validation rules that you can copy create more sophisticated rules specific your business and the country in which it operates. We'll look at all these aspects of rule creation in the rest of this section.

Creating Initial Rules

To create a validation rule, follow these steps:

1. Initiate the Configuration Workbench in the Payroll Control Center using Transaction PYC_CONF_WB.

2. Switch to edit mode by clicking on the **Display/Edit** button.
3. Copy a standard validation rule or another existing validation rule or create a new one.
4. Click the **Basic Information** tab, shown in Figure 10.23.

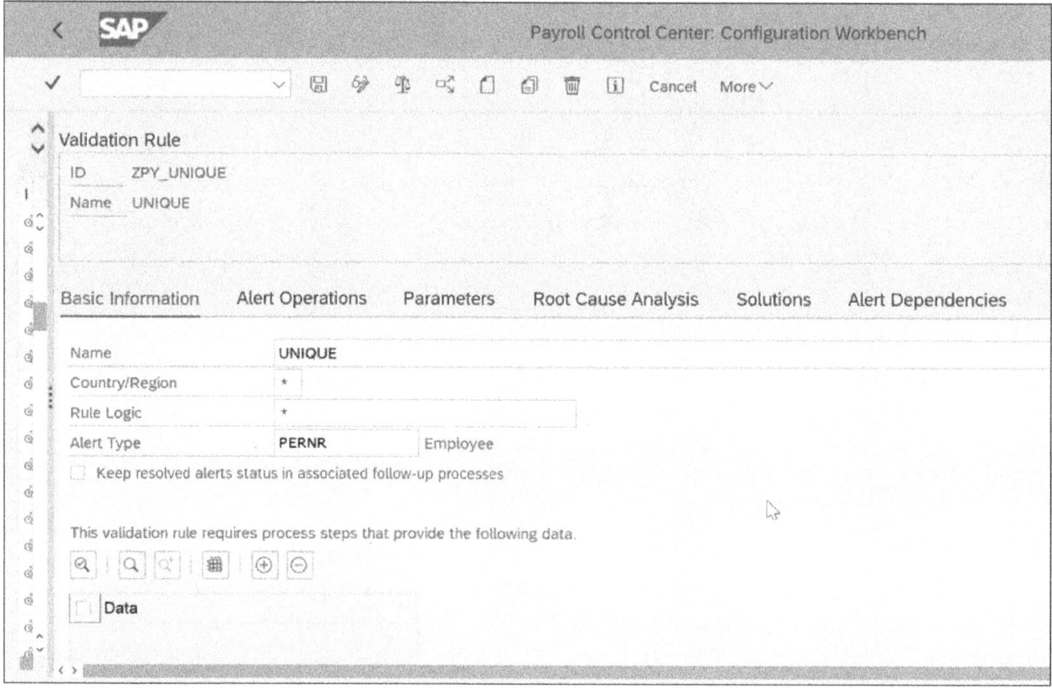

Figure 10.23 Basic Information Tab of the Validation Rule

5. Maintain the **Name** and the **Country/Region** fields for the rule.
6. Now, let's add some logic. The **Rule Logic** field must be maintained with an existing class. In this context, *classes* are code snippets that perform the validation. Classes need to be created in advance using Transaction SE24. The rule logic for your class must be inherited from the base class `CL_PYD_CHK_FP4_BASE`, as shown in Figure 10.24. Creating rule logic requires knowledge of ABAP, and thus, you might need a developer to help you.
7. The **Alert Type** is the entity for which the validation rule will provide an alert if the rule logic detects an anomaly. Typically, this entity is the `PERNR` (personnel number).
8. The **Keep resolved alerts status in associated follow-up process** checkbox is a little intricate to understand. As mentioned before, these validation errors are not syntactical; this validation process identifies possible error situations. If not actually an error, then the payroll administrator can mark this alert as **Resolved**. This checkbox means the administrator's "ok" flag for this rule will override the error status each time it executes and finds the same issue for the same payroll run.

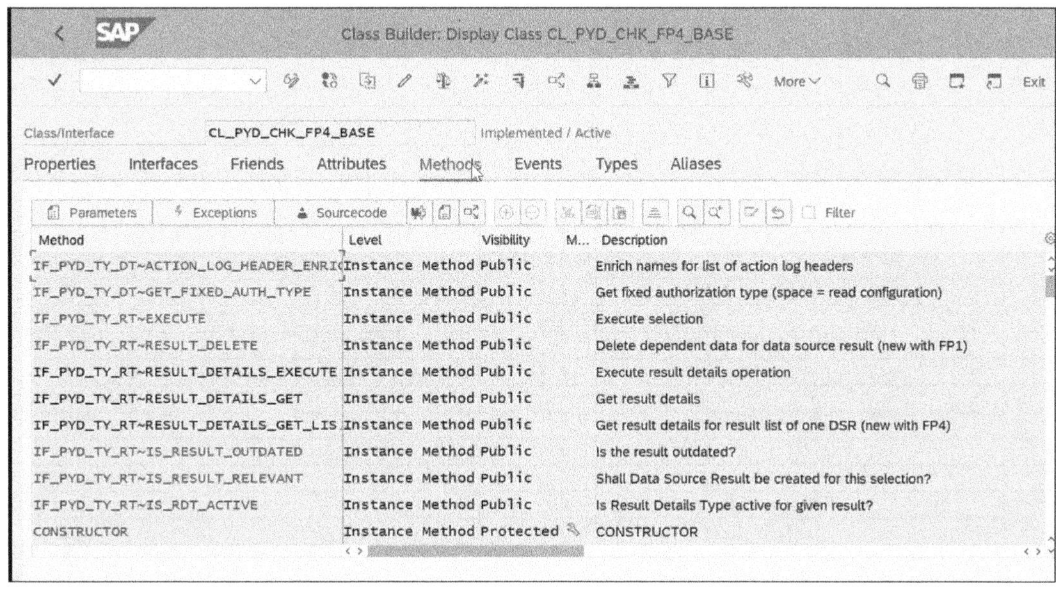

Figure 10.24 Base Class for Validation Rules

9. The **This validation rule requires process steps that provide the following data** field identifies data that may need to come from other process steps that precede the step that initiates the policy (validation) process. You'll need to click the **Insert Row** icon, shown in Figure 10.25, and select the **Payroll Result** row. Then, click **Continue**.

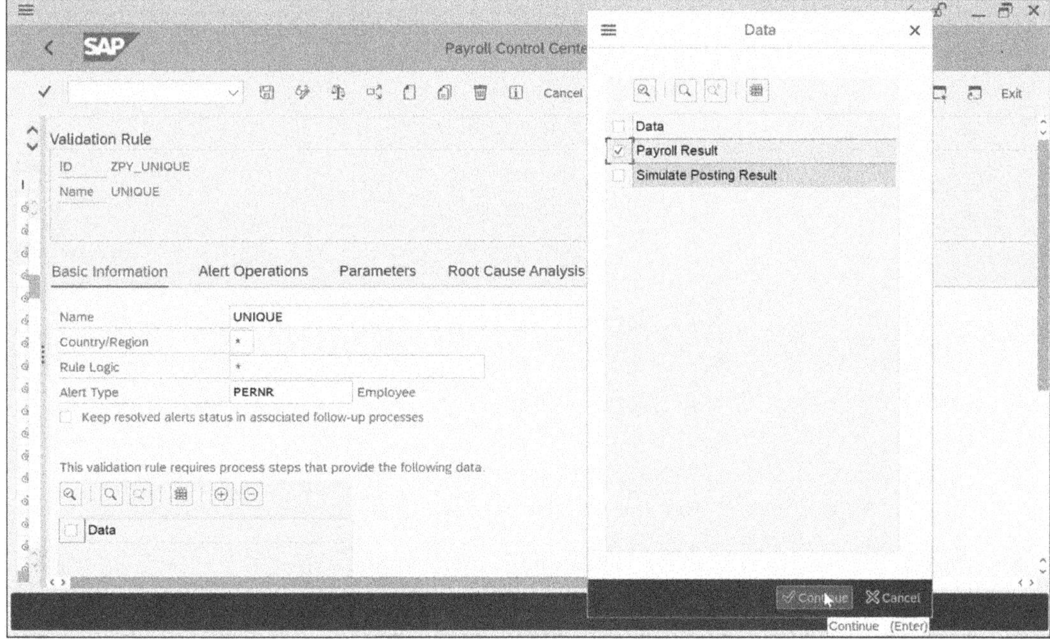

Figure 10.25 Choosing Data for Validation Rule Steps

10. Click the **Alert Operations** tab and select the required operations from the list of operations displayed. In fact, you'll deselect the operations that are not needed since the system brings up every operation for every rule, as shown in Figure 10.26.

    Four operations are available:

    - **Set to Resolved**: A payroll administrator can set the issue to resolved by selecting a reason code.
    - **Reset Status**: A payroll administrator can set the issue back to unresolved by selecting a reason code.
    - **Solution Applied**: A payroll administrator can indicate the solution is already applied, which will move the error from unresolved to resolved.
    - **Remove Solution**: A payroll administrator can indicate the solution is removed, which will move the error from resolved to unresolved.

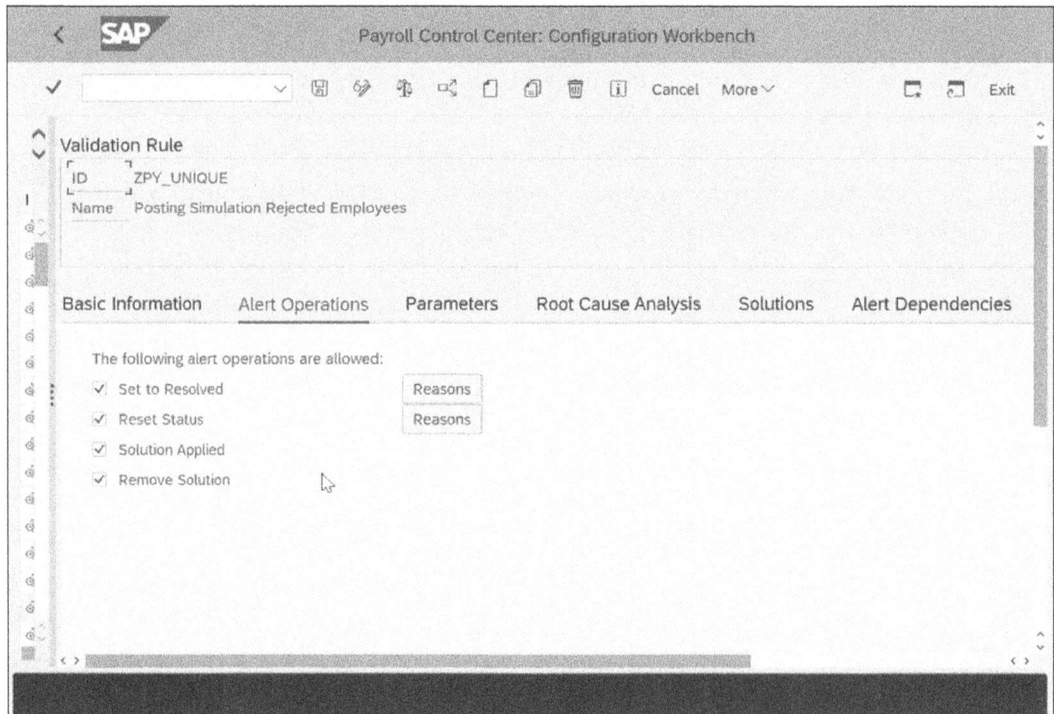

Figure 10.26 Alert Operations in Validation Rules

To set up reason codes, click the **Reasons** button for each action. Note that this button is available only for the first two actions. Insert a row in the **Alert Operations** dialog box by clicking the plus icon (+). Enter a reason code and a reason text, and then click **Continue** to save these settings. Reason codes can be added on the fly, as shown in Figure 10.27.

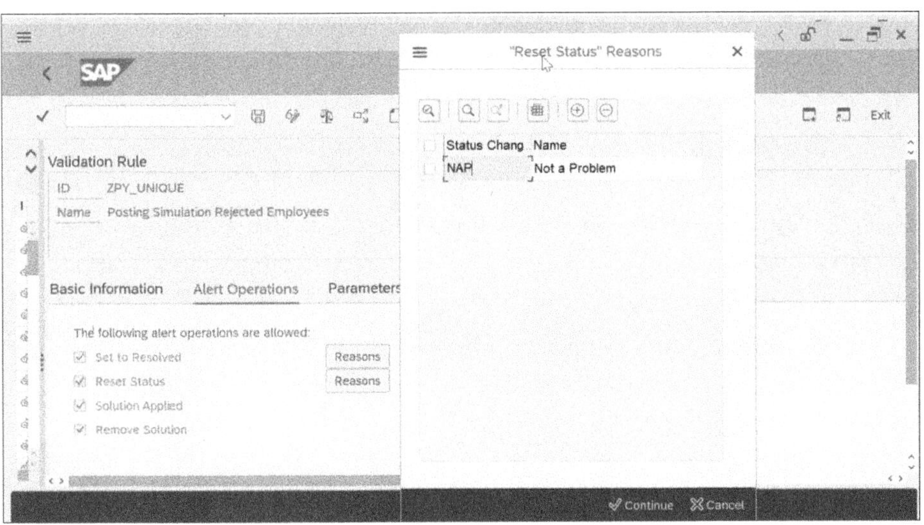

Figure 10.27 Reason Codes for Reset Statuses

11. Parameters are variables that can provide runtime values to the validation rule for the logic and the alerts to use. Next, select the **Parameters** tab and insert the parameters that will feed data into this validation rule by clicking the + icon (shown second from the right in Figure 10.28). Parameters PYP_PROC (Payroll Process) and PYP_PROC_INST (Payroll Process Instance) are automatically added to every rule. The process context includes the selection type (such as the specific payroll area) and the recurrence type (such as the specific payroll period). The parameters maintained on this screen will be also included into the context during execution of the validation rule.

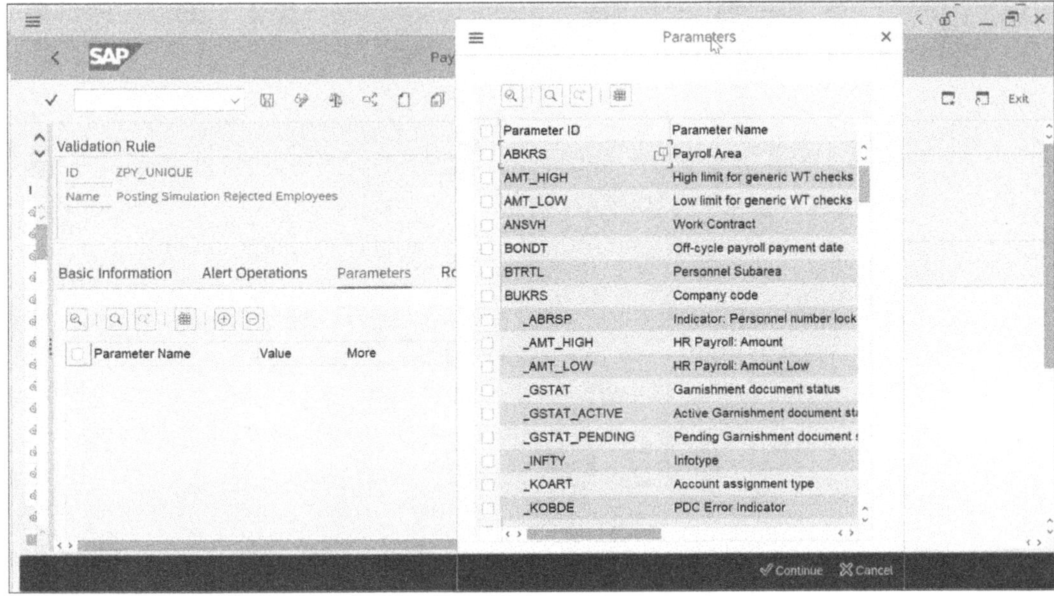

Figure 10.28 Choosing Parameters for the Validation Rule

12. Click **Insert Row** (the + button) and, in the **Parameters** dialog box, enter a value. Then, click **Continue**, as shown in Figure 10.28.
13. Select the **Root Cause Analysis** tab and configure the root cause.
14. Select the **Show key value** checkbox to display the **Key Indicators** column for alerts in My Alerts.
15. Select the **Show summary** checkbox to display the **Details Column** next to **Key Value** in My Alerts.
16. Add result details by clicking on the + icon.
17. In the **Root Cause Analysis** dialog box, choose a result detail by selecting its checkbox, as shown in Figure 10.29. With this step, the result detail will be displayed on the My Alerts page.

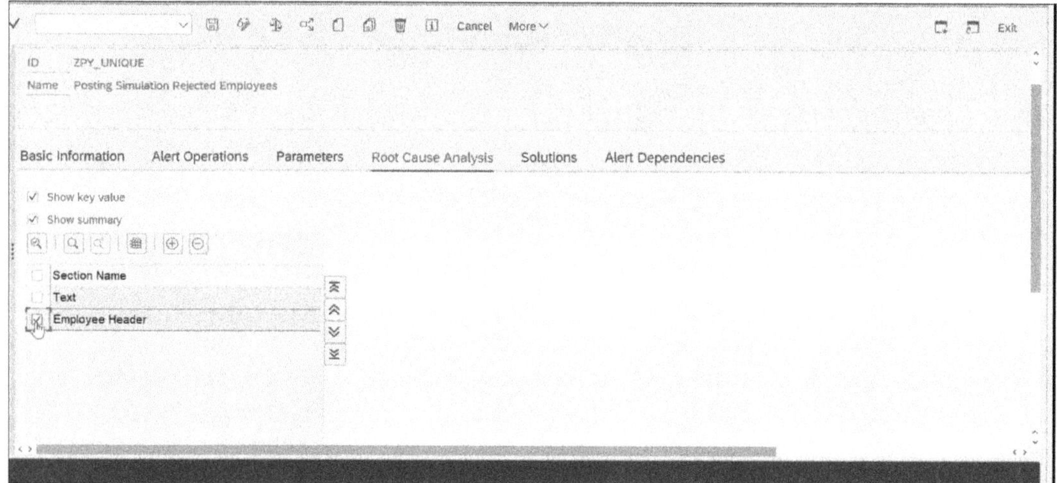

Figure 10.29 Validation Rule Root Cause: Choosing Result Details

18. You can add result details by clicking the **Insert Row** (+) icon, shown in Figure 10.30. Note that you can select result details by selecting the checkbox next to the relevant section name, and you can choose more than one section.
19. Choose the **Solutions** tab and add a solution to the validation rule. A *solution* is typically a hyperlink that the payroll administrator can click from both the **Alerts** and **Alert Details** pages in the My Alerts application to correct the alert for the employee. A typical example of a solution is **Add missing cost center in infotype 0001**. When the payroll administrator clicks this solution, the administrator is taken directly to the Employee Central field that holds the cost center. Section 10.3.4 provides more details on how to define solutions.
20. You can add solutions that have already been configured, by clicking the **Insert Row** (+) icon. In the **Solutions** dialog box, select a solution and click **Continue** to save, as shown in Figure 10.31.

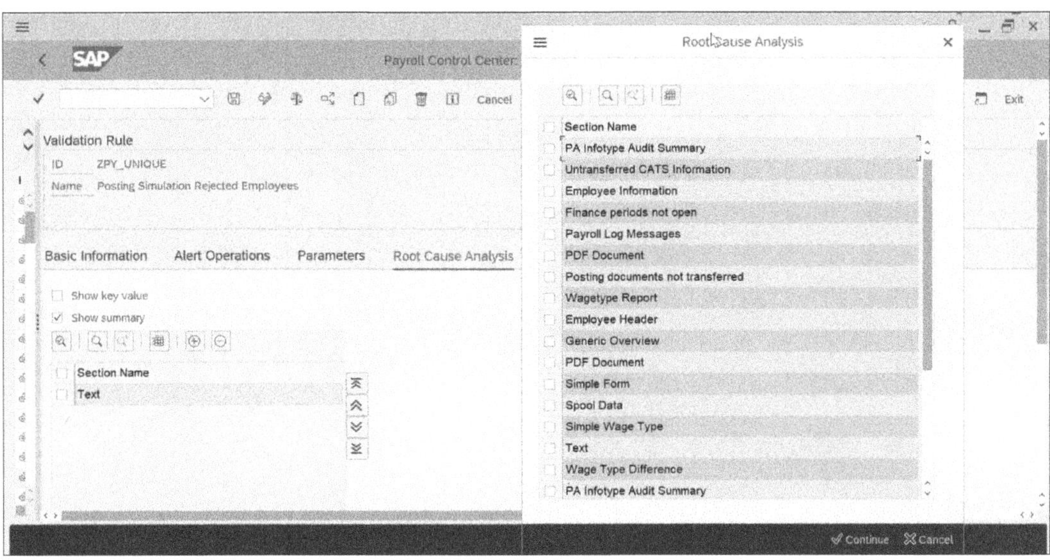

Figure 10.30 Validation Rule: Adding a Result Detail Section Name to a Root Cause

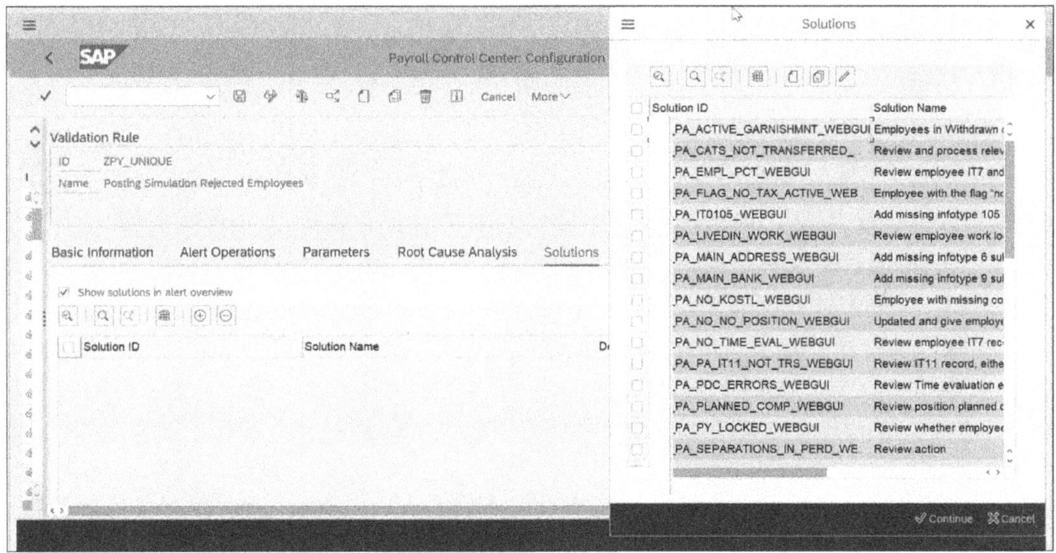

Figure 10.31 Adding Solutions to a Validation Rule

21. Click the **Alert Dependencies** tab and configure the dependency for this validation rule, as shown in Figure 10.32. Two main actions are accessed under this tab, as follows:

   - **This validation rule can be used by other validation rules to suppress raised alerts**: If validation rule X detects an alert, validation rule Y can be made dependent so that validation rule Y does not generate an additional alert. This setting enables one rule to be a filter for any other rule.

– **Don't raise an alert if one of the following validation rules has already raised an alert**: You can suppress the rule if any of the validation rules selected in this dialog box raises an error. Note that this dialog box will only display rules that have the **This validation rule can be used by other validation rules to suppress raised alerts** checkbox activated.

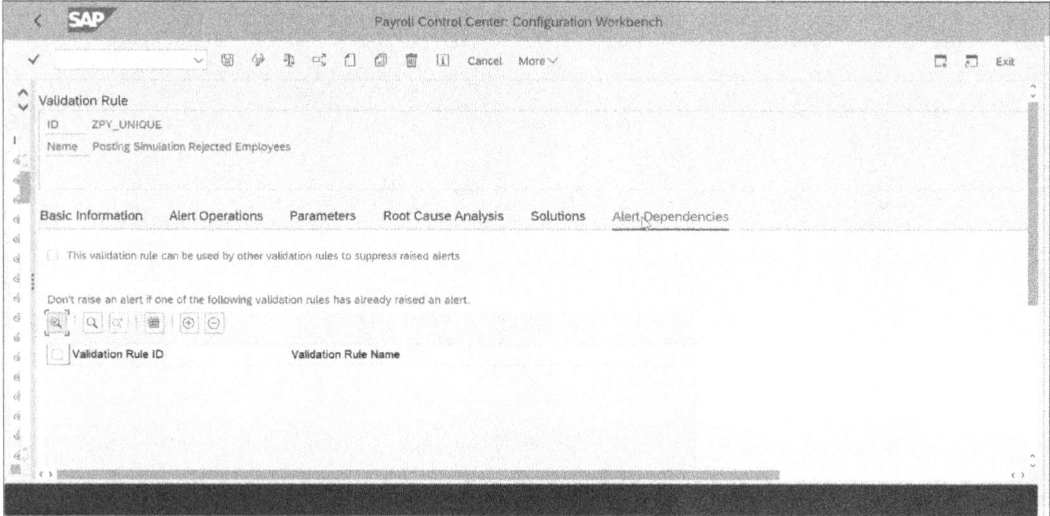

Figure 10.32 Validation Rule Alert Dependencies

22. Finally, click on the **Validate** button and correct any errors and then save.

Standard Rule Logic for Validation Rules

SAP provides a standard set of rule logic that you can use to construct useful validation rules. The most prominent are listed in Table 10.7. Note that SAP continues to add to this list. You can also create your own validation rules, either from scratch or with minor modifications of standard ones. In either case, expertise in ABAP programming is required.

Number	Technical Name of Rule Logic	Validation Rule Example	Scenario Description
1	CL_SBP_PYD_CHK_PA_EMPL_PCT	Employment percentage (infotype 0007) and utilization (infotype 0008) do not match	This validation rule shows employees for which utilization and employment percentage do not match.

Table 10.7 Standard Rule Logic

Number	Technical Name of Rule Logic	Validation Rule Example	Scenario Description
2	CL_SBP_PYD_CHK_PA_LIVEDIN_WORK	Lived-in (infotype 0001) and worked-in (infotype 0006) do not match	This validation rule shows employees that work and live in different countries in the current payroll period (infotypes 0001 and 0006).
3	CL_SBP_PYD_CHK_PA_ZER8IT14IT15	Zero base payment (infotype 0008) inconsistency with payment infotypes 0014 or 0015	This validation rule shows employees that have no base payment in infotype 0008 (infotype 0008 exists but without any amount) but do have recurring/additional payments in infotype 0014 and 0015.
4	CL_SBP_PYD_CHK_PA_CATS_TRANS	Cross-Application Timesheet System (CATS) transfers not completed	This validation rule shows CATS items that have not yet been transferred to Employee Central but before the end date of this pay period.
5	CL_SBP_PYD_CHK_PA_PDC_ERRORS	Plant data collection (PDC) errors	This validation rule shows employees who have a time evaluation PDC error flag.
6	CL_SBP_PYD_CHK_PA_MAIN_BANK	Employees missing main bank details	This validation rule shows employees who are missing a main bank subtype in infotype 0009.
7	CL_SBP_PYD_CHK_PA_PY_LOCKED	Personnel number locked for payroll	This validation rule shows employees who are locked for payroll on infotype 0003.
8	CL_SBP_PYD_CHK_PA_MAIN_ADDRESS	Employees missing a main address	This validation rule shows employees who are missing their main address subtype in infotype 0006. Without this information, in many countries, a bank transfer (before Data Medium Exchange [DME]) requires a main address to create an electronic funds transfer (EFT) file.

Table 10.7 Standard Rule Logic (Cont.)

Number	Technical Name of Rule Logic	Validation Rule Example	Scenario Description
9	CL_SBP_PYD_CHK_PA_IT11_NOT_TRS	Unpaid infotype 0011 records	This validation rule shows employees that have an infotype 0011 record within the current pay period, but the bank transfer date has not been populated (i.e., the amount has not been paid via an ad hoc method). This check is useful where records may have been created for employees but have not been paid where they should have been.
10	CL_SBP_PYD_CHK_PA_ZTERF_ZERO	Time management status is 0 (no time evaluation)	This validation rule shows employees that are excluded from time evaluation. Some business use this status to put employees on hold when they are on secondment, for example.
11	CL_SBP_PYD_CHK_PA_POS_99999999	Active employees in position 99999999	This validation rule shows employees that are in the default position 99999999. Some customers use positions from SAP ERP HCM Organizational Management (OM) as well as from Employee Central Payroll. If SAP ERP HCM OM is used, we recommend that employees should only be in position 99999999 if they are terminated. Having them in position 99999999 has implications for the payroll workflow, for self-service capabilities, and costs.

Table 10.7 Standard Rule Logic (Cont.)

Number	Technical Name of Rule Logic	Validation Rule Example	Scenario Description
12	CL_SBP_PYD_CHK_PA_IT0105	Employees missing infotype 0105 records	This validation rule shows employees who are missing an SAP user ID subtype in infotype 0105. Employees without this subtype cannot access self-service in SAP. Also, this user ID is often used in downstream network/active directory systems within the customer infrastructure so all employees must have this ID.
13	CL_SBP_PYD_CHK_PA_PLANNED_COMP	Position planned compensation does not match infotype 0008 (Basic Pay)	Businesses using planned compensation in SAP ERP HCM OM means that (among other things) employees inherit the pay rate of the position. This check flags instances where the position pay scale and basic pay scale do not match (and where the basic pay scale is at a level above the level within planned compensation).
14	CL_SBP_PYD_CHK_PA_COMPLETENESS	Completeness check for employees appointed in the pay period	This validation rule shows newly appointed employees that do not have the required infotypes.
15	CL_SBP_PYD_CHK_FI_BLOCKED_VEND	Blocked vendor list	This validation rule shows where a vendor has been blocked. This normally occurs when an employee has been absent from work for an extended period of time or when accounts payable locks a vendor due to it not having been used for a long time.

Table 10.7 Standard Rule Logic (Cont.)

Number	Technical Name of Rule Logic	Validation Rule Example	Scenario Description
16	CL_SBP_PYD_CHK_PY_WT_CASH_GT	Overview of cash payments	This validation rule shows employees that have a high net pay, a common check used by payroll administrators. Specify country-specific thresholds (e.g., currency, amount, etc.) in the **Country** columns, for example, "400 ZAR" for South Africa.
17	CL_SBP_PYD_CHK_PY_WT_GRS_GT	Gross pay over an amount	This validation rule is used to show employees with the gross pay wage type (/101) exceeding a threshold. Specify country-specific thresholds (e.g., currency, amount, etc.) in the **Country** columns, for example, "400 ZAR" for South Africa.
18	CL_SBP_PYD_CHK_PY_WT_RET_GT	Retroactive pay over an amount	This validation rule is to show employees with a retroactive wage type (/552 - Difference prev. Period) exceeding a threshold. Specify country-specific thresholds (e.g., currency, amount, etc.) in the **Country** columns, for example, "400 ZAR" for South Africa.
19	CL_SBP_PYD_CHK_PY_WT_BET_LIMIT	Positive deductions within a range	This validation rule shows employees with positive deductions (e.g., deductions are positive in value, but they should always be negative). This check looks for the wage types within the range entered in the parameters.

Table 10.7 Standard Rule Logic (Cont.)

Number	Technical Name of Rule Logic	Validation Rule Example	Scenario Description
20	CL_SBP_PYD_CHK_PY_CHK_WT_VAR_A	Variance in net pay over an amount	This validation rule shows employees where the specified wage type is different in value by more than the threshold specified when comparing the current pay period amount to the previous pay period amount. Specify country-specific thresholds (e.g., currency, amount, etc.) in **Country** columns, for example, "400 ZAR" for South Africa.
21	CL_SBP_PYD_CHK_PY_CHK_WT_VAR_P	Variance in net pay over a percentage	This validation rule shows employees where the specified wage type is different by more than the threshold (in percent) specified for comparing the current pay period amount to the previous pay period amount. Specify country-specific thresholds in percentages.
22	CL_PYC_CHK_AMT_GT_SAMP	Overtime payment over a certain amount	This validation rule shows employees where the specified wage types in the range are greater than the threshold specified. Specify country-specific thresholds (e.g., currency, amount, etc.) in the **Country** column, for example, "300 GBP" for the UK.
23	CL_SBP_PYD_PY_WT_COMP_PCT	Employees with overtime exceeding a percentage of total pay	This validation rule show employees where the specified wage type is more than the specified percentage of the total pay wage type (/101). Specify country-specific thresholds in percentages.

Table 10.7 Standard Rule Logic (Cont.)

Number	Technical Name of Rule Logic	Validation Rule Example	Scenario Description
24	CL_SBP_PYD_PY_LARGE_RETRO	Employees with large retroactive calculations	This validation rule is to show employees with retroactive calculations in excess of the number of days specified in the threshold. Specify country-specific thresholds (e.g., 90 days).
25	CL_SBP_PYD_PY_WT_NET_GT	Employees with net pay greater than the threshold	This validation rule shows employees with net pay in excess of the specified threshold. Specify country-specific thresholds (e.g., currency, amount, etc.) in the **Country** columns, for example, "400 ZAR" for South Africa.
26	CL_SBP_PYD_PY_SEP_W_NET	Separated employees with net pay	This validation rule is to show employees that are separated (employment status is 0) and have net pay.
27	CL_SBP_PYD_PY_NEG_GROSS	Negative gross pay	This validation rule is to show employees with gross pay (/101) less than zero.
28	CL_SBP_PYD_PY_NONCAS_WO_PY	Active, positive time recording employees without pay	This validation rule show employees with a time evaluation status that is neither 0 nor 1 and also do not have a net pay. The reason for this check is that negative time recording employees would be expected to have net pay (unless they are on unpaid leave).
29	CL_SBP_PYD_PY_WT_POS_DEDS	Employees with positive deductions	This validation rule show employees that have a positive wage type /110. Employees should not get a positive deduction. Deductions should be taken from their pay, not added to it.

Table 10.7 Standard Rule Logic (Cont.)

Number	Technical Name of Rule Logic	Validation Rule Example	Scenario Description
30	CL_SBP_PYD_PY_POS_TAX	Employees with tax refunds	This validation rule show employees whose tax wage types that are posted to finance (according to table T52EL) total a positive amount. Employees would not be expected to get a tax refund. This case tends to occur when a retroactive reduction in their pay impacts their overall tax credit.
31	CL_SBP_PYD_PY_DED_GT_GROSS	Employee with deductions greater than gross earnings	Employees should not have deductions (/110) greater than their gross earnings (/110). This validation rule picks up the (rare) cases where this has occurred.
32	CL_SBP_PYD_PY_CLAIMS	Employees with claims	Retroactive reductions in pay can cause claims (for example, for back-dated leave without pay). These situations can result in employees not being paid until the claim is recovered. Therefore, with an understanding of where these cases have occurred, action can be taken before the employee is impacted (by getting zero pay).
33	CL_SBP_PYD_PY_LOG	Employees with messages in the payroll log	Messages that occur in the payroll log may be important for payroll administrators to address any issues. This validation rule provides a particular solution to an error in the payroll log. Businesses should review the errors captured in the payroll log historically and create solutions for common issues.

Table 10.7 Standard Rule Logic (Cont.)

Number	Technical Name of Rule Logic	Validation Rule Example	Scenario Description
34	CL_SBP_PYD_CHK_TAX_ALLOWANCES	Employees with tax allowances exceeding the threshold	This validation rule shows employees who have exceeded a specified tax allowance threshold limit (e.g., over 10 tax allowances), which will result in a review/audit.
35	CL_SBP_PYD_CHK_TAX_FLAG_ACTIVE	Employees with the no tax flag active	This validation rule will display employees who have a flag set to indicate no taxes should be calculated and deducted from the gross pay. This rule reads infotype 0234, flag OZERO.
36	CL_SBP_PYD_PY_NET_ANNU_DED	Employees with deductions that have exceeded the annual deduction limit	This validation rule shows employees whose deductions have exceeded the annual limits.
37	CL_SBP_PYD_CHK_NO_KOSTL	Employees with a missing cost center	This validation rule show employees with a missing cost center.
38	CL_PYC_CHK_CHK_ACT_GARNISHMENT	Employees in withdrawn or leave of absence status with an active garnishment	This validation rule shows employees who are in a withdrawn or inactive employment status but have an active or pending garnishment.
39	CL_PYC_CHK_OC_AMT_GT_SAMP	Employees in an off-cycle run whose net pay is greater than a threshold	This validation detects employees in an off-cycle run whose net pay is greater than a threshold.
40	CL_PYC_CHK_REJ_EES_SP_SAMP	Employees who were rejected in a post-simulation run	This validation rule identifies employees who were rejected in a production run so that an active step can be undertaken to ensure they are paid.
41	CL_SBP_PYD_CHK_PA_ACTIONS	Actions changed or entered in pay period	This validation is useful to identify employees who have had some master data change, to flag them for more scrutiny to determine pay accuracy.

Table 10.7 Standard Rule Logic (Cont.)

Number	Technical Name of Rule Logic	Validation Rule Example	Scenario Description
42	CL_SBP_PYD_CHK_PY_WTRA_OVER_LI	Wage types in a range over a threshold	This generic validation can be used for validating different types of earnings or deductions if they cross a limit collectively.

Table 10.7 Standard Rule Logic (Cont.)

Standard Validation Rules

SAP also provides a set of sample validation rules, listed in Table 10.8.

Sample Validation Rule	Description
PYD_SAMPLE_AMT_GT_CHK (Check Wage Type /559 Greater than 100000 (Process Context Enable))	Raises an alert for each employee whose payment (wage type /559) is greater than 100,000.
PYD_SAMPLE_CHECK_REJECTED_EES (Sample Check Implementation - Find Payroll-Rejected Employee List)	Raises an alert for each employee who is rejected from the payroll run.
PYD_SAMPLE_COMPARE_GT_CHK (Compare Current and Previous Wage Type /101 Difference Greater than 1000 (Sample))	Raises an alert for each employee whose total gross (wage type /101) has a difference of greater than 1,000 between the current payroll period and the previous wage type.
PYD_SAMPLE_COUNTRY_KEY_CHK (Check the Country Key between IT0006 and Personnel Area (Process Content Enable))	Raises an alert for each employee whose country/region attribute in infotype 0006 (Addresses) is inconsistent with the personnel area of the payroll process.
PYD_SAMPLE_NET_GT_CHK (Wage Type /550 Check Greater than 10000 (Sample))	Raises an alert for each employee whose statutory net salary (wage type /550) is greater than 10,000.
PYD_SAMPLE_OC_AMT_GT_CHK (Wage Type /559 Check Greater than Values in Off-Cycle Process (Sample))	Raises an alert for each employee whose statutory net salary (wage type /559) is greater than a specified value in the off-cycle process.
PYC_SAMPLE_SIM_POSTING_REJ_EES (Posting Simulation Rejected Employees)	Raises an alert for each employee who was rejected from the posting simulation step.

Table 10.8 Standard Validation Rules

Standard Root Cause Analysis Details

SAP provides standard root cause analysis details, as shown in Table 10.9.

Result Detail Type	Result Detail Type Name
SAP_EEH	Employee header
SAP_GOV	Generic overview
SAP_PDF	PDF document
SAP_SFO	Simple form
SAP_SWT	Simple wage type
SAP_SPO	Spool data
SAP_T	Text
SAP_WTD	Wage type difference

Table 10.9 Standard Root Cause Analysis

Custom details can be added using Transaction SPRO by following the menu path **Payroll • Payroll International • Payroll Control Center • General Settings • Define Parameter Types**. However, a custom detail must be assigned to a parameter type to be effective.

You'll create custom parameter types on the screen shown in Figure 10.33.

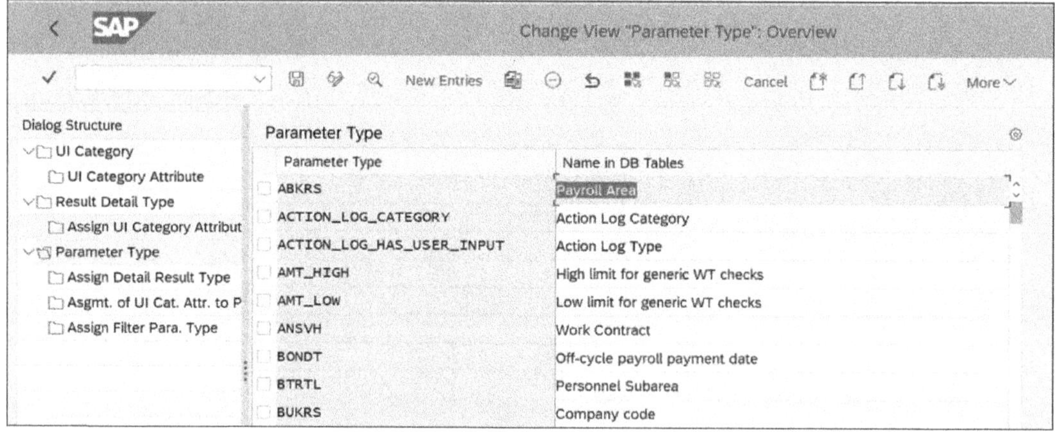

Figure 10.33 Configuring Parameter Types

To configure a custom parameter type, follow these steps:

1. Select **Parameter Type** from the menu on the left and the click **New Entries.** Fill in the details shown in Figure 10.34.

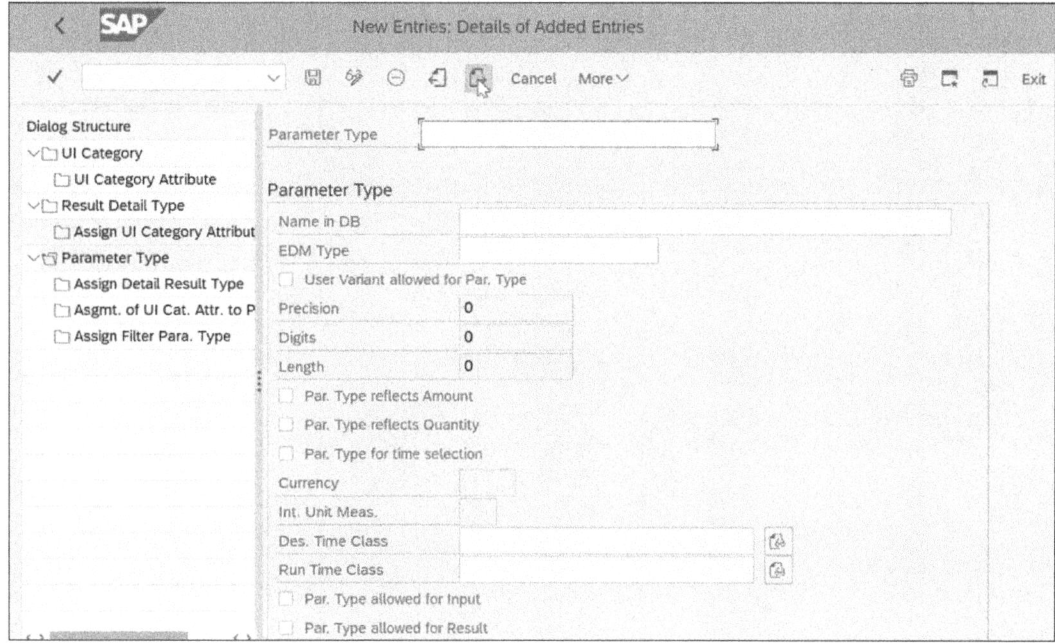

Figure 10.34 Creating a New Parameter Type

2. Enter a name in the **Parameter Type** field.
3. Enter a parameter in the **Name in DB** field.
4. Pick a data type from the **EDM Type** dropdown list.
5. Enter a value in the **Length** field.
6. Select the **User Variant allowed for Par. Type** checkbox if this parameter can be saved as a variant.
7. The **Run time class** field must be maintained with a class that has been already programmed to run the function that does the work for this parameter (such as getting an input parameter, value list, etc.). From this dropdown list, you can select from among preexisting classes, but you'll need the help of an ABAP developer if you require a new class to be created.
8. The last two checkboxes (**Par. Type Allowed for Input** and **Par. Type Allowed for Output**) determine if this parameter is an input or output parameter.
9. Leave the other fields blank and save.
10. Once this configuration is complete, you can select the checkbox next to your new parameter type and click on **Assign Detail Result Type**, as shown in Figure 10.35.

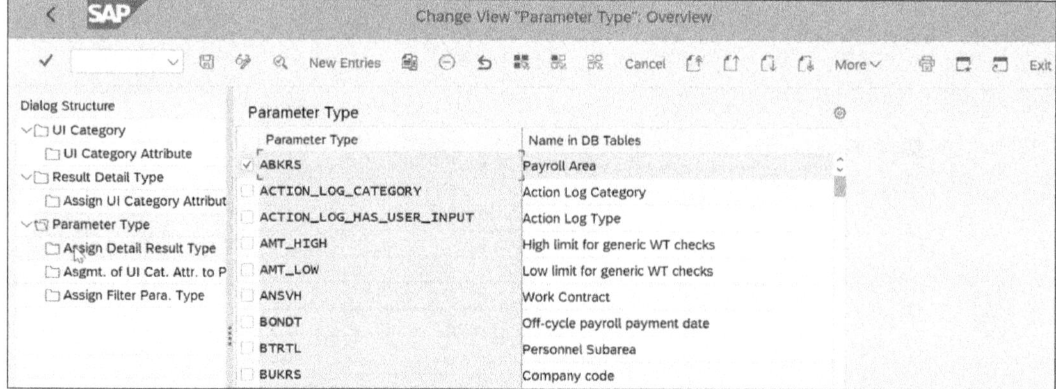

Figure 10.35 Assigning a Result Type to a Parameter

A new screen will open, where you'll click on **New Entries**, as shown in Figure 10.36.

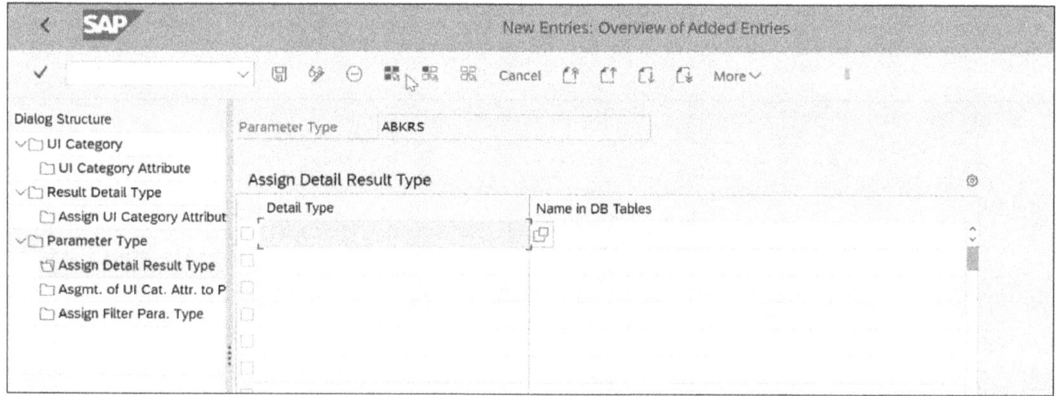

Figure 10.36 Assigning a Detail Type to a Parameter Type

Choose a **Detail Type** from the dropdown list. Enter a name in the **Name in DB** field and save.

Again, creating a new class with new parameter types requires help from an ABAP programmer.

10.3.4 Defining Solutions

A solution is a hyperlink that is available in My Alerts. A solution can be clicked by a user to navigate to the specific area in the system that allows the user to correct the issue. A solution, of course, is part of the validation rule we created in the previous section. The validation rule, like many other components, has a **Solution** tab. A solution is designed for a particular situation and can be valid for multiple validation rules—a solution is meant to be reusable. A solution consists of a few components, such as a solution variable and a solution body, which, when put together, form the complete hyperlink. The reason for this approach is because the subcomponents of a solution are reusable as well.

Our first step is to define the solution variable. SAP provides several standard solution variables, listed in Table 10.10.

Variable	Description
SYSTEM	Service URI of the Employee Central Payroll system, for example, *<protocol>://<host>[:<port>]*
SYSTEM_EC	Service URI of the Employee Central system that replicates data to the Employee Central Payroll system, for example, *https://apiexample.successfactors.com*.
EMPLOYEE_ID	Employee personnel number
EMPLOYEE_ID_EC	Employee ID in Employee Central system, mapped with infotype 0105, subtype ECUS (Communication)
PERSON_ID	Person ID of the employee, which uniquely identifies a person in the system

Table 10.10 Standard Solution Variables

Typically, you would copy a solution variable and use it with minor changes.

To define a solution variable, use Transaction SPRO to navigate to Customizing and follow the menu path **Payroll • Payroll International • Payroll Control Center • Payroll Control Center Configuration Simplification • Maintain Additional Information for Check Types and Solutions • Define Solution Variables**. Solution variables will be displayed on the screen, as shown in Figure 10.37.

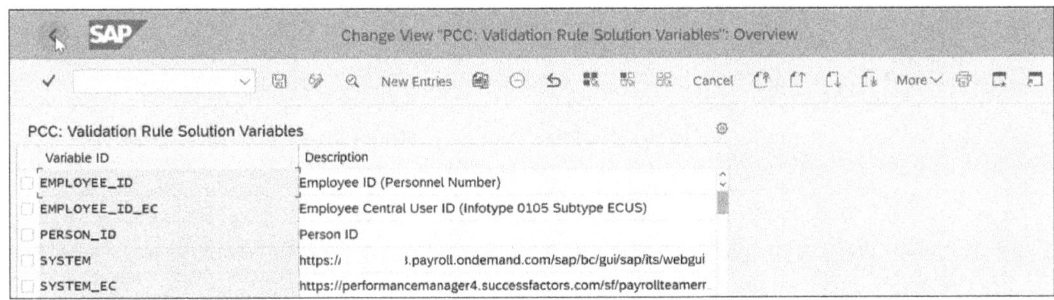

Figure 10.37 Solution Variables

Click the **New Entries** or **Copy** button and fill in the required fields (**Variable ID** and **Name**) and then select either the **Fixed Value** or **Variable Class** option, as shown in Figure 10.38.

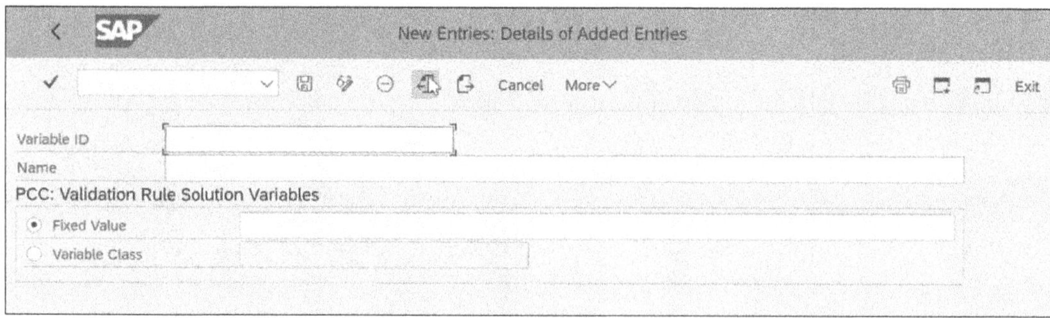

Figure 10.38 New Entries in Solution Variables

If the variable has a fixed value, enter this value into the **Fixed Value** field.

If the variable is expected to be filled in during runtime, you must create a class implementing interface IF_PYD_SOLUTION_VAR and ensure that method GET_VALUE is implemented. You'll need an ABAP developer to help you create this class, and the parameters listed in Table 10.11 will be needed. You'll then add the class name to the **Variable Class** field.

Input/Output	Parameter	Reason
Input	IT_PAR	A value list that represents the current process contexts (for example, payroll areas, payroll period, and process ID)
Input	IS_RESO_REL_KEY	The current parameter type and ID of the alert (For example, PERNR could be the parameter and the ID could be 10000002.)
Input	IV_SEMANTICAL_ID	The solution ID
Input	IV_VARIABLE_ID	The variable ID
Output	RV_RESULT	The result decided by the input parameters

Table 10.11 Parameters Required to Implement Solution Variables in a Variable Class

Our next major step is to define the solution using Transaction SPRO to navigate to Customizing and then following the menu path **Payroll • Payroll International • Payroll Control Center • Payroll Control Center Configuration Simplification • Maintain Additional Information for Check Types and Solutions • Define Solutions.**

Click the **New Entries** or **Copy** button and fill in the required fields, as shown in Figure 10.39. You may freely define the **Solution ID** field, but you should start with a "Z" or "Y" to avoid SAP namespaces. The text you enter into the **Title** field is displayed as the link text in the alert, while the text in the **Description** field is shown as a tooltip.

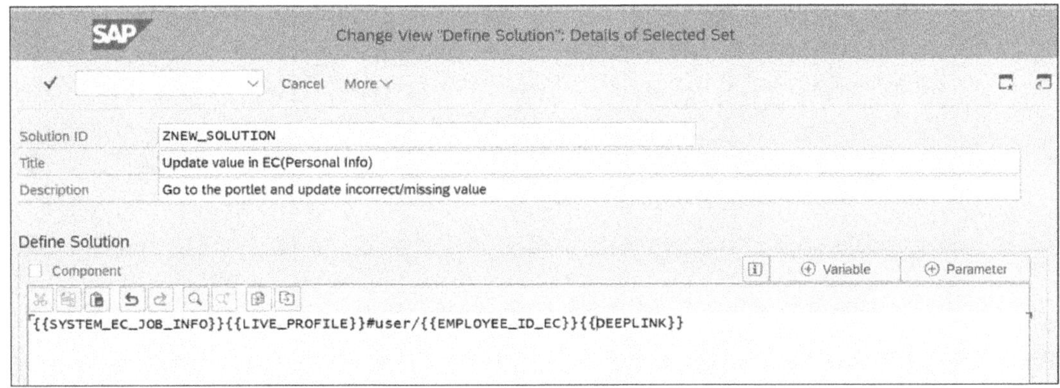

Figure 10.39 Defining a Solution

Select the **Component** checkbox if this component is an SAPUI5 component. Selecting this checkbox means that this link connects to the Employee Central Payroll system, available for use externally through a web browser. This link, also known as a *mashup*, can be accessed from the Employee Central screen.

Click the **+ Variable** button to add a variable and the **+ Parameter** button to insert parameters. A variable is assigned its value during runtime. A parameter is assigned during design time, when you assign the solution to a validation rule. Both variables and parameters are inserted into the solution like a string that is chained together. You can use a combination of multiple variables, parameters, and constants to piece together the link that forms the solution.

Table 10.12 lists the solution link fields. Parameters are shown with double square brackets, while variables with double curly brackets, which is the convention followed on this configuration screen.

Field	Example
Solution ID	For a people profile link, you might use YL_PEOPLE_LINK
Title	People Profile Maintenance
Description	Maintain data in Employee Central system
Component	Not checked, since this is not an SAPUI5 component
Solution Link	{{Z_LIVE_PROFILE}}#user/{{EMPLOYEE_ID_EC}} {{Z_DEEPLINK}} *#user is a fixed constant that is assigned the user's Employee Central User Id using the last solution variable {{EMPLOYEE_ID_EC}}*

Table 10.12 Explanation of Solution Link

The **Solution Link** field is the link that will be displayed in the Payroll Control Center when master data is missing. This link will take the user directly to the page of the specific employee in the correct portlet where the master data needs to be rectified. Since this can happen to any employee, the employee ID field is dynamic. As shown in Table 10.13, the structure of the link is in three parts—one fixed value + one variable + one fixed value. Thus, we decided to split the link into three variables and assigned fixed values for the first and the third fields. We assigned a class variable to the second field that will bring in the Employee ID field at runtime.

Variable ID	Value/Explanation
Z_LIVE_PROFILE	■ Name: Service Path for People Profile ■ Fixed Value: <host>/sf/liveprofile#blockview/ ■ An example of <host> is *https://performancemanager4.successfactors.com* but could be different in your case depending on the datacenter you're using for your Employee Central instance
EMPLOYEE_ID_EC	■ Name: Employee ID in Employee Central ■ Variable class: CL_PYC_SOL_VAR_EMPLOYEE_ID_EC ■ If the parameter type in IS_RESO_REL_KEY uses the data element PERNR_D, the class returns the employee's employee ID at run time. In other words, the EMPLOYEE_ID can only be derived if the alert type refers to personnel number
Z_DEEPLINK	■ Name: Deep Link ■ Fixed Value: /blockid/default_block22

Table 10.13 Variable Structure of the Solution Link Field of the Solution Object

The Z_DEEPLINK variable in Table 10.13 requires further explanation. This relatively new feature allows you to position the screen directly on the portlet that needs to be modified. You can identify the portlet by its **Block ID** in the **Configure People Profile** screen in Employee Central, as shown in Figure 10.40.

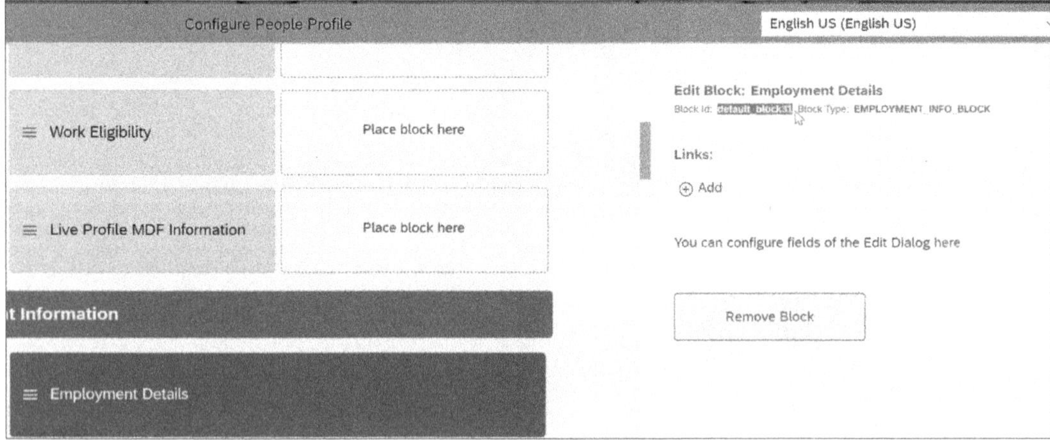

Figure 10.40 Blocking an ID from the Configure People Profile Screen

10.3.5 Creating Analytics Charts

An analytics chart is a visual representation of pay-related statistics to help you see the impact of the payroll being processed. This tool is most useful in the monitoring step so a payroll manager can spot issues that are not necessarily at an employee level. An example is a comparison between the number of employees processed last month versus this month.

To create an analytics chart, follow these steps:

1. Initiate the Configuration Workbench in the Payroll Control Center using Transaction PYC_CONF_WB.
2. Switch to edit mode by clicking on the **Display/Edit** button.
3. From the left panel, choose **Analytics Chart**, as shown in Figure 10.41. Copy a standard or existing analytics chart or create a new one. To copy, double-click an existing analytics chart to copy, click the **Copy** icon in the application toolbar. Then, enter a unique ID for the new analytics chart and click **Continue**. All the attributes and values, except for the ID, will be copied to the new chart. To create a new analytics chart from scratch, click the **Create** icon in the application toolbar. Then, enter a unique ID and a name for your new analytics chart and then click **Continue**.

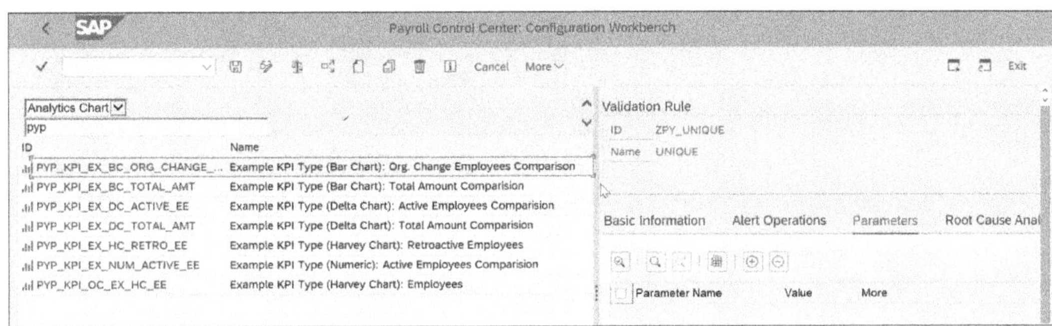

Figure 10.41 Configuring Analytics Charts

4. In the new analytics chart, click the **Basic Information** tab, as shown in Figure 10.42.

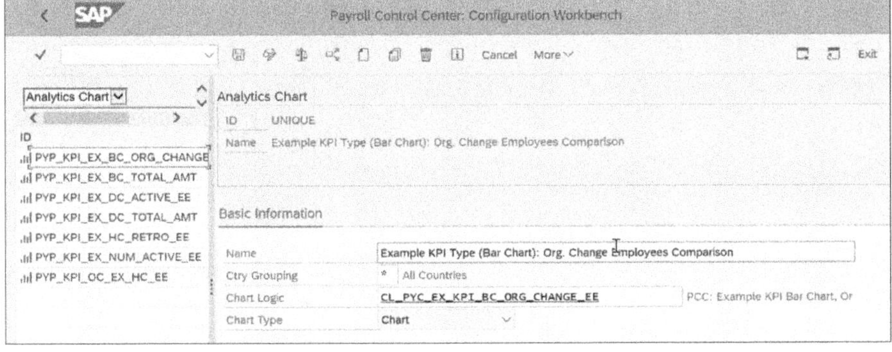

Figure 10.42 New Analytics Chart

5. Add the name (**Name**) and the country (**Ctry Grouping**) for which the rule is required.
6. Now, let's add the chart logic. In the **Chart Logic** field, you must enter an existing class. Recall that classes are code snippets that create the chart and must be created in advance in Transaction SE24. Creating rule logic requires knowledge of ABAP, and thus, you might need a developer to help you.
7. Valid classes for the **Chart Logic** field are limited to inheriting from one of the following base classes, which determine the type of chart:
   - `CL_PYC_KPI_BASE` (Payroll Process KPI Base Class)
   - `CL_PYC_KPI_DELTA_CHART_BASE` (Delta Chart)
   - `CL_PYC_KPI_HARVEY_CHART_BASE` (Harvey Chart)
   - `CL_PYC_KPI_NUMERIC_BASE` (Numeric Chart)
8. The **Chart Type** field is a read-only field showing you the type of chart you just selected.
9. Click the **Validate** icon to see if any errors exist.
10. Save these settings.

Analytics charts are saved in a PYAC transport object. Analytics charts are grouped into groups called "Analytics" and assigned to processes. The following list includes some sample analytics charts provided by SAP that you can copy from, with their technical names in parenthesis:

- Total Amount Comparison (`PYP_KPI_EX_BC_TOTAL_AMT`), a bar chart comparison of total payroll between last and current period.
- Total Amount Comparison (`PYP_KPI_EX_DC_TOTAL_AMT`), a delta chart comparison of total payroll between last and current period.
- Org. Change Employees Comparison (`PYP_KPI_EX_BC_ORG_CHANGE_EE`), displays the number of employees with organizational changes between last and current periods.
- Active Employees Comparison (`PYP_KPI_EX_NUM_ACTIVE_EE`), displays a percentage increase or decrease in the growth of employees between last and current pay periods.
- Retroactive Employees (`PYP_KPI_EX_HC_RETRO_EE`), a Harvey Ball chart for the number of employees with a retroactive payroll run between the last and the current payroll periods.
- Active Employees Comparison (`PYP_KPI_EX_DC_ACTIVE_EE`), the number of active employees compared between last and current pay periods.

10.3.6 Creating Analytics Groupings

Grouping analytics charts is the next step, and you'll organize your charts into logical groups that are meaningful to be displayed in the executed processes for payroll, such as simulation, live run, and subsequent processing.

To create an analytics grouping, follow these steps:

1. Initiate the Configuration Workbench in the Payroll Control Center using Transaction PYC_CONF_WB.
2. Switch to edit mode by clicking on the **Display/Edit** button.
3. From the left panel, choose **Analytics**. Copy a standard analytics grouping or existing analytics grouping or create a new one, as shown in Figure 10.43. To copy, double-click an existing analytics grouping to copy. Then, click the **Copy** icon in the application toolbar, enter a unique ID for the new analytics grouping, and then click **Continue**. All the attributes and values, except for the ID, will be copied to the new chart. To create a new analytics grouping from scratch, click the **Create** icon in the application toolbar. Then, enter a unique ID and a name for the new analytics object.

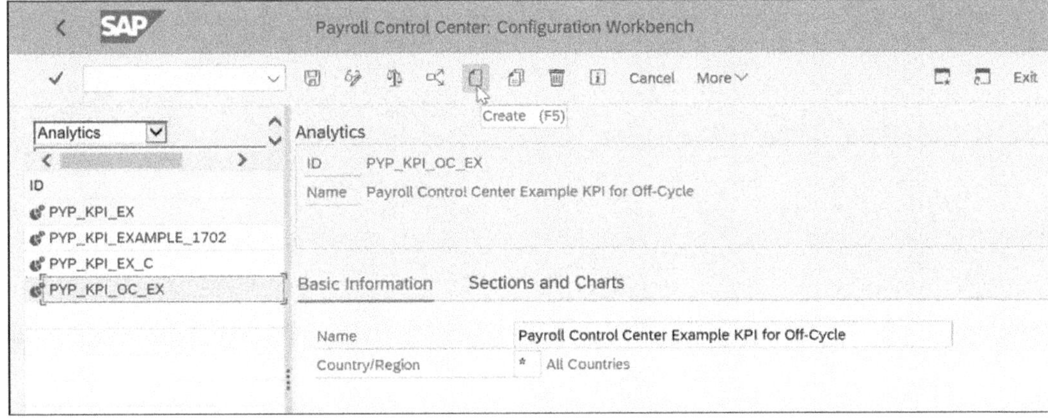

Figure 10.43 Analytics Create Button

4. In the new analytics object, click the **Basic Information** tab, as shown in Figure 10.44.

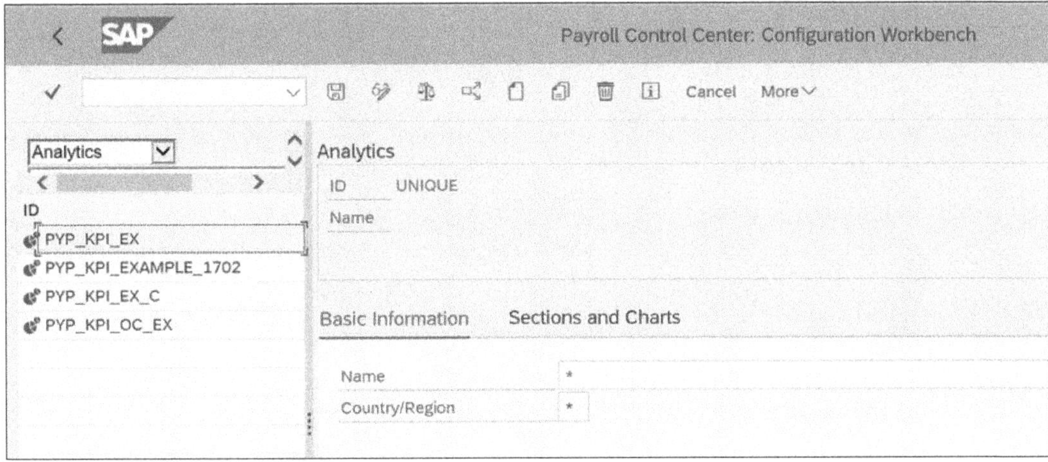

Figure 10.44 Analytics New Entry

5. Add the **Name** and the **Country/Region** for which the rule is required.
6. Now, choose the **Sections and Charts** page and group your analytics charts.
7. Create a new section by clicking the **New Section** button.
8. Click **Continue**.
9. Add a new chart by clicking on the **New Chart** button.
10. Select the analytics charts you need and click **Continue**, as shown in Figure 10.45.

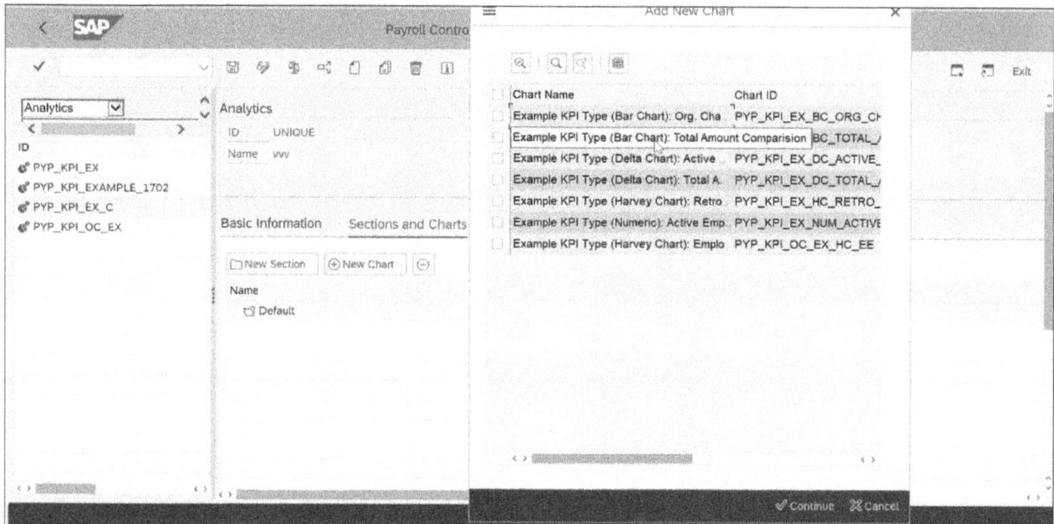

Figure 10.45 Adding a Chart to Analytics

11. Repeat these steps until you've created all the groups you need. You can adjust the positions of the charts by clicking on the up and down arrows.
12. Click the **Validate** icon to see if any errors exist.
13. Save these settings.

Analytics are transported using a PYAN transport object.

10.3.7 Creating Step Templates

A step template is a basic building block for processes and includes name of the program that must be run for the process, along with the name of the variant that stores the parameters required to complete the run successfully.

To create a step template, follow these steps:

1. Initiate the Configuration Workbench in the Payroll Control Center using Transaction PYC_CONF_WB.
2. Switch to edit mode by clicking on the **Display/Edit** button.

3. From the left panel, select **Step Template**. Copy a standard step template or existing step template or create a new one, as shown in Figure 10.46. To copy, double-click an existing step template to copy. Then, click the **Copy** icon in the application toolbar, enter a unique ID for the new step template, and then click **Continue**. All the attributes and values, except for the ID, will be copied to the new chart. To create a new step template from scratch, click the **Create** icon in the application toolbar. Then, enter a unique ID and a name for the new step template and click **Continue**.

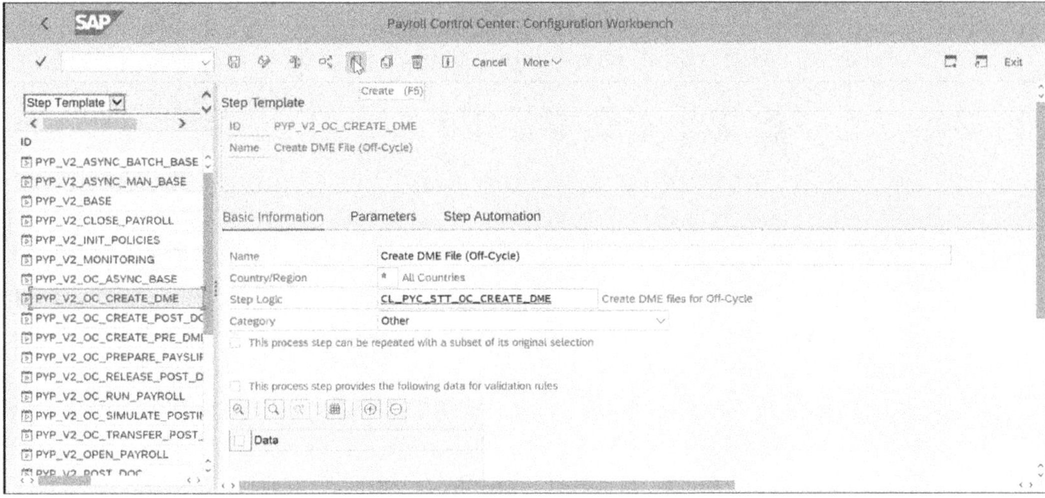

Figure 10.46 Creating a Step Template

4. In the new step template, click the **Basic Information** tab, as shown in Figure 10.47.

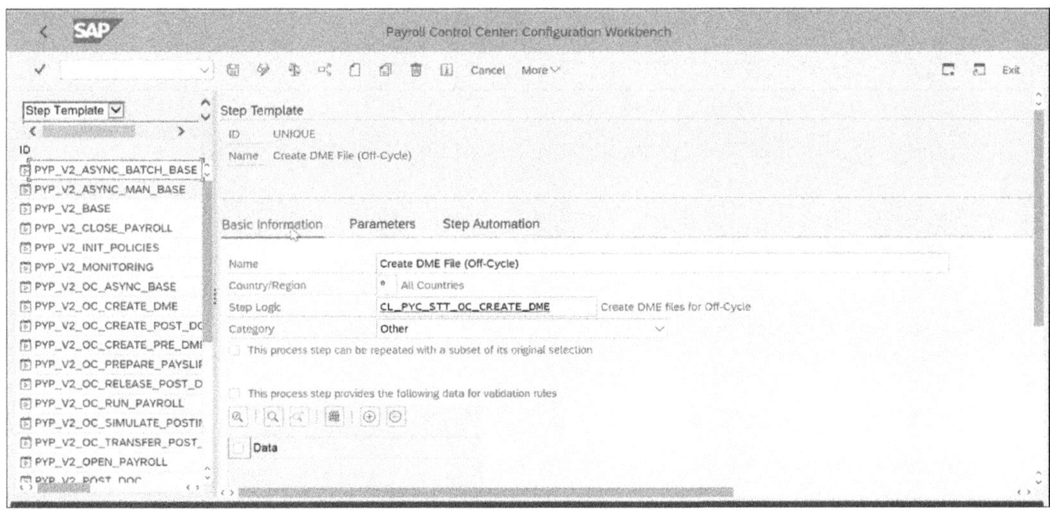

Figure 10.47 Basic Information in the New Step Template

5. Add the **Name** and the **Country/Region** for which the rule is required.
6. Now, choose the **Step Logic**. If you want to create new step logic, you must use Transaction SE24.
7. Select one of the following three categories of steps:
   - **Monitoring**: If this step is for monitoring, it must inherit from class CL_PC_SIT_MONITORING.
   - **Initiate Policies**: If this step initiates a policy, it must inherit from class CL_PC_SIT_EXECUTE_POLICIES.
   - **Other**: If another kind of step that involves running ABAP programs in background mode, it must inherit from class CL_PC_SIT_ASYNC_BATCH_BASE. Otherwise, it must inherit from class CL_PC_SIT_BASE.
8. Selecting the **This process step can be repeated with a subset of its original selection** checkbox means this step can run for a subset of the data. For example, the manager may want to run a payroll for the entire payroll area, or just the employees who errored out, or one single employee whose error has just been fixed.
9. Selecting the **This process step provides the following data for validation rules** checkbox means the step must run to provide data to a validation rule. The data must be identified by clicking on the **Insert Row** (+) icon, as shown in Figure 10.48. Then, select the applicable data by selecting among the checkboxes and click **Continue**. This step ensures the rule is executed only with the latest data.

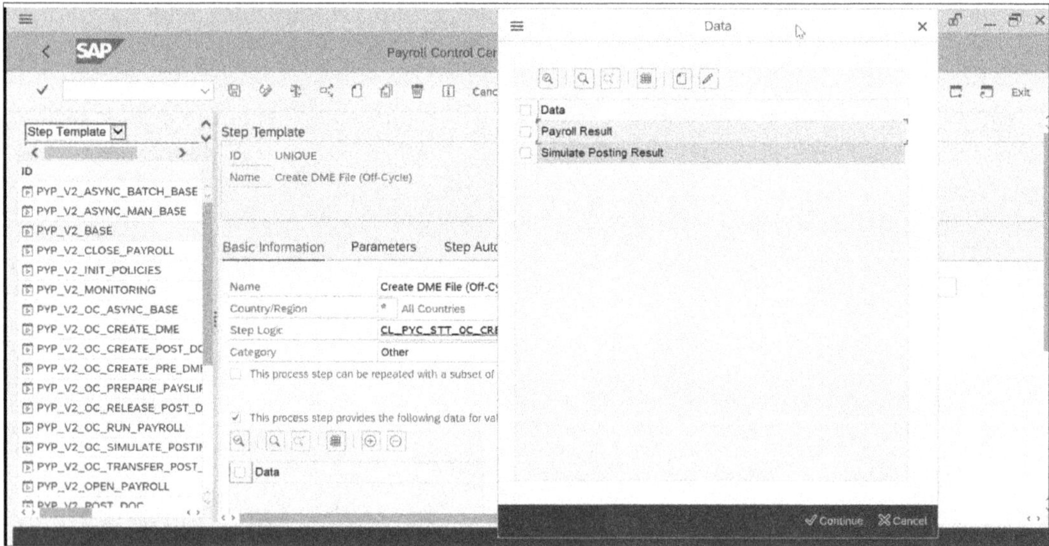

Figure 10.48 Choosing Data for a Step Template

10. Choose the parameters required to run this step by navigating to the **Parameters** tab. Note that the following parameters are always selected by default to keep the context of the process alive:

- PYP_PROC (Payroll Process)
- PYP_PROC_INST (Payroll Process Instance)
- PYP_PROC_GRP_TEMPLATE (Payroll Process Step Group Template)
- PYP_PROC_TEMPLATE (Payroll Process Template)
- PYP_STEP_SORT_FIELD (Payroll Process Step Sort Field)

11. Click on the **Insert Row** icon. Select the parameters, such as the program name. Click on the **Mandatory** checkbox if the step cannot run without this parameter, for example, the payroll area, which must be chosen when configuring the process in the next step.
12. Click the **Step Automation** tab and configure the following fields:
    - **This Process Step Can Be Started Automatically:** This option means you do not need to manually start the process. The daemon will kick it off.
    - **This Process Step Can Be Confirmed Automatically:** This option means this job will be automatically confirmed if no errors arise.
    - **Automatically Confirm Even If This Process Step Did Finish with Errors:** This option means this job will be automatically confirmed even if errors arise.
13. Validate your settings by clicking the **Validate** icon and then save.

To avoid creating new step templates, several step templates are provided by SAP, listed in Table 10.14. The first three templates are models that can be used for customer steps that support payroll but are not actually part of running payroll. The remaining are all steps to run various components of payroll including pre- and post-processing.

Process	Step Template ID	Step Template Name
Generic step template	PYP_V2_ASYNC_BATCH_BASE	This step template can be the base for customer steps executing background jobs asynchronously.
Generic step template	PYP_V2_ASYNC_MAN_BASE	This step template can be the base for customer steps that require some offline activities manually.
Generic step template	PYP_V2_OC_ASYNC_BASE	This step template can be the base for customer steps in off-cycle processes executing background jobs.
(Team) monitoring payroll	PYP_V2_INIT_POLICIES	Initiate Policies

Table 10.14 Standard Step Templates by Process

Process	Step Template ID	Step Template Name
(Team) monitoring payroll	PYP_V2_MONITORING	Monitoring
(Team) monitoring payroll	PYP_V2_RUN_PAYROLL_TEST	Create Test Payroll Data
(Team) monitoring payroll	PYP_V2_SIMULATE_POSTING	Simulate Posting Run
Productive payroll	PYP_V2_OPEN_PAYROLL	Start Payroll
Productive payroll	PYP_V2_RUN_PAYROLL	Run Payroll
Productive payroll	PYP_V2_SIMULATE_POSTING	Simulate Posting Run
Productive payroll	PYP_V2_INIT_POLICIES	Initiate Policies
Productive payroll	PYP_V2_MONITORING	Monitoring
Productive payroll	PYP_V2_CLOSE_PAYROLL	End Payroll
Pay Slip	PYP_V2_ST_PRE_PAYSLIP	Create Pay slip
Pay Slip	PYP_V2_ST_PRINT_PAYSLIP	Print Pay Slip
Posting	PYP_V2_POST_DOC	Create Posting Document
Posting	PYP_V2_REL_POST_DOC	Release Posting Document
Posting	PYP_V2_TRANSFER_POST_DOC	Transfer Posting Document
Bank transfer	PYP_V2_ST_PRE_DME_PROC	Create Pre-DME File
Bank transfer	PYP_V2_ST_CREATE_DME	Create DME File
Bank transfer	PYP_V2_ST_SEND_DME	Send DME File
Planned off-cycle: productive payroll	PYP_V2_OC_RUN_PAYROLL	Run Payroll (Off-Cycle)
Planned off-cycle: productive payroll	PYP_V2_OC_SIMULATE_POSTING	Posting Simulation (Off-Cycle)
Planned off-cycle: productive payroll	PYP_V2_INIT_POLICIES	Initiate Policies
Planned off-cycle: productive payroll	PYP_V2_MONITORING	Monitoring

Table 10.14 Standard Step Templates by Process (Cont.)

Process	Step Template ID	Step Template Name
Planned off-cycle: pay slip	PYP_V2_OC_PREPARE_PAYSLIP	Create Pay Slip (Off-Cycle)
Planned off-cycle: pay slip	PYP_V2_ST_PRINT_PAYSLIP	Print Pay Slip
Planned off-cycle: posting	PYP_V2_OC_CREATE_POST_DOC	Create Posting Documents (Off-Cycle)
Planned off-cycle: posting	PYP_V2_OC_RELEASE_POST_DOC	Release Posting Document (Off-Cycle)
Planned off-cycle: posting	PYP_V2_OC_TRANSFER_POST_DOC	Transfer Posting Document (Off-Cycle)
Planned off-cycle: bank transfer	PYP_V2_OC_CREATE_PRE_DME	Create Pre-DME File (Off-Cycle)
Planned off-cycle: bank transfer	PYP_V2_OC_CREATE_DME	Create DME File (Off-Cycle)
Planned off-cycle: bank transfer	PYP_V2_ST_SEND_DME	Send DME File
Ad hoc off-cycle: subsequent activities	PYP_V2_OC_CREATE_POST_DOC	Create Posting Documents (Off-Cycle)
Ad hoc off-cycle: subsequent activities	PYP_V2_OC_RELEASE_POST_DOC	Release Posting Document (Off-Cycle)
Ad hoc off-cycle: subsequent activities	PYP_V2_OC_TRANSFER_POST_DOC	Transfer Posting Document (Off-Cycle)
Ad hoc off-cycle: subsequent activities	PYP_V2_OC_PREPARE_PAYSLIP	Create Pay Slip (Off-Cycle)
Ad hoc off-cycle: subsequent activities	PYP_V2_ST_PRINT_PAYSLIP	Print Pay Slip
Ad hoc off-cycle: subsequent activities	PYP_V2_OC_CREATE_PRE_DME	Create Pre-DME File (Off-Cycle)
Ad hoc off-cycle: sub-sequent activities	PYP_V2_OC_CREATE_DME	Create DME File (Off-Cycle)
Ad hoc off-cycle: subsequent activities	PYP_V2_ST_SEND_DME	Send DME File

Table 10.14 Standard Step Templates by Process (Cont.)

10.3.8 Creating Policy Types and Process Types

A policy type defines the vital attributes of a policy—such as country or region. A process type collects the different steps, step contexts, recurrence types, and selection types of a process. Depending on the process category, a process type also collects the following information:

- Event handlers for monitoring processes and team monitoring processes
- Team dimensions for team monitoring processes
- Off-cycle selection for planned off-cycle productive payroll processes
- Policy types for monitoring processes, team monitoring processes, productive and planned off-cycle processes, and productive payroll processes

We'll discuss the creation process for both policy types and process types in the following sections.

Policy Types

To create a policy type, follow these steps:

1. Initiate the Configuration Workbench in the Payroll Control Center using Transaction PYC_CONF_WB.
2. Switch to edit mode by clicking on the **Display/Edit** button.
3. From the left panel, choose **Policy Type**. Copy a standard or existing policy type or create a new one, as shown in Figure 10.49.

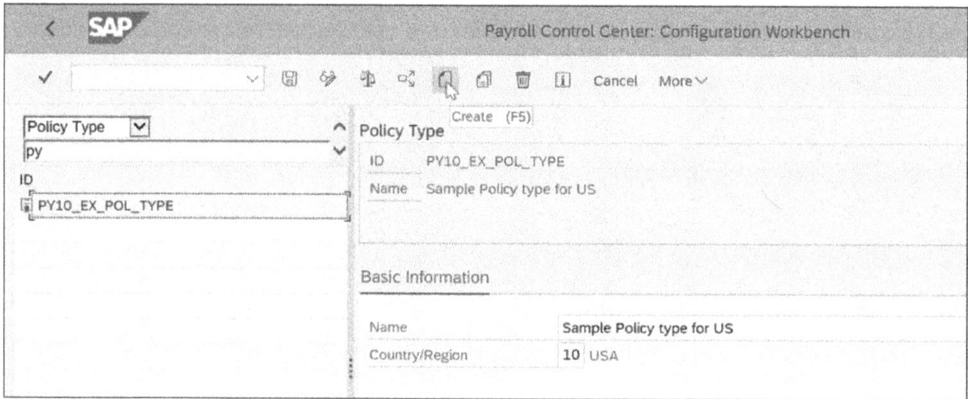

Figure 10.49 Creating New Policy Type

4. To copy, double-click an existing policy type to copy and click the **Copy** icon in the application toolbar. Then, enter a unique ID for the new policy type and click

Continue. All the attributes and values, except for the ID, will be copied to the new policy type. To create a new policy type from scratch, click the **Create** icon in the application toolbar. Then, enter a unique ID and a name for the new policy type and click **Continue**.

5. In the new policy type, click the **Basic Information** tab, as shown in Figure 10.50.

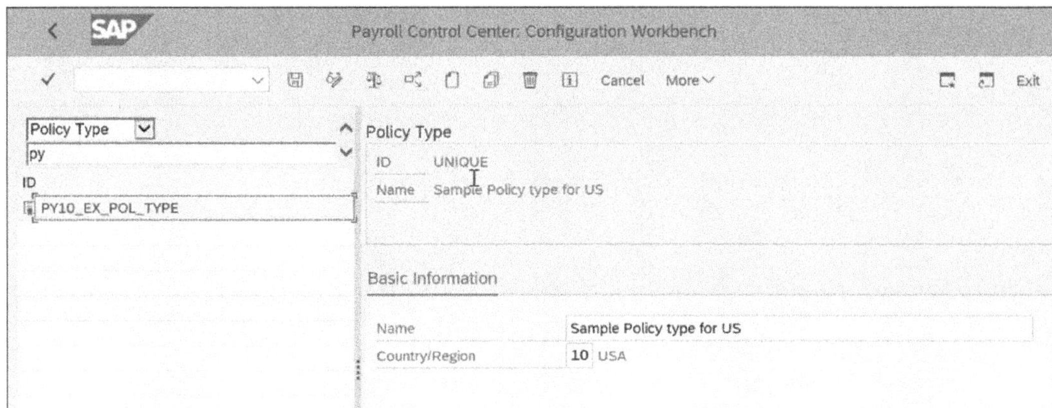

Figure 10.50 Basic Information for a Policy Type

6. Add the **Name** and the **Country/Region** for which the rule is required.
7. Click the **Validate** icon to see if any errors exist.
8. Save these settings.

Note that a policy type is transported using a PYTP transport object.

Process Types

To create a process type, follow these steps:

1. Initiate the Configuration Workbench in the Payroll Control Center using Transaction PYC_CONF_WB.
2. Switch to edit mode by clicking on the **Display/Edit** button.

   From the left panel, choose **Process Type.** Copy a standard process type or an existing process type or create a new one, as shown in Figure 10.51. To copy, double-click an existing process type to copy. Then, click the **Copy** icon in the application toolbar, enter a unique ID for the new process type, and click **Continue.** All the attributes and values, except for the ID, will be copied to the new chart. To create a new process type from scratch, click the **Create** icon in the application toolbar. Then, enter a unique ID and a name for the new process type and click **Continue**.

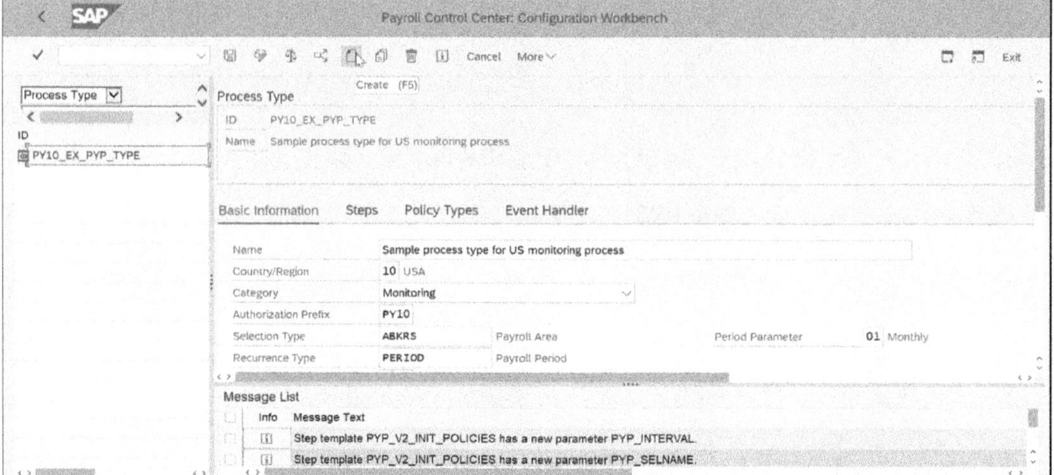

Figure 10.51 New Process Type

3. In the new process type, click the **Basic Information** tab, as shown in Figure 10.52.

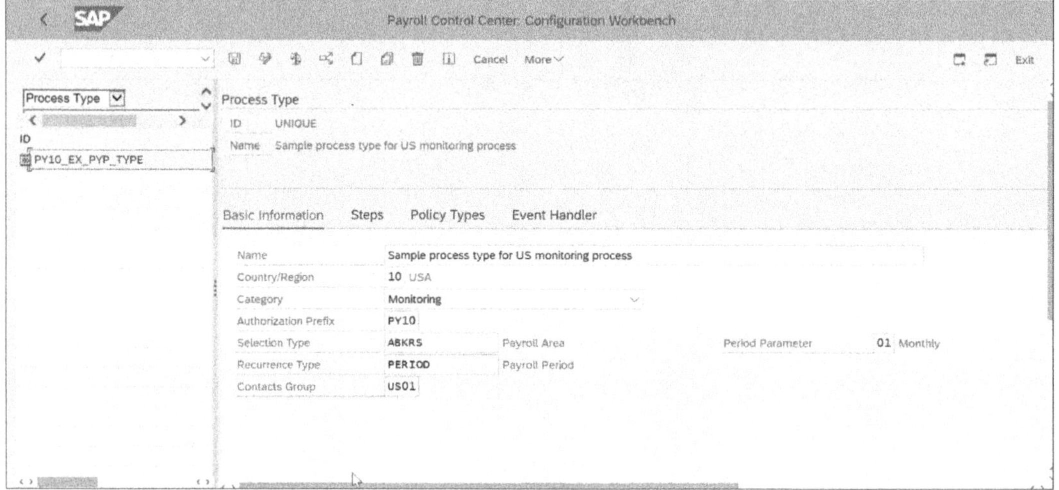

Figure 10.52 Basic Information for a Process Type

4. Add the **Name** and the **Country/Region** for which the rule is required.
5. Choose from the following options in the **Category** dropdown list:
   - **Team Monitoring**
   - **Monitoring**
   - **Productive Payroll**
   - **Productive Off cycle**
   - **Planned off cycle**

6. Add an **Authorization Prefix**, up to 4 characters (which will be combined with the process ID and thus will become the authorization ID for the process).
7. Add a **Selection Type**, such as **Payroll Area**.
8. Add **Period Parameters** (run frequency), as shown in Figure 10.53. This run frequency must be configured in advance. To configure payroll periods, use Transaction SM30 with view V_T549Q or use Transaction SPRO, as discussed in Chapter 4, Section 4.2.6.

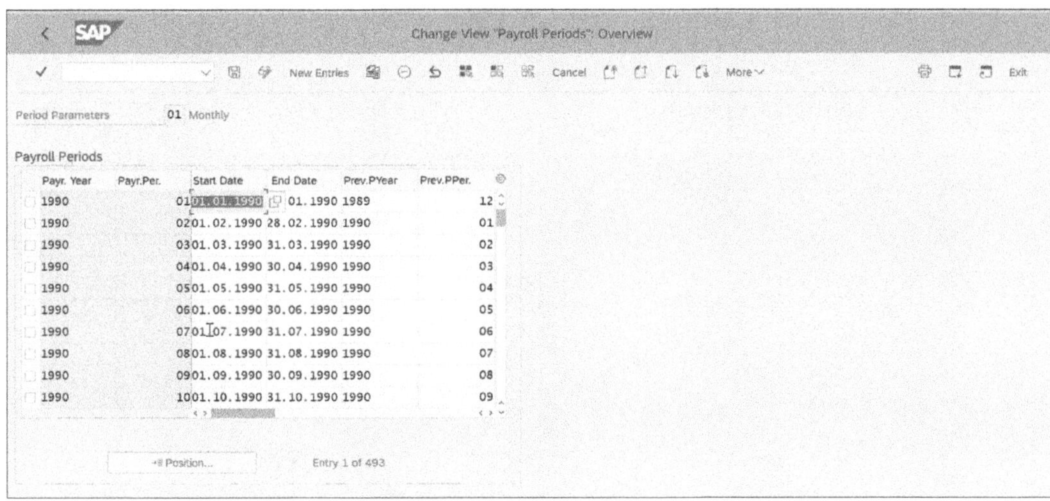

Figure 10.53 Period Parameter Configuration

9. Add a **Recurrence Type**, as follows:
   - For regular process, use the payroll period (field PERIOD).
   - For off-cycle process, use the payment date (field BONDT).
10. Add a **Contacts Group** (the administrator group, as shown in Figure 10.54). This administrator group must be configured in advance, as outlined in Chapter 5.

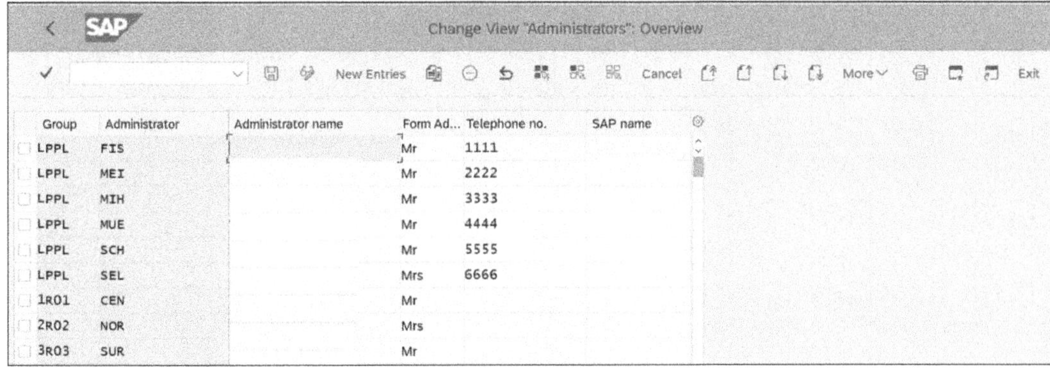

Figure 10.54 Configuring an Administrator Group

11. Under the **Steps** tab, choose the steps and then click **Continue**. Step selection can be performed in one of two ways:
    - Manual assignment, as shown in Figure 10.55:
        - Click the **Insert Row** button.
        - Select one or more steps.
        - Click **Continue**.
    - Import process:
        - Click the **Import** button.
        - Select one or more process types.
        - Adjust the sequence.

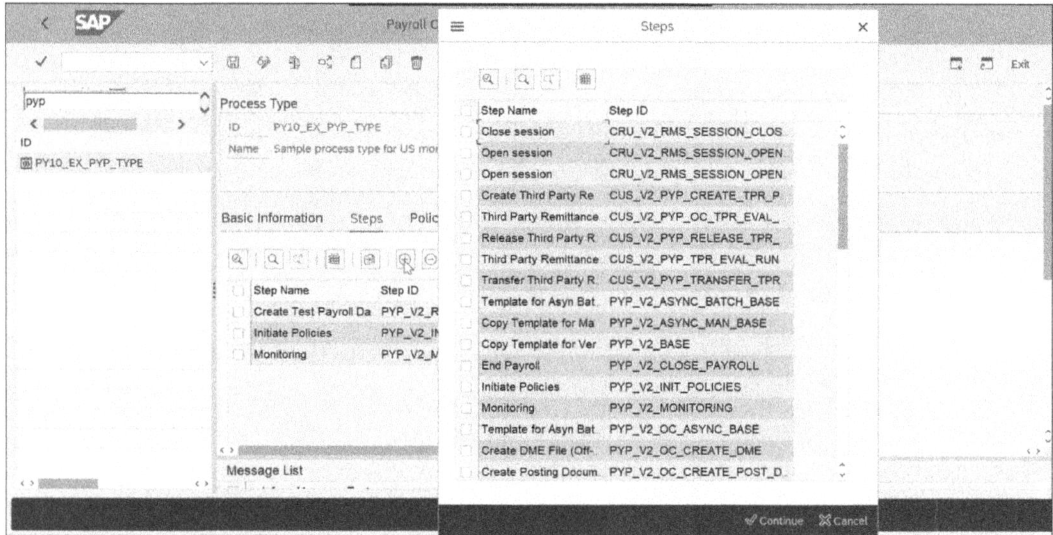

Figure 10.55 Manual Assignment

12. For planned off-cycle productive process, click the **Off-Cycle Selection** tab, as shown in Figure 10.56, and specify the value of the selection type. (Note that this tab appears for off-cycle processes only.) For the **Selection Logic** field, enter an existing class. You can build a new class in Transaction SE24 (Class Builder). The rule logic must inherit from base class `CL_PYC_SELECTION_OC_RT` (Off-Cycle Selection Runtime).

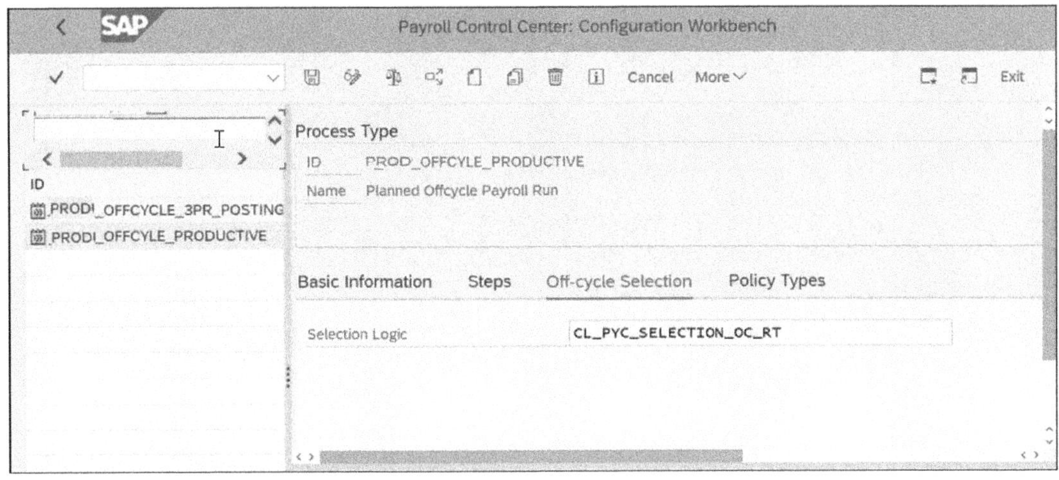

Figure 10.56 Off-Cycle Selection Tab

13. For monitoring and team monitoring processes, click the **Event Handler** tab, as shown in Figure 10.57 and specify two key fields:
   - **Event Handler Enabled**: If selected, the system confirms re-execution if a step has been performed before.
   - **Relevance Check Logic**: Choose a relevant check logic.

   Note that the policies available in the Manage Processes application depend on the policy types assigned to the process types in the same country.

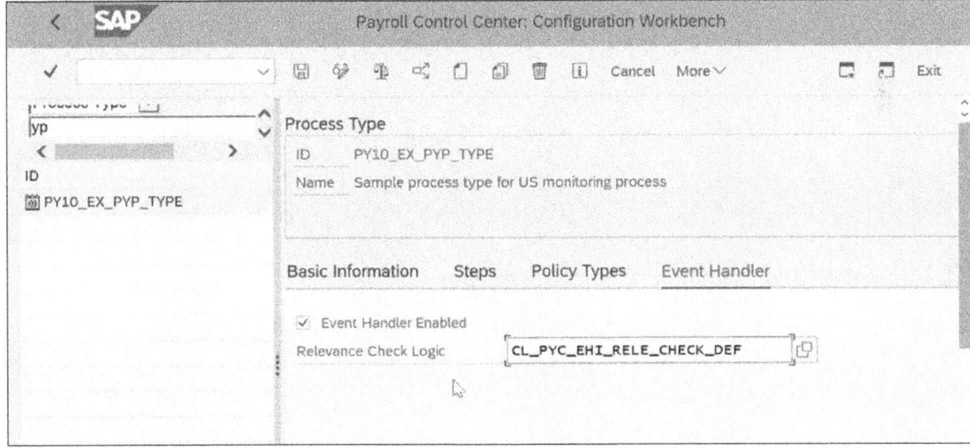

Figure 10.57 Event Handler Tab

14. For team monitoring processes, click the **Team Dimensions** tab, as shown in Figure 10.58, and proceed as follows:
   - Click on **Insert Row**.
   - Choose one or more dimensions from table PYC_D_TMDI_DE.

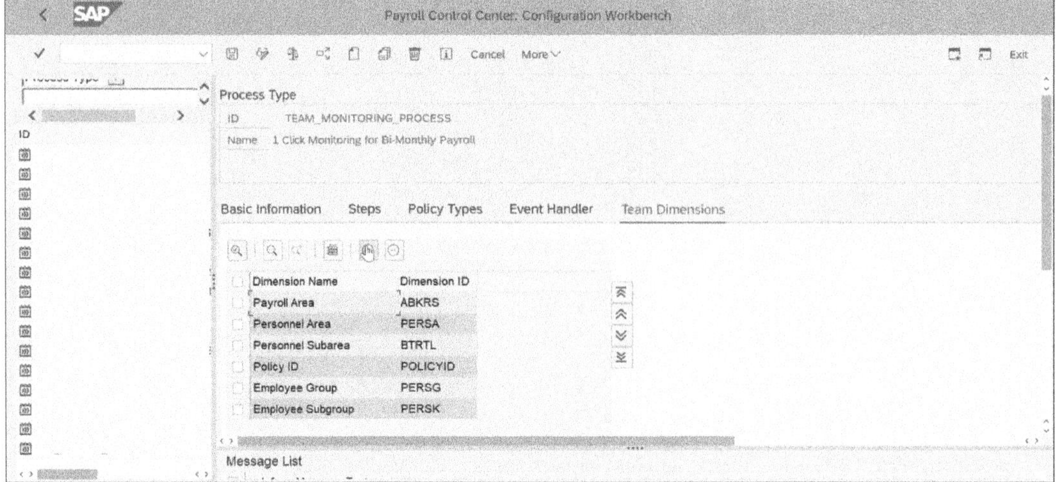

Figure 10.58 Team Dimensions Tab

15. An authorized business user can use the **Dimension** field in the Manage Teams application to set up teams for handling alerts.
16. Click the **Validate** icon to see if any errors exist.
17. Save these settings.

The following sample process types are available:

- PY10_EX_PYP_TYPE: Monitoring (US)
- PY99_EX_PYP_TYPE_TM: Team Monitoring (International)
- PY10_EX_PYP_TYPE_PROD: Regular Productive (US)
- PY10_EX_PYP_TYPE_PO: Planned Off-cycle - Productive Payroll (US)
- PY10_EX_PYP_TYPE_OO: Planned Off-cycle - subsequent Posting (US)
- PY10_EX_PYP_TYPE_AO: Ad-hoc Off-cycle (US)

10.3.9 Creating Policies and Processes

Policies and processes are the fundamental nuts and bolts of an Employee Central Payroll system. While policies govern the data used to generate pay results through appropriate validations and analytics, processes actually perform the various steps required to read the data, calculate payroll, produce results, and send these results to receiver systems, such as banking and finance.

Policies and processes are created differently based on the following two fundamental scenarios:

- *Simplified configuration* is where the policies and processes are directly created in the production Employee Central Payroll system from the Employee Central frontend.

The big advantage of this method is that a business user can configure and control the layout of the processes without depending on the technical team. This approach requires the simplified configuration capability (business function HCM_LOC_CI_92) to be *activated*.

- *Traditional configuration* is where you use the Configuration Workbench in the backend (through Transaction PYC_CONF_WB) to configure the processes. Since this activity is configuration, a business user will need technical help to get this configuration done. However, once created, the setup can be transported to other clients using the standard transport management system. This approach requires the simplified configuration capability (business function HCM_LOC_CI_92) to be *deactivated*.

The recommended scenario is to use the simplified configuration, for its ease of use and availability for the business user. But SAP lets you choose the method that is most suited to your business situation. Two items to note, however, are as follows:

- You should never mix and match between the two setups, that is, have some processes using the simplified configuration and others using the traditional configuration. This approach would lead to inconsistencies that are difficult to overcome. For this reason, the Manage Process and Manage Policies applications must be disabled in Employee Central if you want to use the traditional configuration.
- The processes and policies configured through the simplified configuration approach appear in the Configuration Workbench as display only. These settings cannot be changed using Transaction PYC_CONF_WB.

We'll discuss both methods of configuration in the following sections as well as provide instructions for generating recurrent payroll processes.

Simplified Configuration

The simplified configuration method needs the required authorization in Employee Central, which we'll discuss in Section 10.4. Let's now look at how to create policies and processes using simplified configuration.

Creating Policies

To create a policy, follow these steps:

1. Log on the Employee Central system and go to the **Payroll Control Center** screen through the **Manage Payroll** screen. Start the Manage Policies application, as shown in Figure 10.59.

   You can also reach this configuration screen via the following URL: *https://<host>:<port>/sap/bc/ui5_ui5/sap/hrpy_pcc_conf_2/index.html?sap-client=<client>&sap-ui-language=<language_code>*.

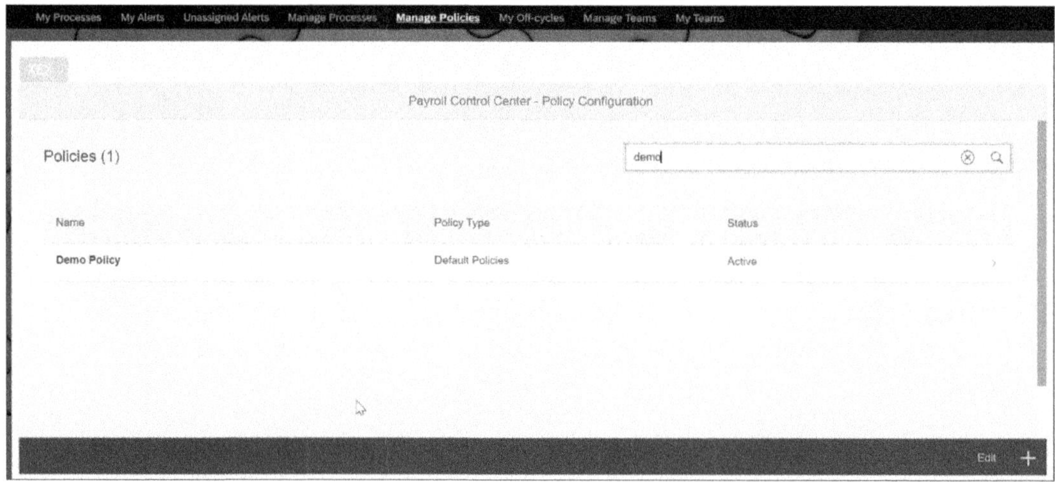

Figure 10.59 Manage Policies Screen in Employee Central

2. Review the policies listed on this page. Each policy can have one of three different statuses, as follows:
   - **New** (not yet activated, being developed)
   - **In Process By** (being edited by a user)
   - **Active** (validated and saved, ready to use)
3. Click the + icon to add a new policy.
4. Select **Policy Type**, which brings up a list of policy types including the one we configured earlier in Section 10.3.8, as shown in Figure 10.60. Select the policy type that is relevant for your scenario.

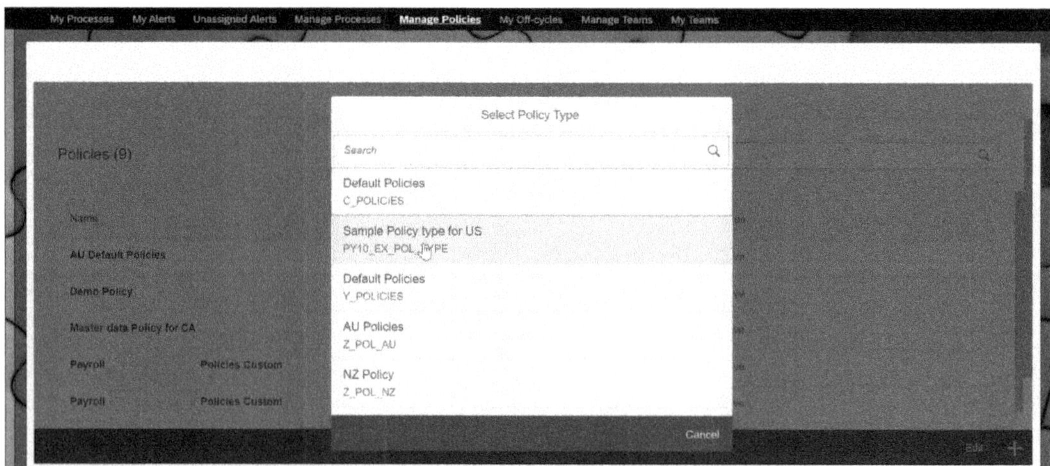

Figure 10.60 Creating a Policy: Choosing a Policy Type

5. Enter a policy name in the **Policy Name** field.
6. Choose one or more check types (called "validation rules" in the Configuration Workbench) by clicking on the + button, as shown in Figure 10.61.

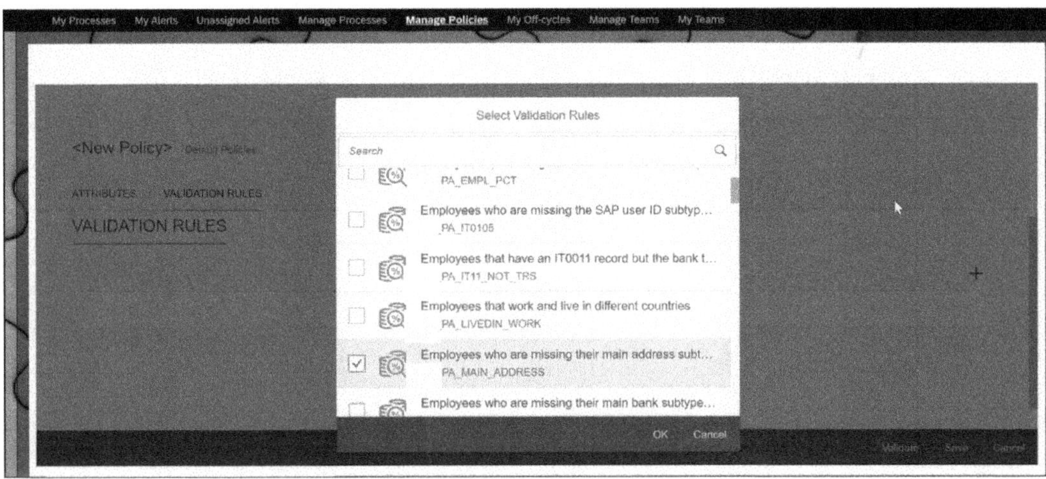

Figure 10.61 Creating a Policy: Choosing a Check Type/Validation Rule

7. Click **Validate** to check for errors and then save.

Creating Processes

To create a process, follow these steps:

1. Log on the Employee Central system and go to the **Payroll Control Center** screen through the **Manage Payroll** screen. Start the Manage Processes application, as shown in Figure 10.62.

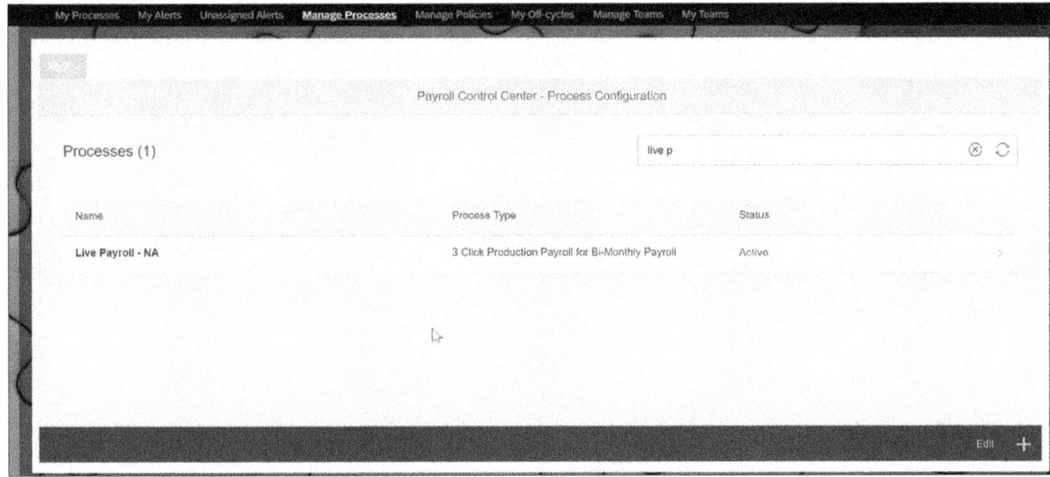

Figure 10.62 Manage Processes Configuration in Employee Central

You can also reach this configuration screen via the following URL: *https://<host>:<port>/sap/bc/ui5_ui5/sap/hrpy_pcc_conf_1/index.html?sap-client=<client>&sap-ui-language=<language_code>*.

2. Review the processes listed on this page. Each process can have one of four different statuses, as follows:
   - **New** (not yet activated, being developed)
   - **New In Process By** (new process being edited by a user)
   - **Active** (validated and saved, ready to use)
   - **Active in process by** (validated and saved process, being edited by a user)
3. Click the + button to add a new process.
4. Select **Process Type**, which will bring up a list of process types, including the one we configured earlier in Section 10.3.8, as shown in Figure 10.63. Select the process type that is relevant for your scenario.

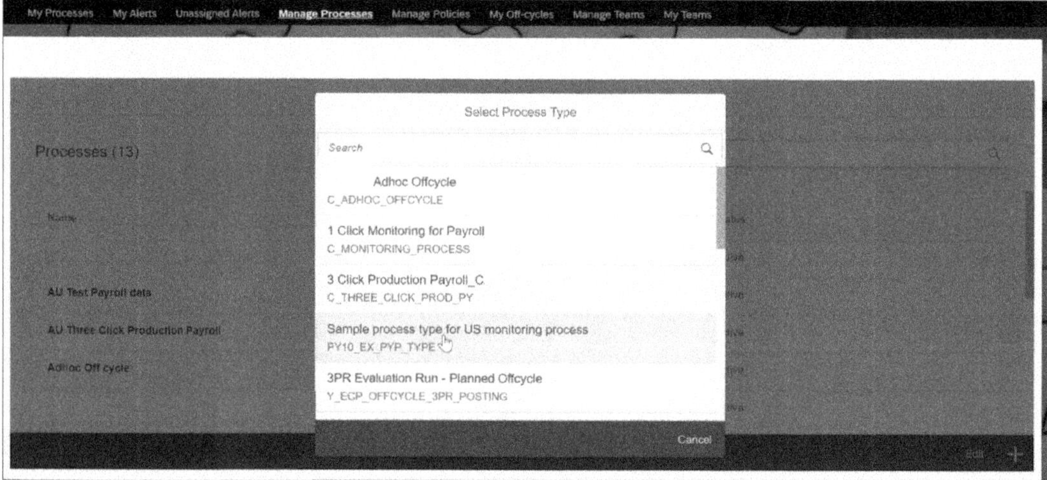

Figure 10.63 Creating a Process: Choosing a Process Type

5. In the **Name** field, enter the process name.
6. Under the **Selection** tab, enter one or more parameters by clicking on the + button for each field applicable for this process. Choose the parameter value, as shown in Figure 10.64, and then click **OK**.

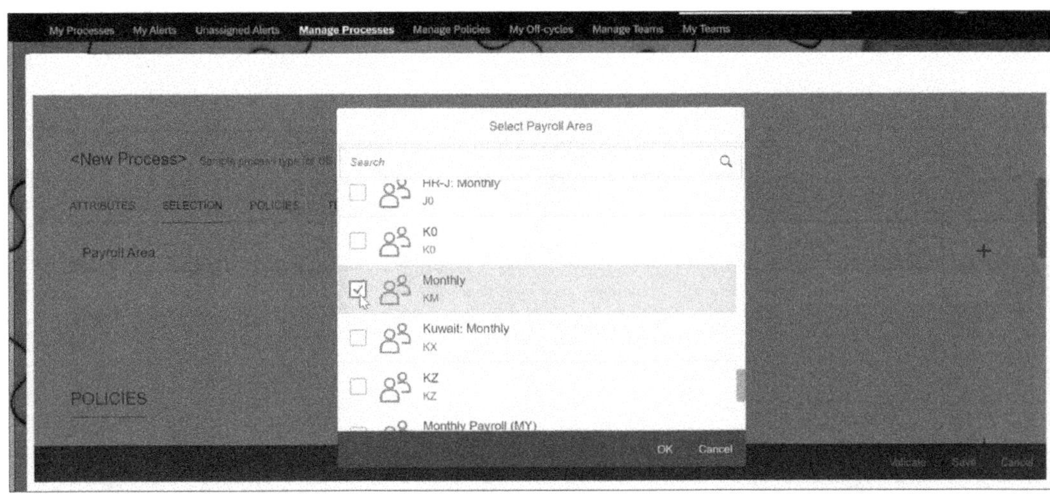

Figure 10.64 Creating a Process: Choosing a Parameter Value

7. Under the **Policies** tab, choose one or more policies relevant for this process by clicking the + icon. This list will already be filtered down based on the selection of process type and country and may not have a value to choose, as shown in Figure 10.65.

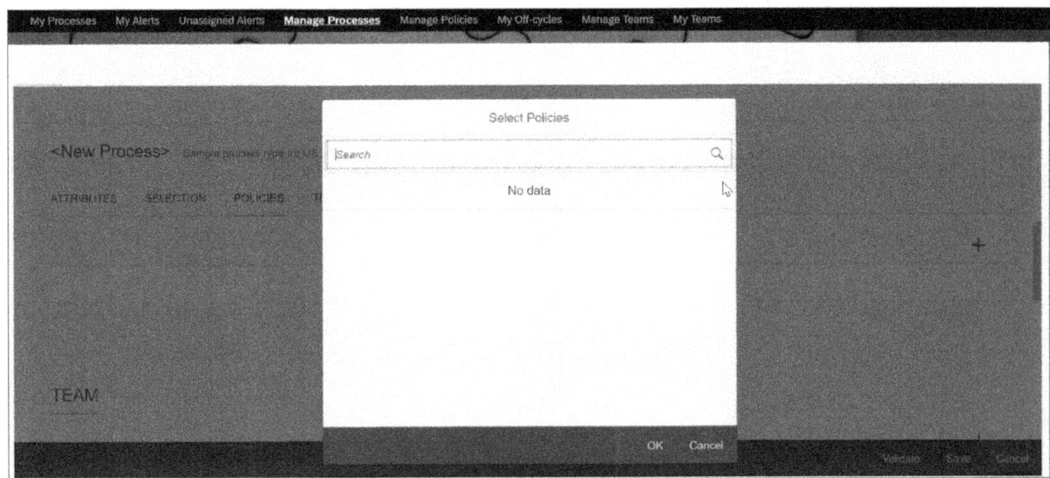

Figure 10.65 Creating a Process: Choosing a Policy Type

8. If you're creating a planned off-cycle process, enter the off-cycle reason by clicking on the + icon in the new tab called **Off Cycle Reason**.
9. If you're creating a planned off-cycle (other) process, choose the preceding off-cycle process by clicking on the + icon in the new tab called **Parent Off-Cycle Process**.

10. Under the **Team** tab, choose one or more payroll administrators by clicking on the + icon if the process contains a monitoring step, as shown in Figure 10.66. These team members will have access to the results of the process and will work on alerts as they occur.

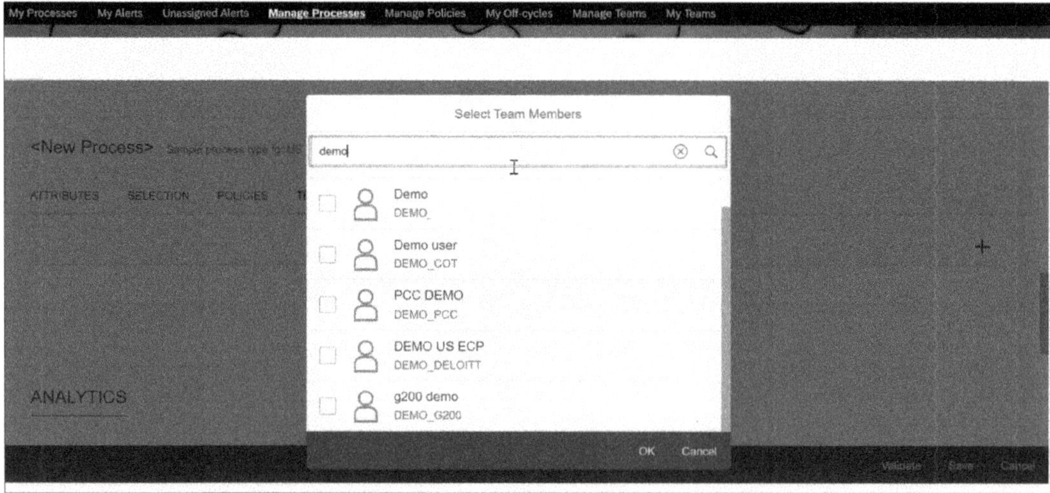

Figure 10.66 Creating a Process: Choosing Team Members

11. Under the **Analytics** tab, from the **Analytics** dropdown list, select the analytics chart that you would want to display, as shown in Figure 10.67.

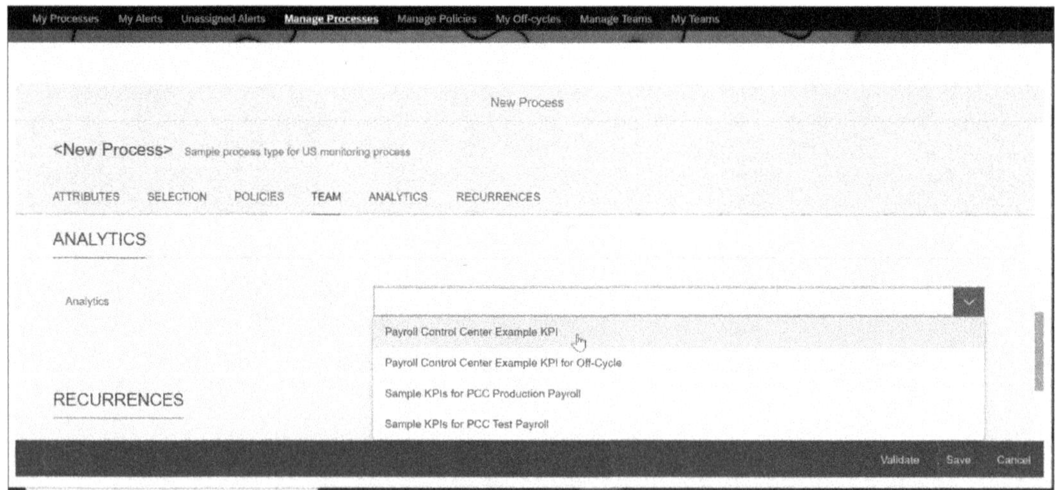

Figure 10.67 Creating a Process: Choosing Analytics

12. On the **Recurrences** tab, click on the + button and select the periodic cycle for which you would want to generate this process again and again, as shown in Figure 10.68.

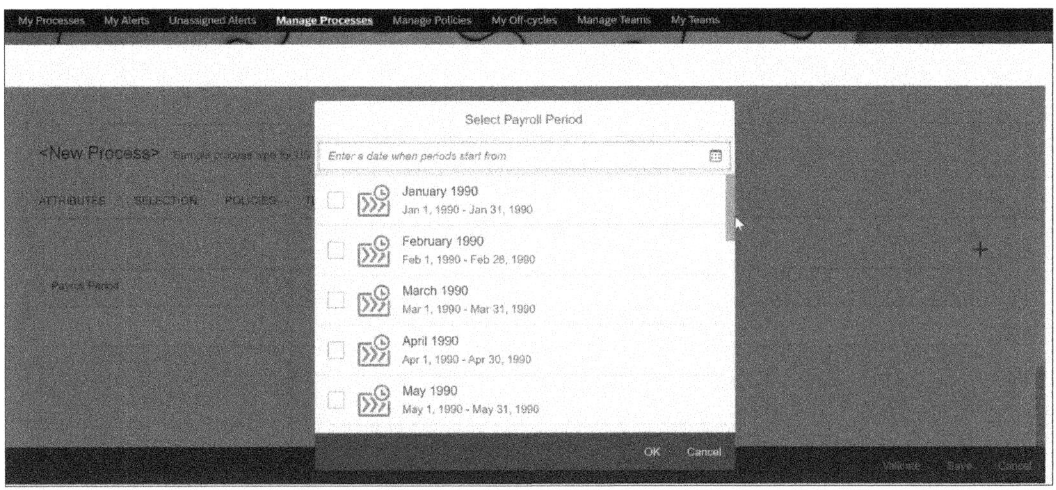

Figure 10.68 Creating a Process: Recurrences

13. Click **Validate** to check for errors and then save.

Traditional Configuration

To use traditional configuration on the Employee Central Payroll side to create policies and processes, make sure that business function HCM_LOC_CI_92, which activates the simplified configuration for the Payroll Control Center, is *not* activated. Let's now look at how to create policies and processes using traditional configuration.

Creating Policies

To create a policy, follow these steps:

1. Initiate the Configuration Workbench in the Payroll Control Center using Transaction PYC_CONF_WB.
2. Switch to edit mode by clicking on the **Display/Edit** button.
3. From the left panel, choose **Policy**. Copy a standard policy or an existing policy or create a new one. To copy, double-click an existing policy to copy and click the **Copy** icon in the application toolbar. Then, enter a unique ID for the new policy and click **Continue**. All the attributes and values, except for the ID, will be copied to the new policy. To create a new policy from scratch, click the **Create** icon in the application toolbar. Then, enter a unique ID and a name for the new policy and click **Continue**.
4. Click the **Basic information** tab.
5. Add the **Name** and the **Country/Region** for which the rule is required.
6. Go to the **Validation Rules** tab and add validation rules by clicking on the + icon (shown earlier in Figure 10.61).
7. Click the **Validate** icon to see if any errors exist and then save.

Policies are transported using a PYPL transport object with the relevant policy ID.

Creating Processes

To create a process, follow these steps:

1. Initiate the Configuration Workbench in the Payroll Control Center using Transaction PYC_CONF_WB.
2. Switch to edit mode by clicking on the **Display/Edit** button.
3. From the left panel, choose **Process**. Copy a standard process or an existing process or create a new one.
4. To copy, double-click an existing process to copy. Then, click the **Copy** icon in the application toolbar, enter a unique ID for the new process, and click **Continue**. All the attributes and values, except for the ID, will be copied to the new policy. To create a new process from scratch, click the **Create** icon in the application toolbar. Then, enter a unique ID and a name for the new process and click **Continue**.
5. Go to the **Basic information** tab.
6. Add the **Name** and the **Country/Region** for which the rule is required.
7. Choose the **Process Category**.
8. Enter the **Authorization Prefix**. The authorization prefix, combined with the process ID, controls who has access to this process via authorization object P_PYD_INST.
9. For the **Selection Type**, choose the parameters relevant for this process.
10. For the **Recurrence Type**, specify how often this process needs to be run.
11. For the **Contacts Group**, choose an administrator group that includes the administrators responsible for this process. This group must be present in table T526.
12. Click the **Steps** tab and specify the steps for this process. This activity can be performed in one of two ways, as follows:
    - Manual method:
        - Click the **Insert Row** button.
        - Select one or more steps that show up. (The list is driven by the step templates you configured earlier.)
        - Adjust the sequence using the up or down arrows next to this table.
        - Double-click each step and fill in the parameter values if required (for example, ABAP program name).
    - Copy method:
        - Click the **Import** button.
        - Choose one or more processes.
        - Adjust the sequence and double-click each step copied to ensure the parameters have the correct values.
13. Click the **Selections** tab.

14. Add parameter values by clicking on the **Insert Row** icon. The process will now select only employees that fit these criteria (e.g., payroll area).
15. Click the **Policies** tab.
16. Add policies by clicking on the **Insert Row** button.
17. If for a monitoring or productive payroll process, you'll see the **Teams** tab. Click the **Teams** tab to select an administrator group.
18. If for a team monitoring process, you'll see the **Team Dimensions** tab. Click the **Team Dimension** tab. Then, select the field(s) that you want to use as a team dimension and click **Continue**. This setting enables the Manage Teams application and lets a payroll manager set up the values to identify the teams that can manage alerts and monitor payroll.
19. If for a productive payroll process, you'll see the **Preceding Monitoring Processes** tab. This tab is for identifying if the preceding process has the setting **Keep resolved alerts status in associated follow-up processes** selected under the **Basic Information** tab. If so, if a validation is set to **Resolved** in the preceding monitoring process, that status will remain as **Resolved** for this process even if the condition that triggered the validation rule in the first place has not changed. Click the **Insert Row** button and select the preceding monitoring processes that are applicable.
20. If for an off-cycle process, you'll see the **Off Cycle Selection** tab. Select the selection logic that should be used to identify the off-cycle payment infotype 0267 that needs to be processed. This infotype has a default rule logic, which can be changed using Transaction SE24. If changed, the rule logic must be based on base class `CL_PYC_SELECTION_OC_RT`, as shown in Figure 10.69.

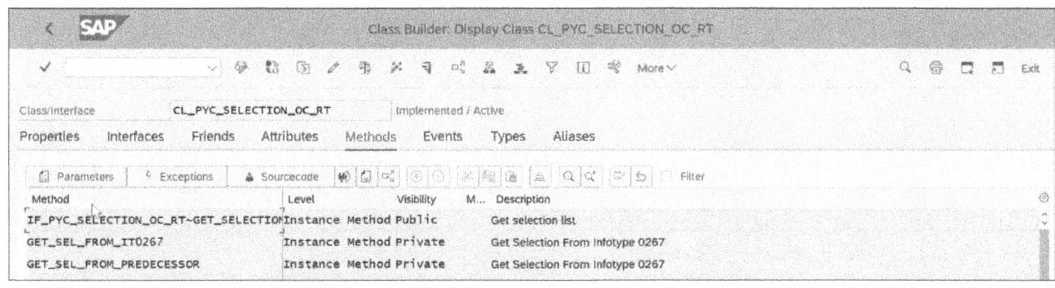

Figure 10.69 Class CL_PYC_SELECTION_OC_RT

21. If for an off-cycle process, you'll also see an **Off Cycle Reason** tab. Select the off-cycle reason under this tab.
22. Click the **Analytics** tab and select the analytics you want for this process.
23. Click the **Event Handler** tab.
24. The **Event Handler Enabled** checkbox specifies, after a test payroll run, whether to open all control records or only the control records for errors.

25. The **Relevance Check Logic** field specifies what type of master data changes will require revalidation. The default logic is `CL_PYC_EHI_RELE_CHECK_DEF`.
26. Click the **Validate** icon to see if any errors exist and then save.

The process will be transported with the following logical transport objects:

- PYPR with process ID
- PYPT with generated process template ID
- PYDI with generated artifacts (for example, steps, process context, etc.)

Generating Recurrences of Payroll Processes

Using the process configuration directly in the production system using simplified configuration generates process recurrences automatically. But if the process configuration is performed in the Configuration Workbench, the recurrences must be set up using a separate transaction—Transaction PYC_PROCESS_INSTANCES. Recurrences can be executed using Transaction PYC_INSTANCE_GES, as shown in Figure 10.70.

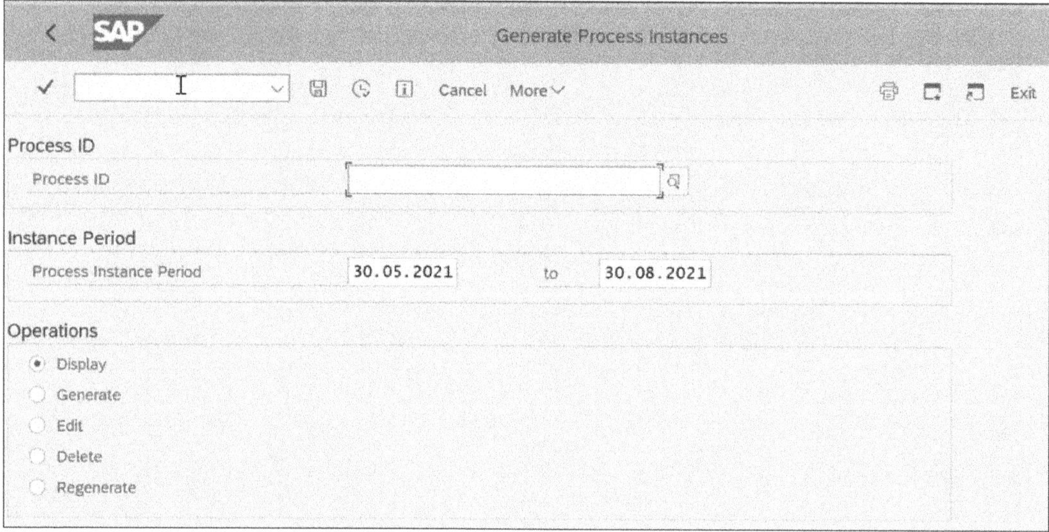

Figure 10.70 Setting Up Recurrences Using a Report

On this screen, you'll fill in the following information:

1. Enter the **Process ID**.
2. Identify the periods for generation in the **Process Instance Period**.
3. Select **Generate** in the **Operations** section.
4. For planned off-cycle processes, maintain the **Specify Pay Date**, **Pay ID**, and **Period Information** fields.
5. For planned off-cycle subsequent activities processes, choose **Create from predecessor**.

6. For ad hoc off-cycle process, click on **New Entry** and maintain the rest of the information for the process instance.
7. Click on **Execute**.

A table of information lists all the processes that are generated.

10.3.10 Setting Up Authorizations for the Payroll Control Center

The Payroll Control Center requires authorizations to be set up in the backend. This task is performed in the traditional way that security is set up in SAP systems, using Transaction PFCG. A user must be provided the traditional HR authorizations based on the need to view sensitive pay data in the system. This access is managed using traditional authorization objects, such as authorization object P_ORGIN.

Note

Detailed instructions for setting up authorizations is beyond the scope of this book. For additional information on this topic, see *http://s-prs.co/v538603*.

In addition, specific roles are required for operating the Payroll Control Center. SAP provides the template roles listed in Table 10.15 for this purpose.

Business Role	Activity	Sample Technical Role
Auditor	Uses the Audit Trail application and (read-only access to) My Processes application	SAP_HR_PYC_ANALYST
Payroll Control Center administrator	Uses the backend transactions and supporting tools for the Payroll Control Center	SAP_HR_PYC_BACKEND_ADMIN
Policy creator	Uses the Manage Policies application	SAP_HR_PYC_CONFIG_POLICY
Process creator	Uses the Manage Processes application	SAP_HR_PYC_CONFIG_PROC
Process manager	Uses the My Processes application	SAP_HR_PYC_PROC_MANAGER
Payroll administrator	Uses the My Alerts and Unassigned Alerts applications	SAP_HR_PYC_PY_ADMIN
Team leader	Uses the My Teams application and has (read-only access to) the My Processes application	SAP_HR_PYC_TM_MNG

Table 10.15 Payroll Control Center Roles and Their Technical Implementations

Business Role	Activity	Sample Technical Role
Team creator	Uses the Manage Teams application	SAP_HR_PYC_TM_SETUP
Configuration Workbench user	Configures the Payroll Control Center through the Configuration Workbench	None

Table 10.15 Payroll Control Center Roles and Their Technical Implementations (Cont.)

The following authorization objects are available to adjust these roles:

- P_PYC_POL
- S_SERVICE
- P_PYD_INST
- S_SPO_ACT
- P_PYD_IAUT
- P_PYD_RDT
- P_PYC_CACT

Activities and authorization objects required for these business roles are listed in Table 10.16.

Business Roles	Activities	Authorization Object
Policy creator	■ Create policies ■ Edit policies ■ Delete policies	P_PYC_POL
	■ Access the OData service for the Manage Policies application	S_SERVICE
Process creator	■ View policies ■ Add policies to processes	P_PYC_POL
	■ Create processes ■ Edit processes ■ Delete processes	P_PYC_PYP
	■ Access the OData service of the Manage Policies application	S_SERVICE

Table 10.16 Business Roles and Their Available Authorization Objects

Business Roles	Activities	Authorization Object
Process manager	■ View processes ■ Manage processes and steps ■ Create notes in steps ■ Read notes in steps ■ Assign alerts ■ Start or pause payroll administrators and teams assigned to process	P_PYD_INST
	■ View spool	S_SPO_ACT
	■ Access the OData service of the Manage Policies application	S_SERVICE
Payroll administrator	■ View alerts ■ Validate alerts ■ Change or forward alerts ■ Filter alerts	P_PYD_INST
	■ Access the OData service of the Manage Policies application	S_SERVICE
Team creator	■ Display team setup ■ Edit team setup	P_PYD_INST
	■ Access the OData service of the Manage Policies application	S_SERVICE
Team leader	■ View processes ■ Create notes in steps ■ Manage teams ■ Forward alerts to teams ■ Activate or deactivate team members ■ Assign alerts to payroll administrators	P_PYD_INST, P_PYD_IAUT
	■ Access the OData service of the Manage Policies application	S_SERVICE
Auditor	■ View processes ■ Read action log	P_PYD_INST
	■ Access the OData service of the Manage Policies application	S_SERVICE
Payroll Control Center administrator	■ Carry out administrative tasks for the Payroll Control Center	P_PYD_AAUT
Configuration Workbench user	■ Carry out administrative tasks for the Payroll Control Center	P_PYD_AAUT

Table 10.16 Business Roles and Their Available Authorization Objects (Cont.)

10.3.11 Technical Tasks for the Payroll Control Center

In this section, we'll discuss the technical processes and jobs, along with their various characteristics, and provide guidance on how to set these processes up.

Technical Processes and Jobs

For the Payroll Control Center, you'll need to understand several technical concepts. For the Payroll Control Center to launch processes and policies, the solution requires the help of background processes that run constantly to pick up and service these requests. The diagram shown in Figure 10.71 provides an overview of these background processes and how they work together to steer the payroll process forward.

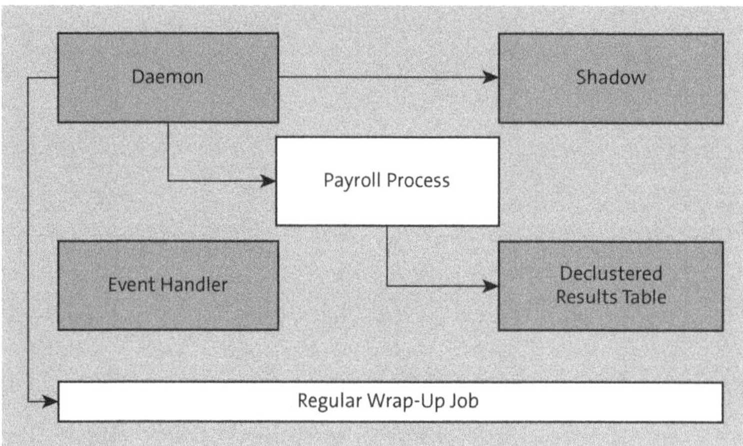

Figure 10.71 Technical Tasks in the Payroll Control Center

A job that runs in the background without requiring any user input is known as a *background job*. A specific type of background job is a *daemon job*. Daemon jobs run in the background at a predetermined frequency to ensure that the payroll is running. A daemon job works like a listener—once it receives a request to process something, it calls the specific payroll job to process it. The details of this job reside in the step specifications of the payroll step that makes this request.

Payroll processes can leave a number of table entries, logs, and inconsistent data structures that can clog up the available space in your technical environment. A *wrap-up job* is a process that cleans up after the payroll processes have completed. The report PYD_REGULAR_WRAP_UP is the program for the main wrap-up job, which in turn launches several other wrap-up jobs, which are listed in Table 10.17.

Report Name	Purpose of the Report
PYC_EVENT_HANDLER_REORG	Deletes event handler data or event handler logs that are older than a predefined period.

Table 10.17 Wrap-Up Jobs in the Payroll Control Center

Report Name	Purpose of the Report
PYD_DAE_REGS_REORG	Deletes daemon registrations (use cases and events) that have already been processed by the regular daemon jobs.
PYD_SHADOWS_REORG	Deletes finished and inconsistent shadow processes that are older than a predefined period. Shadows are used to hold information about selection objects (for example, PERNRs) and can be used to start a process step instance for a subset of the original selection.
PYD_RC_SHADOW_REORG	Cleans up the status of the recheck shadow processes and provides input for report PYD_SHADOWS_REORG.
PYD_DELETE_EXPIRED_SESSIONS	Deletes expired user sessions.

Table 10.17 Wrap-Up Jobs in the Payroll Control Center (Cont.)

A payroll process executes a number of steps for a selected number of employees as defined by the run criteria selected during the launch of the run request. After executing this process for the first time, sometimes, you'll need to process a subset of the full scope of this process and communicate the results to the administrator. These processes that are "shadows" of the original process with reduced scope, are known as *shadow processes*. A few different types of shadow processes may be involved, depending upon the scenario, as follows:

- **DS: Deviation selection**
  This shadow process is generated out of a repeat run of payroll (or test payroll) for erroneous employees (that is, employees with alerts detected during the initiate policies process). When the payroll process manager tries to repeat and run the test payroll (or to run payroll), the options for executing the step for all employees or for erroneous employees includes running it only for employees that errored out.

  This job runs the same steps but only for those employees that errored, reducing the time taken to process them drastically.

- **EH: Event handler**
  This shadow process is triggered when there are changes detected by another background process called the *event handler*. This job is always listening for a change made to the payroll master data after the payroll has been run. Since this change could result in a recalculation of payroll, the process is automatically deployed for only those employees whose master data changed.

- **RC: Recheck/validate**
  During monitoring the payroll process, an error may be discovered. Once it is fixed, the administrator can start the validation process again by clicking on the **Validate** icon from the Alerts page. This can spawn the RC type of shadow process to execute the validation steps just for that employee.

Payroll Test Runs

The payroll results in test mode typically for traditional SAP ERP HCM Payroll used to be displayed in the payroll log only after the processing completes. The declustering feature of Employee Central Payroll makes it possible to store the test payroll results in the de-clustered table structure. This makes it possible to run reports that the administrator is familiar with to perform further analysis on the payroll results.

However, the payroll test run must be done from the Payroll Control Center. The payroll driver is executed via the step **Create Test Payroll Data** (technical name PYP_V2_RUN_PAYROLL_TEST in the standard delivery) in the **Payroll Process for Test-Payroll** step (technical name PYP_V2_TEST_PAYROLL in the standard delivery).

Running payroll from the backend using Transaction SE38 in simulation mode does not enable the writing of test results into the declustered tables. The daemon jobs and associated shadow processes along with event handler interventions are controlled through a single report called PYC_ADMIN_TRANSACTION. Authorization to create batch jobs is necessary for this task. This authorization involves authorization object S_BTCH_NAM.

To start the payroll test run, follow these steps:

1. Either run the report using Transaction SE38 or follow the menu path **Human Resources • Payroll • International • Tools • Payroll Control Center • Admin Transaction Report.**
2. Click on **Maintain the user name for batch processing** menu on the left and enter the batch user name in the **Batch User** field, as shown in Figure 10.72. The batch user is used for scheduling all the batch processing programs started by the daemon report (PYD_DAE_PROCESS) that is scheduled by this report in the next step. The user needs to be a non-dialog user (a technical or a batch user).

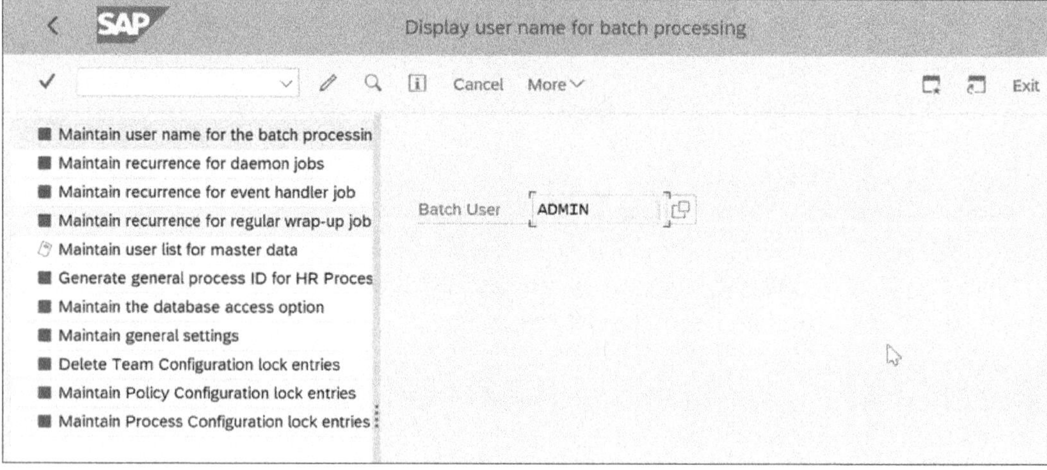

Figure 10.72 Maintaining a Batch User Name in the Admin Transaction Report

3. Click on **Maintain recurrence for daemon jobs** from the menu on the left and then fill in the **Recurrence** field (usually a small interval, such as 1 minute), as shown in Figure 10.73.

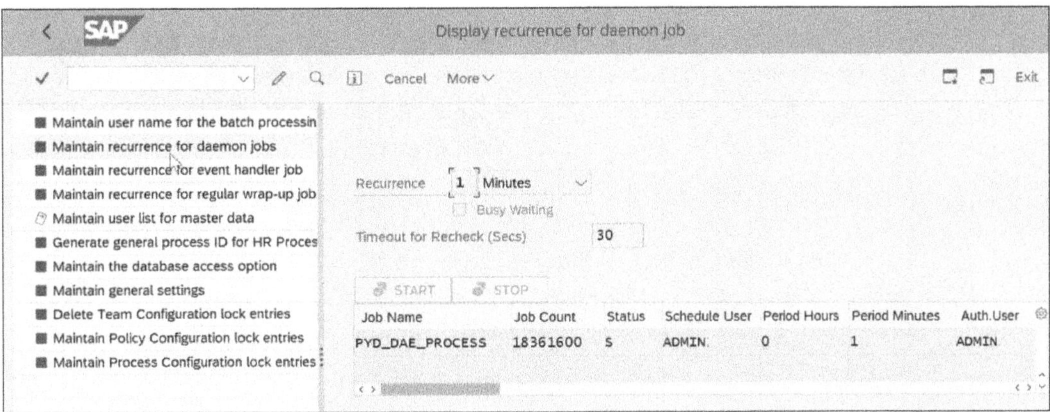

Figure 10.73 Recurrence for Daemon Jobs

On this screen, you'll also fill out the following:

- **Busy Waiting**: This setting is required to improve system response so that the daemon can immediately respond to a request without waiting for the wait time to be over. This setting does block one job batch process to the application server.
- **Timeout for Recheck (Secs)**: This setting is to prevent a shadow process started by the **Validate** button from My Alerts from taking too long. The setting should be slightly more than the time typically taken to process a single validation. The timeout setting applies to the batch user, not the administrator who initiated it by clicking the **Validate** button.
- Daemon job table (with the **Job Name** and **Job Count** fields, among others, as shown in Figure 10.73): This table needs to be monitored to ensure processes that are not started by the daemon report PYD_DAE_PROCESS are not in this table. Manually scheduled Transaction SM36 or SE38 processes can clutter this table, which has the potential of degrading performance.
- **START** and **STOP** buttons: Use these buttons to start or stop a daemon process.

4. Click on **Maintain recurrence for event handler jobs** from the menu on the left and then fill in the **Recurrence** field. You'll define how often this process is executed. Notice again the **START** and **STOP** buttons: These buttons start or stop the event handler job SAP_STD_EH_FOR_ALL_PROCESS, which picks up all items that have been flagged with a master data change after the payroll process has been run, as described at the beginning of this section.
5. Click on **Maintain recurrence of regular wrap up jobs** from the menu on the left. In the **Recurrence** field, indicate the time interval at which the wrap-up job should execute to clean up inconsistent or unnecessary data structures.

6. Click on **Maintain user list for master data** from the menu on the left. This list shows users who can change master data when the payroll control record is locked. This information may be necessary for certain exceptional cases during payroll processing when a few employees fail to generate correct payroll due to master data errors.

   By default, payroll administrators have this privilege. You can add others to this list in the **User Name** field if necessary. One user that should be added is the batch user that brings master data from Employee Central through replication, so that the corrections done in Employee Central can flow through Payroll and be used for the payroll process.

7. Click on **Generate general process ID for HR Process Workbench** from the menu on the left. This setting facilitates the integration of the HR Process Workbench (Transactions PEST and PUST) with the Payroll Control Center. All HR Process Workbench reports can pass their information to the Payroll Control Center using this ID. A dummy process model SAPXXPCC is used for the interaction between the Payroll Control Center and the HR Process Workbench. The process ID, generated by clicking on the **Generate** button, is used for the run instance for this dummy process model.

8. Click on **Maintain the database access option** from the menu on the left. Two settings are available in the **Database Access Option** field: **Use EXISTS instead of JOIN in SQL** and **Default: Use JOIN in SQL**, as shown in Figure 10.74. These settings help improve the performance of certain types of databases that may be used for the payroll environment. For most use cases, the **Use EXISTS instead of JOIN in SQL** option provides better performance.

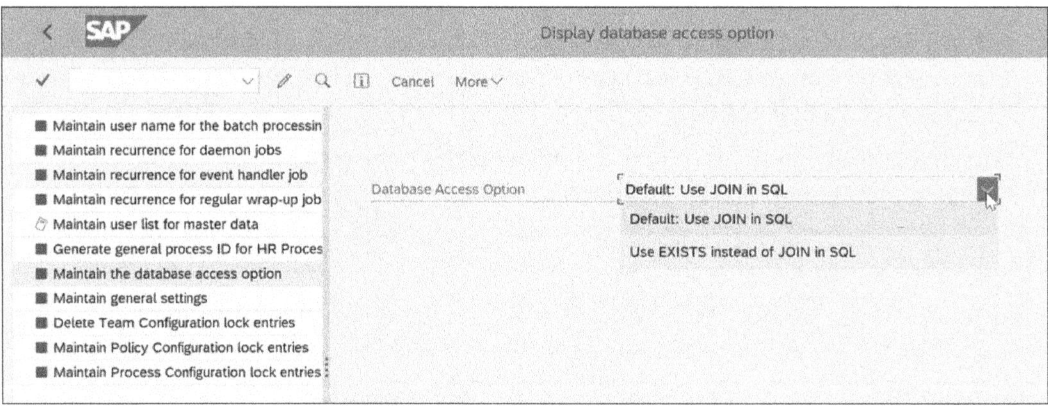

Figure 10.74 Database Access Options

9. Click on **Maintain general settings** from the menu on the left. This setting is to allow the upload of documents in the My Processes application while running payroll by selecting the **Attachment allowed** checkbox. You can define the **Maximum file size.** To restrict the types of files that can be uploaded, use the separate configuration **Payroll Process Management • Define Allowed Mime Types for Document Upload** in the Payroll Control Center configuration in Transaction SPRO.

You can also specify the number of lines that can be downloaded from the alert history (Excel download) using the **Maximum number of rows for EXCEL download** field, as shown in Figure 10.75.

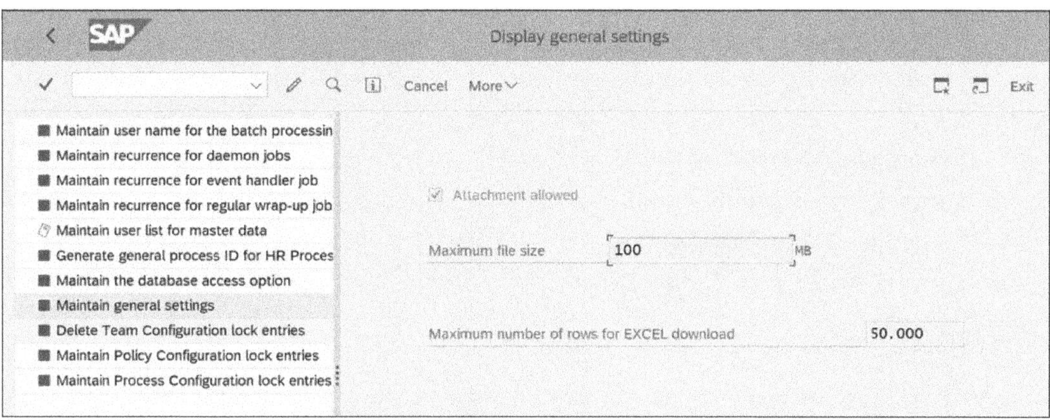

Figure 10.75 Maximum File Size for Attachments

10. Click on **Delete Team configuration lock entries** from the menu on the left and then click on the pencil icon to access the screen. Another user updating an item in the Manage Teams application and not saving it can create a lock situation. This setting is to release that lock if it happens. Select the checkbox next to the item that needs to be unlocked and then click on the **Delete Row** icon to delete this lock if required, as shown in Figure 10.76.

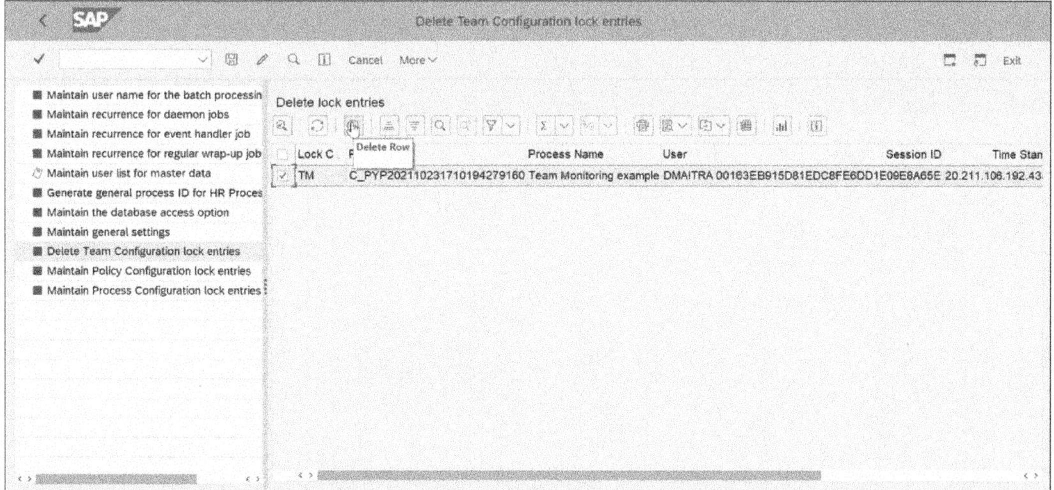

Figure 10.76 Deleting Team Configuration Lock Entries

11. Click on **Maintain Policy Configuration lock entries** from the menu on the left. Another user updating an item in the Manage Policies application and not saving it

can create a lock situation. This setting is to release that lock if it happens. Use the checkbox on the screen to delete this lock if required.

12. Click on **Maintain Process Configuration lock entries** from the menu on the left. Another user updating an item in the Manage Processes application and not saving it can create a lock situation. This setting is to release that lock if it happens. Use the checkbox on the screen to delete this lock if required.

Deleting Completed Processes

When the payroll administrator initiates the My Processes application, it loads all processes including completed ones. To improve response time for loading, it may be beneficial to delete those runs that have been completed for some time. For this task, use the report PYC_SUPPORT_DEL_COMPLETED_PI by launching Transaction SE38, as shown in Figure 10.77.

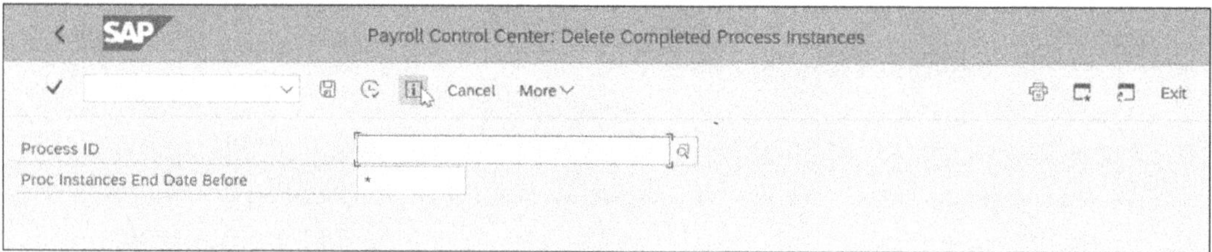

Figure 10.77 Deleting Completed Processes

You can enter the relevant **Process ID** information and click on **Execute**. You can also use the **Process Instances End Date Before** field to restrict the deleted processes to a certain date in the past.

Before you delete a completed process, you might want to download the audit log, which would no longer be available after the deletion operation is completed. Download the audit log by running report PYC_SUPPORT_DOWNLOAD_PI_ALH using Transaction SE38, as shown in Figure 10.78. Enter the process ID in the **Process Instance ID** field. Enter the maximum number of records per file in the **Maximum Row Number per File** field. Then, enter or choose a **Directory Path** and select a file separator from the **CSV File Separator** dropdown list. Finally, execute.

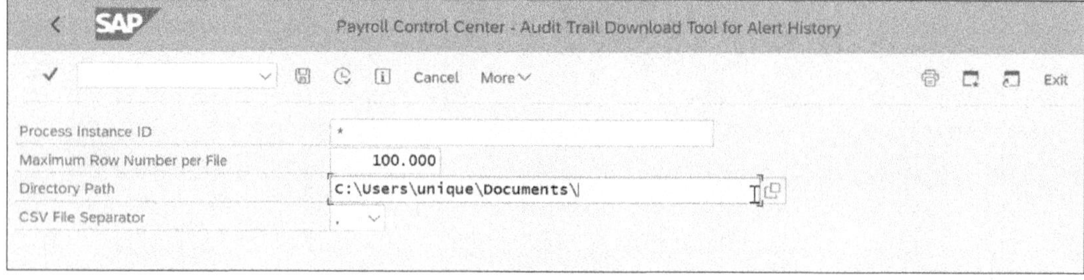

Figure 10.78 Transaction to Download Audit Logs

10.4 Configuring the Payroll Control Center in Employee Central

We've covered a lot of ground so far, but we can summarize the configuration required to enable the Payroll Control Center in Employee Central in six steps. Note that some of these steps may have already been completed when the replication between the two systems was enabled (see Chapter 7 and Chapter 8):

1. Configure system-level objects (Section 10.4.1):
   - SAP system configuration
   - Payroll Control Center configuration
2. Set up role-based permissions for system administrators (Section 10.4.1).
3. Configure the Employee Central Payroll system (Section 10.4.1).
4. Set up role-based permissions for users using the different tabs (Section 10.4.2).
5. Assign the Employee Central Payroll system to users (Section 10.4.3).
6. Create a custom tile for the audit trail (Section 10.4.4).

Table 10.18 lists the applications in Employee Central that enable the Payroll Control Center.

Employee Central Application	Employee Central Payroll Application
Manage Policies	Policy Configuration
Manage Processes	Process Configuration
Manage Teams	Team Configuration
My Alerts	Alert Management
My Off-cycles	Manage Off-Cycle Payrolls
My Processes	Process Management
My Teams	Team Management
Unassigned Alerts	Team Alerts

Table 10.18 Payroll Control Center Applications in Employee Central

Before starting the configuration process, you should activate the **Enable Payroll Control Center** option in provisioning, using the steps covered in Chapter 3, Section 3.1. In the following sections, we'll look at the steps needed to configure the Payroll Control Center, omitting the steps we've already covered in earlier chapters.

10.4.1 Configuring System-Level Objects

This section describes how to set up the Employee Central Payroll system that an Employee Central application will connect to. We'll also introduce you to the Payroll

Control Center configuration page and explain the settings in Employee Central required to make it work.

SAP System Configuration

Let's now set up the metadata for the permission category of the payroll system so that the system can be assigned correctly to an administrator. Follow these steps:

1. Log on to Employee Central and go to the Admin Center.
2. Enter "Configure Object Definitions" into the search bar to pull up the **Configure Object Definitions** page, as shown in Figure 10.79.

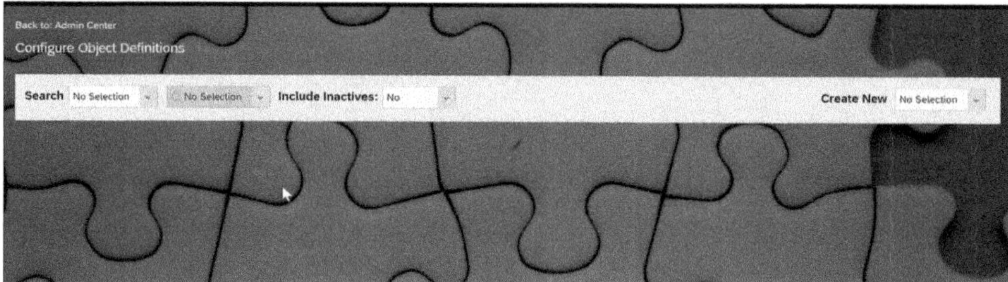

Figure 10.79 Configure Object Definitions Page in Employee Central

3. Select **Object Definition** and **SAP System Configuration** in the two dropdown lists next to the word **Search**, as shown in Figure 10.80. Then, click on **Take Action • Make Corrections** in the top-right corner.
4. On the page that appears, fill in the **Effective Dating** and **Status** fields, as shown in Figure 10.80.

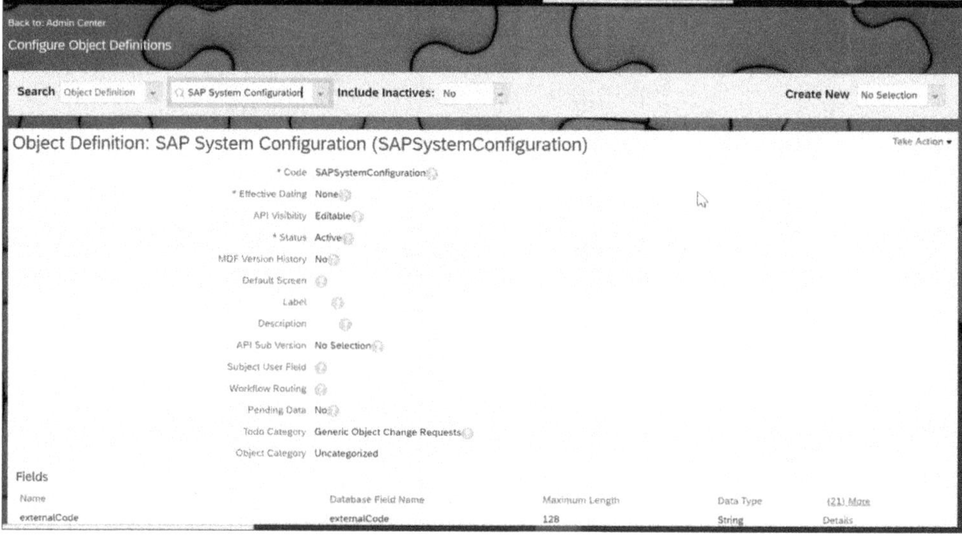

Figure 10.80 SAP System Configuration Object

5. On the same screen, scroll down to the **Security** section. Then, select **Yes** from the **Secured** dropdown list.
6. Choose **SAP System Configuration** from the **Permission Category** dropdown list, as shown in Figure 10.81, and then save.

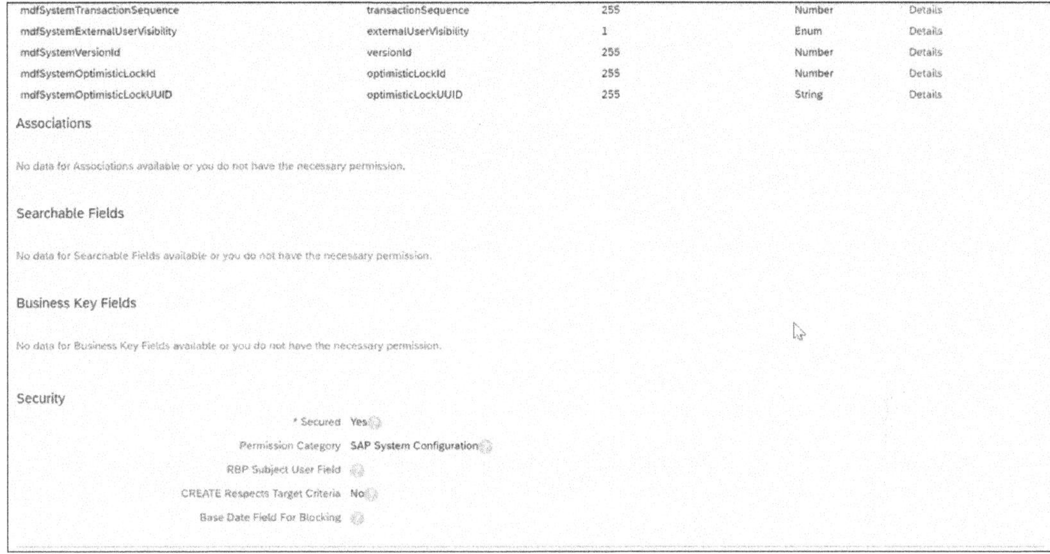

Figure 10.81 Security Settings for SAP System Configuration

Your next task will be to add the **SAP System Configuration** in the **Role-Based Permission** menu for the administrator who is configuring the system. This step is necessary in this sequence so that the next steps can be completed. Follow these steps:

1. Go to **Manage Roles** in the Admin Center and turn on the permission categories listed in Table 10.19 for the roles assigned to the person performing the configuration.

Category	Permission
Manage SAP System Configuration	Access to SAP System Configuration
SAP System Configuration	View and edit
Payroll Integration – Payroll System Assignment	View and edit
Payroll Integration – Payroll Control Center Configuration	View and edit

Table 10.19 Permission Categories for the SAP System

2. Go to the Admin Center and pull up the **SAP System Configuration** screen by entering "Manage Data" into the search bar and then choosing **SAP System Configuration**

in the **Create New** dropdown list, which opens the screen shown in Figure 10.82. Maintain the following fields and then save:

- **System Name**
- **System URL** (you would have received this URL from SAP in the initial welcome document.)
- **Client** (SAP provides one client in each environment by default. A user can request additional clients based on licensing agreements.)

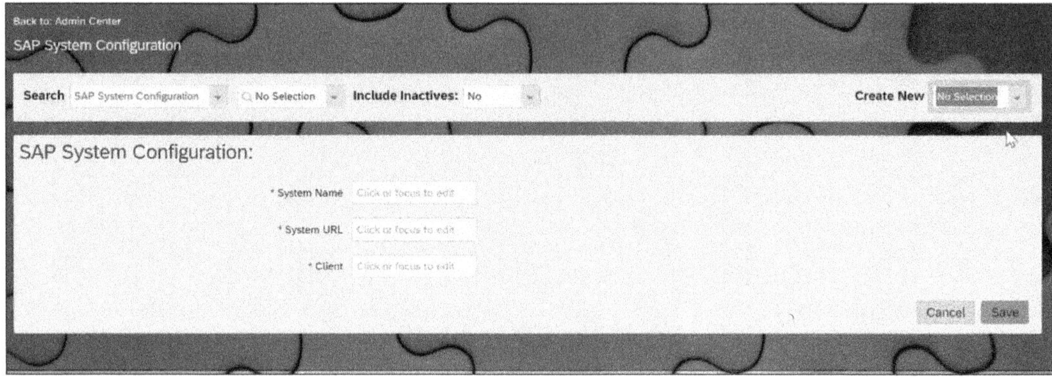

Figure 10.82 SAP System Configuration: Required Fields

Payroll Control Center Configuration

The **Payroll Control Center Configuration** page connects Payroll Control Center applications in Employee Central to the specific Employee Central Payroll systems that are required. To set up this connection, follow these steps:

1. Go to the Admin Center and enter "Payroll Control Center Configuration" into the search bar to pull up the **Payroll Control Center Configuration** page, as shown in Figure 10.83.

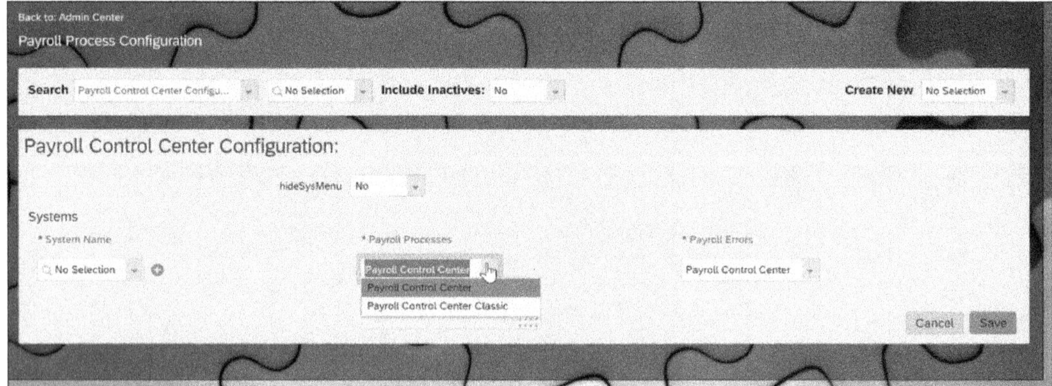

Figure 10.83 Payroll Control Center Configuration

2. Select **Payroll Control Center Configuration** in the **Create New** dropdown menu.
3. Enter the necessary details for the following fields and then save:
   - **System Name**: Enter the name of the payroll system you provided in the previous setting.
   - **Payment Processes**: Two options exist, choose one based on the desired appearance:
     - **Payroll Control Center Classic** (classic SAP Fiori UX)
     - **Payroll Control Center** (enhanced SAP Fiori UX)
   - **Payroll Errors**: Two options exist, choose one based on the desired appearance:
     - **Payroll Control Center Classic** (classic SAP Fiori UX)
     - **Payroll Control Center** (enhanced SAP Fiori UX)

10.4.2 Setting Up Role-Based Permission for Users

The next step is to add permissions in the menu for role-based permissions for the business users who will be using the system, using an appropriate business user role for the Payroll Control Center. An example is shown in Figure 10.84.

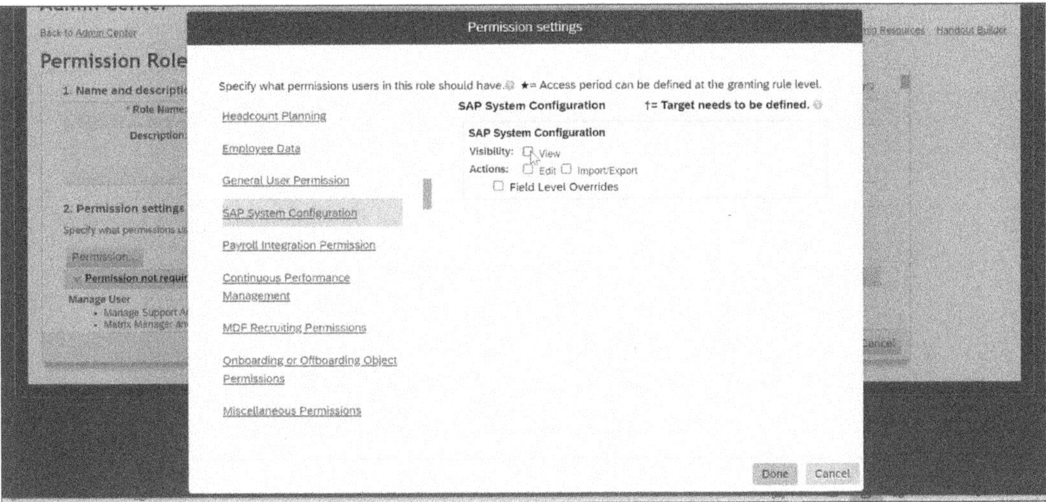

Figure 10.84 Role-Based Permission Settings for the SAP System Configuration Category

For this activity, go to Manage Roles from the Admin Center by entering "Manage Permission Roles" in the search bar. Turn on the appropriate permission categories for the roles assigned to the business user(s), as listed in Table 10.20.

Category	Permission
SAP System Configuration	Optional - View
Payroll Integration – Payroll System Assignment	Optional - View
Payroll Control Center	My Processes Access My Alerts Access Unassigned Alerts Access Manage Processes Access Manage Policies Access My Off-cycles Access Manage Teams Access My Teams Access

Table 10.20 Business User Permissions

10.4.3 Assigning the Employee Central Payroll System to Users

Go to the Admin Center and enter "Payroll System Assignment" into the search bar to open the **Payroll System Assignment** page, as shown in Figure 10.85. Select **Payroll System Assignment** from the **Create New** dropdown list.

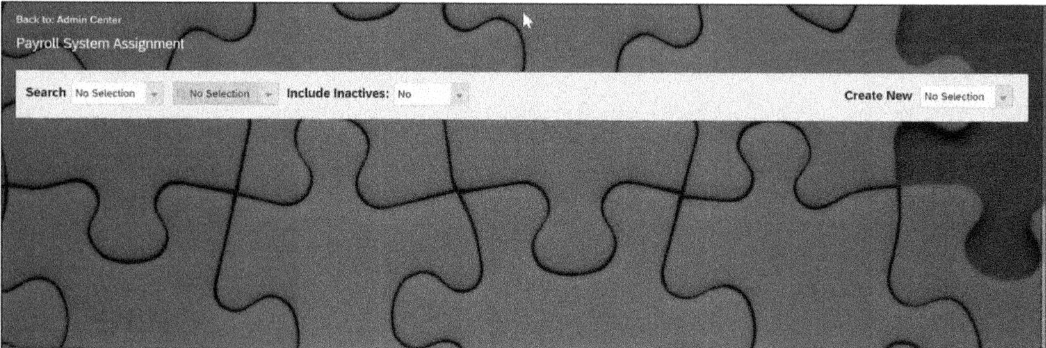

Figure 10.85 Payroll System Assignment

Enter the **User Name** that you want to assign, as shown in Figure 10.86.

Choose the systems and select the system that should be the default for that user and then save these settings. This system will be the default system for the user, but the user can always move from the application if given access to more than one system.

Figure 10.86 Assigning a System to a User

10.4.4 Creating a Custom Tile for Audit Trails

A custom tile for audit trails is a good way to give your payroll users easy access to audit information. An audit trail shows the status of active and completed jobs in the Payroll Control Center. The tile provides direct access to this screen, without going through the Employee Central Payroll menu path.

Go to the Admin Center and pull up the **Manage Homepage** menu, as shown in Figure 10.87. On this screen, you'll configure your custom tile.

Figure 10.87 Manage Homepage

To configure your custom tile, follow these steps:

1. Click the **Add Custom Tile** button.
2. For the **General** step, maintain the **Title Name** and **Description** fields, as shown in Figure 10.88, and then click **Next**.

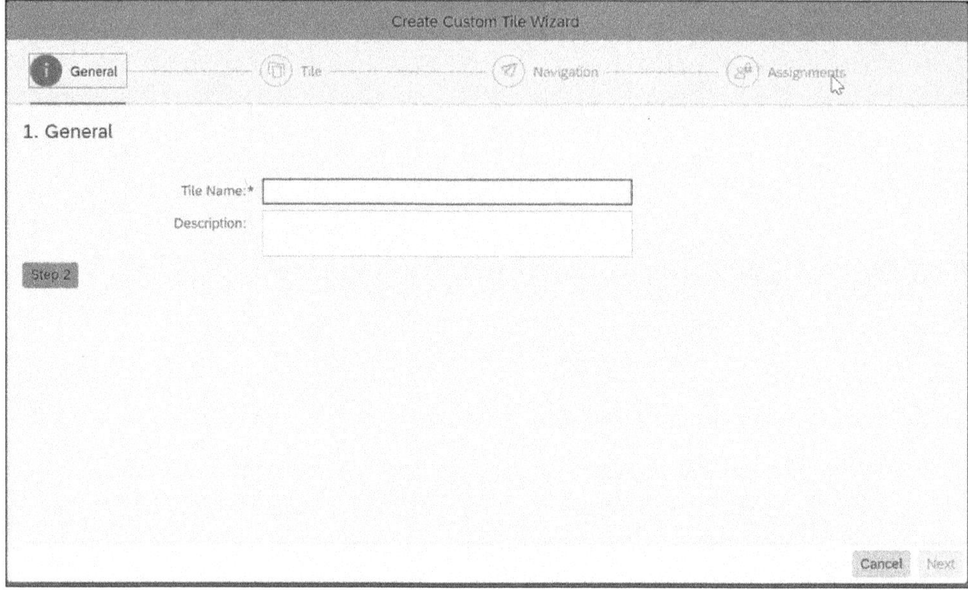

Figure 10.88 Custom Tile Configuration

3. For the **Title** step, shown in Figure 10.89, fill in the following fields and then click **Next**:
   - **Type**: Leave as **Static**.
   - **Title**: Enter the same name used in the previous screen.
   - **Icon**: Choose an icon from the available list.
4. For the **Navigation** step, choose **URL** from the **Target** dropdown list, as shown in Figure 10.90.
5. In the URL field, the URL = Payroll System URL + Service Path + Client ID + Language. Each element is as follows:
   - You enter the relevant payroll system URL followed by the service path and the client ID.
   - The service path is *sap/bc/ui5_ui5/sap/hrpy_pcc_al_2/index.html*, which is fixed.
   - If you don't specify a language ID in the URL, the default language is the user's logon language in Employee Central.

   An example audit trail URL might be *https://www.server.com:12345/sap/bc/ui5_ui5/sap/hrpy_pcc_al_2/index.html?sap-client=000*.

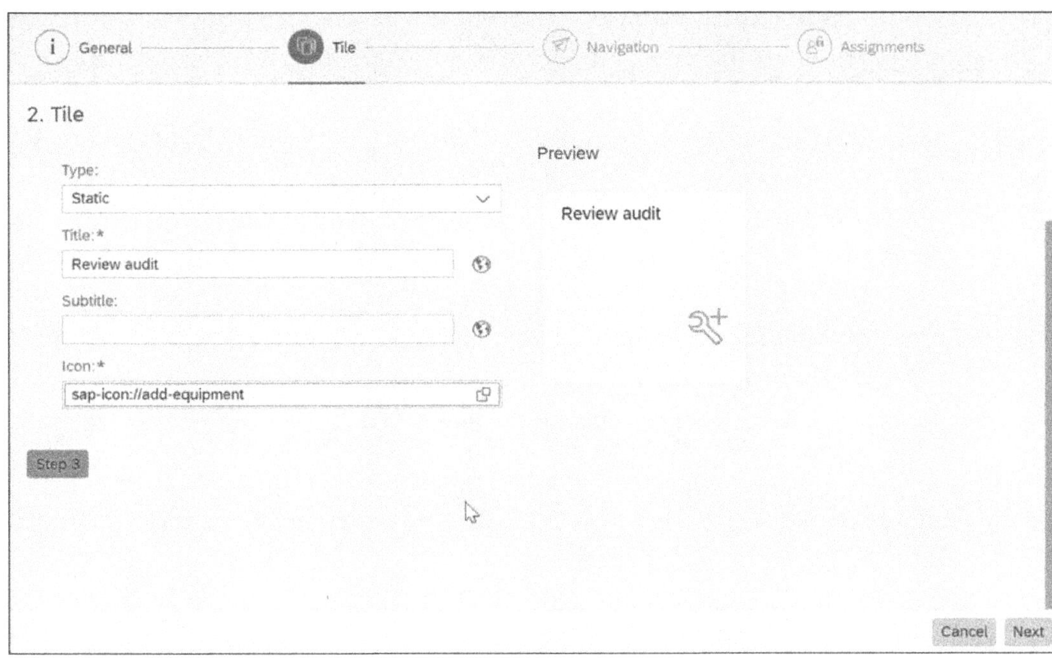

Figure 10.89 Title of the Tile

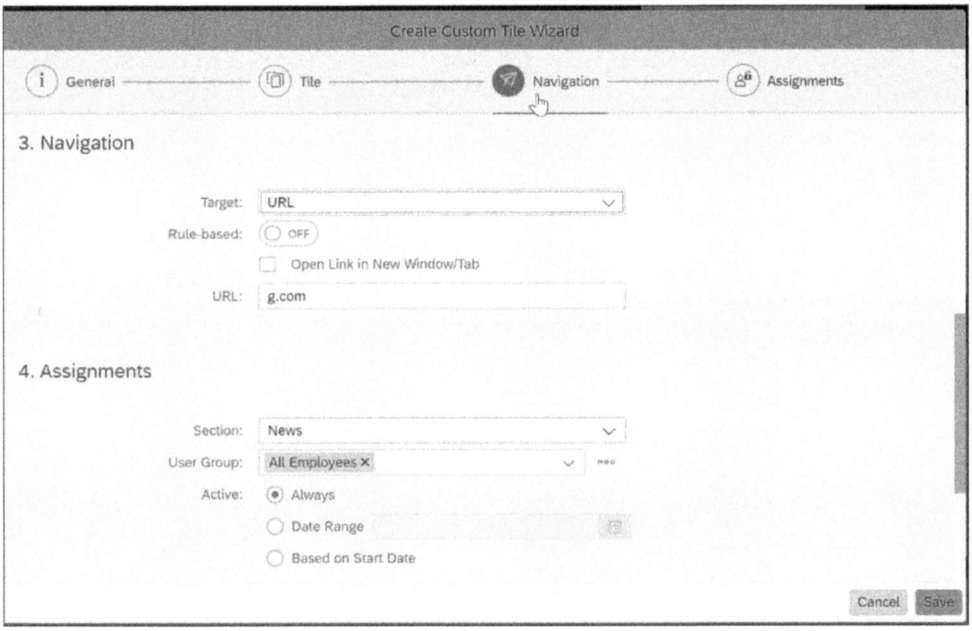

Figure 10.90 Navigation Settings for a Tile

6. For the **Assignments** step, from the **Section** dropdown list, choose a section on the homepage where the tile should appear. In our example shown in Figure 10.91, we've chosen **News**.
7. From the **User Group** dropdown list, select the user group that should have access to this tile.
8. The **Active** radio buttons enable the tile on specific dates; for instance, you can restrict the tile to only appear on the dates specified on this screen. If you select the **Always** option, you can define a date range or tie the tile to the start date of the user who is viewing this tile.
9. Save these settings.

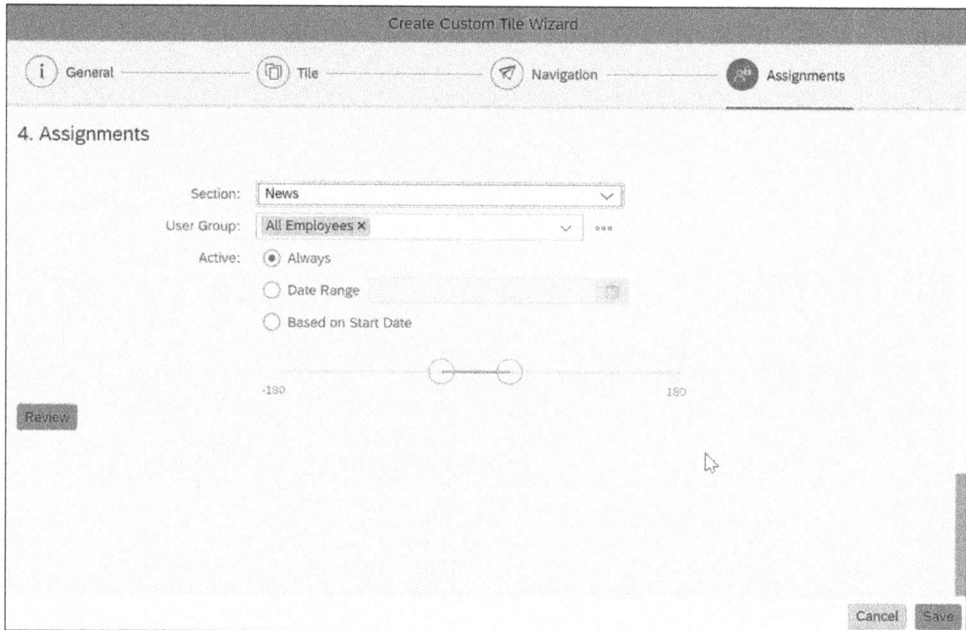

Figure 10.91 Assignment Settings for a Tile

10.5 Making the Payroll Control Center Operational

Now that we've configured the Payroll Control Center, let's look at how the configuration can be deployed for operational use. The Payroll Control Center is an end-to-end tool that enables data validation and the running of payroll as well as subsequent processes to pay employees and complete accounting and other downstream processes related to employee wages and benefits.

The following categories of activities summarize the business setup tasks required for payroll processing:

- **Policy creation**
  We reviewed how to configure policies in the previous section using the Manage Policies application The policy creation process builds the polices that are applied to monitoring processes to validate data before a productive payroll run. Policies are created at the beginning of implementation. Remember to update them as new requirements pop up through acquisitions or legislative changes.
- **Process creation**
  We reviewed how to configure the steps of a payroll process using the Manage Processes application in the previous section. Once the configuration is complete, these steps will automatically show up in the My Processes and My Off-cycles applications from which you can start, stop, and review the various steps that have been set up for your payroll processes.
- **Monitoring**
  Monitoring setup involves ensuring the monitoring processes are started at the correct times so that data validation errors can be reviewed before the live payroll run, with ample time to fix issues. Performed through the My Processes application and conducted by a payroll process manager, a monitoring process involves the following steps:
  - Produce test payroll results for all employees
  - Produce check results for employees who belong to the payroll areas for which the process is executed
  - Assign alerts to payroll administrators and monitor the resolution of the alerts until the start of the productive payroll

  The ability to find issues with master data as early as possible ensures that the real productive payroll can be relatively error free.
- **Team creation**
  Teams are created to help manage the timely review and correction of master data for a large employee base. Accessed through the Manage Teams application (see next section), the team creation process creates teams with their authorizations and attributes for effective collaboration on the tasks involved to complete payroll. This task goes through continuous updates to add and delete team members since the payroll team composition generally changes over time.
- **Team monitoring**
  This activity is to ensure effective collaboration between teams to get the review and fix tasks completed on time. This activity involves reassigning tasks when required to distribute them optimally among team members. Accessed through the My Processes and My Teams applications and performed by payroll process managers as well as team leads, a team monitoring process allows the segregation of assignments to teams. It facilitates team collaboration through automatic distribution of alerts.

In addition to the monitoring discussed in the earlier bullet point, it involves the following:

- Distribution of alerts automatically based on predefined criteria
- Ability for team leads to reassign alerts to other team members manually

- **Productive payroll**
  Productive payroll should be already configured, as described in the previous section. Use the My Processes application to ensure the timing and steps are correctly set up and available. Productive payroll is the final run of payroll and involves the following steps:
  - Open payroll (release payroll control record for payroll)
  - Productive payroll run
  - Generation of check results
  - Posting simulation
  - Monitoring
  - Exit payroll

  Use the My Alerts application to review the errors produced by the SAP payroll engine and take action on them.
- **Others**
  Other tasks are required before and after running payroll. The configuration of these tasks would have been performed already through Manage Processes application, as described in the previous section. This other category of processes typically includes the following tasks:
  - Importing data from vendors
  - Exporting data to vendors
  - Bank transfers
  - Processing pay statements
  - Posting to the general ledger
  - Making third-party remittances to pay creditors, vendors, and regulatory agencies

  As a setup activity, you must ensure these tasks are set up with their correct dependencies and with timing considerations in mind.
- **Planned off-cycle (productive) processes**
  Planned off-cycle processes are productive payroll processes that are preplanned for delivering a payment that does not coincide with the normal pay cycle. A good example of this scenario is the annual bonus. The configuration of these tasks would have been performed already through the Manage Processes application, as described in the previous section. From a setup perspective, you must ensure that this process is visible in the My Processes application at the time the off-cycle payroll run is required.

- **Planned off-cycle (others) processes**
  Other planned off-cycle processes include the subsequent and preceding steps of a planned off-cycle productive process. This kind of process has the same steps as the "Others" category. The configuration of these tasks would have been performed already through the Manage Processes application, as described in the previous section. As a setup check, you must ensure that these processes are also scheduled correctly and appear in the right sequence in the My Processes application.

10.6 Operating the Payroll Control Center

In this section, we'll review the features available for payroll business users to manage their daily activities using the setup that was performed in the previous section. First, we'll discuss the My Processes application and describe the various operations that arise during a day-to-day payroll process cycle. Next, we'll briefly review the My Teams application and show you how a payroll manager can leverage this application to set up an efficient collaboration framework. Finally, we'll review the My Off-cycles application to understand how to run payroll for individual employees if required urgently.

10.6.1 My Processes Application

Several useful operations are available for a payroll manager via the My Processes application. The available actions can be categorized in the following way:

- Managing payroll processes
- Assigning individual alerts to processors
- Assigning default processors for all alerts of by a policy
- For team monitoring processes, pausing or starting the teams that are working on the alerts
- Processing alerts
- When a change to an existing policy or process occurs, carrying out the necessary manual activities so that the change takes effect on time

Managing Payroll Processes

The Manage Payroll application allows you to manage all regular payroll processes and off-cycle processes, except for the ad hoc off-cycle process that is managed by the My Off-cycles application, which we'll discuss later in Section 10.6.3.

In the Payroll Control Center, invoke the My Processes application by choosing **Payroll** from the Employee Central menu path and then clicking **My Processes**, as shown in Figure 10.92.

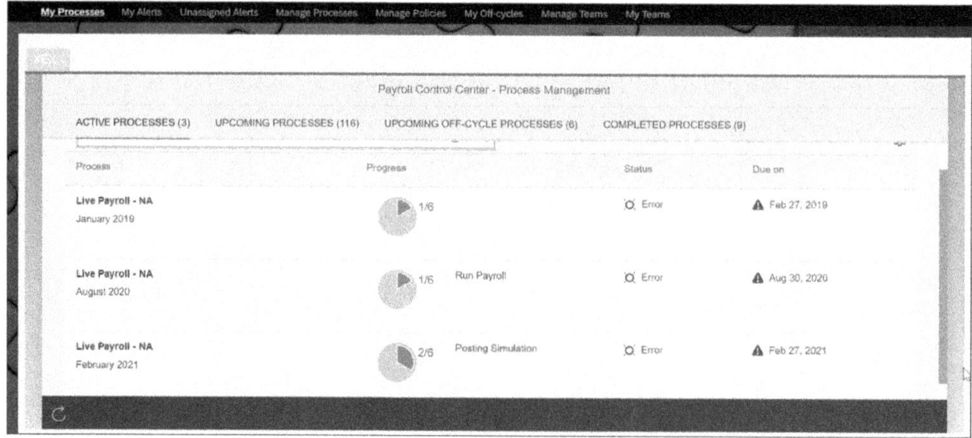

Figure 10.92 My Processes

Processes will be classified into the following three tabs:

- **Active Processes**
- **Upcoming Processes**
- **Completed Processes**

You may see a fourth tab, labeled **Upcoming Off Cycle Processes**, if you've created planned off-cycle processes for your organization.

As soon as you start a process from the **Upcoming Processes** tab, that process is removed from that tab and moved to the **Active Processes** tab. Once completed, the process automatically moves to the **Completed Processes** tab.

Under the **Active Processes** tab, you can see the progress of a process, its error status, and its due date by clicking on it, which also takes you to the current active step of the process, as shown in Figure 10.93.

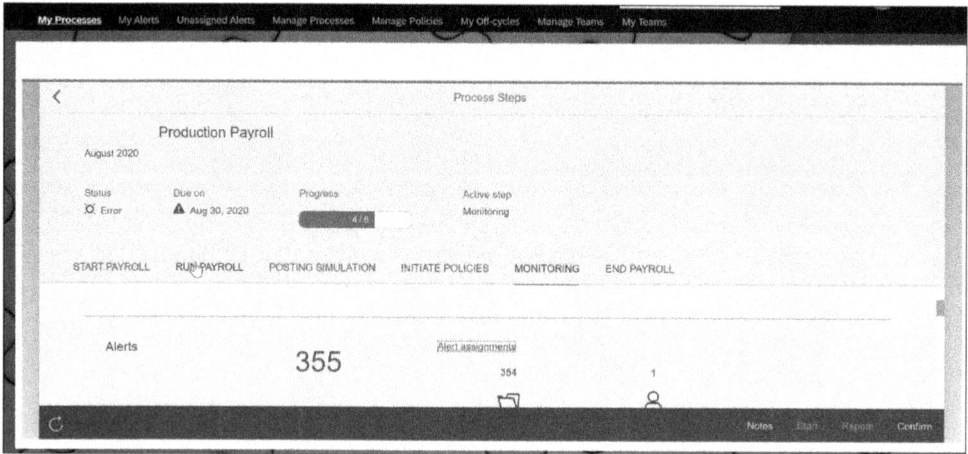

Figure 10.93 Current Process Step

To control the processing of payroll, you can take the actions listed in Table 10.21 from this screen.

Action	Procedure
Manually start the step	Click on the **Start** button.
Add notes	Click on **Notes**.
Upload attachment	Click on the **Add Attachment** button after clicking on **Notes**, as shown in Figure 10.94.
Restart	Click on **Repeat**.
Confirm	Click on **Confirm**. You won't be able to restart this process once you click this button. Also, the process will not move to the next step until you confirm. One more point about confirmation—you can confirm a step even if alerts exist. So, pay special attention to the alerts to ensure that they do not need any correction before the next step.

Table 10.21 Actions Available to Control Payroll Processing

Figure 10.94 shows how to add attachments to your notes by clicking on the **Notes** button, entering a **Title**, and then choosing **Browse...** to select the **File** you want.

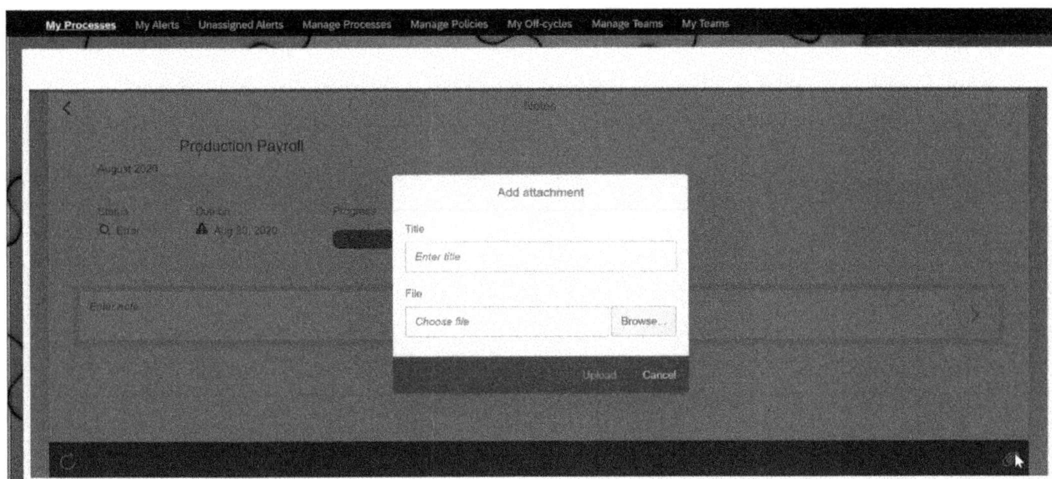

Figure 10.94 Adding Attachments in Notes

Similar buttons to the buttons listed in Table 10.21 exist under the **Upcoming Processes** and **Upcoming Offcycle Processes** tabs so you can start or see the details of the process. In addition, you can also see the planned start date of the process.

Under the **Completed Processes** tab, the date displayed is the date the process completed, along with its status. The error status is also available under the status bar.

Under the **Upcoming Offcycle Processes** tab, the following categories of processes are visible:

- **Planned off-cycle**: For productive payroll
- **Planned off-cycle**: For other, subsequent activities
- **Ad hoc off-cycle**: For subsequent activities of ad hoc off-cycle productive payroll

The status for the upcoming off-cycle process will have some additional details that are relevant for off-cycle payroll, as follows:

- **Number of employees assigned to the planned offcycle productive payroll process**, which means it can be started, or **No employees assigned to the planned offcycle productive payroll process**, which means it can be canceled.
- **Parent not yet finished for planned off cycle – other processes**, which means the subsequent process that this activity runs cannot be started yet since payroll for this employee isn't complete.
- **Number of requests for an adhoc off cycle**, which means it is ready to be started, OR **No requests for an adhoc off cycle**, which means it is ready to be canceled.

Assigning Individual Alerts to Processors

A payroll process manager can maintain the administrator team that will focus on errors. For this step, the manager simply clicks on **Alert Assignments** in the **Monitoring** step in the My Processes application. Current assignments will be displayed, and alerts that are not yet assigned will have the **Not assigned** status, as shown in Figure 10.95.

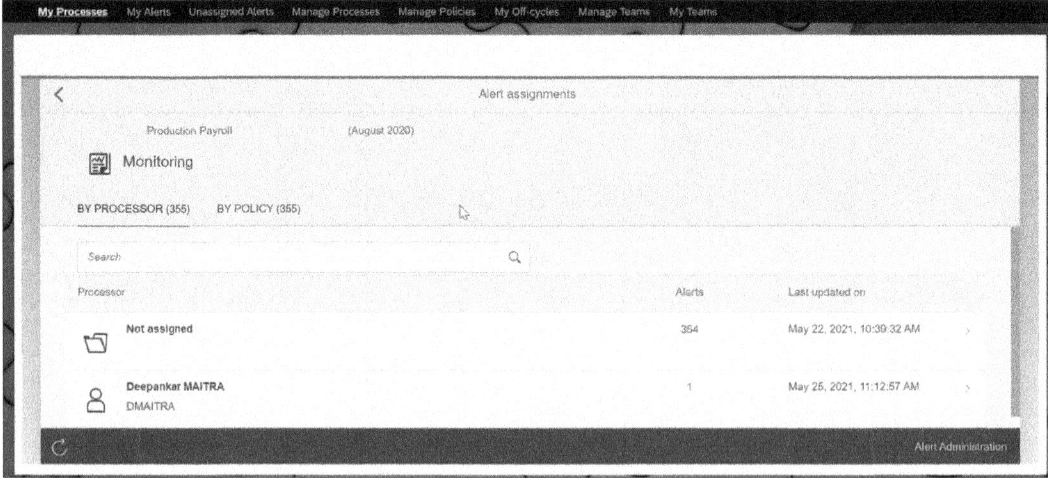

Figure 10.95 Alert Assignment

Double-click on the **Not assigned** folder, and now, alerts that are not assigned can be assigned to a processor by clicking the **Assign Processor** button, as shown in Figure 10.96.

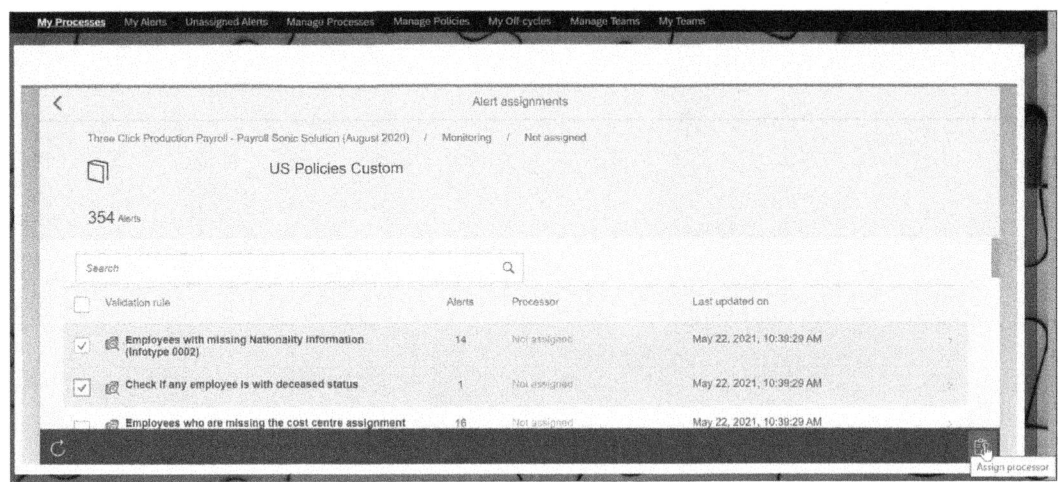

Figure 10.96 Assign Processor Button

Clicking this button lets you assign a member of the team already registered in the My Teams application, as shown in Figure 10.97. Simply choose the processor and confirm.

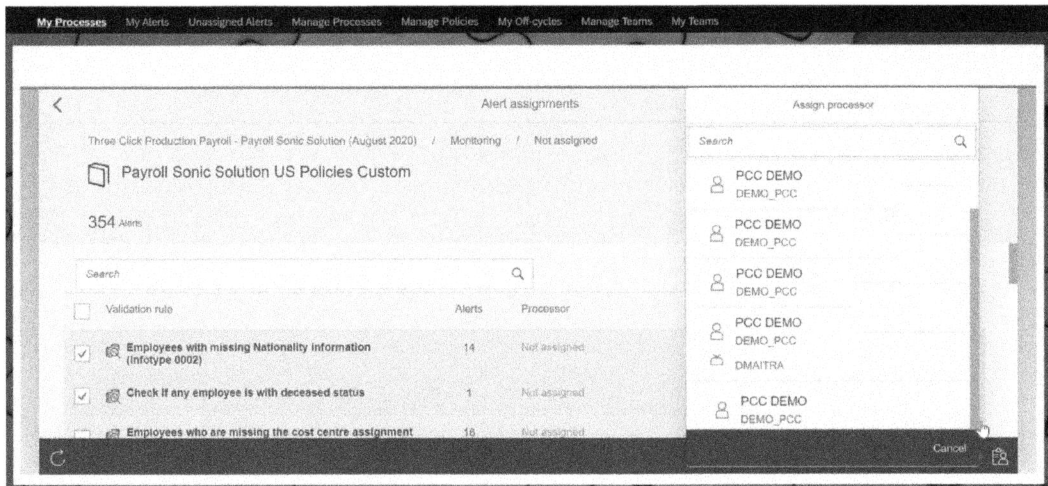

Figure 10.97 Assign Processor Screen

Assigning Default Processors for All of a Policy's Alerts

From the **Alert Assignments** page, you can set the default processor for all validations of a policy by clicking on the **By Policy** tab. Then, click on the **Settings** button and choose **Default processor,** as shown in Figure 10.98.

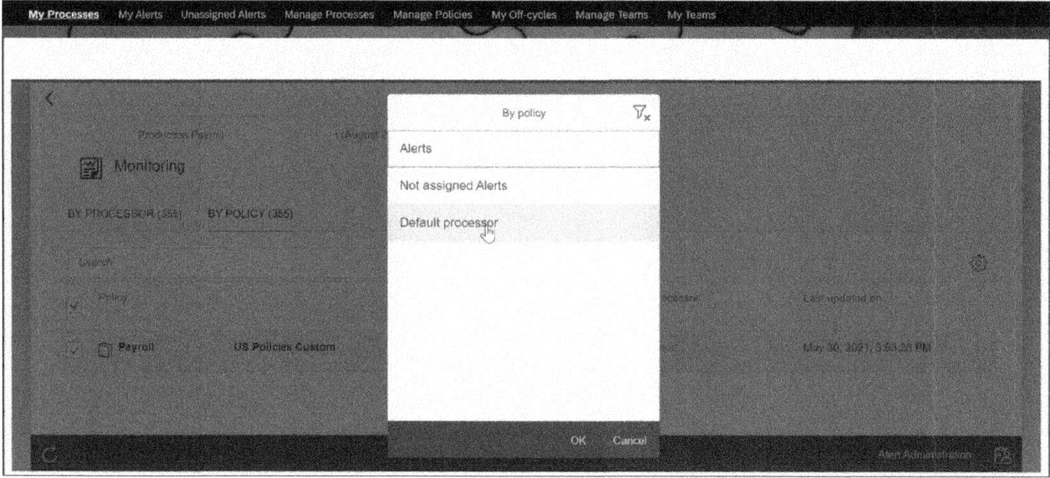

Figure 10.98 Default Processor for Policy

Pausing/Starting the Teams Working on Alerts

For a team monitoring process, you cannot change assignments since these assignments are based on preset criteria. However, a manager can start or pause teams via the Team Assignments application in the Payroll Control Center. With this application, toggling the **Processing** column for each team displayed can start or stop that team from functioning.

Processing Alerts

As a payroll administrator, you can access an alert assigned to you from the My Alerts page and process the alert. When you're in an individual alert, you'll see five different statuses, as follows:

- **Unresolved**, which means the alert requires action.
- **Solution Applied**, which means the solution proposed by the application was applied already and will be revalidated the next time you initiate policies during the monitoring process.
- **Resolved**, which is essentially flagging an alert as a false alarm and indicates to the application to proceed without waiting for a correction.
- **Completed**, which means the solution was applied and is now revalidated.
- **Not Assigned**, which means you can pick up and assign the alert to yourself.

You can also forward the alert to another team by clicking the **Forward** button, as shown in Figure 10.99.

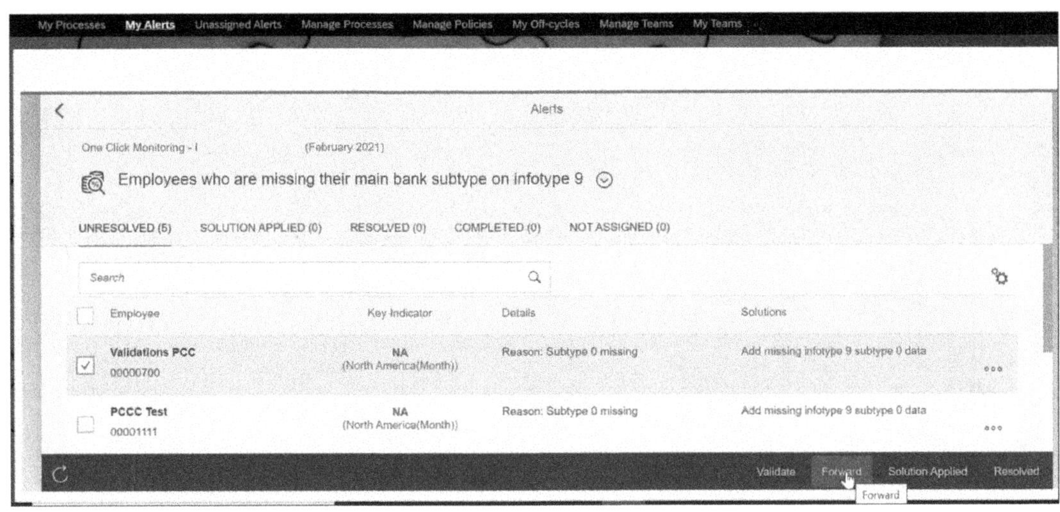

Figure 10.99 Forward Alert Button

Manual Activities

Table 10.22 lists a few scenarios where manual effort may be required, along with the action required for each case.

Scenario	Action
Adding a new check or policy	If the policy to which the check is added is part of a started process, it needs to be restarted to activate the new check
Adding or deleting team members or change the schedule of a process	The My Processes application needs to be refreshed.

Table 10.22 Scenarios Requiring Manual Action

10.6.2 Manage Teams Application

To create teams, use the Manage Teams application. In the Payroll Control Center, choose the **Manage Teams** application, as shown in Figure 10.100. The team monitoring processes will automatically show up.

If you click a process, the team for that process can be accessed by clicking on the **Team Setup** button. Use the **Add Team** button to add a team member. Teams can be identified either by name or by entering a dimension. A dimension is a value in the infotype 0001 of the employee. This information can be used to group the team together, similar to the administrator fields in infotype 0001 in traditional processing through SAP ERP HCM Payroll.

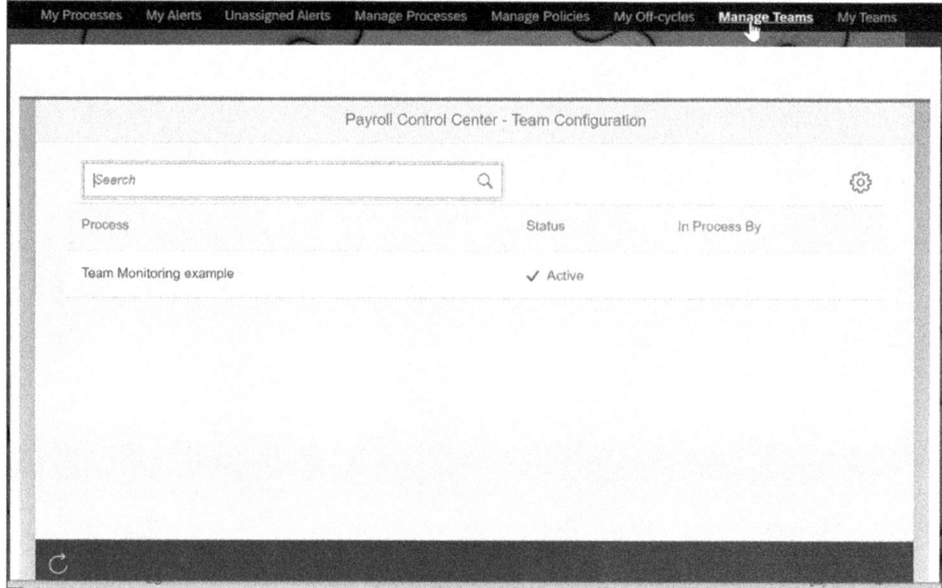

Figure 10.100 Manage Teams Application

After teams are set up, they must be activated from the **Edit Team Setup** page by following these steps:

1. Click **Edit** in the bottom corner, as shown in Figure 10.101.

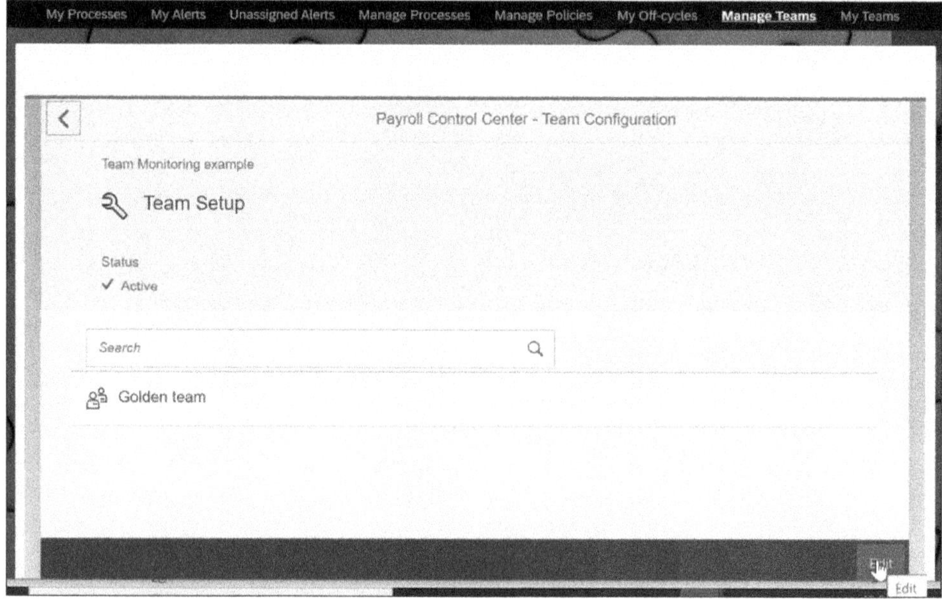

Figure 10.101 Edit Button in the Manage Teams Application

2. Click on **Add New Team** and enter a team name in the **Team** field under **Attributes**, as shown in Figure 10.102.

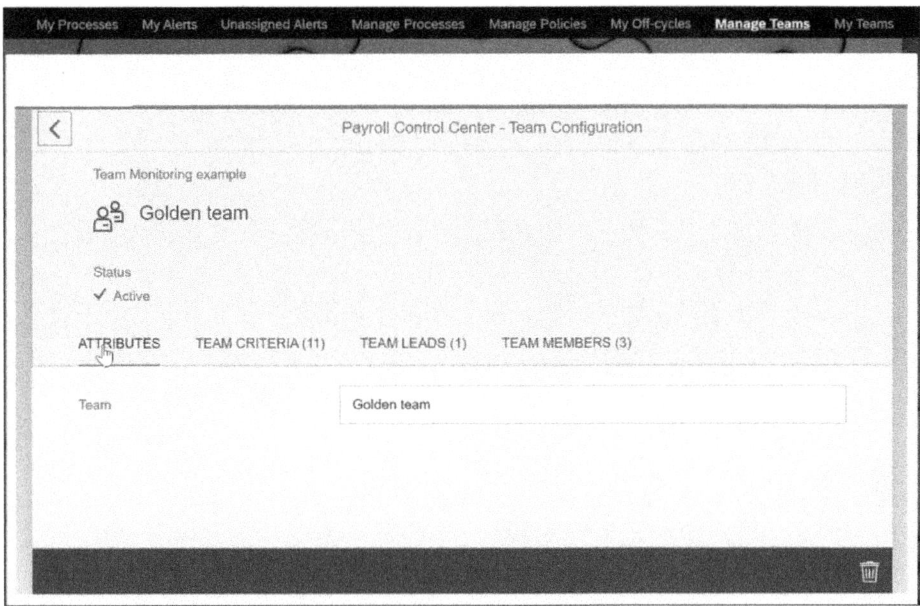

Figure 10.102 Attributes Tab in the Manage Teams Application

3. Under the **Criteria** tab, enter criteria in the team criteria field based on dimensions created earlier, as shown in Figure 10.103.

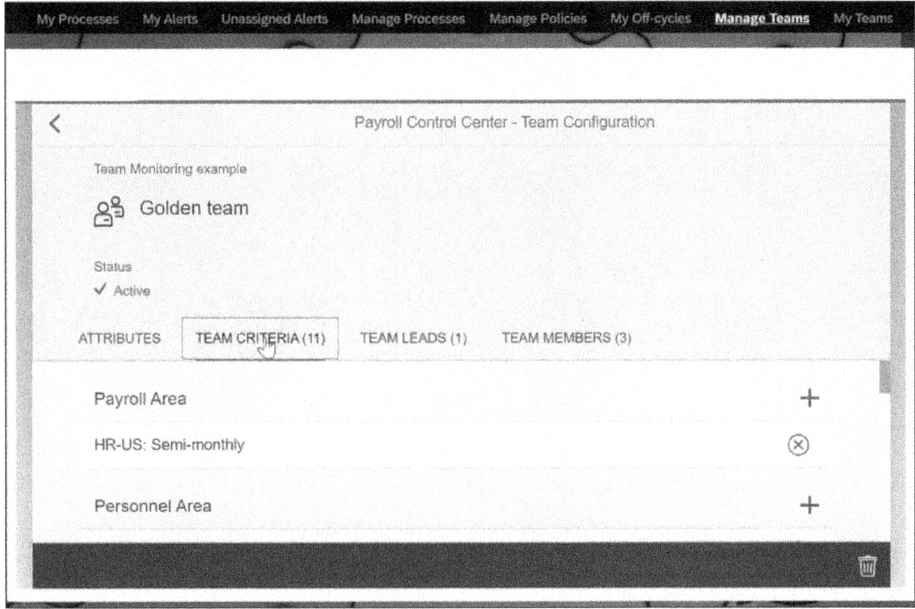

Figure 10.103 Criteria Tab in the Manage Teams Application

4. Define a team lead under the **Team Lead** tab, as shown in Figure 10.104.

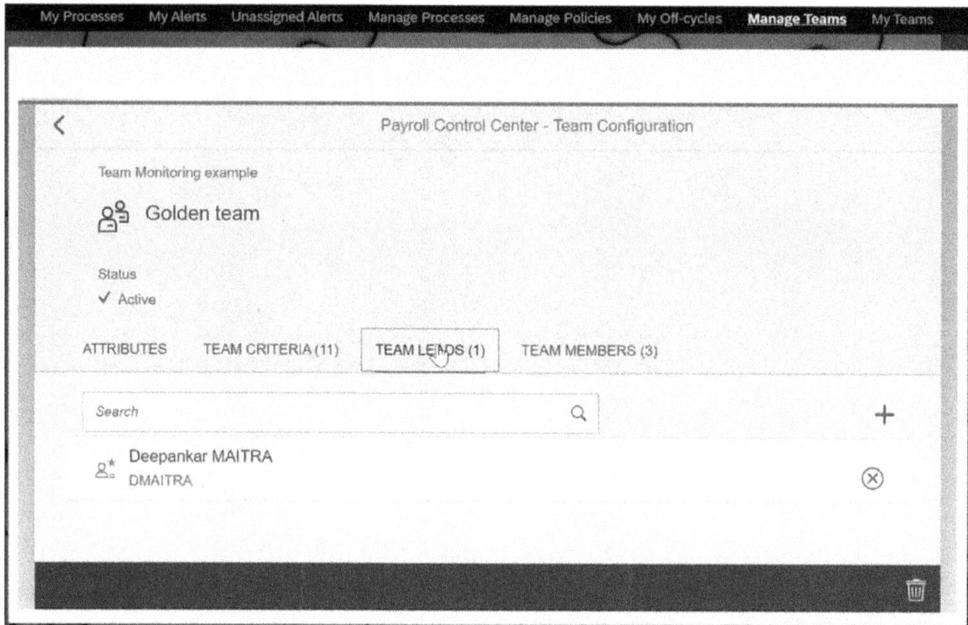

Figure 10.104 Team Lead Tab in the Manage Teams Application

5. Define team members under the **Team Members** tab, as shown in Figure 10.105.

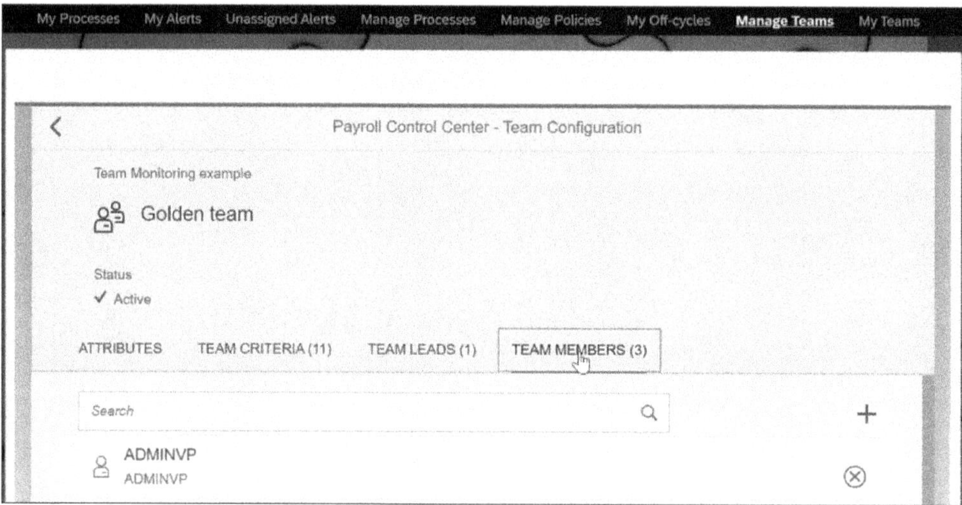

Figure 10.105 Team Members Tab in the Manage Teams Application

6. Go back to the **Teams Setup** page and click on **Activate** to activate this team for the payroll process instance chosen.

10.6.3 My Off-Cycles Application

Ad-hoc off-cycle processes are run without much planning, generally to satisfy an urgent but unpredictable need, such as an employee shorted on pay due to a glitch in an interface. This kind of payroll run involves the productive payroll and subsequent processing steps, but only for a small selection of employees that require it. The steps involved in an off-cycle process include the following:

1. Run productive off-cycle process.
2. Create posting documents (off-cycle).
3. Release posting document (off-cycle).
4. Transfer posting document (off-cycle).
5. Create pay slip (off-cycle).
6. Print pay slip.
7. Create pre-DME file (off-cycle).
8. Create DME file (off-cycle).
9. Send DME file.

An ad-hoc off-cycle process is created using the My Off-cycles application, as shown in Figure 10.106.

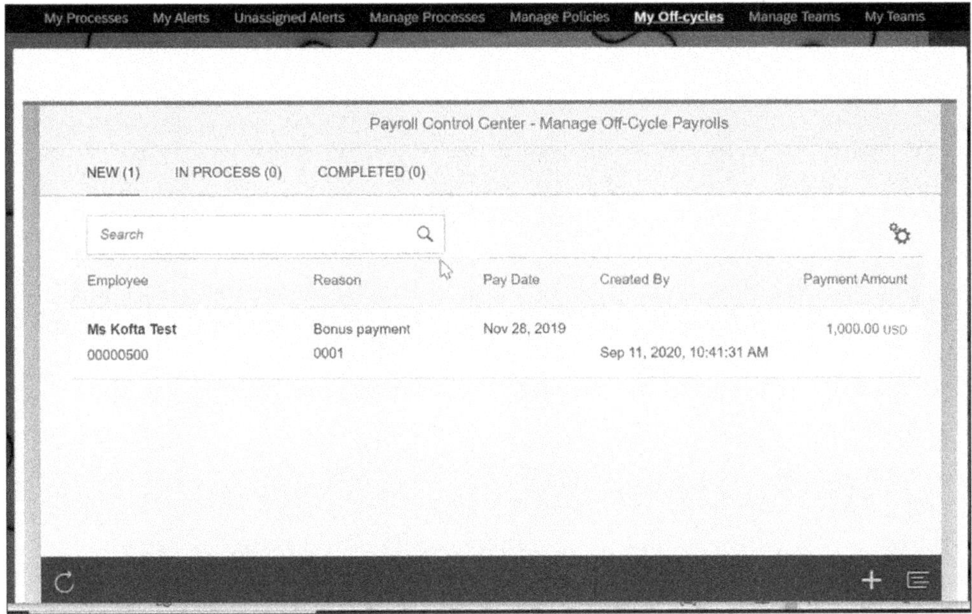

Figure 10.106 Ad-Hoc Off-Cycle Process in My Off-Cycles

The following steps are required to run an off-cycle payroll in the Payroll Control Center:

1. Go to the My Off-cycles application.
2. Click the **Add** icon (+) in the bottom-right corner.
3. The next entries will depend on the off-cycle reason. The different reasons that typically occur are listed with the fields that must be filled in, as listed in Table 10.23.

Off-Cycle Reason	Fields Required
Correction	■ **Employee** ■ **Off Cycle reason** ■ **Adjustment results** (difference between current and existing payroll runs) ■ **Pay date**
Advance pay with original result	■ **Employee** ■ **Off Cycle reason** ■ **Pay date** ■ **Payment information**
Bonus	■ **Employee** ■ **Off Cycle reason** ■ **Pay date** (must be later than accounted to date in infotype 0003) ■ **Payment information** from infotype 267
Advance absence	■ **Employee** ■ **Off Cycle reason** ■ **Pay date** (must be between current and next pay dates) ■ **Absence** from infotype 2001 (later than accounted to date in infotype 0003)

Table 10.23 Off-Cycle Reasons and Their Required Fields

4. Click **Check Request** to simulate off cycle run. The variant used will be in table T520CV.
5. Save these settings.

The ad-hoc off cycle starts, which is then managed from the My Processes application by a payroll manager.

In certain situations, an off-cycle process must be run straight after a regular cycle. Usually, this need is driven by a master data change, such as a late submission of an absence or an adjustment to the amount of a bonus. The My Off-cycles application can recognize some of these instances and provide a notification. A payroll administrator can then start an off-cycle process by clicking on the notification. One advantage of this method is that no further input is required. A system can inherit the characteristics of the master data change to start the off-cycle process with the correct parameters. This method works for bonus and advanced vacation payments only.

Sometimes, an ad-hoc payroll is performed for the wrong employee and must be amended. Ad-hoc off-cycle requests can be canceled if required, but only if they meet certain conditions. If any of the following conditions are met, the button for cancellation appears automatically on the bottom right.

- This payroll is the last off-cycle for this employee.
- This is the employee's last payroll result.
- The payroll control record has the **Exit payroll** status.
- Follow-on activities, such as postings, haven't started yet for this employee.

10.7 Summary

In this chapter, we explored configuring the Payroll Control Center in detail so the solution can be used for processing payroll with accuracy and timeliness. You're now familiar with how to configure the various features that make the Payroll Control Center indispensable for completing payroll efficiently. As with any configuration activity, decisions must be made, and each decision impacts choices about other elements of the system. Using the Payroll Control Center is entirely optional—you can run the Employee Central Payroll engine in the traditional way with the same result. One efficiency feature that existed in traditional payroll—the process model, which allowed the execution of multiple payroll steps in sequence—still exists. But the Payroll Control Center also provides this functionality and much more, and thus, users of the Payroll Control Center do not typically use the process model anymore.

Chapter 11
Setting Up Pay Statements

When payroll runs smoothly, the employee receives money in the bank. But that amount is not usually what the employee is paid. To understand the components of pay and the amounts withheld from pay, the employee requires a pay statement. The bank statement merely shows the net amount that was deposited.

This chapter introduces the most popular report in a payroll system—the pay statement. We'll discuss the available types of pay statements, explain specific features of each type, and provide a grounding in how to configure pay statements. Part of the configuration may require turning on key figures for payroll results, which is discussed in Chapter 12. The pay statement requires specific permissions to be visible not only to the employee but also to the payroll administrators who process these statements and validate their accuracy. Finally, some specific features will need to be turned on to make off-cycle statements visible in a user-friendly card-based format.

11.1 Types of Pay Statements

The pay statement itself needs to be developed using HR Forms in Employee Central Payroll, which is a developer task. The discussion in this section focuses on the rendering of the pay statement: The types represent the different ways they appear on your screen. An administrator can enable a pay statement to appear in the following places:

- Employee Central
- SAP SuccessFactors Mobile app
- The latest homepage
- Custom access points in Employee Central via deep links

The type of pay statement determines where the pay statement can appear. Four types of pay statement display options are configurable:

1. **Arrow-based pay statement (legacy pay statement)**
   Also known as the *legacy pay statement*, this type of pay statement is the simplest to configure. As shown in Figure 11.1, the navigation of this pay statement is basic. You must move from one pay statement to another using the arrow keys. The system will automatically position the screen on the latest pay result, from which you can

move to past pay statements using the left arrow key on the statement. You can then go forward to more current pay statements using the right arrow key.

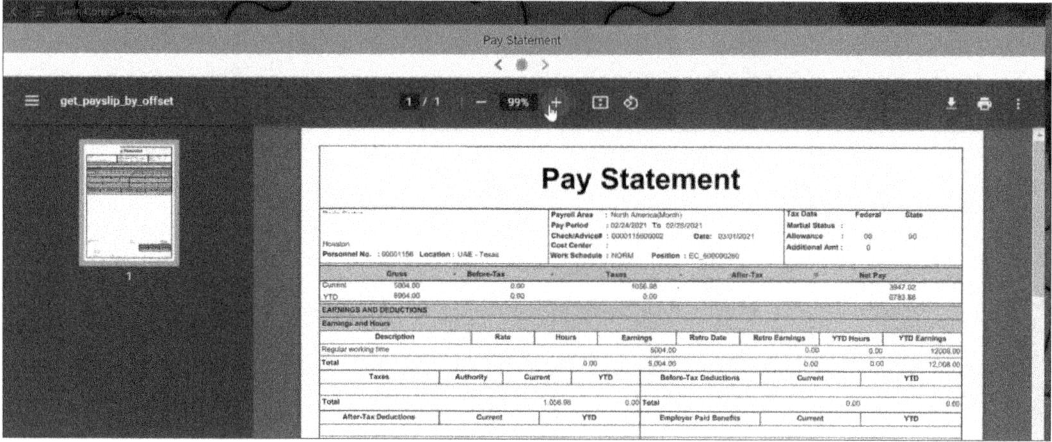

Figure 11.1 Legacy Pay Statement (Arrow Based)

2. **Direct pay statement**
   This variety of pay statement appears with an index on the left and the statement on the right. The index is a running list of all pay periods that have been processed for this employee in ascending chronological order. The system will automatically position the screen on the latest pay result, from which you can move to past pay statements by clicking on the period you're interested in from the pane on the left, as shown in Figure 11.2.

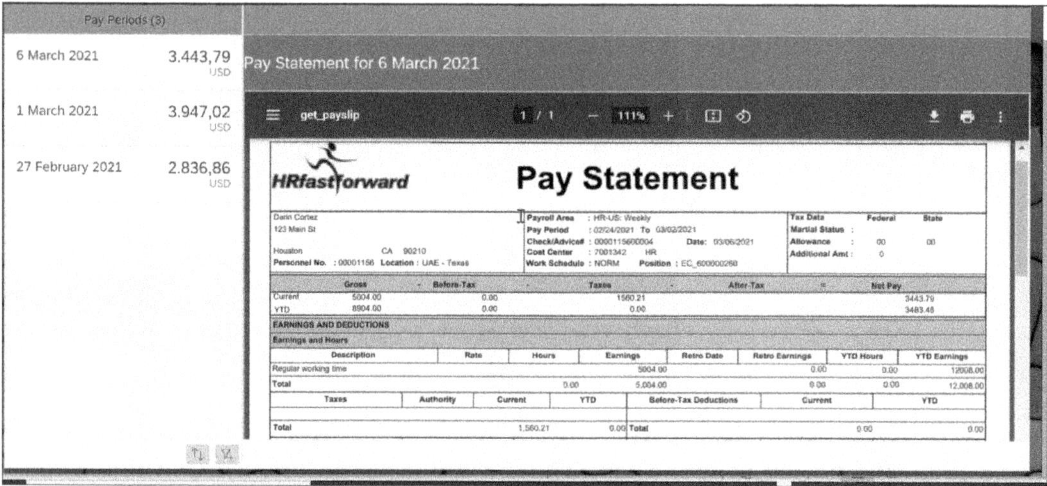

Figure 11.2 Direct Pay Statement

3. **SAP-Fiori like pay statement**
   SAP Fiori-like pay statements rely on pay data replicated from Employee Central Payroll to Employee Central. Using this type has the principal advantage of mobile enablement. Employees can view their pay statements from their smartphones. From an appearance perspective, not much is different from a direct pay statement. The system will automatically position the screen on the latest pay result, from which you can move to past pay statements by clicking on the period you're interested in from the pane on the left.

4. **SAP Fiori-like pay statement with key figures**
   A variation of the previous type, the SAP Fiori-like pay statement with key figures has the added capability of adding summary-level information next to the pay statement links in the pane on the left. For example, in addition to Period 10, 2020, you'll see the net amount of $2,000 next to it. If you click on this summary information, you can access the pay statement in the pane on the right.

11.2 Configuring Pay Statements

The configuration steps for pay statements vary based on the type you choose. Table 11.1 compares the different steps.

Step	Legacy Pay Statement	Direct Pay Statement	SAP Fiori-Like Pay Statement	SAP Fiori-Like Pay Statement with Key Figures
Employee Central Payroll: Set up HR Forms.	√	√	√	√
Employee Central Payroll: Set up feature HRFOR.	√	√	√	√
Employee Central Payroll: Implement Business Add-In (BAdI) IF_HRESS_PAYSLIP_BADI~PROVIDE_FILTERED_RGDIR to control when a pay statement is visible.		√		
Employee Central Payroll: Implement BAdI HR_EC_PAYSLIP_BADI_DEF to control when a pay statement is visible.	√			

Table 11.1 Features Applicable for Each Type of Pay Statement

Step	Legacy Pay Statement	Direct Pay Statement	SAP Fiori-Like Pay Statement	SAP Fiori-Like Pay Statement with Key Figures
Employee Central: Configure OAuth 2.0 to integrate Employee Central authorization to read payroll results directly from Employee Central Payroll.		√		
Employee Central: Allow admins to access the OData application programming interface (API) through basic authentication.	√		√	√
Employee Central: Allow access to payroll integration/employee payroll run results (current and history), and WageTypeAccumulationConfig.			√	√
Employee Central: Configure payroll key figures.				√
Employee Central: Payroll unified configuration.		√		
Set up pay data in **Payroll Information** block for Employee Self-Service (ESS).			√	√
Replicate payroll results from Employee Central Payroll to Employee Central.			√	√
Configure key figures.				√
Configure deep link */sf/latestpayperiod* if required.	√	√		

Table 11.1 Features Applicable for Each Type of Pay Statement (Cont.)

Let's now look at all these steps in more detail:

1. This step is required for all pay statements. Setting up HR Forms is a development activity, as discussed in Section 11.1.
2. This step is also required for all pay statements and makes use of a simple feature where you tell the system that HR Forms is being used to display the pay statement. Use Transaction PE03 open feature HRFOR. Enter the name of the HR Forms form created in the previous step. Save.

3. This step is an optional development activity involving the BAdI `IF_HRESS_PAYSLIP_BADI~PROVIDE_FILTERED_RGDIR`, implemented with Transaction SE18. This step is required only if the timing of the display needs to be controlled. For example, let's say you don't want to show the employees their pay statements before the pay date, so that you have room for last-minute corrections.
4. This step is the same as step 3, only with BAdI `HR_EC_PAYSLIP_BADI_DEF`.
5. This step is the configuration of OAuth 2.0 and is a prerequisite for the direct pay statement. We'll review this topic in Chapter 13, Section 13.7.
6. This step covers the permissions required to display the statement if you're not using the direct pay statement. See Chapter 4, Section 4.1, for more guidance on role-based permission settings.
7. This step covers additional permissions. See Chapter 4, Section 4.1, for more guidance on role-based permission settings.
8. This step is required for showing key figures—we'll review this step in Chapter 12, Section 12.1.6.
9. This step is the unified configuration for payroll, required for almost everything payroll-related in Employee Central, but not for arrow-based or SAP-Fiori-like statements. Refer to Chapter 4, Section 4.2.3, for more information.
10. This step prepares Employee Central to receive data from Employee Central Payroll. Refer to Chapter 12, Section 12.1, for more information.
11. This step involves getting the data pulled over from payroll to Employee Central. This connection is essential for SAP Fiori-based statements. Refer to Chapter 12, Section 12.4, for more information.
12. This step is required only if you're using key figures to drive the pay statement in the SAP Fiori-based statement. Although a direct statement shows basic key figures, it does not require this configuration step. However, direct pay statement key figures require that you install support pack EA-HR SP86. Installing support pack EA-HR SP80 will provide for direct pay statements but not key figures. Your technical team must download the support packs from the SAP site using their customer accounts.
13. This step is an optional item if you want to launch pay statements from other areas of Employee Central. The deep link gives you the ability to launch the pay statement page without having to go through any other tile or menu path. A deep link can be used in multiple places at various points of the process. We saw one example in Chapter 10 when we discussed solutions and how a deep link can position the solution automatically to the correct page with the desired information without having to navigate through unnecessary data pages.

Enabling pay statements in the SAP SuccessFactors Mobile app is possible by enabling the **Pay Summary** feature and configuring the corresponding permissions. All the pay statement types are available on mobile, except for the arrow-based (legacy) statement.

11.3 Viewing Pay Statements

Pay statements can be viewed in multiple ways in Employee Central. The following sections discuss the most common methods of viewing them.

11.3.1 Pay Statement on the My Employee File Page

Pay statements can be viewed from the **My Employee File** page. Navigate to the **Payroll Information** block, as shown in Figure 11.3 and click on the **Pay Statement** link under the **Earnings and Deductions** heading.

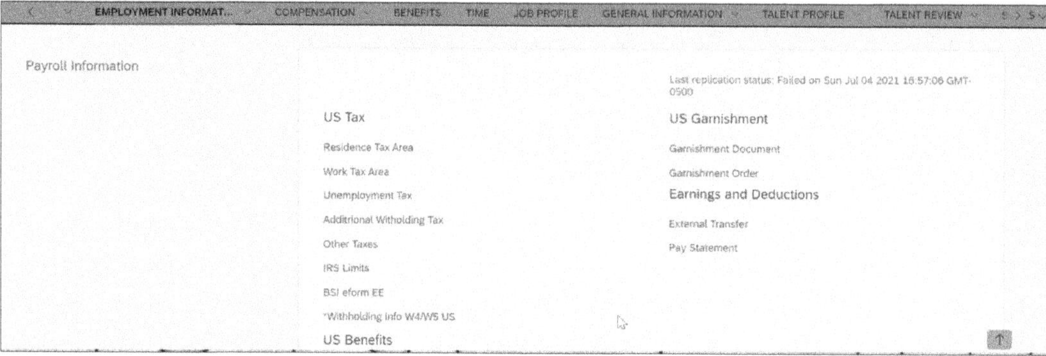

Figure 11.3 Pay Statement from My Employee File

11.3.2 Viewing Pay Statements via Quick Actions on the Latest Homepage

The latest Employee Central homepage has a set of cards for navigating to the pay statement, as shown in Figure 11.4. Navigate to the **Quick Actions** section and click on the **View Payslip** card. Clicking on **Pay History** takes the user to the navigation overview page.

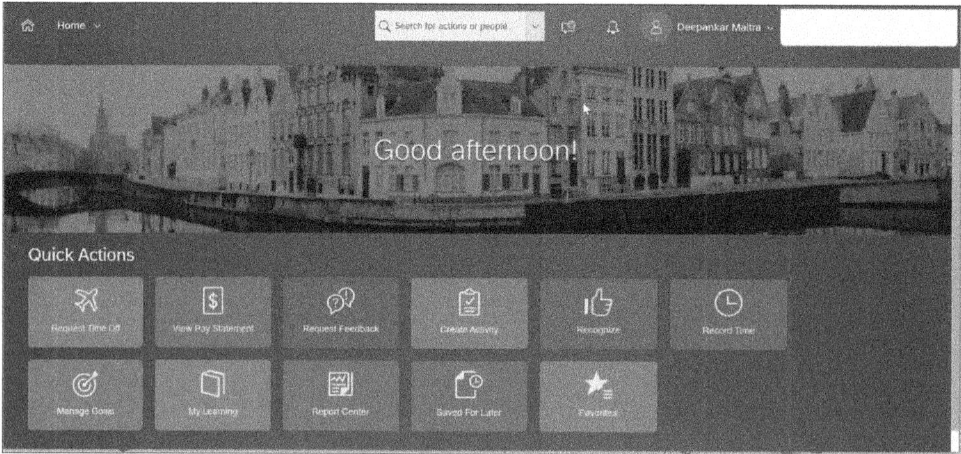

Figure 11.4 Pay Statement from Quick Actions

The homepage must be enabled in provisioning or from the Upgrade Center, and permissions must be granted to access the latest homepage.

11.3.3 Off-Cycle Payment Cards and Reports

If an employee receives a bonus payment, that payment can be displayed using the payment card, as shown in Figure 11.5. If multiple payments exist, multiple cards will be visible to display the payments.

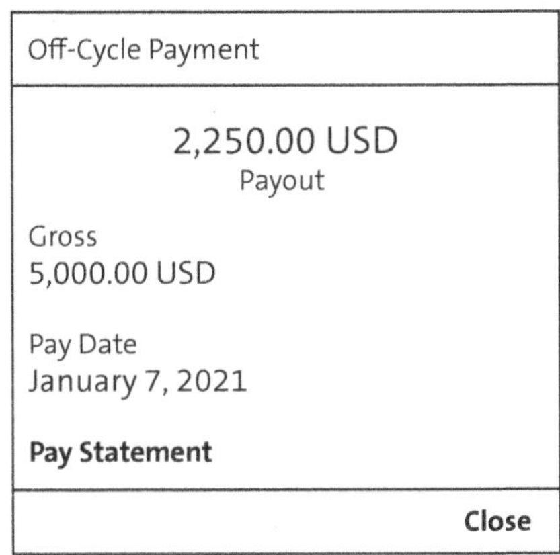

Figure 11.5 Off-Cycle Pay Card

To make these cards visible, the off-cycle report must be executed from Employee Central Payroll. To run this report, point-to-point replication with OData must have been configured already (see Chapter 7 and Chapter 8), and you must have configured the OAuth protocol (see Chapter 13, Section 13.7).

To run the report, execute Transaction SE38 and run report RP_HRSFEC_PTP_UXR_OFF-CYCLE, which will result in the screen shown in Figure 11.6.

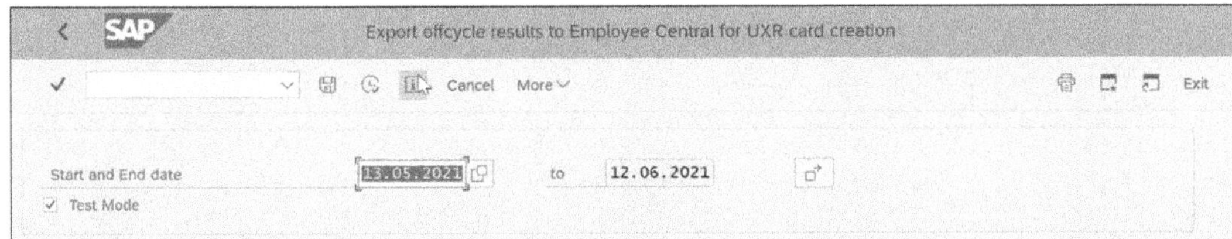

Figure 11.6 Report to Export Off-Cycle Results from Payroll

The report can be scheduled to run at regular intervals using the **Start and End date** fields. Care must be taken to run it only after the pay processing is complete.

11.4 Summary

This chapter concentrates on the various types of pay statements that are available and provides the guidance required to understand which pay statement type to use and when. As a customer-facing deliverable, pay statements are read by every employee in the organization, so getting this setup correct and delivering pay statements in a timely manner are as important as running the periodic payroll itself.

Chapter 12
Using Payroll Results in Employee Central

In the last chapter, we discussed setting up and using pay statements. One of the types of statements was the SAP Fiori-like pay statement, which requires that payroll results be available in Employee Central. This chapter discusses how to get those results over into the Employee Central system. In some respects, this process is a reverse replication where we're getting information from Employee Central Payroll to Employee Central, instead from Employee Central to Employee Central Payroll.

This chapter discusses the concept of bringing payroll results back to Employee Central. In certain scenarios (e.g., SAP Fiori-like pay statements, mobile pay statements, or Employee Central Advanced Reporting), the payroll result should be carried over to Employee Central. In this chapter, we'll discuss how to set up the configuration in Employee Central Payroll for the required wage types and configure the import parameters so that this data can be made available in Employee Central. We'll also discuss how to set up the required permissions and run the replication program to start the process.

12.1 Configuration

The configuration for bringing payroll results back to Employee Central involves several steps:

1. Picking the right wage types.
2. Configuring the import mechanism.
3. Creating business rules for the payroll run type picklist, for the wage type picklist, and for the grouping reason picklist.
4. Setting up key figures.
5. Setting up permissions.

After this configuration is complete, you still must replicate the payroll results to transfer the results from your Employee Central Payroll system to your Employee Central system, every time a payroll has been run. In this section, we'll show you how to perform the first four of these tasks. Setting up permissions and replicating the payroll results will be covered in Section 12.2 and Section 12.3, respectively.

12.1.1 Specifying the Relevant Wage Types in Employee Central Payroll

A prerequisite for the replication of pay results is the determination of which wage types should be brought into Employee Central. Since Employee Central is not the pay system, only 10 to 15 wage types can be accommodated, so you must choose carefully.

> **Note**
> Business function HCM_SFEC_MDEC2HR must be activated in Employee Central Payroll for the export of wage types to work.

To specify the relevant wage types in Employee Central Payroll, follow these steps:

1. Navigate to Customizing in Transaction SPRO and follow the menu path **Personnel Management • Integration Settings for SuccessFactors Employee Central • Data Extraction Of Payroll Results.**
2. Define your subapplications, as shown in Figure 12.1. Each subapplication relates to a specific set of wage types as defined by the application. For example, ERNU describes the wage types for earnings earned in a pay period. More information is available in the SAP documentation at *http://s-prs.co/v538605*. Once you decide which subapplications should be shown in Employee Central, select those applications by selecting the **Stat...** checkbox on the right, select the radio button for the frequency (**Year/ Month/Day**), and save these settings.

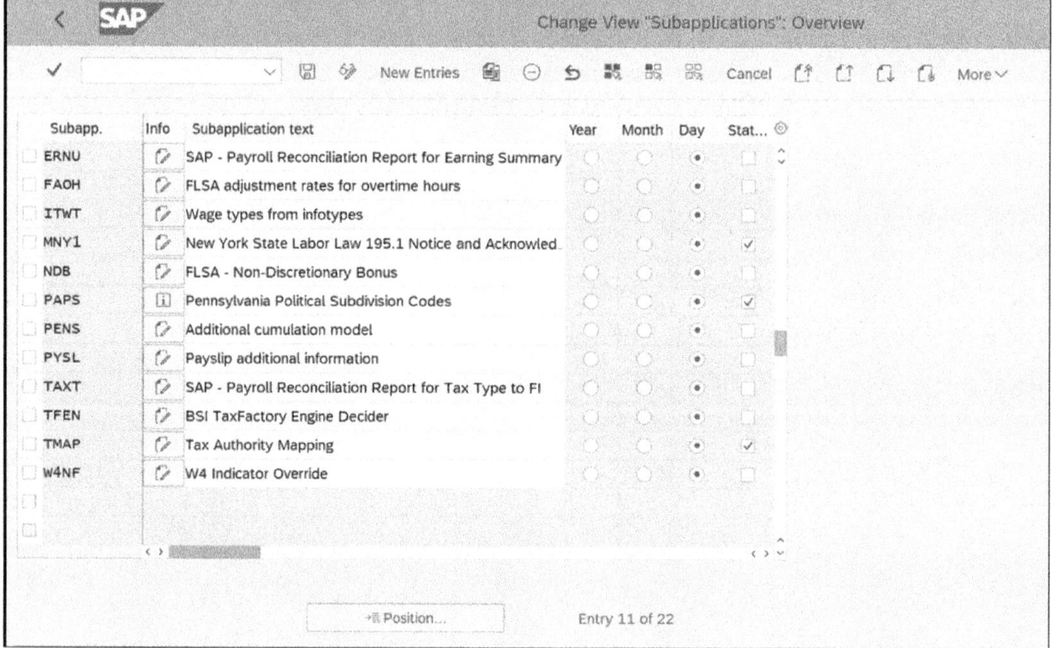

Figure 12.1 Defining Subapplications

3. Define subapplication attributes, as shown in Figure 12.2. In this step, you can select a condition as well that is relevant for certain countries only, like Belgium. Generally, you would choose **Single employee** in the **Type** column to get the results for an individual employee. You can also maintain the **Start Date** and **End Date** fields to control the timing of when the data is pushed to Employee Central. Once selections are made, click the **Save** icon.

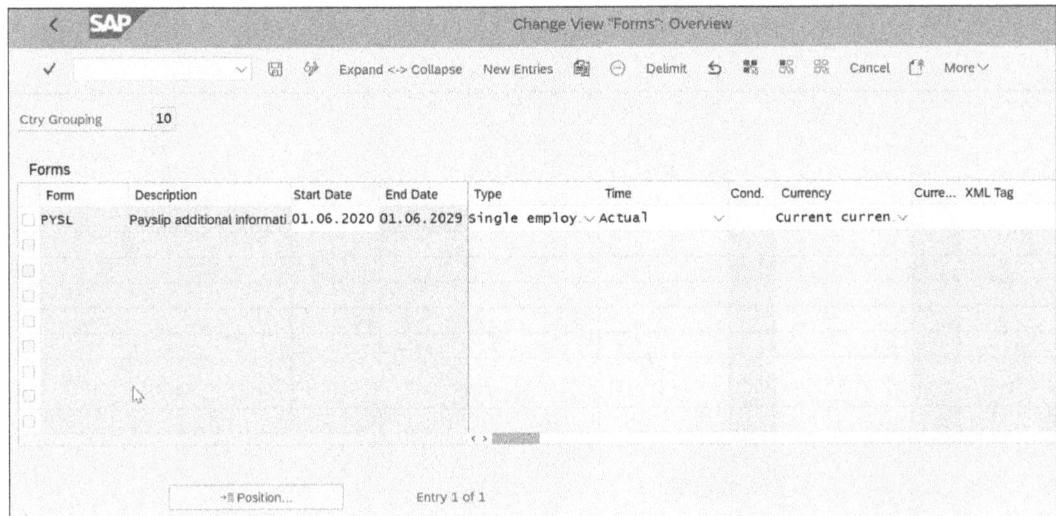

Figure 12.2 Defining Subapplication Attributes

4. Next, we'll define wage type groups by choosing the wage types that should be included in the replication of the subapplication. This decision could be based the country you're dealing with—if trying to configure a multicountry implementation, each country must be treated like a subapplication since the wage types for each country will be different. First, define the group, as shown in Figure 12.3. Click on **New Entries**. Enter a name in the **Group** field and enter a description. Then, click the **Save** icon.

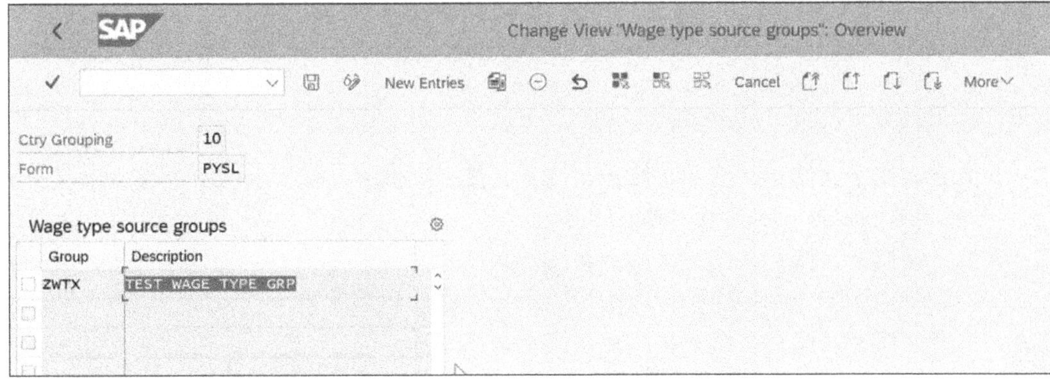

Figure 12.3 Defining a Wage Type Group

5. Finally, set the wage types for this group, as shown in Figure 12.4. Click on **New Entries**. Select the group we just created. Now, enter the wage type value in the **Value** field. Notice that you can also select a processing or evaluation class in the **Class** field. Specifying a class will have the effect of selecting all the wage types that belong to this evaluation class. Once done, click on the **Save** icon.

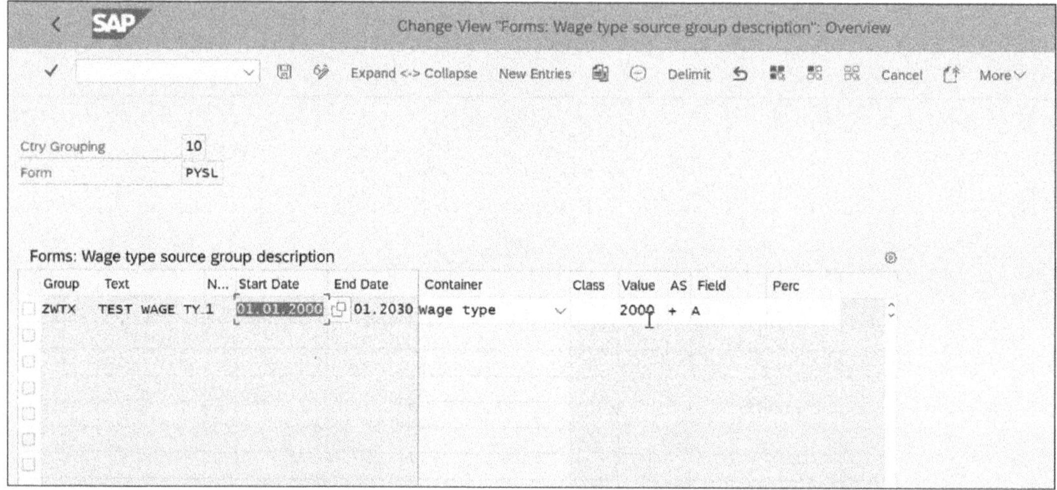

Figure 12.4 Wage Types for the Group

12.1.2 Setting Up Import Related Configuration in Employee Central

To configure the replication of payroll results, you'll need to maintain the following three fields:

- **Payroll Run type** (for example, regular, off-cycle, etc.)
- **Wage type**
- **Grouping Reason** (optional): Used for person-related totals mostly in concurrent employment scenarios for grouping related payroll attributes for the same employee (e.g., tax company in the US for maintaining totals at the employer level). Concurrent employment is not covered in this book, but you can review grouping reasons under the personnel management configuration by following the menu path **Personnel Management • Personnel Administration • Basic Settings Basic Settings for Concurrent Employment • Editor for Personnel Assignment Groupings.**

The code values for these fields do not exist in Employee Central. Thus, you must define these values in custom picklists.

Begin with creating a picklist in Employee Central for each of these three fields by clicking the **Create a picklist** icon, as shown in Figure 12.5. Enter a code and name and then save.

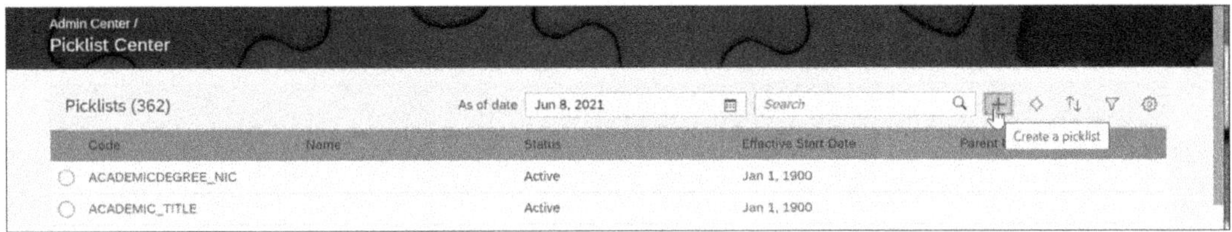

Figure 12.5 Creating a Picklist

Next, add the values you would like to use for this picklist. For example, "R" for regular run and "O" for Offcycle." You can do this by selecting the newly created picklist from the same screen and clicking on the + icon to add values.

Once done, you'll need to assign the picklist to each field and define a business rule for each field, which we'll discuss next.

12.1.3 Creating a Business Rule for Payroll Run Type Picklist

To create a business rule for the payroll run type picklist, in Employee Central, enter "Configure Object Definitions" in the search box. then, select **Object Definition** and **Employee Payroll Run Results**, which will open the screen shown in Figure 12.6.

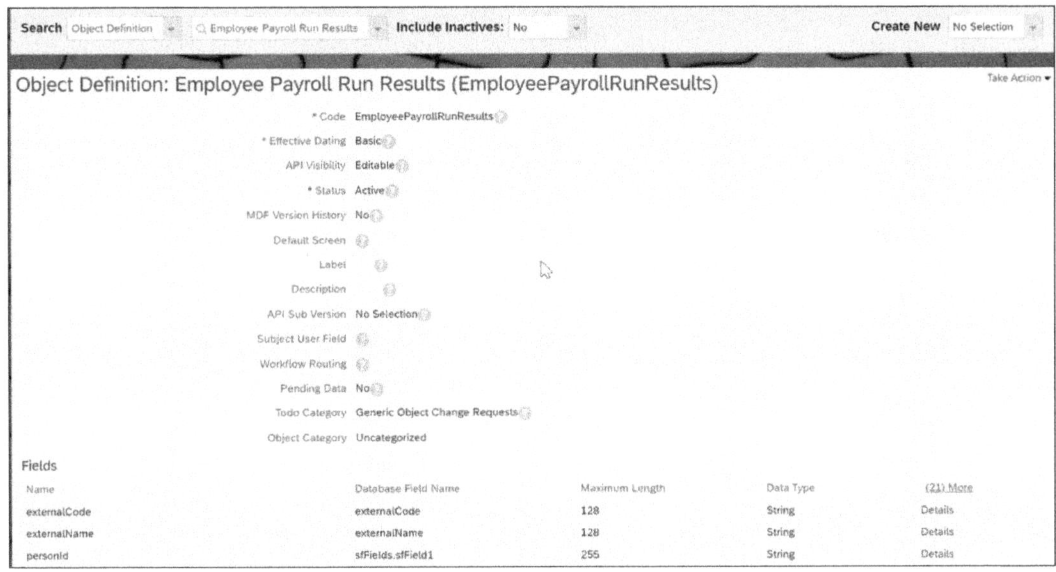

Figure 12.6 Payroll Result Run Type Object Definition

Choose **Take Action • Make Correction** in the top-right corner. On the screen that appears, select **Details** next to the **payrollRunType** field, which will open the screen shown in Figure 12.7. Enter the name of the picklist in the **Valid Values Source** field and then click **Done**.

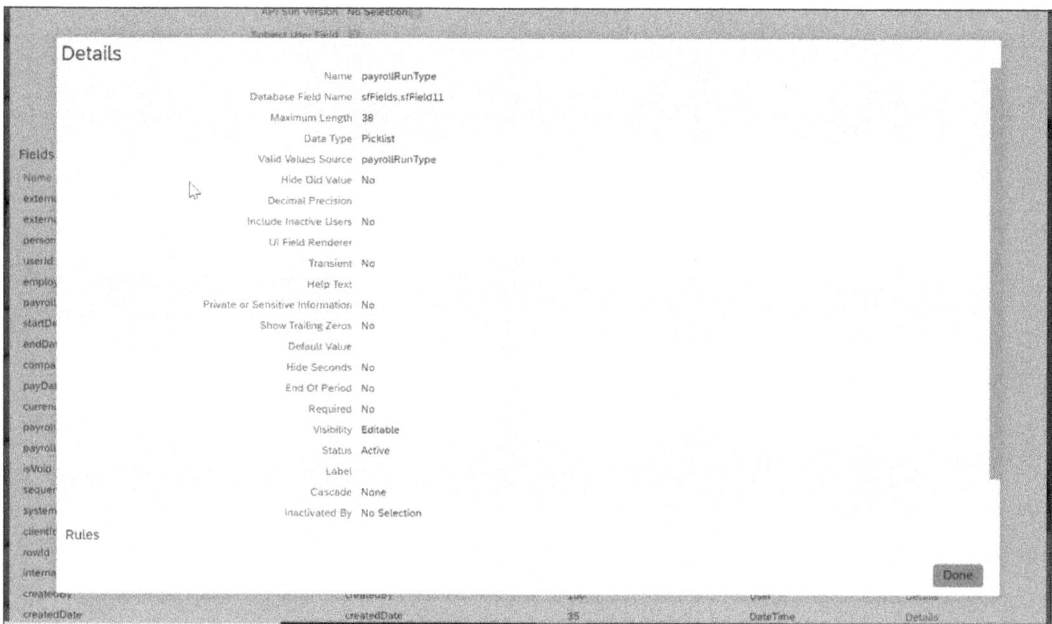

Figure 12.7 Payroll Run Type Results Picklist

Now, you'll need to populate the picklists with the correct values by following these steps:

1. Go to the Admin Center and enter "Configure Business Rules" into the search bar. Next, perform these steps:
   - Click **Create new rule**, which will result in the screen shown in Figure 12.8. Select **Basic**.

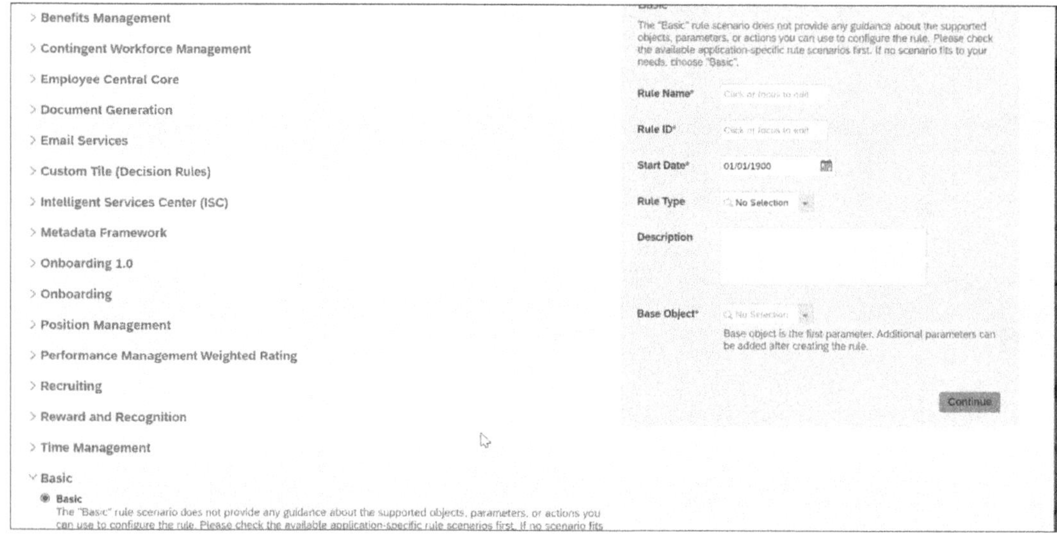

Figure 12.8 Creating a Business Rule

- After maintaining the **Rule Name** and **Rule ID** fields, select **Employee Payroll Run Results** from the **Base Object** dropdown list.
- Click **Continue**, which will open the screen shown in Figure 12.9.

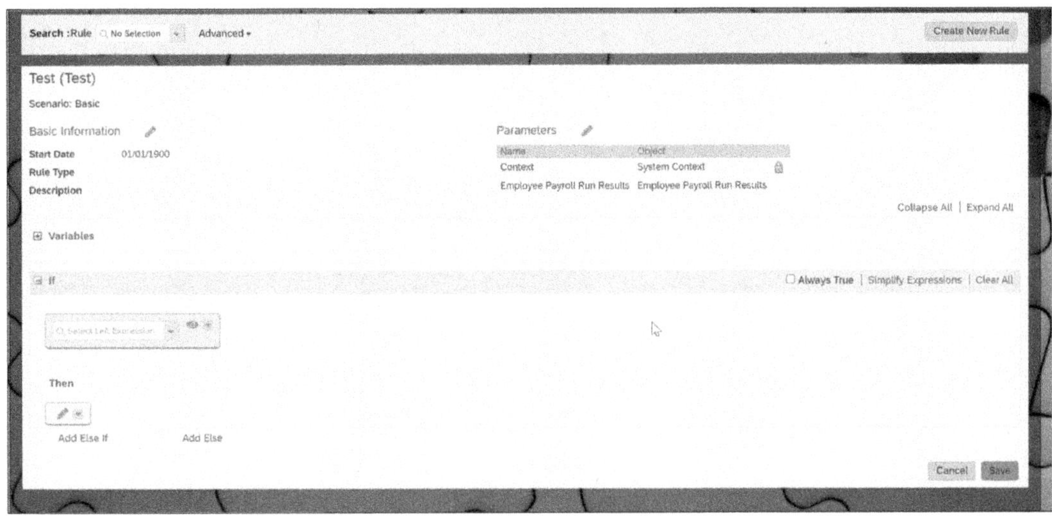

Figure 12.9 Details of a Business Rule

- Now, go to the **If** statement and choose **Employee Payroll Run Results** – **Source: Payroll Run Type.** Check that it is equal to the payroll type that you're configuring, as represented in Employee Central Payroll (typically, blank for a regular run, "A" for bonus off-cycle, etc.). You can check these values from the **Payroll Type** dropdown list in the RP_HRSFEC_ExportPayrollResults report, described in Section 12.3.
- Go to the **Then** statement, click on the pencil icon and set the picklist value corresponding to this type of run that you set up in the picklist values earlier. For this step, choose **Set** from the first field, **Payroll Run Type** from the second field, and the corresponding value from the last field.
- Save these settings.

2. Now, you'll assign this rule to the **Save Rules** section in the **Employee Payroll Run Results Item** object, shown in Figure 12.9, and then save. You can get to the screen shown in Figure 12.10 by typing "Configure Object Definition" in the search bar and then selecting the object, as shown earlier in Figure 12.6, except enter "Employee Payroll Run Results Items" instead of "Employee Payroll Run Results" in the second search field.

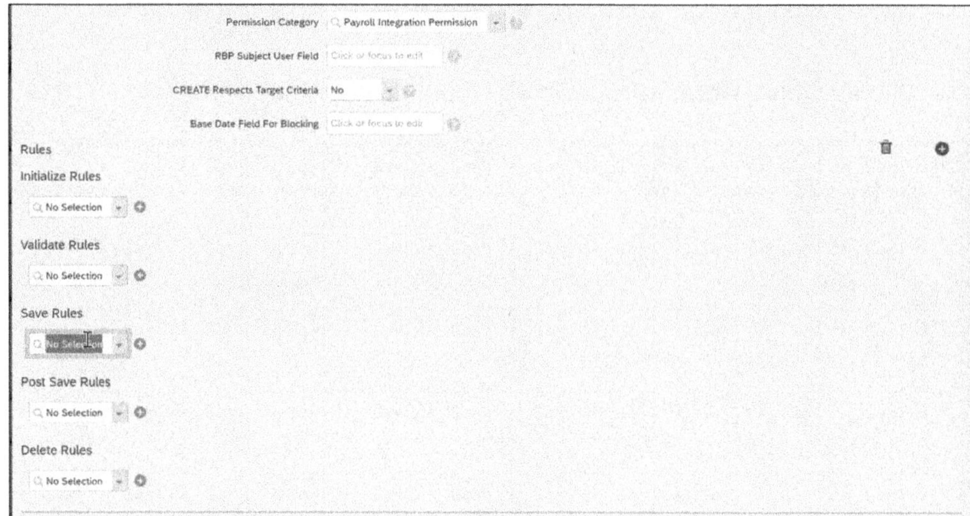

Figure 12.10 Assigning the Business Rule to Employee Payroll Run Results Items

12.1.4 Creating a Business Rule for Wage Type Picklist

To create a business rule for the wage type picklist, in Employee Central, enter "Configure Object Definitions" in the search box. Then, select **Object Definition** and **Employee Payroll Run Results Items** in the dropdowns lists next to **Search**. Next, choose **Take Action • Make Correction** in the top-right corner.

Select **Details** at the end of the **wageType** field row and enter the name of the picklist in the **Valid Values Source** field, as shown in Figure 12.11, and then click **Done**.

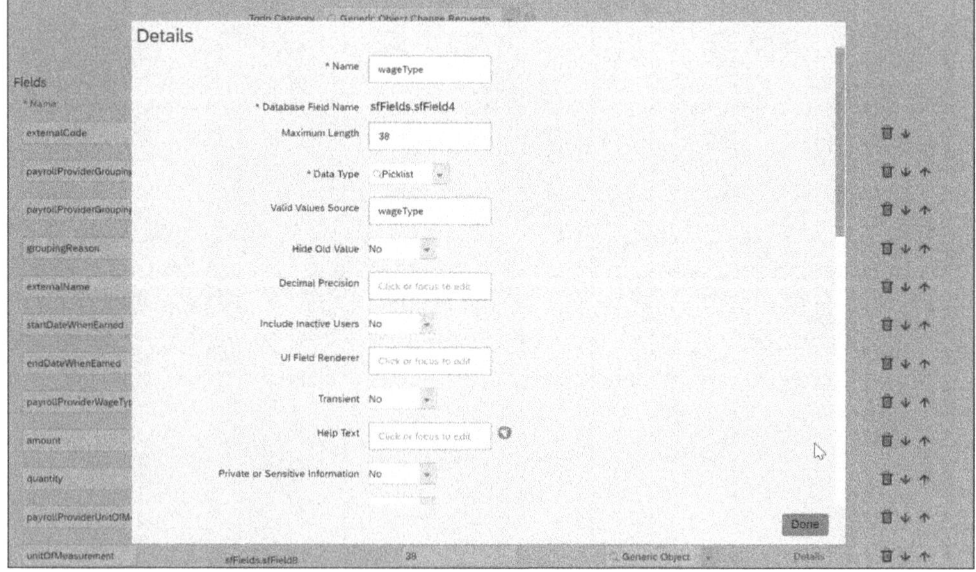

Figure 12.11 Wage Type Detail Picklist Assignment

Now, you'll need to populate the picklists with the correct values by following these steps:

1. Go to the Admin Center and enter "Configure Business Rules" by into the search bar.
   - Click **Create new rule**, which will open the screen shown in Figure 12.8. Select **Basic**.
   - After maintaining the **Rule ID** and **Rule name** fields, choose **Employee Payroll Run Results Items** from the **Base Object** dropdown list.
   - Click **Continue**, which will open the screen shown in Figure 12.12.

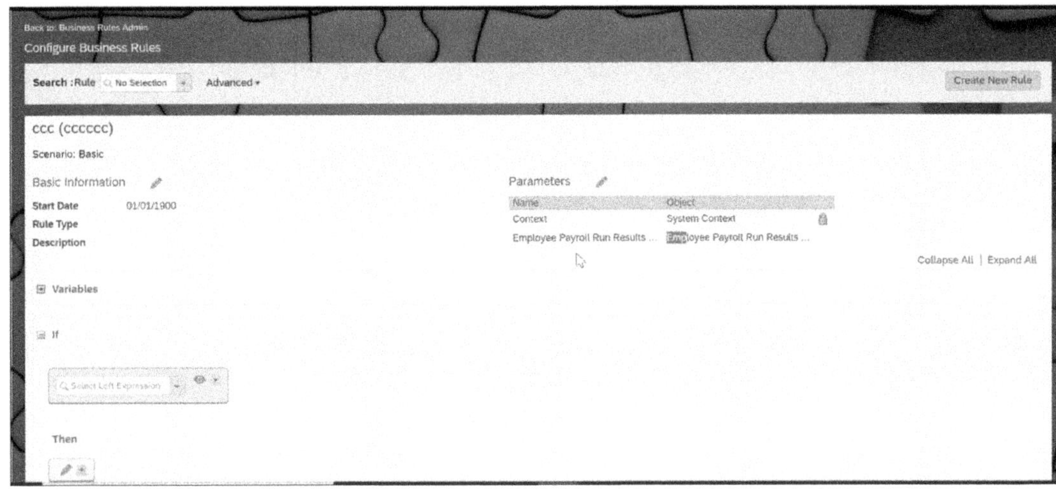

Figure 12.12 Business Rule Details

   - Now, go to the **If** statement and choose **Employee Payroll Run Results – Source: Wage Type**. Check that it is equal to the wage type that you're configuring, as represented in Employee Central Payroll.
   - Go to the **Then** statement and choose the picklist value corresponding to the wage types you desire and save. Then, click on the pencil icon and set the picklist value corresponding to this wage type that you set up in the picklist values earlier. Select **Set** from the first field, select **Wage Type** from the second field, and then choose the corresponding value from the last field.
2. Assign this rule to the Employee Payroll Run Results Items object using the **Configure Object Definition** screen, shown earlier in Figure 12.10, and then save. You can get to this screen by entering "Configure Object Definition" in the search bar and then selecting the object, as shown earlier in Figure 12.6. This time, however, enter "Employee Payroll Run Results Items" instead of "Employee Payroll Run Results" into the second search field.

12.1.5 Creating a Business Rule for Grouping Reason Picklist

To create a business rule for the grouping reason picklist in Employee Central, enter "Configure Object Definitions" in the search box.

Next, select **Object Definition** and **Employee Payroll Run Results Items** from the dropdown lists next to **Search**. Then, choose **Take Action • Make Correction** in the top-right corner.

Select **Details** at the end of the **groupingReason** field row and enter the name of the picklist in the **Valid Values Source** field, as shown in Figure 12.13. Click **Done**.

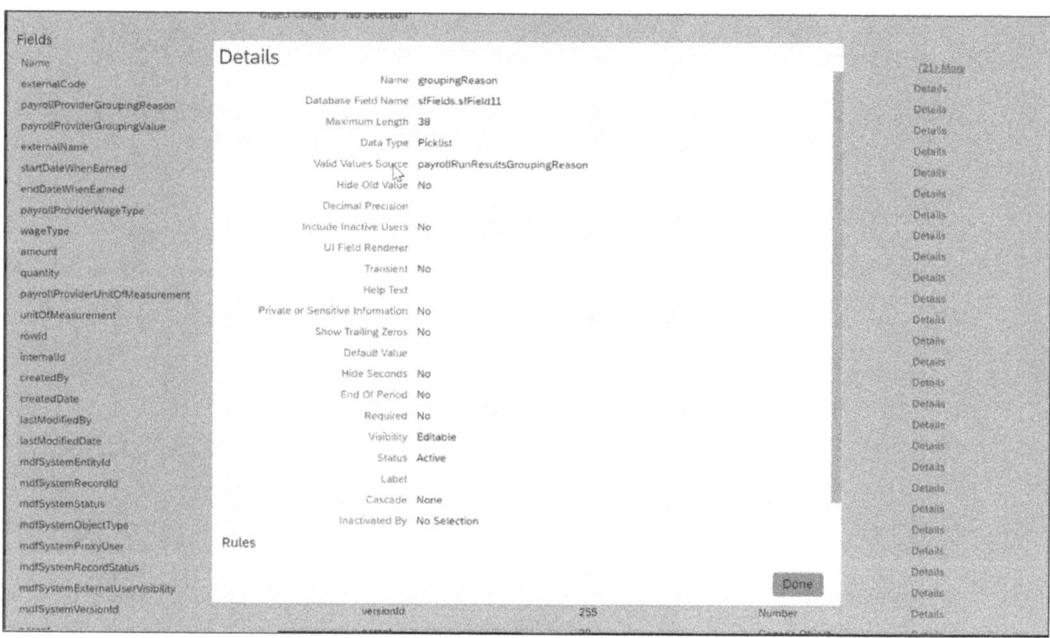

Figure 12.13 Valid Values for Grouping Reason

Now, you'll need to populate the picklists with the correct values by following these steps:

1. Go to the Admin Center and enter "Configure Business Rules" into the search bar.
   - Click **Create new rule**, which will result in the screen shown earlier in Figure 12.8. Select **Basic**.
   - After maintaining the **Rule ID** and **Rule name** fields, choose **Employee Payroll Run Results Item** from the **Base Object** dropdown list.
   - Choose **Continue**, which will open the screen shown earlier in Figure 12.12.
   - Now, go to the **If** statement and choose **Employee Payroll Run Results – Source: Grouping Reason**. Check that it is equal to the grouping reason that you're configuring, as represented in Employee Central Payroll.

  - Go to the **Then** statement and choose the picklist value corresponding to the grouping reasons you desire.
  - Save these settings.

2. Assign this rule to the **Save Rules** section in the Employee Payroll Run Results Item object using the **Configure Object Definition** screen, shown earlier in Figure 12.10, and then save. You can get to this screen by entering "Configure Object Definition" in the search bar and then selecting the object in the same way, as shown earlier in Figure 12.6. However, this time, enter "Employee Payroll Run Results Items" instead of "Employee Payroll Run Results" in the second search field.

12.1.6 Setting Up Payroll Key Figures

Setting up payroll key figures involves configuring the WageTypeAccumulationConfig object. In this process, you can indicate which wage types should be used for Employee Central interfaces.

Start with the Admin Center in Employee Central and enter "Manage Data" into the search bar. Select **WageTypeAccumulationConfig** from the **Create New** dropdown, as shown in Figure 12.14.

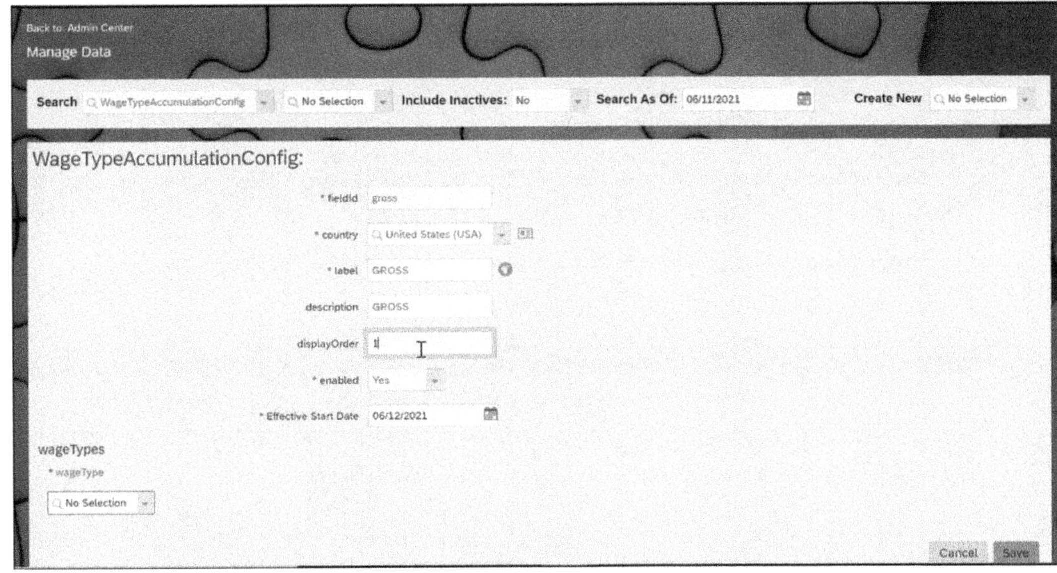

Figure 12.14 Wage Type Accumulation Configuration

On this screen, maintain the information listed in Table 12.1 and then save these settings.

Field Name	Description
fieldID	Two standard fields are already provided and cannot be changed: **Gross** and **Net**. Add any other fields for which you want to create key figures.
displayOrder	This field controls the order in which key figures are displayed.
Description	Text description of the key figure. This information will not be displayed to the user.
Label	The text that will be displayed to the user.
Enabled	This activation field allows you to deactivate key figures that you no longer want to use but want to retain for the future.
externalCode	Generated after saving.
EffectiveStartDate	Date when the key figure comes into effect.
wageTypes	Choose the wage types that you want to combine for this key figure.

Table 12.1 Fields in the Wage Type Accumulation Object

12.2 Setting Up Specific Permissions

The permissions required for the displaying the various kinds of pay statements were discussed in Chapter 11, Section 11.2.

The second scenario for the Employee Central Advanced Reporting tool equires the following payroll integration permissions to be turned on:

- **Employee Payroll Run Results (View Current and View History)**
- **Employee Payroll Run Result Items (View)**

The permissions are turned on using the **Manage Permission Roles** screen in Employee Central. You can access this screen by entering "Manage Permission Roles" into the search bar. For details on setting permissions, see the Employee Central Master guide in the SAP Documentation portal at *http://s-prs.co/v538606*.

12.3 Replicating Payroll Results from Employee Central Payroll to Employee Central

The final step after the configuration in Employee Central is to set up the process for replication from Employee Central Payroll. To continuously receive payroll-related key figures on time, the job for replication must be scheduled to synchronize with the pay processing cycle.

Run the RP_HRSFEC_ExportPayrollResults report using Transaction SE38, as shown in Figure 12.15.

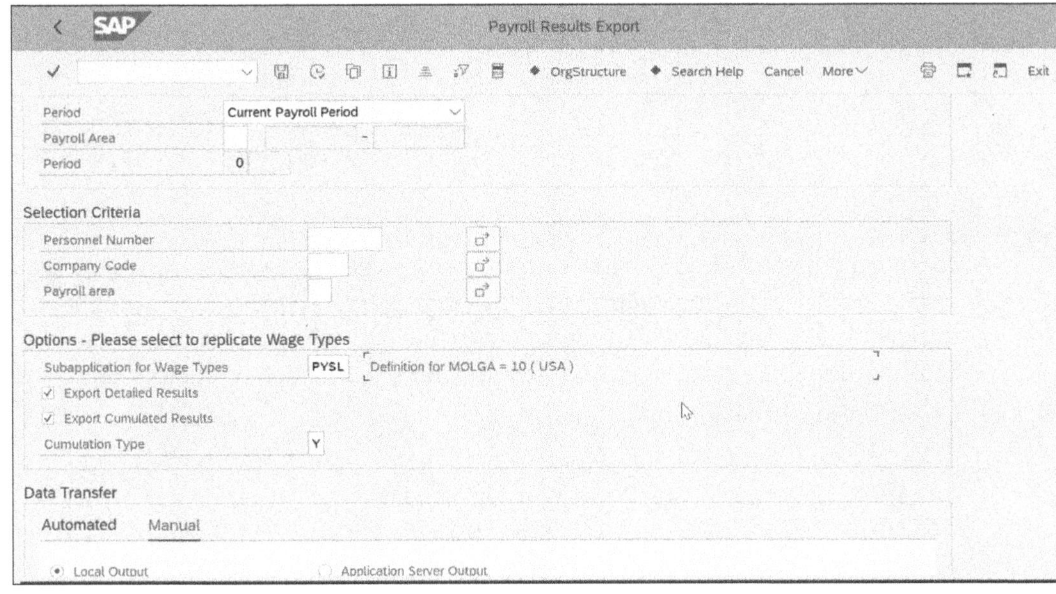

Figure 12.15 Report RP_HRSFEC_ExportPayrollResults

On this screen, you'll fill out the following fields:

- **Period**: The period selection is to control which period should be synchronized (**Current Payroll Period** is the default, but you can select other options).
- **Selection Criteria**: Selection criteria are provided to choose the organizational units that should be synchronized.
- **Options**: The **Options** group is used to choose the subapplication scope, where you identify the wage types to be replicated. Options include cumulated views, detailed views, or both.
- **Data Transfer**: Two options are available: **Automatic** and **Manual**. You would only choose **Manual** if you want to export the results to a file. You can use the **Import Data** feature in Employee Central to import these key figures manually from these files. Otherwise, the automatic setting will move the results automatically to the Metadata Framework (MDF) tables for key figures in Employee Central that you configured earlier in this chapter.

12.4 Limitations for Using Payroll Results

In certain scenarios, payroll results cannot be replicated in the current version of Employee Central. Watch for the following scenarios:

- Voids (when a pay result is voided for an employee)
- Reversals (when a payment is reversed)
- Off-cycles
- Wage types that are numeric and have no amount
- Wage types that are indirectly assigned with variable values

12.5 Summary

In this chapter, we became familiar with the concept of reverse replication—bringing payroll results back to Employee Central. This replication will prepare you with payroll results ready for reporting in various situations, including Employee Central Advanced Reporting. Bringing employee master data and pay results together unlocks new possibilities for insightful reporting.

Chapter 13
Setting Up Single Sign-On and Logout

While using Employee Central Payroll, keep in mind the fact that this environment is a different environment than your Employee Central system, thus creating multiple logon points that require different user IDs and/or passwords. Resolving this multiple-point logon issue is what we'll discuss in this chapter.

Single sign-on (SSO) is immensely popular in many business environments because of the ease with which users can navigate from one system to another. In most offices, users only need to remember one user ID and password—the logon for the corporate LAN, input when the user first logs on to the office computer. From that point on, the other applications can derive authentication for the user through effective deployment of commercial software.

In our case, we're dealing with two systems: Employee Central and Employee Central Payroll. These two systems both have the ability to authenticate users and, if the user has logged on to the Employee Central system, you'll also need to provide the user the required access to Employee Central Payroll without asking the user to log on again. Note, however, that the reverse is not true, if an employee logs on to Employee Central Payroll, they are not also immediately logged on to Employee Central.

This chapter discusses how SSO can be achieved since this capability is fully configurable. We'll review Security Assertion Markup Language (SAML) and how this language is used to deliver this functionality. We'll cover the configuration required in both systems—Employee Central and Employee Central Payroll. Then, we'll look at user setup and identity federation, which is how we tell each system in advance how to authenticate a request from the other system. We'll also look at OAuth 2.0—the standard that you must configure to allow the reverse communication from Employee Central Payroll to Employee Central for certain business scenarios. This chapter does not provide the SSO details for connecting the office LAN to Employee Central, since this link is not possible with just SAP software—configuration using other commercial software will be required.

13.1 Single Sign-On Basics

Before we dive into SSO to improve payroll processing, you must be familiar with the following terms:

- Identity provider (IdP): This system checks the identity and authenticates the user. In our case, the IdP is the Employee Central system.
- Service provider (SP): This system provides the requested service. The SP can be either the Employee Central Payroll system or the Employee Central system. Since Employee Central is co-located with the IdP, services requested from Employee Central do not need SSO in this scenario. So, when we talk about an SP in this context, we're mostly referring to the Employee Central Payroll system.
- SAML: This language provides standards-based logon and logout services. SAML 1.1 was replaced by SAML 2.0 in the early 2000s, which has become the standard has been for SSO, accepted worldwide for exchanging authentication information between software as a service (SaaS) systems.

As shown in Figure 13.1, the SSO process all starts with the user initiating a request from the browser. Let's say the user has already logged on to Employee Central and is now clicking the link to request infotype 0014 (Recurring Earnings and Deductions) for an employee, as shown in Figure 13.2. The click sends a request to the IdP. The IdP provides authentication and logon credentials for the browser to access the SP (i.e., Employee Central Payroll). This step starts a session in Employee Central Payroll and initiates the service that the user is looking for.

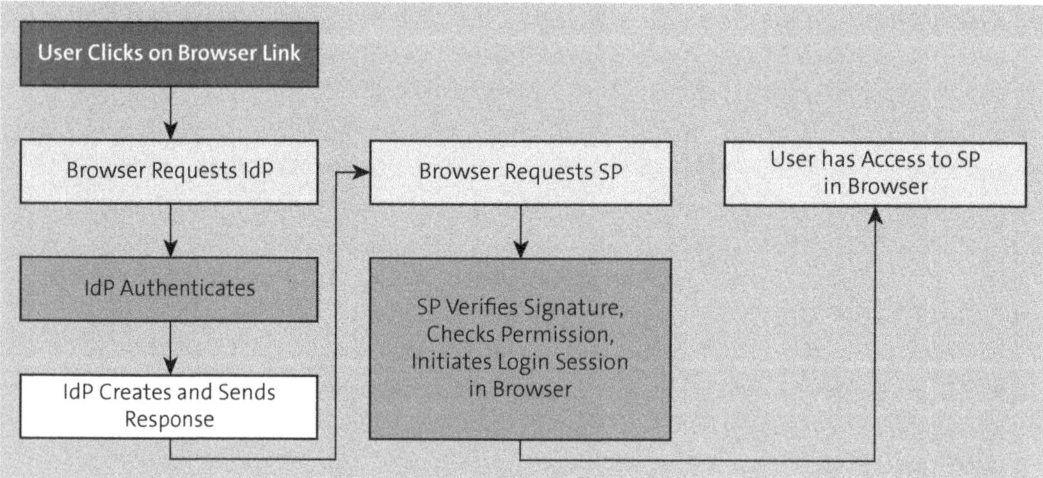

Figure 13.1 Simple SSO Scenario: Accessing the SP (Employee Central Payroll) from the IdP (Employee Central)

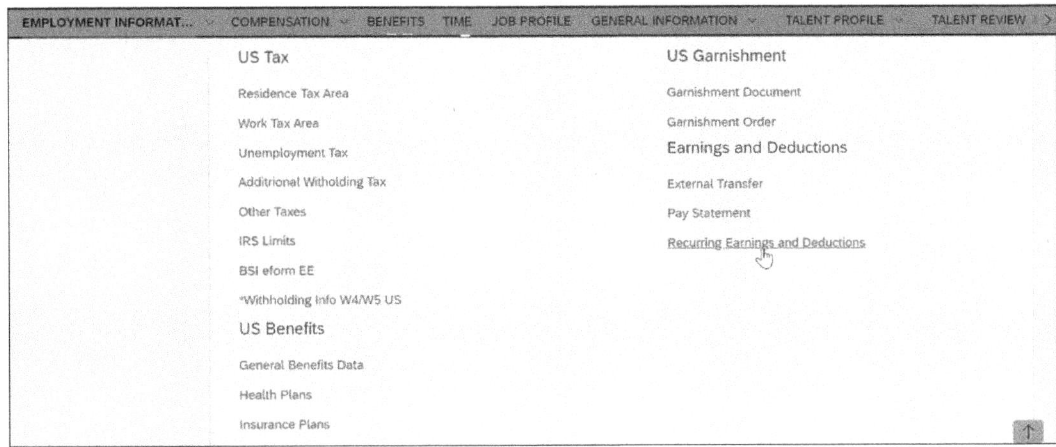

Figure 13.2 Recurring Earnings and Deductions Link

We'll discuss how to prepare your systems for SAML 2.0, both Employee Central Payroll and Employee Central. We'll configure the SP, discuss the setup of user-related attributes, and describe the steps for establishing identity federation. To use SAML 2.0, you'll interact with Transaction SAML2.

13.2 Preparing for SAML 2.0

To prepare for SAML 2.0, preparation is required in two places:

- Employee Central Payroll
- Employee Central

13.2.1 Employee Central Payroll

To prepare your Employee Central Payroll system for SAML 2.0, you must create the local provider for the SAML protocol, as follows:

1. Start Transaction SAML2.
2. Click on the **Enable SAML 2.0 Support** button.
3. Enter a system name in the **Provider Name** field.
4. Under the **Service Provider Settings** tab, select **Automatic** from the **Selection Mode** dropdown list under the **Identity Provider Discovery: Common Domain Cookie (CDC)** heading.
5. Click **Save** to enable the SP, as shown in Figure 13.3.

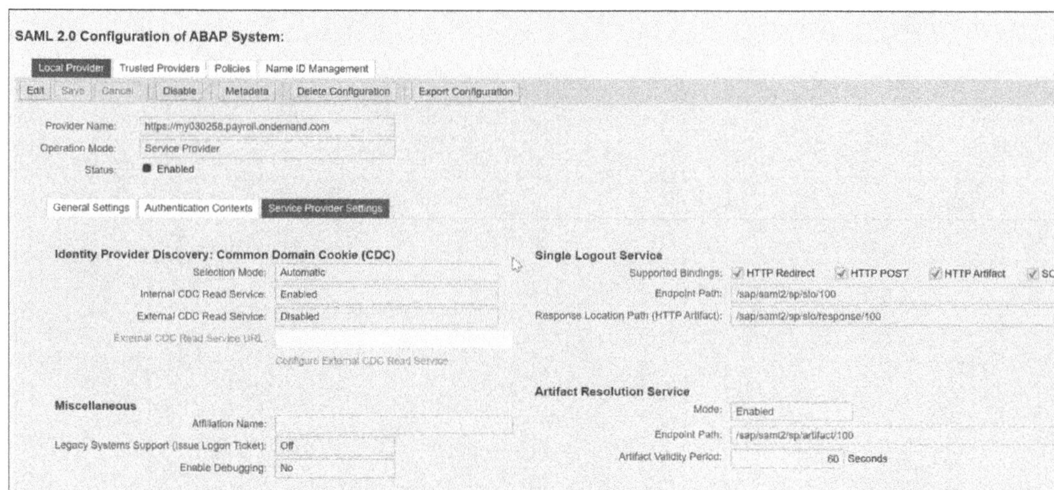

Figure 13.3 SP Settings for Transaction SAML2

Next, you'll need to activate SAML 2.0, by following these steps:

1. Start Transaction SICF and click **Execute.**
2. Activate the following services, as shown in Figure 13.4, by right-clicking and selecting **Activate Service**:
   - **/default_host/sap/public/bc/sec/saml2**
   - **/default_host/sap/public/bc/sec/cdc_ext_service**

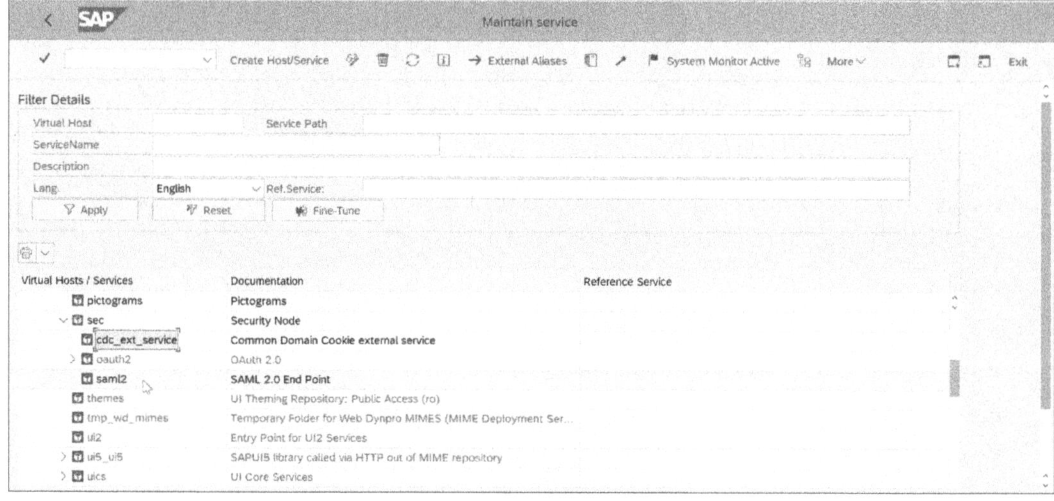

Figure 13.4 Activating SAML 2.0 Services

Finally, run Transaction SICF_SESSIONS and confirm **Security Sessions Management** is enabled by checking that a green square is displayed under the **Current Status** field.

13.2.2 Employee Central

To prepare your Employee Central system for SAML 2.0, follow these steps:

1. Download the metadata by starting Transaction SAML2 in Employee Central Payroll again. Open the **Local Provider** tab and click on the **Metadata** button. As shown in Figure 13.5, in the popup window, select all the entities to be included in the metadata and then click the **Download Metadata** button. In this example, we've selected all entity descriptors: **Service Provider**, **Application Service Provider**, and **Security Token Service**. An example file is shown in Figure 13.6.

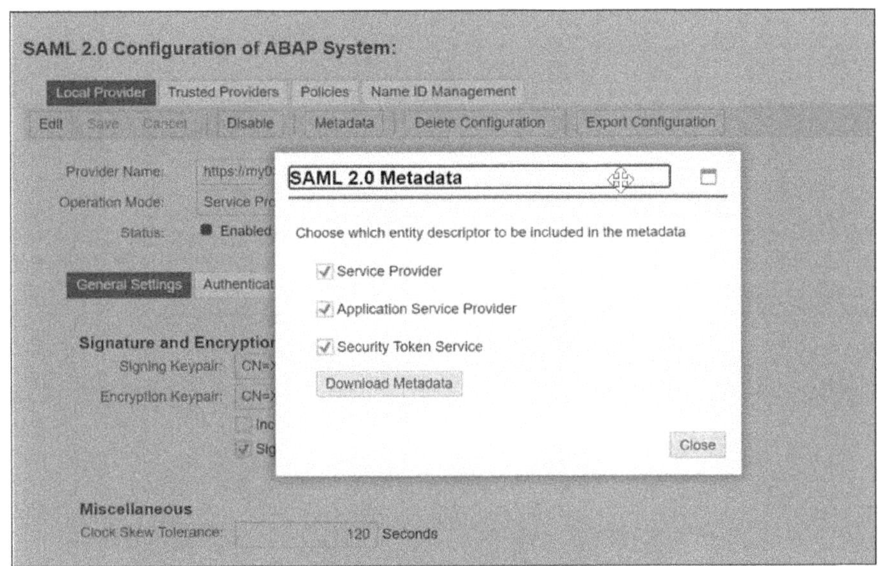

Figure 13.5 Metadata Download in the SAML 2.0 Local Provider Tab


```
<?xml version="1.0"?>
- <m:EntityDescriptor xmlns:m="urn:oasis:names:tc:SAML:2.0:metadata" ID="S00163eb9-15d8-1edb-9fbc-3dddaf37ecf5" cacheDuration="P16Y9M26DT1H30M"
  validUntil="2038-01-01T00:00:01Z" entityID="https://            payroll.ondemand.com">
  - <ds:Signature xmlns:ds="http://www.w3.org/2000/09/xmldsig#">
    - <ds:SignedInfo>
        <ds:CanonicalizationMethod Algorithm="http://www.w3.org/2001/10/xml-exc-c14n#"/>
        <ds:SignatureMethod Algorithm="http://www.w3.org/2000/09/xmldsig#rsa-sha1"/>
      - <ds:Reference URI="#S00163eb9-15d8-1edb-9fbc-3dddaf37ecf5">
        - <ds:Transforms>
            <ds:Transform Algorithm="http://www.w3.org/2000/09/xmldsig#enveloped-signature"/>
            <ds:Transform Algorithm="http://www.w3.org/2001/10/xml-exc-c14n#"/>
          </ds:Transforms>
          <ds:DigestMethod Algorithm="http://www.w3.org/2000/09/xmldsig#sha1"/>
          <ds:DigestValue>z9c9hdSTsX/oqaKNvtB5WWaYihE=</ds:DigestValue>
        </ds:Reference>
    </ds:SignedInfo>
    <ds:SignatureValue>lRUH0yYALGDqN+AwlqfzceXvKwAvxM9tJUEOsvNtsJ5MsrJrFTrPUGA7RS5PB8N0c+waG7xqn64u
      7LsImZ6B7vGy/cghRhVrnjh+bPRyew/2JN1cezN6J4LFvN/v1mahtOlxZdRDaIaBsK0TNQVoL50n
      LLSMCIt/dZOre+U8ubCcCpiSV9PKKPp1Jmner8tEWuhVR7tCgkwEpbf8Ck1l4MOYb7M/ZAlRdiOS
      0CghKP59SV13VAT4ki58/4ClQeZDneQx6ZLFqoiSGAmKSidim3atmeoytUHFsYBHYiQ/+HXOmuWB
      MroMtHuUzxJqClfgRyl4C+thu4i0+wILo9vQiA==</ds:SignatureValue>
  - <ds:KeyInfo>
    - <ds:X509Data>
        <ds:X509Certificate>MIIDWzCCAkMCCAogGAgnFDUBMA0GCSqGSIb3DQEBCwUAMHAxCzAJBgNVBAYTAkRFMRwwGgYDVQQK
          ExNTQVAgVHJ1c3QgQ29tbXVuaXR5MRMwEQYDVQQLEwpTQVAgV2ViIEFTMRQwEgYDVQQLEwtJMDAy
          MDg5Mjc4NjEYMBYGA1UEAwwPWEVYX1NTRkFfUzJTVlBTMB4XDTE4MDgyNzE0MzUwMVoXDTM4MDEw
          MTAwMDAwMVowcDELMAkGA1UEBhMCREUxHDAaBgNVBAoTE1NBUCBUcnVzdCBDb21tdW5pdHkxEzAR
          BgNVBAsTClNBUCBXZWIgQVMxFDASBgNVBAsTC0kwMDIwODkyNzg2MRgwFgYDVQQDDA9YRVhfU1NG
          QV9TMlNWUFMwggEiMA0GCSqGSIb3DQEBAQUAA4IBDwAwggEKAoIBAQCcjn8eJTu0KnYcEpnoMGGa
          Q8Lj9Vrr41WetaFVS2Ts/IV13WBKEj9C51KVapHBFEXoDnWrxDdIaS03gaN0WI4eDGunUHFvkThA
          WNSvJ1EcufBNHVaiUkM8iT4Z3aoRtzW1VqOsPV9Rqx9ZEBkdZmvoE/yKz5I74HllrFZacSFwQCO6
          Su4Z7Hj+K0PPma09o1On7CG6UIn2YS0C3kYsGbBrNfe8f5MY6hMlvbiVR8FXef9dxz+LjLloHWTo
          XNxqjYHOeMM/uWbbS6LzQ4zwbINZyVkq3oSYHcayWi+but6pBeXi3V70i9Ddk5VIlg86FEPypbEN
```


Figure 13.6 Metadata File

2. Sign on to provisioning and select your company ID.
3. Under **Service Provider Settings**, look for the group "Authorized SP Assertion Consumer Settings" by clicking the find icon.
4. Then, from the metadata file, enter the URLs highlighted in the screen shown in Figure 13.7, located under the **Assertion Consumer Service** column and the **Logout URL** column on the provisioning screen shown in Figure 13.8. Note that you'll need assistance from your SAP partner to get this configuration set up in provisioning.

Figure 13.7 URLs for Single Logout and Assertion Services

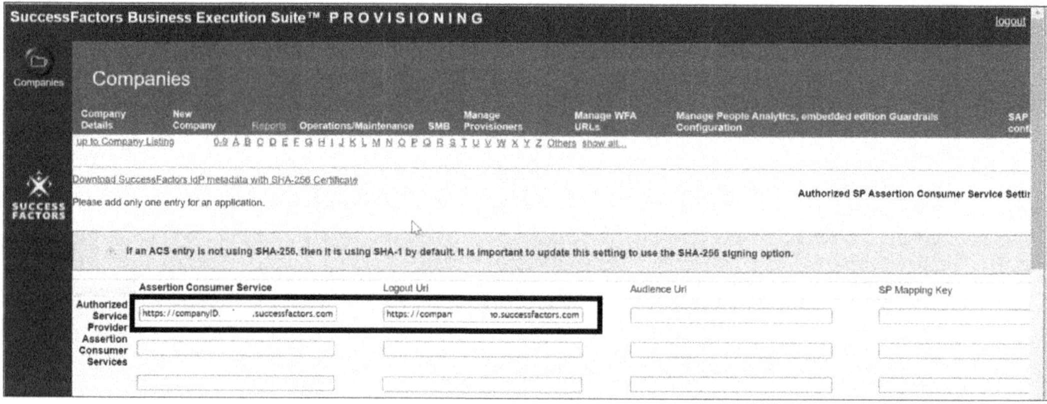

Figure 13.8 Assertion and Logout URLs in Provisioning

13.3 Configuring the Employee Central Payroll Service Provider for SAML 2.0

To begin the configuration of Employee Central Payroll as the SP for SAML 2.0, follow these steps:

1. Start Transaction SAML2 in Employee Central Payroll.
2. Select the **Trusted Providers** tab and upload the metadata that you downloaded earlier by clicking on **Add • Upload Metadata File**, as shown in Figure 13.9.

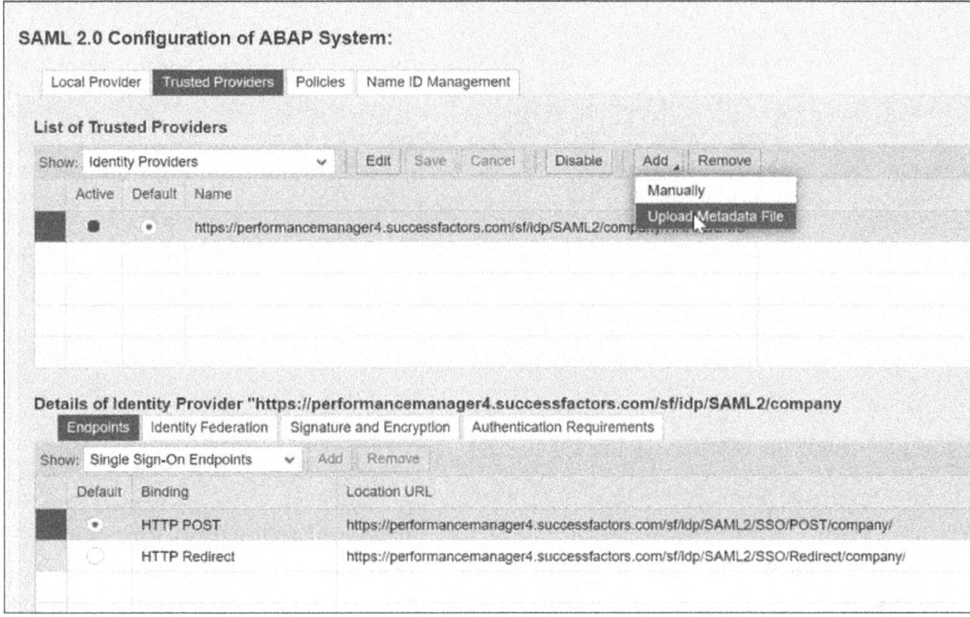

Figure 13.9 Uploading Metadata

3. To get the metadata from the Employee Central system that is be loaded in the previous step, enter this link into a new web browser: *https://<server URL>/idp/samlmetadata?company=<companyID>*.

   In this context, the *<server URL>* is your SAP SuccessFactors URL root (e.g., *performancemanager4.successfactors.com*). The metadata will automatically download. Save this file in a folder.

4. Click on the **Add from Metadata** option in step 1, choose the file we downloaded, click **Open**, and then click **Next**, as shown in Figure 13.10.

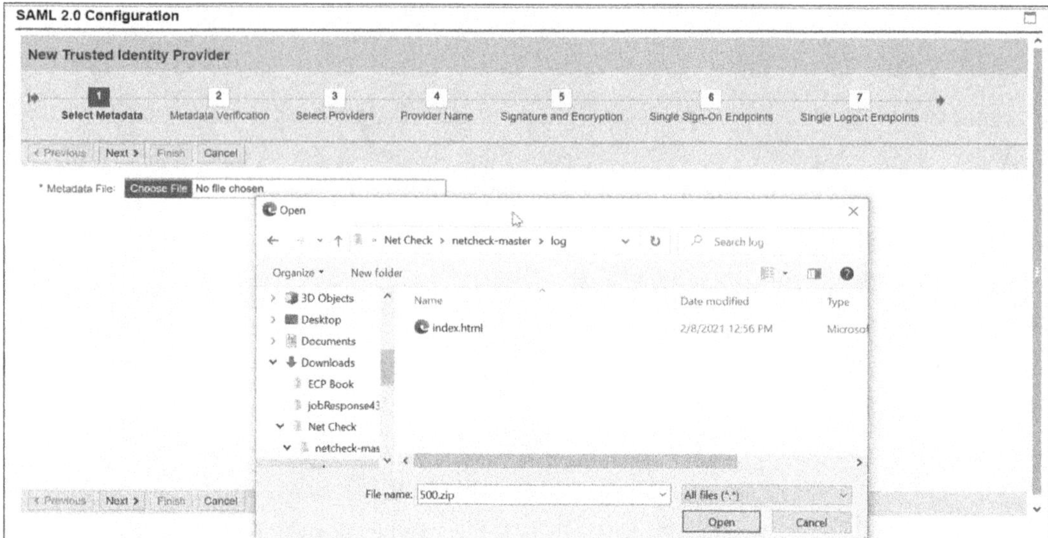

Figure 13.10 Loading the Metadata File for Trusted Provider Settings in Employee Central Payroll

5. Define a name for trusted provider in the **Name** field.
6. Under the **Signature Encryption** tab, click **Next**.
7. Go to the **Endpoints** tab, as shown in Figure 13.9.
8. In the **Binding** column, choose **HTTP Post**.
9. Go to the **Signature and Encryption** tab and click **Next**.
10. Go to the **Authentication Requirements** tab. In the **Authentication Response** area, select **Application URL** from the **Assertion Consumer Service** dropdown list and select the **HTTP POST** from the **Binding** dropdown list, as shown in Figure 13.11.

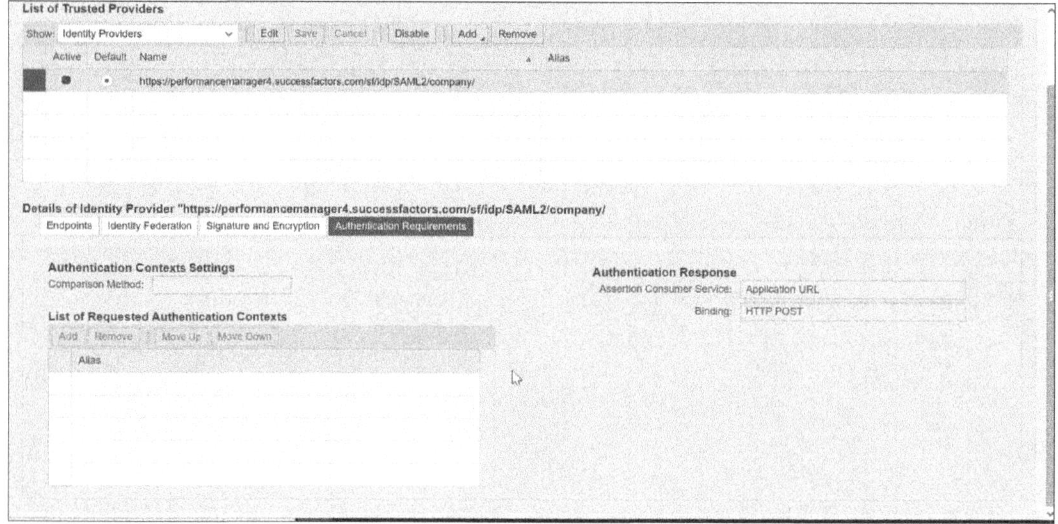

Figure 13.11 Authentication Requirements

11. Click **Finish**.
12. Under the **Identity Federation** tab, click the **Edit** button and then the **Add** button. A list of **Supported Name ID Formats** will be displayed. Select **Unspecified**.
13. Set **Allow Identity Provider to Create NameID** to **No**, as shown at the bottom of Figure 13.12.

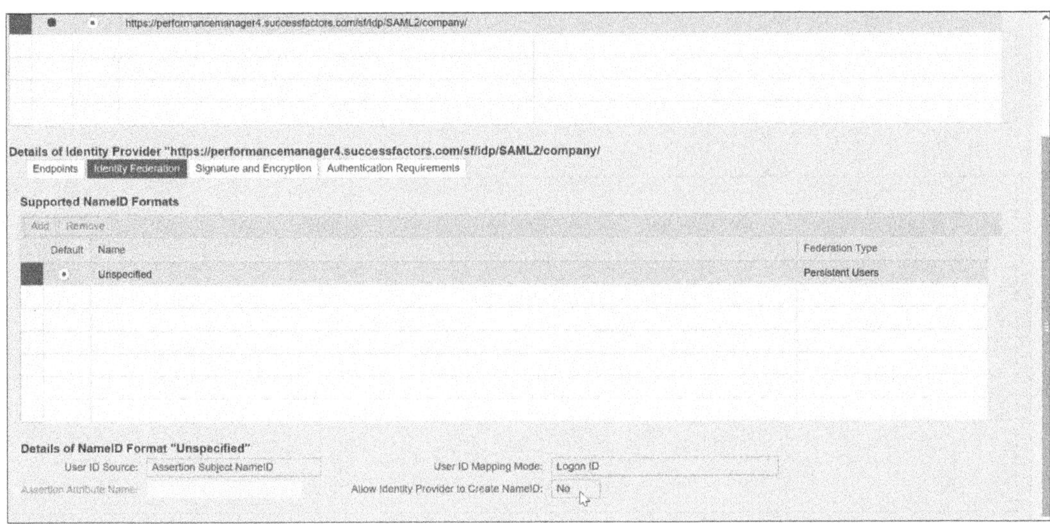

Figure 13.12 Identity Federation

14. Save to enable the trusted provider.
15. Now, under the **Local Provider** tab, turn on the automatic IdP selection under **Service Provider Settings** by selecting **Automatic** from the **Selection Mode** dropdown list, as shown in Figure 13.13.

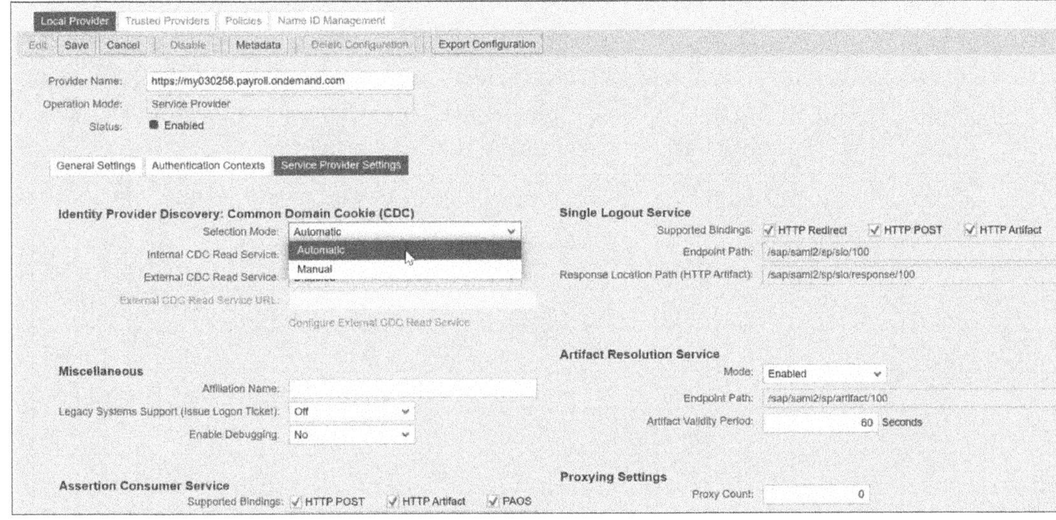

Figure 13.13 Automatic IdP Selection

13.4 User Mapping and Identification

The IdP and SP need a common identifier to uniquely identify the person to be authenticated. This authentication process requires that the user ID be the same in both the Employee Central and Employee Central Payroll systems. A session is initiated by the assertion consumer service in Employee Central for the user with this user ID, which is taken from the SAML 2.0 assertion.

Users are identified in the Employee Central and Employee Central Payroll systems using different technologies and authentication methods. We'll examine these methods briefly in this section.

In the Employee Central Payroll system, users are identified by user ID or by an alias in Transaction SU01, as shown in Figure 13.14.

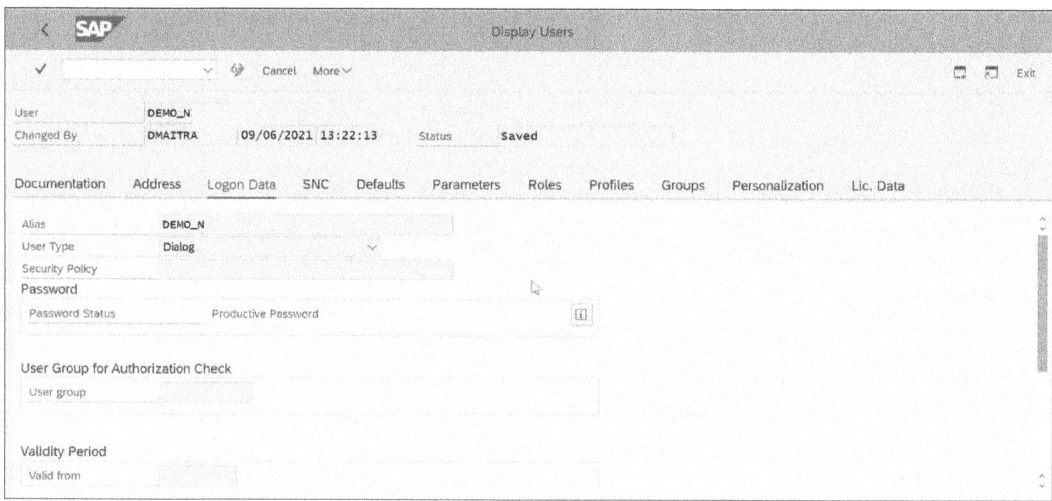

Figure 13.14 Transaction SU01 to Set Up Users in Employee Central Payroll

In contrast, users in Employee Central are identified by user ID, which is typically not visible directly in the system. To find a user ID in Employee Central, from the Admin Center, enter "Employee Export" into search bar, as shown in Figure 13.15.

Click on the **Export User File** button and the report will be produced automatically, as shown in Figure 13.16.

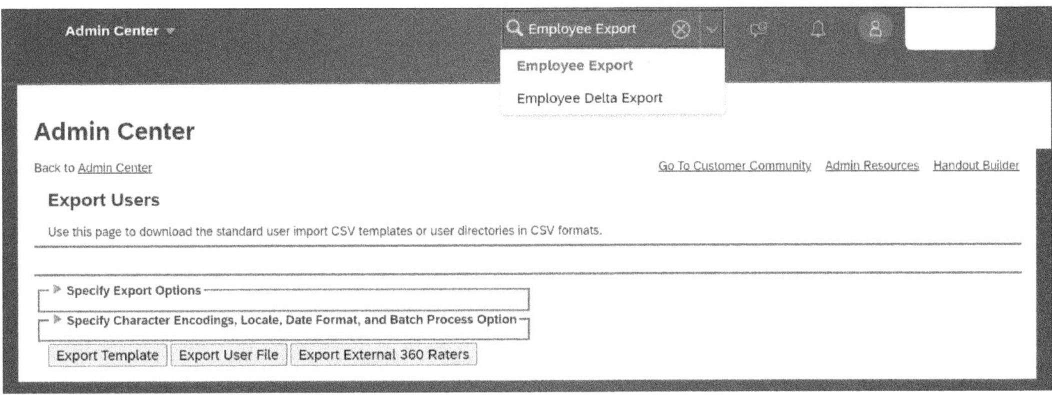

Figure 13.15 User Data Export Screen

STATUS	USERID	USERNAM	FIRSTNAM	NICKNAMI	MI	LASTNAMI	SUFFIX	TITLE	GENDER	EMAIL	MANAGEF	HR	DEPARTM	JOBCODE	DIVISION	LOCATION	TIMEZONE	HIREDATE	EN
STATUS	USERID	Username	First Name	Nickname	Middle Na	Last Name	Suffix	Title	null	Email	Manager	Human Re	Departme	Job Code	Division	Location	Time Zone	Hire Date	En
active	100	100				Director		Sr. Manag	M		99	110	Campus R	HRFF Sr. N	Recruitme	Orlando (I	US/Eastern		
active	100001	100001	WD EC			Test User1		Cloning In			NO_MAN/	NO_HR	N/A	Accountar	N/A	Minneapo	US/Easter	1/6/2020	
active	100002	100002	WD EC			Test User1		Cloning In		WD_EC_T	NO_MAN/	NO_HR	N/A	Accountar	N/A	Minneapo	US/Easter	1/6/2020	
active	100003	100003	WD EC			Test User3		Cloning In		WD_EC_T	NO_MAN/	NO_HR	N/A	Accountar	N/A	Minneapo	US/Easter	2/4/2021	

Figure 13.16 User Data Export Report

If the user IDs do not match, change the user ID in the Employee Central Payroll system so that it matches the Employee Central user ID. An alternative approach is to set up an alias in Employee Central Payroll, through identity federation, which we'll discuss next.

13.5 Establishing Identity Federation

So, how does the Employee Central Payroll system know that a requesting user is the same person that has access to the requested information since the requestor identifier is coming from the Employee Central system? (Similar questions may arise in the reverse situation if the requested service is from the Employee Central system and the requestor is from the Employee Central Payroll system.)

Identity federation refers to the "brotherhood" of technical user IDs in two systems. Two ways exist to establish identity federation, as follows:

1. In the first approach, user IDs are set up to be identical. This approach requires the ICF nodes HRPAO_PAOM_MASTERDATA, PAYSLIP, and NWBC to be configured to use the logon of standard SAP users, as shown in Figure 13.17. Use Transaction SICF, enter these nodes in the **Service Name** field and click on **Apply** to access the node directly. (Return to Section 13.2.1 for instructions for using Transaction SICF.) Once you're in the service, click on the **Logon Data** tab and scroll to the **Authentication** section. Select the **Standard SAP User** radio button, as shown in Figure 13.17. Repeat these steps for the other two services.

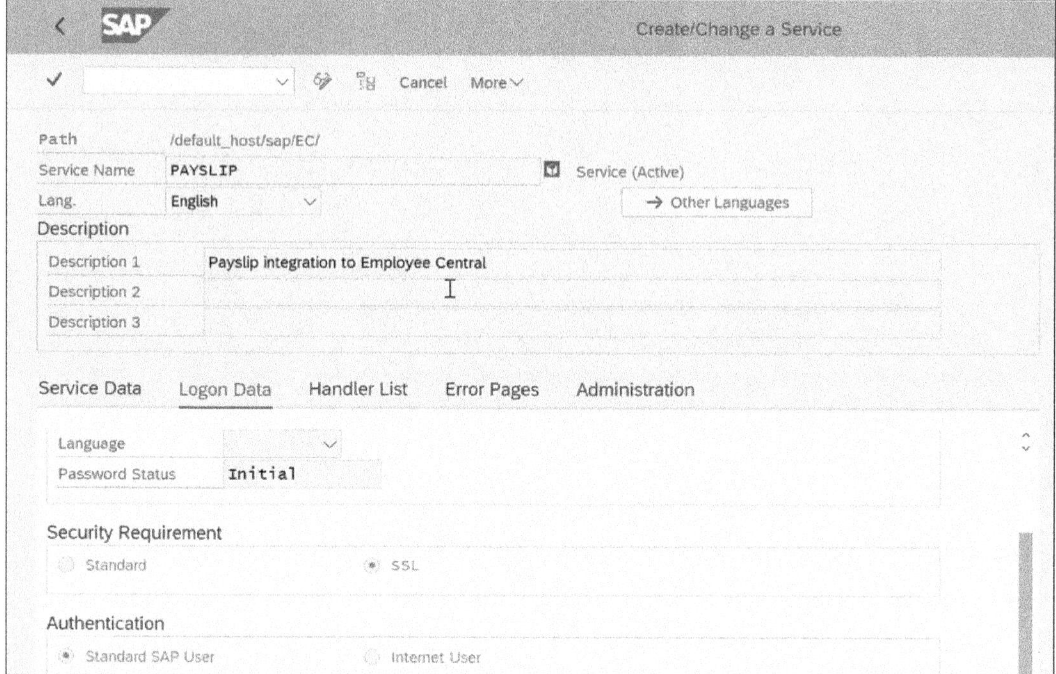

Figure 13.17 Logging On as a Standard SAP User or an Internet User for the PAYSLIP Node

2. The second method involves mapping the Employee Central user ID to an alias. This approach requires the ICF nodes HRPAO_PAOM_MASTERDATA, PAYSLIP, and NWBC to be configured in a certain way to use the logon of an internet user, as shown in Figure 13.17. This configuration involves the same steps as before, except that the **Internet User** radio button should be activated, instead of the **Standard SAP User** radio button.

13.6 System Settings

Some settings are necessary in browsers and mobile apps to ensure that SSO works smoothly in these applications.

13.6.1 Browser

A user will need to add the SAP SuccessFactors and SAP SuccessFactors Employee Central Payroll URLs to the trusted sites settings in web browser options. How you manage these options depends on the browser you use (Google Chrome, Microsoft Edge, etc.).

If your browser includes a feature that is enabled to always submit a user certificate, this feature needs to be turned off. You should also allow popups, which are needed for the NWBC logout function.

13.6.2 Mobile Operating System

To enable the use of mobile smartphones for viewing pay slips, use Transaction RZ10 to set the profile parameters enabling the Transport Security app. Select the **Extended maintenance** radio button and click on **Change**, as shown in Figure 13.18.

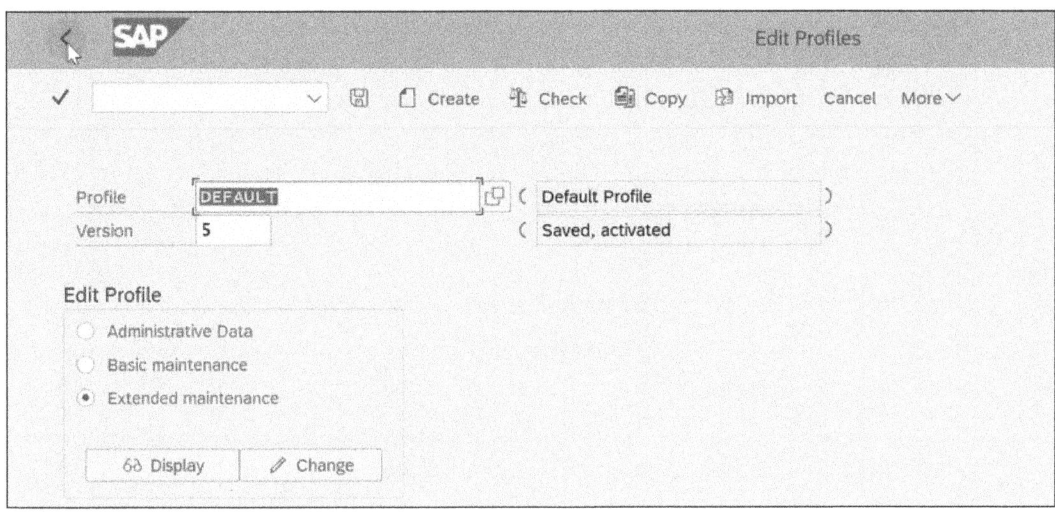

Figure 13.18 Transaction RZ10

Enter the following details, as shown in Figure 13.19:

- For parameter **ssl/ciphersuites**, enter "135:PFS:HIGH::EC_P256:EC_HIGH."
- For parameter **ssl/client_ciphersuites**, enter "150:PFS:HIGH::EC_P256:EC_HIGH."

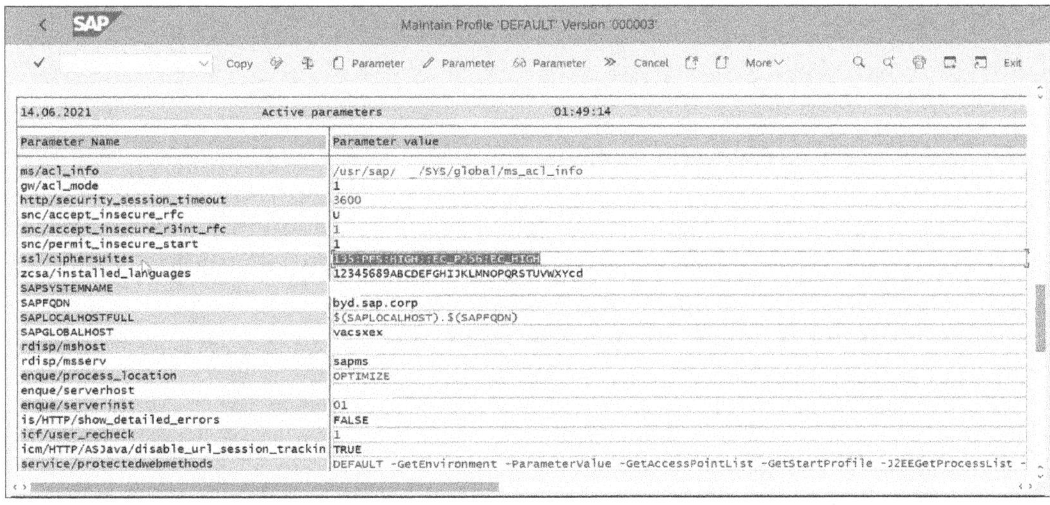

Figure 13.19 Mobile OS Profile Setting for Pay Slips

13.7 OAuth 2.0

OAuth is another protocol like SAML 2.0 that SAP uses to connect Employee Central Payroll with Employee Central. While SAML 2.0 is great for SSO, OAuth 2.0 has better performance for data transfers.

Enabling OAuth 2.0 is a prerequisite for the direct pay statements, as described in Chapter 11, Section 11.2. In a direct pay statement scenario, Employee Central Payroll reaches out to Employee Central to determine if a user has permission to view the relevant periods and payroll key figures. Periods and key figures are necessary to drive the direct pay statement, which is indexed by these fields. In this section, we'll have a brief discussion on how to set up OAuth 2.0. These instructions must be followed in sequence.

13.7.1 Initial Settings in Employee Central Payroll

Using Transaction SPRO, complete the following steps to perform the initial settings in Employee Central Payroll:

1. Activate SAP Gateway by following the menu path **SAP NetWeaver • SAP Gateway • OData Channel • Configuration • Activate or Deactivate SAP Gateway.**
2. Create an SAP system alias called "LOCAL" by following the menu path **SAP NetWeaver • SAP Gateway • OData Channel • Configuration • Connection Settings • SAP Gateway to SAP system • Manage SAP System Aliases.** As shown in Figure 13.20, click on **New Entries** and select the **Local SAP** checkbox. Then, click the **Save** icon.

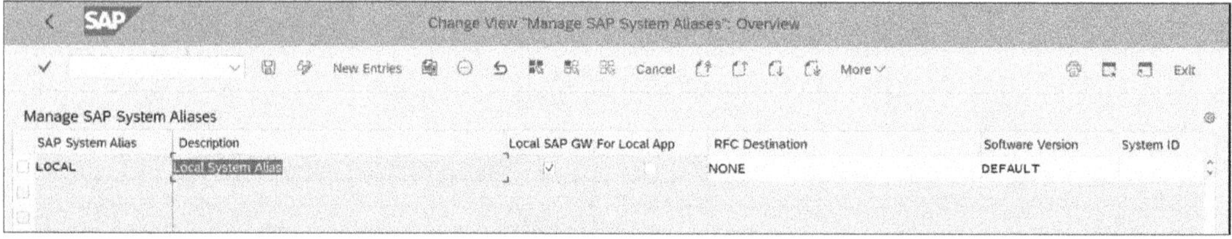

Figure 13.20 Managing SAP System Aliases

3. Activate corresponding service by following the menu path **SAP NetWeaver • SAP Gateway • OData Channel • Administration • General Settings • Activate and Maintain Services**, as shown in Figure 13.21. Click on the service and, from the **ICF Node** dropdown list (the pencil icon), select **Activate**. The service you'll need for direct pay statements is called HRSFEC_PAY_OVERVIEW_SRV.

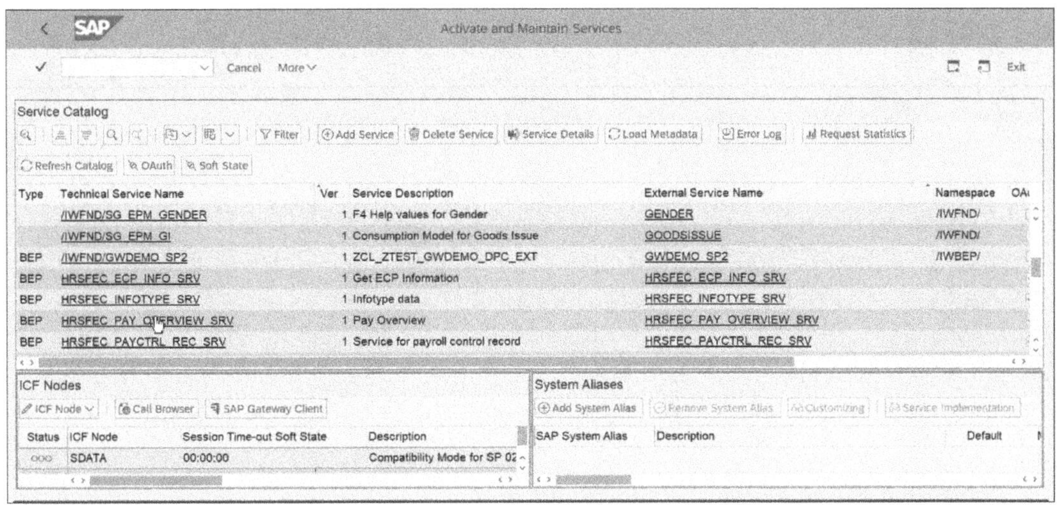

Figure 13.21 Activating Services

4. Assign the LOCAL alias to this service by following the menu path **SAP NetWeaver • SAP Gateway • OData Channel • Administration • General Settings • Assign SAP System Aliases to OData Service.** As shown in Figure 13.22, select the service, click on **Add System Alias**, and choose **LOCAL**.

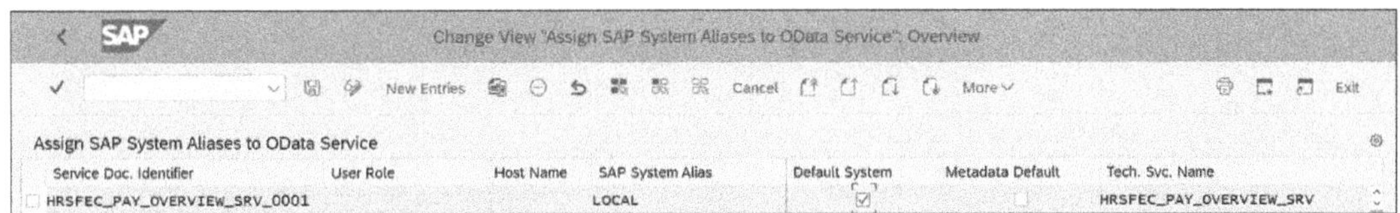

Figure 13.22 Assigning the LOCAL Alias to a Service

13.7.2 Setting Up X.509 Keys in Employee Central

An X.509 key is a certificate that is widely accepted by international associations for authenticating communications between systems. This certificate is generated from Employee Central and loaded into Employee Central Payroll to establish trust between the two systems.

In Employee Central, go to the Security Center. Click on **X509 Certificates** and then select **OAuthX509Key**.

Select **Add** and then choose **New X509 Key**. In the **Configuration Name** field, as shown in Figure 13.23, enter the name of your certificate, without any spaces. Generate and save the key by clicking on the **Download** button. Then, download the key, which you'll need in the next step.

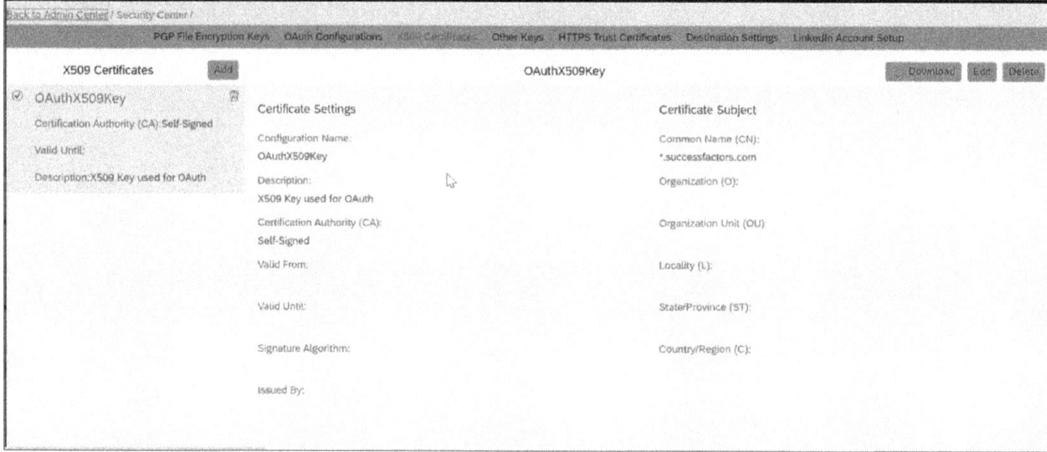

Figure 13.23 X.509 Key

13.7.3 Configuring the OAuth Identity Provider in Employee Central Payroll

To configure the OAuth IdP, follow these steps:

1. Use Transaction SAML2.
2. Go to the **Trusted Providers** tab, as shown in Figure 13.24.

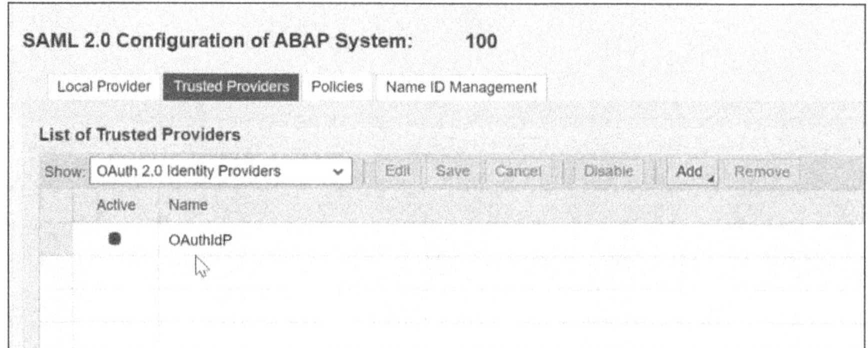

Figure 13.24 OAuth IdP

3. Select **OAuth 2 Identity Providers** from the **Show** dropdown menu.
4. Select **Add • Manually** and provide the X.509 file we downloaded in Section 13.7.2.
5. Choose **Edit** in the newly created IdP. Add the **NameID** format as **Unspecified**.
6. In the **Details** section of the **Name Id Format Unspecified**, select the following:
   - **Assertion Subject Name ID** from the **User ID Source** dropdown list
   - **Logon Alias** from the **User ID Mapping Mode** dropdown list
7. Save and enable.

13.7.4 Creating Service Users in Employee Central Payroll

Service users are used when retrieving configuration information from the system for specific services. To create service users, follow these steps:

1. Open Transaction SU01 and navigate to the **Address** view, as shown in Figure 13.25.

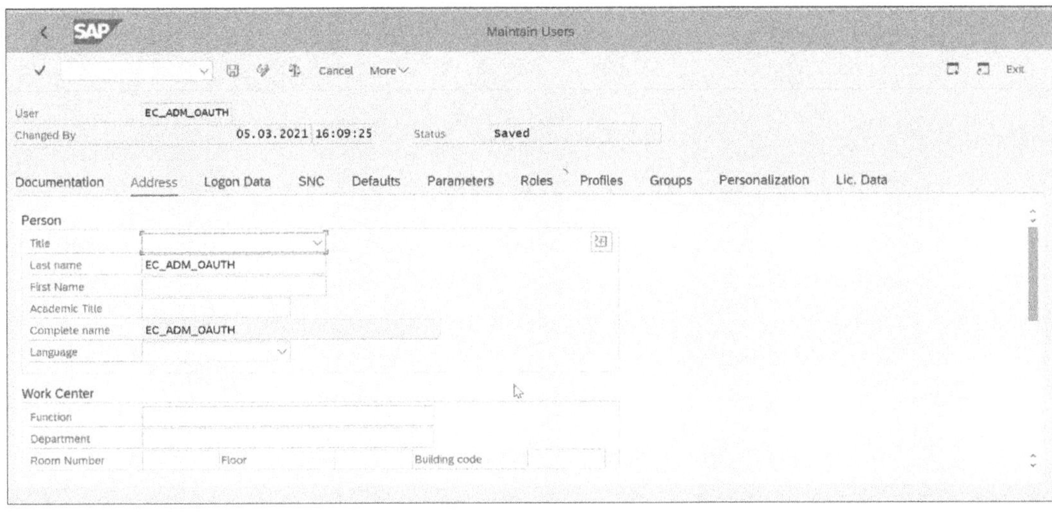

Figure 13.25 Transaction SU01 Address View

2. Create the following users:
   - EC_ADM_OAUTH
   - EC_ESS_OAUTH

For both users, under the **Address** tab, enter a last name. Then, under the **Logon Data** tab, select **System** from the **User Type** dropdown list and enter a password, as shown in Figure 13.26. Then, click the **Save** icon.

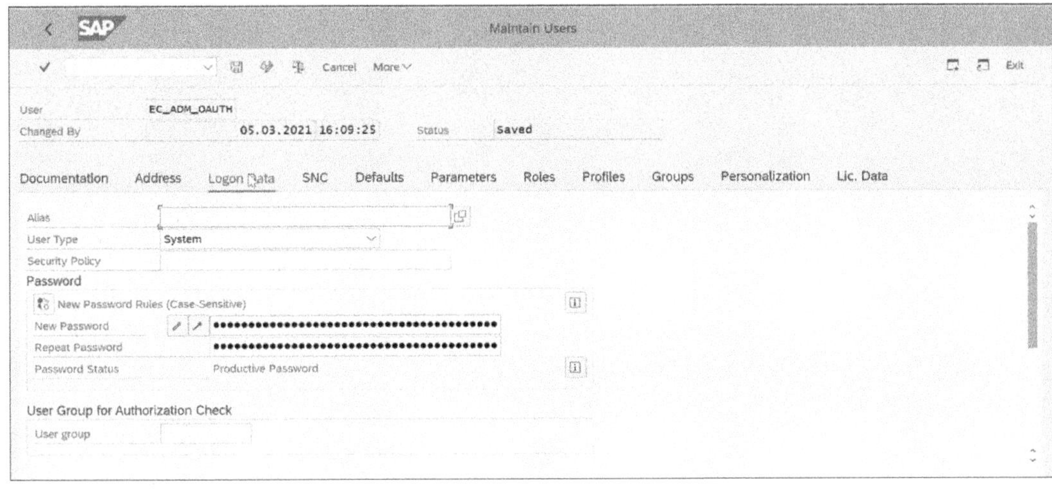

Figure 13.26 Logon Data for Users

13.7.5 Registering the OAuth Client in Employee Central Payroll

To register the OAuth client, first use Transaction SOAUTH2. For each of the IDs we created in Section 13.7.5, EC_ADM_OAUTH and EC_ESS_OAUTH, you'll need to perform the following actions:

1. Ensure the **Password** and **SSL Client Certificate** checkboxes are selected under **Client Authentication**.
2. Ensure the **Grant Type SAML 2.0 Bearer Active** checkbox is selected under the **Resource Owner Authentication** section.
3. For the **Trusted OAuth 2.0 IdP** field, choose the IdP we created earlier in Section 13.7.3.
4. In the **Scope Assignment** section, perform two tasks:
   - For **EC_ADM_OAUTH**, enter "HRSFEC_ECP_INFO_SRV_0001" (in the **Payroll System Info** field) and "HRSFEC_INFOTYPE_SRV_0001" (in the **Infotype Existence** field).
   - For **EC_ESS_OAUTH**, enter "HRSFEC_PAY_OVERVIEW_SRV_0001."

13.7.6 Allowing Outbound Messaging for OAuth in Employee Central

For this last step, log on to Employee Central, go back to the Security Center, and select **OAuth Configurations**, as shown in Figure 13.27.

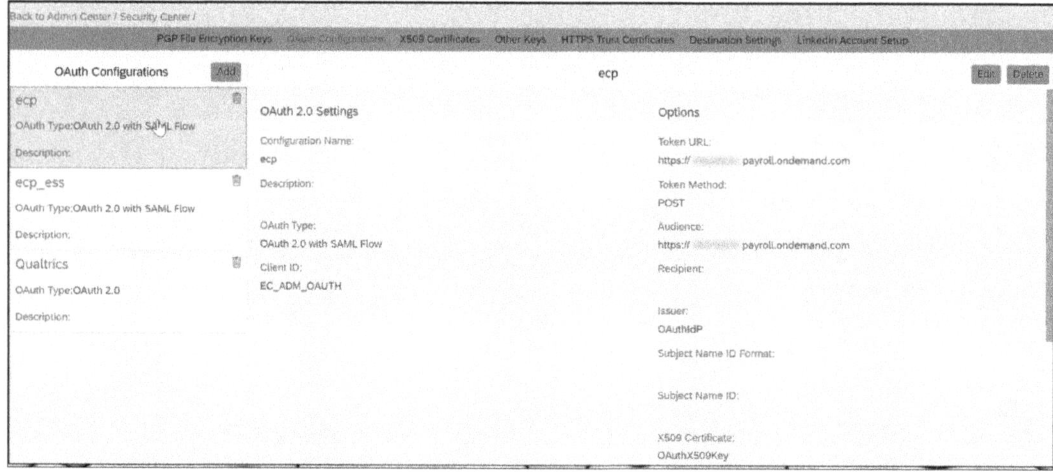

Figure 13.27 Outbound OAuth Configuration in Employee Central

Click **Add** to set up the two different configurations, as shown in Table 13.1. The ADM (administration) configuration connects the infotype and system detail viewing service for administrators while the Employee Self-Service (ESS) configuration enables the viewing of the direct pay statement.

Field Name	ADM Configuration	Employee Self-Service Configuration
Configuration Name	ecp	ecp_ess
OAuth Type	OAuth 2.0 with SAML flow	OAuth 2.0 with SAML flow
Client ID	EC_ADM_OAUTH	EC_ESS_OAUTH
Client Secret	Password you set up earlier for EC_ADM_OAUTH in Employee Central Payroll	Password you set up earlier for EC_ESS_OAUTH in payroll
Token URL	URL of your Employee Central Payroll system	URL of your payroll system
Token Method	POST	POST
Audience	Local provider name (Section 13.2.1)	Local provider name (Section 13.2.1)
Issuer	IdP name (created earlier in Section 13.7.3)	IdP name (created earlier in Section 13.7.3)
Subject Name ID	Blank	Blank
Subject Name ID Format	Unspecified	Unspecified
X509 Keys	Choose the X.509 key you created earlier	Choose the X.509 key you created earlier

Table 13.1 OAuth Configuration Parameters in Employee Central

13.8 Summary

In this chapter, you learned about a difficult technical topic: SSO. While a little tricky to understand, this configuration is entirely technical, one-time, and relieves the user from additional logon processes. If SSO is not turned on, your users will need to enter their credentials once in every Employee Central session to access any transaction in Employee Central Payroll. OAuth is necessary to enable direct pay statements, which is the pay statement most users prefer. You can do more with SSO, such as allow SSO right after authenticating to the office LAN, but this extension will require additional commercial SSO software.

Chapter 14
Monitoring Tools

Surely, you've heard of Murphy's Law—if something can go wrong, it will. This simple realization drives the motivation for this chapter and the next. SAP is great at enabling automation through batch jobs that run on their own at particular times. When they run, the business processes work perfectly. But at times things get stuck and need a little nudge. To fix such situations, you must first realize that a problem exists. This chapter is about that first step—realizing that something has gone wrong in the system.

This chapter provides insight into the monitoring process. Every system has the potential to fail. You cannot escape failure at some point in the system lifecycle. But failures are costly, and detecting a situation on time is almost as good as escaping it. If you detect a potential failure situation early, you can fix the issue before the problem spreads or impacts end users in a significant way. This early detection is where monitoring tools are useful.

The payroll process itself has plenty of checks and balances built into its steps. The payroll process is carefully timed and diligently executed with your payroll administrators reviewing many reports to ensure that the pay computation is accurate. However, this accuracy is dependent on one prime requirement—employee data coming through to the payroll system in a timely and accurate manner. For example, a person hired yesterday must be in the payroll system today so that pay can be calculated for the one day in the week that this person worked.

Since employee data is so important for processing payroll, the scheduled process that brings in the data must be monitored for errors. You saw in Chapter 7 and Chapter 8 how the Employee Central system can be connected to Employee Central Payroll to send data at periodic intervals. The data is extracted by a service from Employee Central, sent down a web connection, and then processed by payroll to convert this data into the infotypes required to calculate pay. Failure can occur at any point in this chain.

This chapter describes the various tools available to monitor this process so that any problem can be detected early. You'll learn how to check the status of replication using tools in Employee Central Payroll as well as the Data Replication Monitor in Employee Central. We'll also introduce you to the versatile application log in Employee Central

Payroll, and you'll learn how to run the Consistency Check report. This chapter ends with a short discussion on read access logging.

14.1 Checking the Status of Replicated Employee Data

SAP provides a standard report that you can use to check the replication status of employee data. This report is initiated by using Transaction HRSFEC_REPL_STATUS, as shown in Figure 14.1.

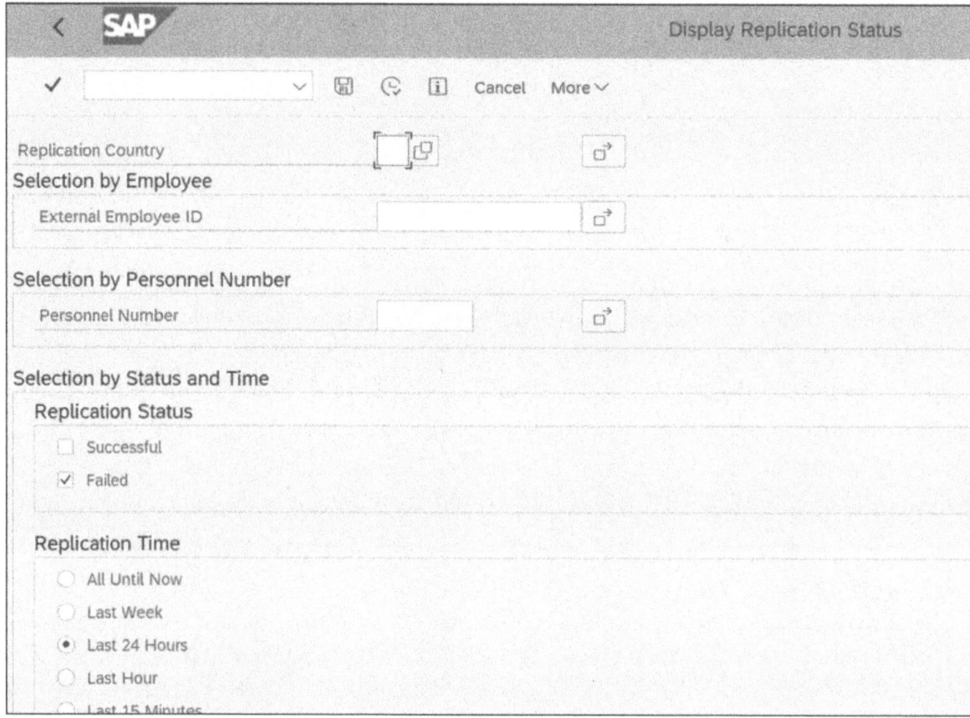

Figure 14.1 Transaction HRSFEC_REPL_STATUS

You can restrict the status to an employee or group of employees through the **Personnel Number** field. You can also restrict the number of status records returned by choosing a **Replication Status option** (**Successful** or **Failed**). This option specifies whether you would like to see all records that came in or only the ones that produced an error.

Sometimes, the data that came in is not in error but is still incorrect. Perhaps, the mapping is wrong, or a default value is not properly coming through. In such situations, you can look at the XML document that comes in from the interface. When you execute the report, by clicking the **Execute** button, you'll see the status of the replication, as shown in Figure 14.2.

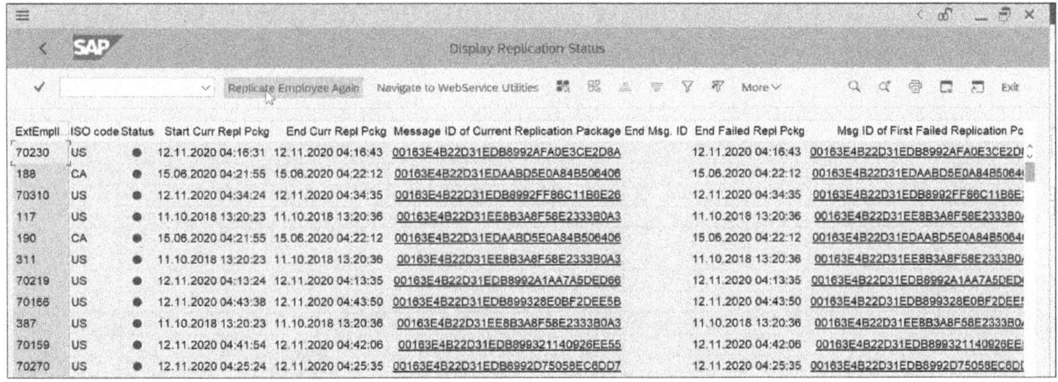

Figure 14.2 Displayed Replication Status Table

If you click on the hyperlink under **Msg ID of Failed Replication Package**, the system provides an option to download the XML file, as shown in Figure 14.3.

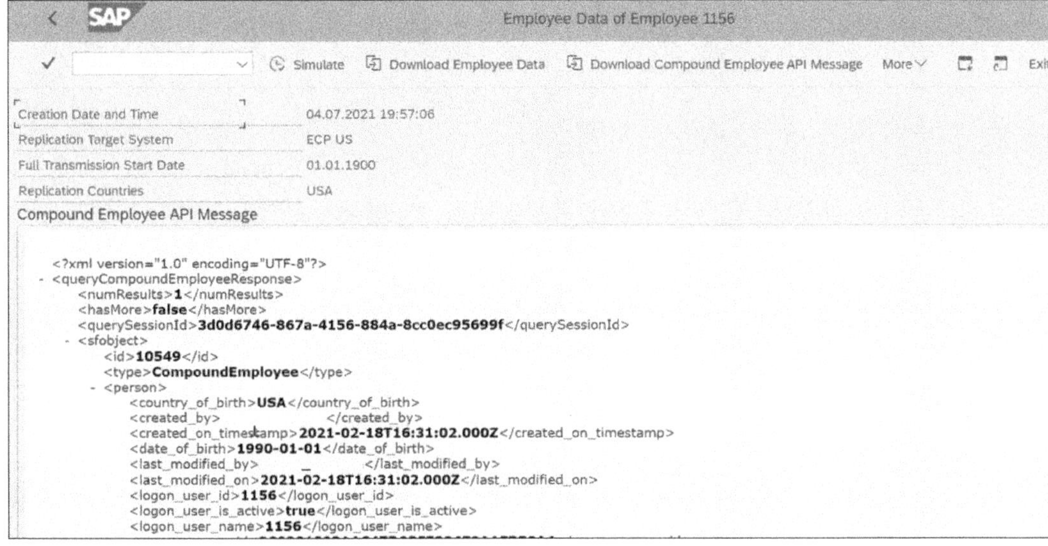

Figure 14.3 XML Message in Log Text

You can also export the report to a spreadsheet for further reference and analysis.

Since this report is a standard report, you can schedule it to execute at a regular interval. Typically, the replication runs several times a day, but the status program can be invoked once a day for someone to check and clear messages that are still in error. The system remembers an unsuccessful transmission and continues to attempt the synchronization in every run until that transmission is successful. However, you'll see the error in the status messages only once.

14.2 Using the Employee Central Data Replication Monitor When Replicating Employee Master Data

Employee Central has a tool for reviewing the status of replication called the Data Replication Monitor. The Data Replication Monitor includes some special features for automatic rescheduling if a failure occurs. To start the Data Replication Monitor, enter "Data Replication Monitor" into the search bar, and the screen shown in Figure 14.4 will appear.

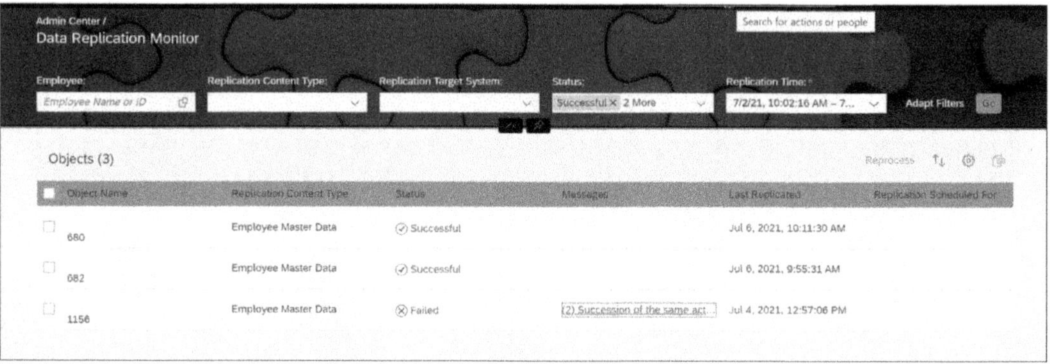

Figure 14.4 Data Replication Monitor

A common error for replication occurs when someone is in the record of an employee to edit an infotype in the Employee Central Payroll system. As you recall, not all employee infotypes are maintained in Employee Central. Specific payroll-related infotypes, such as tax infotypes, are maintained directly in the Employee Central Payroll system. Whenever any infotype for an employee is being edited, SAP places a lock on all infotypes for that employee, to maintain data integrity. As a result, the replication process will fail with the message **Data is locked**. These types of errors can be annoying because there is nothing to investigate—the system clears the lock when the user has finished editing and saves the record. Thus, reprocessing this data is all that you need to do. You can set up the Data Replication Monitor to reprocess these types of errors. The system usually waits for 60 minutes before attempting the next replication.

Some standard errors that fall into this scenario include the following:

- Payroll area is locked for master data maintenance.
- Online maintenance is locked.
- Status of payroll area does not allow maintenance (for instance, when the payroll area is released for processing).

You can also schedule a manual push replication for any employee from the Data Replication Monitor, as shown in Figure 14.4, by clicking the **Reprocess** button.

When the payroll status is in **Released for correction** mode, manual edits can be performed by a payroll administrator. Whether replication can continue to work during this phase is determined by setting the correction phase indicator in Employee Central Payroll. We'll discuss the correction phase indicator in Chapter 15, Section 15.2.

If manual replication is performed using push replication during a payroll correction phase, the system replicates the data even if the correction phase indicator is set turned off.

14.3 Using the Application Log: Transaction SLG1

The application log is a generic monitoring capability with two logical components:

- A business log
- A technical log

The *business log* is the same report that you access via Transaction HRSFEC_REPL_STATUS (Section 14.1). This report specializes in detecting errors that occur when loading the data into Employee Central Payroll infotypes. These errors include the following:

- Code values that do not exist in configuration
- Missing mandatory fields
- Locks for editing placed on the record by other users

The technical log shows other technical errors, including issues with connectivity and the construction of the XML messages used to transmit the data. The technical log is applicable for all types of integrations, not just employee data replication. To bring up the log for employee data replication, launch Transaction SLG1 and enter "HRSFEC" in the **Object** field, as shown in Figure 14.5.

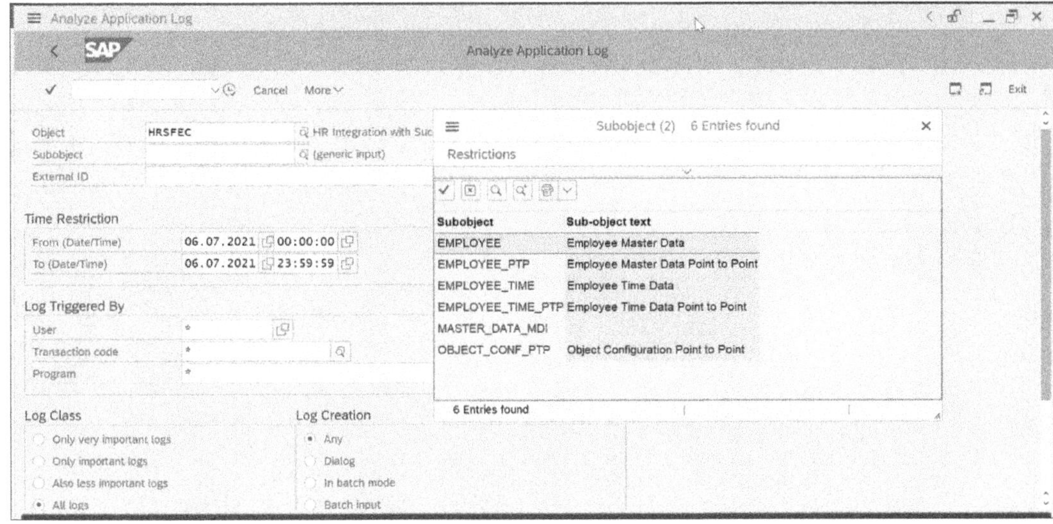

Figure 14.5 Transaction SLG1 Application Log

Depending on the **Subobject** field you choose, you can bring up the business log (**EMPLOYEE**) or the technical log (**EMPLOPYEE_PTP**).

Let's now discuss two important aspects of this log. First, you'll learn how to control the detail shown by these logs—after all, more detail sometimes takes more time to go and is thus is not always helpful. Then, we'll review how to get rid of logs that have fulfilled their purpose but are clogging up your environment.

14.3.1 Application Log: Setting the Log Level

Transaction SLG1 has the unique feature of setting the log level for controlling what you see in the error messages. Higher levels show more details about errors. During the implementation of a new system, log levels are generally set high so that any discrepancy can be investigated thoroughly. Once the system has gone live, and the process has settled down to a smooth operational rhythm, these levels can be lowered so that the monitoring team can avoid sifting through high volumes of messages unnecessarily.

To set the level, use Transaction SM30 and pull up table T77S0 by entering this table number in the **Table/View** field and clicking on **Maintain**. Choose the **LGLVE** and **LGLVP** switches in group **SFEC**, as shown in Figure 14.6.

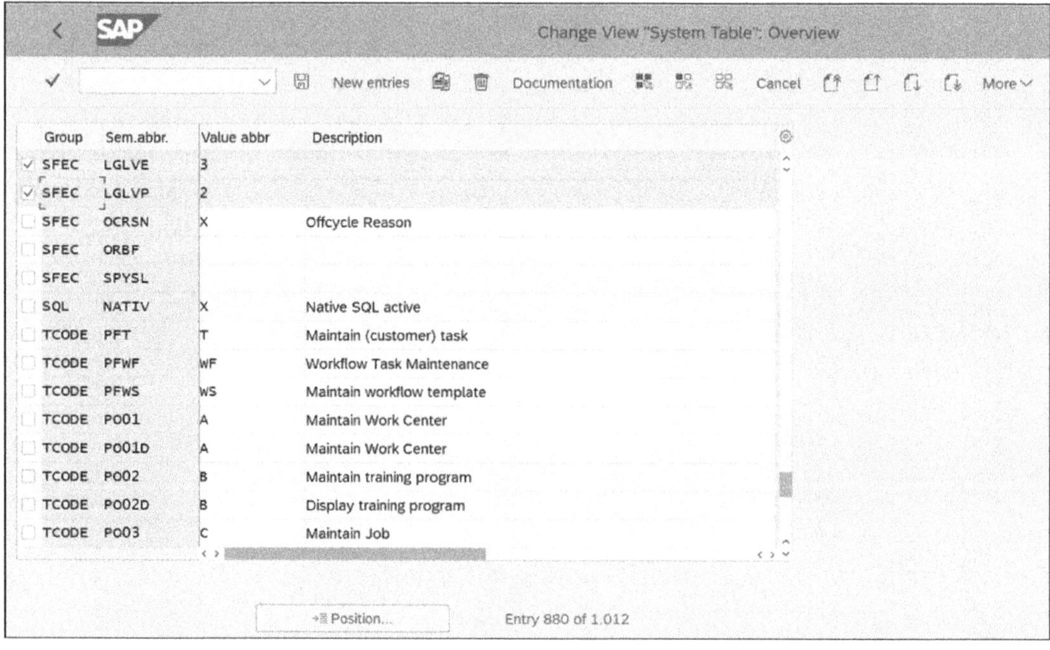

Figure 14.6 Setting the Log Level in Table T77S0 with Transaction SM30

The semantic abbreviation LGLVE refers to the business log, and LGLVP refers to the technical log. Depending on which log you want to set, enter a value for the appropriate abbreviation in the **Value abbr.** field.

The business log has four levels, listed in Table 14.1.

Type of Message	Level 0	Level 1 (Default)	Level 2	Level 3
Error	No	Yes	Yes	Yes
Warning	No	No	Yes	Yes
Success	No	Yes	Yes	Yes
Failure	No	Yes	Yes	Yes
Context info on employee	No	Yes	Yes	Yes
Additional context info	No	No	No	Yes

Table 14.1 Types of Messages and Their Levels in the Business Log

The technical log is simpler, with only three levels, listed in Table 14.2.

Type of Message	Level 0	Level 1 (Default)	Level 2
Error	No	Yes	Yes
Warning	No	No	Yes
Success	No	No	Yes
Failure	No	Yes	Yes

Table 14.2 Types of Messages and Their Levels in the Technical Log

14.3.2 Deleting Expired Business and Technical Logs Using Transaction SLG2

A best practice is to delete old logs periodically to maintain good response times in your system. Not doing so increases clutter and delays both the read time and write time for logs.

Transaction SLG2 brings up the screen shown in Figure 14.7, where you can choose a date range or log numbers. The default setting for logs is 90 days. You can choose to delete logs immediately or schedule a batch job for deleting logs later via this transaction.

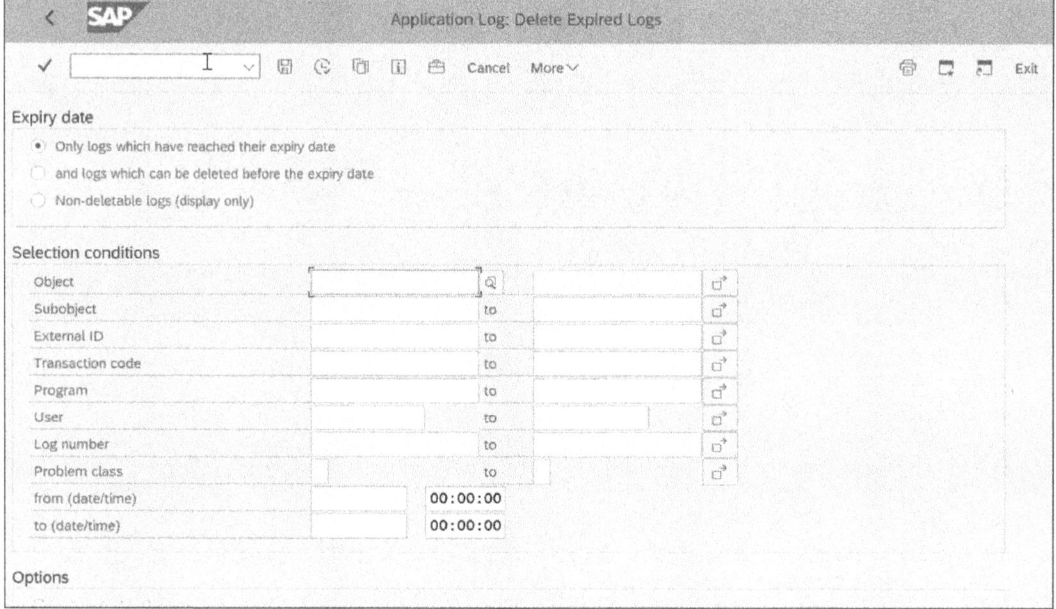

Figure 14.7 Deleting Expired Logs Using Transaction SLG2

14.4 Using Web Services Utilities: Transaction SRTUTIL

Transaction SRTUTIL is a powerful tool for monitoring every message that comes in. This transaction helps detect issues with the inbound service interface and the web services framework. As an entry-point monitoring tool, this tool is the first to see an incoming message. So, if you do not find a message in this log, it was most likely not received by this system.

To launch this tool, use Transaction SRTUTIL and, on the selection screen, under the **Receiver Information**, enter "EmployeeMasterDataBundleReplicationRequest_IN" in the **Interface Name** field, as shown in Figure 14.8.

This tool shows all received replication messages, so be sure to limit the search using appropriate date and time ranges. Double-click on a message and click on **Original XML** to display it in XML format if required, as shown in Figure 14.9.

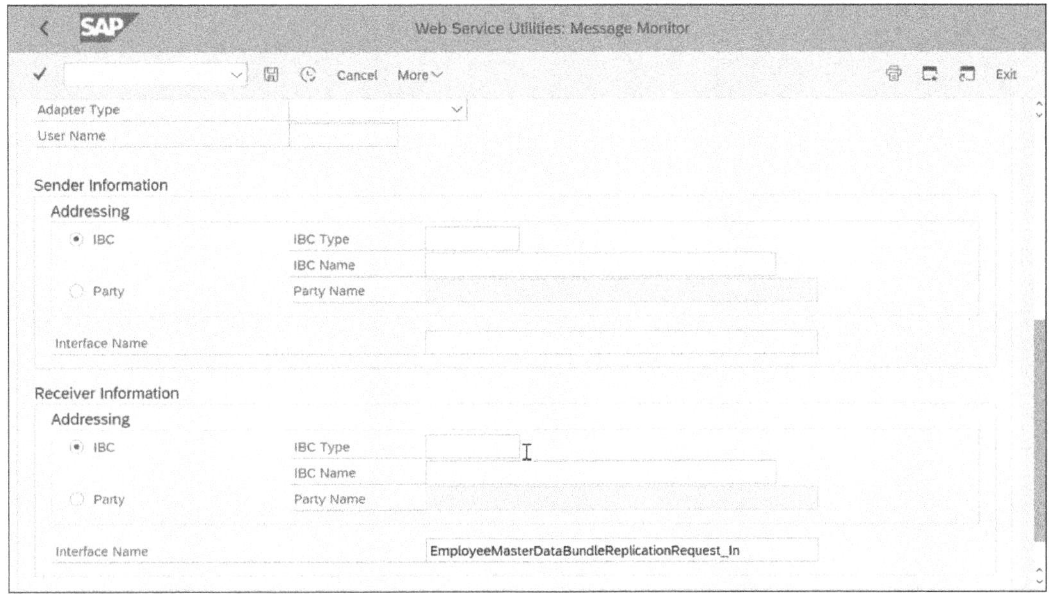

Figure 14.8 Transaction SRTUTIL

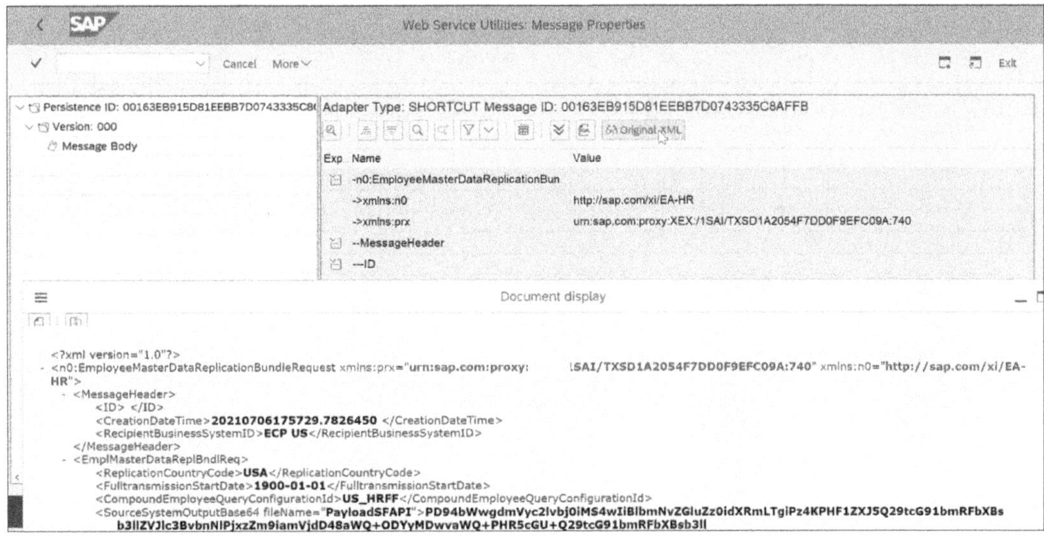

<?xml version="1.0"?>
- <n0:EmployeeMasterDataReplicationBundleRequest xmlns:prx="urn:sap.com:proxy: ISAI/TXSD1A2054F7DD0F9EFC09A:740" xmlns:n0="http://sap.com/xi/EA-HR">
- <MessageHeader>
<ID> </ID>
<CreationDateTime>20210706175729.7826450 </CreationDateTime>
<RecipientBusinessSystemID>ECP US</RecipientBusinessSystemID>
</MessageHeader>
- <EmplMasterDataReplBndlReq>
<ReplicationCountryCode>USA</ReplicationCountryCode>
<FulltransmissionStartDate>1900-01-01</FulltransmissionStartDate>
<CompoundEmployeeQueryConfigurationId>US_HRFF</CompoundEmployeeQueryConfigurationId>
<SourceSystemOutputBase64 fileName="PayloadSFAPI">PD94bWwgdmVyc2lvbj0iMS4wIiBlbmNvZGluZz0idXRmLTgiPz4KPHF1ZXJ5Q29tcG91bmRFbXBs
b3llZVJlc3BvbnNlPjxzZm9iamVjdD48aWQ+ODYyMDwvaWQ+PHR5cGU+Q29tcG91bmRFbXBsb3ll

Figure 14.9 Displaying Original XML in Transaction SRTUTIL

14.5 Using the Consistency Check Report for Replication Data

The Consistency Check report compares the log created by Employee Central with the log created by Employee Central Payroll to detect inconsistencies caused by an interruption in the processing due to a system glitch. This report does not compare the employee data in the two systems.

To run this report, start Transaction HRSFEC_REPL_CONSIST, as shown in Figure 14.10.

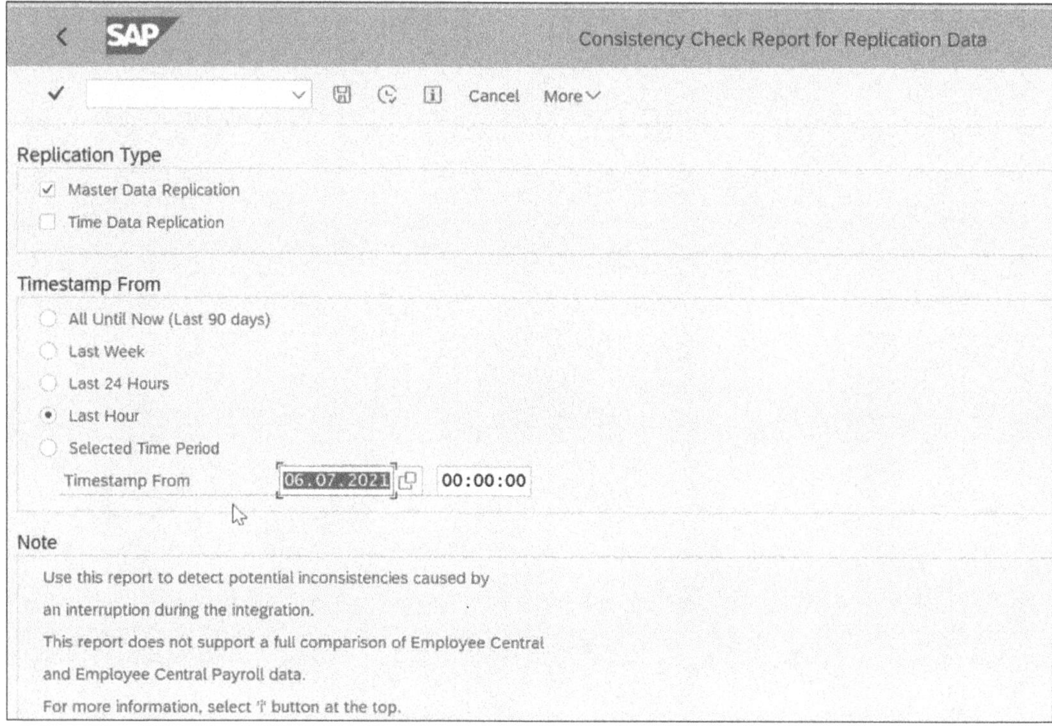

Figure 14.10 Consistency Check Using Transaction HRSFEC_REPL_CONSIST

Select **Master Data Replication** or **Time Data Replication** as required. Select the period required and click on **Execute**.

For **Master Data Replication**, the **Reprocess** option will trigger a resynchronization of the employees reported as inconsistent. If all is good, you'll see a message like the one shown in Figure 14.11.

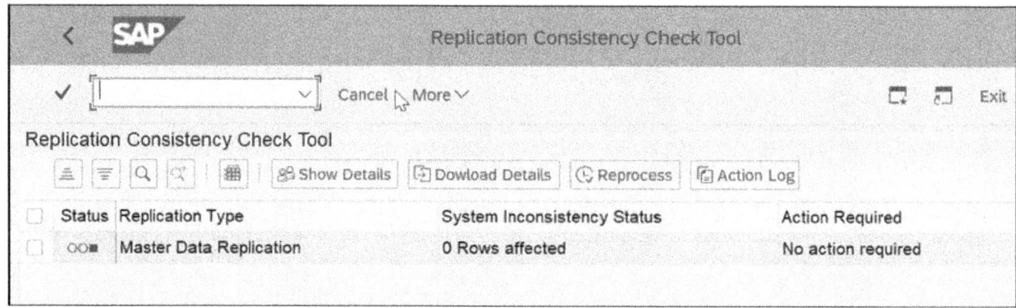

Figure 14.11 All Clear Message from the Consistency Check.

14.6 Read Access Logging for Employee Central Payroll

Another monitoring best practice is to identify users who are reading the tables, mainly to ensure that sensitive data is being accessed only by authorized users. Some countries may even have laws that require this visibility to be available in their human resources (HR)/payroll systems.

Read access logging is how you accomplish this task and involves several steps. In this section, we'll provide you with short guide on how to enable read access logging for an infotype that is read from the mashup screen in Employee Central. Read access logging for HR data is handled in Employee Central and is outside the scope of this book.

Additional Reading

Read the SAP documentation on read access logging for further details, including prerequisites and authorizations, at *http://s-prs.co/v538602*.

Start the read access log by using Transaction SRALMANAGER, as shown in Figure 14.12.

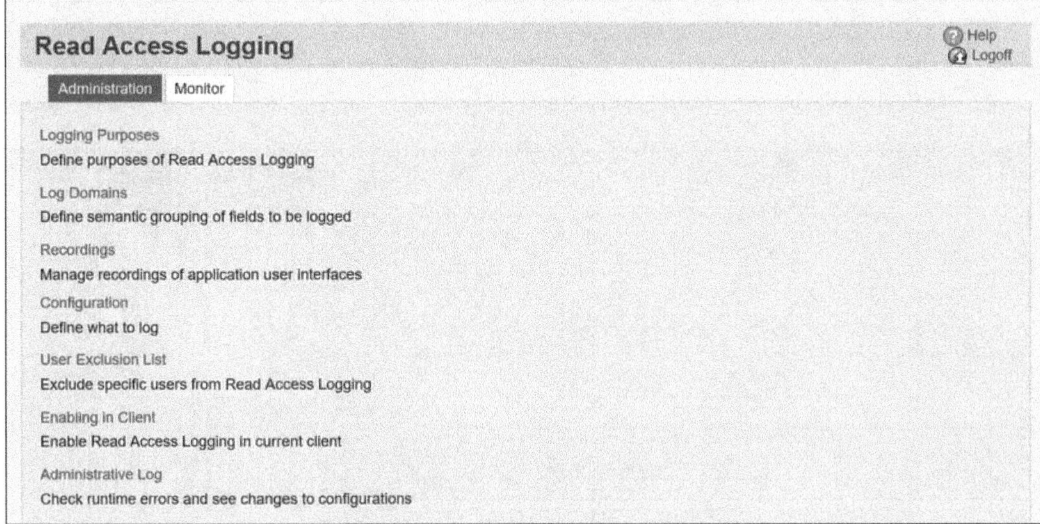

Figure 14.12 Transaction SRALMANAGER for Read Logging

Then, proceed with the following steps:

1. Create a recording, as shown in Figure 14.13, by clicking on the **Create** button in the **Search Result** section. The system will confirm that the recording is in progress.

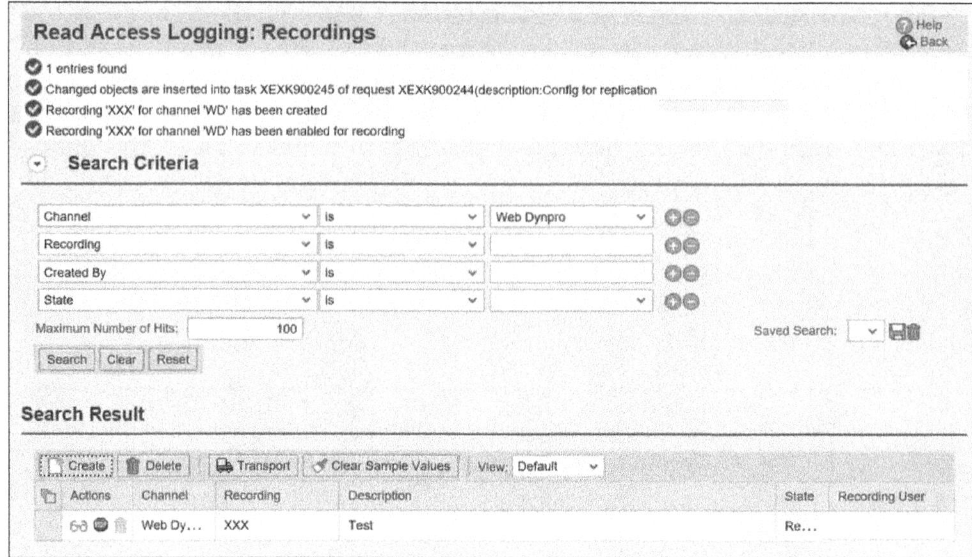

Figure 14.13 Creating a Recording

2. Start recording in Employee Central: Open the mashup screen for this infotype from the **Payroll Information** block. Right-click on the field you want to be logged and select **Read Access Logging • Record Field**, as shown in Figure 14.14.

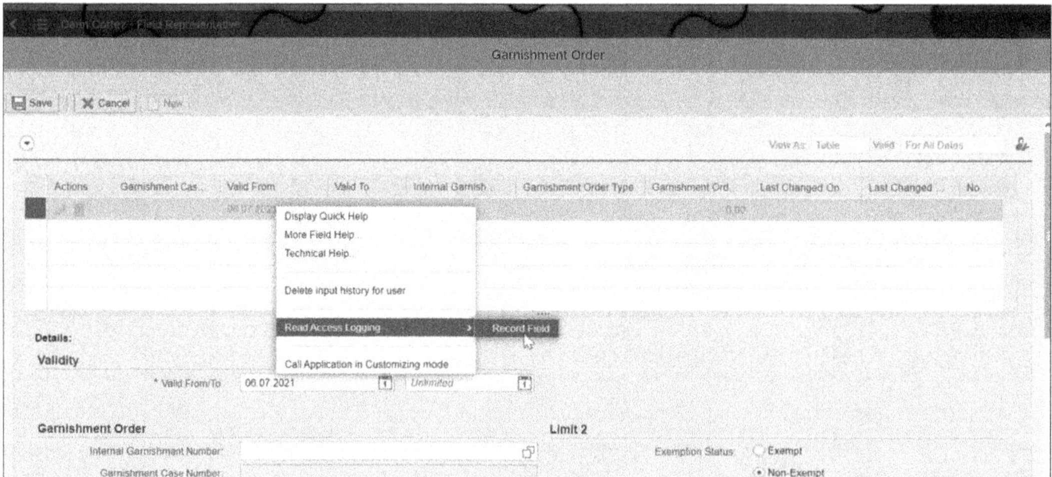

Figure 14.14 Recording in Employee Central Mashup Screen

3. Stop recording in Transaction SRALMANAGER by clicking on the stop icon, shown earlier at the bottom of Figure 14.13.
4. Complete the configuration activity in Transaction SRALMANAGER by going to the **Administration** tab, choosing **Configuration**, and clicking on **Create**. Create a log group for the **Purpose** field, select a group, enter a description, and click **Create**.

5. Drag and drop the recorded fields and assign a domain, as shown in Figure 14.15.

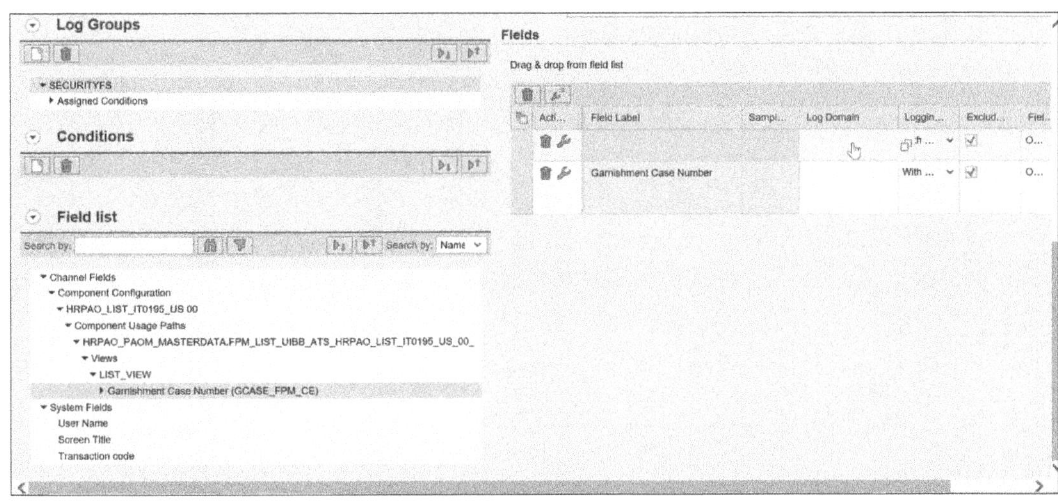

Figure 14.15 Complete Configuration by Dropping Recorded Fields and Saving

6. Save as active.
7. Enable the client by clicking on **Enable Read Access Logging in Client** under **Enabling in Client**, as shown in Figure 14.16. This step will set up the garnishment infotype for logging. Note that this step might have been already completed to log other data items before.

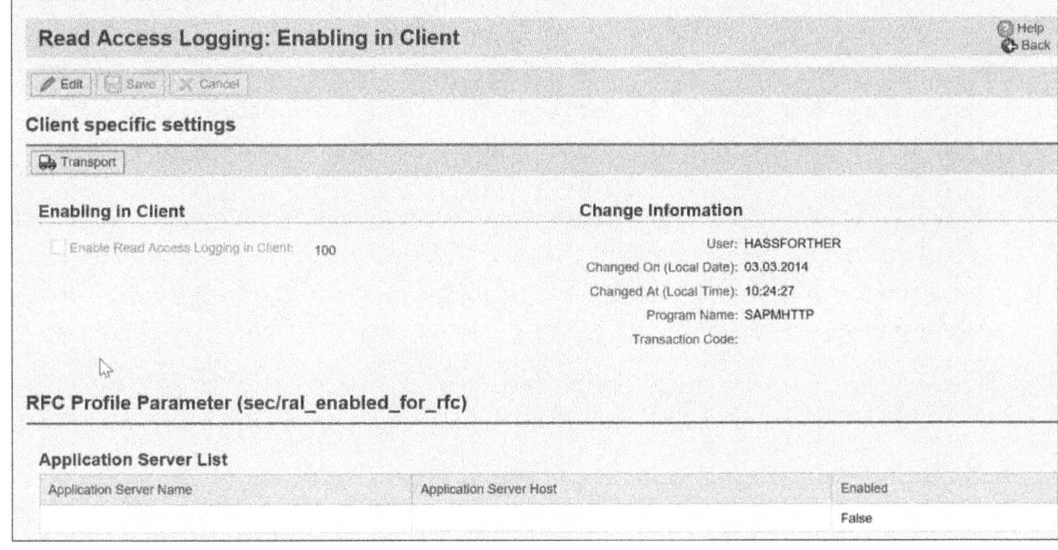

Figure 14.16 Enabling Read Logging in Your Client

You can read the log as it is populated, from the same transaction (Transaction SRAL-MANAGER) under the **Monitor** tab, by filling in your selections and searching for the specific log you need, as shown in Figure 14.17.

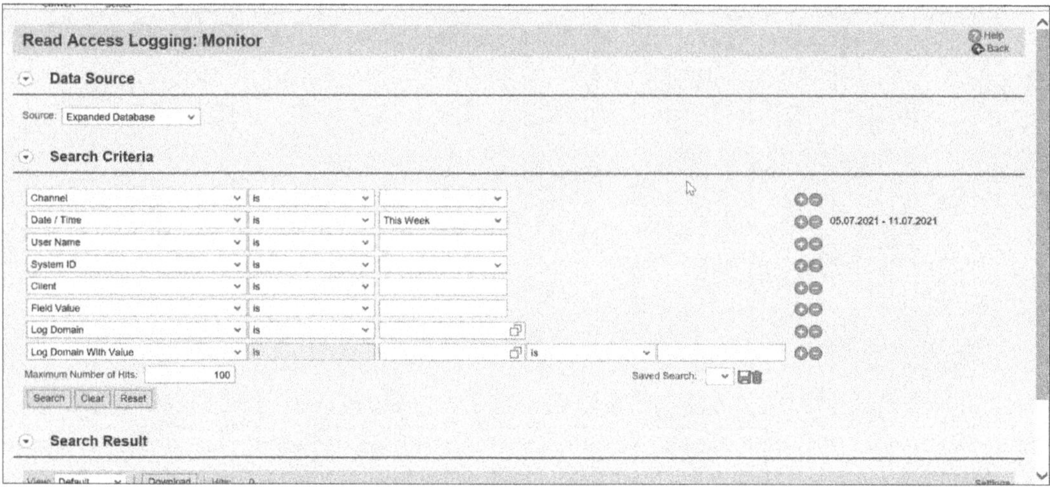

Figure 14.17 Displaying the Read Logs

14.7 Summary

Auditing and reporting on a system's health are an important function for an organization and are required to maintain business continuity. Checking on system functions regularly helps businesses avoid catastrophic problems with long recovery periods. In this chapter, we reviewed the different logs available for monitoring the replication process, which is the main process required to equip your Employee Central Payroll system with latest master data. Accuracy in pay calculations is highly dependent on this data being timely and accurate. We also explored a feature for read logging sensitive data access to alert payroll administrators about who is reading the data so the company can be protected from unwanted data leaks. In the next chapter, we'll focus on fixing some of the tough issues you might encounter with replication and discuss some possible solutions.

Chapter 15
Troubleshooting

In the previous chapter, you saw how to put your ear to the ground and listen for unanticipated problems so they can be resolved quickly. Once you've discovered an issue, the error message is the starting point for its resolution. However, in some scenarios, the error message may not be enough. This chapter is about those tough situations that are difficult to diagnose and fix.

This chapter discusses a few unique scenarios with replication that need special care to resolve. These situations are infrequent but can happen in an organization from time to time. If one of these issues arises, this discussion will at least prepare you for it and perhaps offer a few options to move forward with resolution. In this chapter, we'll discuss how to investigate issues with replication, how to use the correction phase indicator, and how to correct hire dates in tricky situations.

15.1 Troubleshooting Master Data Replication from Employee Central to Employee Central Payroll

The following issues are commonly encountered in data replication:

- Errors arising from intercompany transfers
- Conflicting job information records, for instance, duplicate hires
- Errors related to using **Edit** instead of **Make Correction** to update data
- Errors when a person has already been hired
- Data changes resulting in multiple statuses, for example, when data is changed multiple times on the same day

15.1.1 Intercompany Transfer Errors

The **Employee is active since...** or **Do not add more than 1 employment status changing actions at same day...** errors occur during an intercompany transfer where the employee's last day in the old legal entity and their first day in the new legal entity are the same. Two options exist for resolving this issue:

- Option 1: Investigate if the business allows the dates to be different (if so, update one of the dates, and the issue is resolved).
- Option 2: If dates can't be different, then delete the hiring job information of the previous employment before replicating.

15.1.2 Conflicting Job Information Record Errors

The error message **Start date before initial entry date (31.12.9999)** occurs when two job information records with the same start date, for the same person, are replicated. In this scenario, Employee Central Payroll has a tough time determining which is the correct hire record.

The solution is to delete the second (duplicate) job information record (typically a data change), perform the replication a second time, and then re-create the second record.

15.1.3 Edit Feature Errors

Getting the error message **Succession of the same action of type &1 is not supported** during replication is typically caused by the use of the **Edit** feature to add records in the **Job Information** portlet. Using this function, you can introduce records without an event, but doing so causes inconsistencies on the Employee Central Payroll side since Employee Central Payroll does not understand what event to assign.

The best way to prevent this error message is to use the **Make Correction** feature instead of **Edit**, which will enforce the use of an event and event reason.

15.1.4 Person Already Hired Error

If you receive the **Person already hired** error message during replication, you may be using an enhancement for personnel number mapping. Check the logic in the `HRSFEC_B_CE_EXT_PERNR_MAP` Business Add-In (BAdI) and modify it to ensure that a unique personnel number is always assigned.

15.1.5 Multiple Statuses Error

The error message **Don't add more than 1 employment status changing actions at same day dd/mm/yyyy (HRSFEC_SERVICES 203)** is caused if two job information records exist for the same date with two different statuses. Employee Central allows this inconsistency, but Employee Central Payroll does not.

The easiest method to resolve this issue is to get rid of the second record and merge its changes with the first one. For example, if the first record is a transfer and the second is a salary change, include the salary change in the transfer record itself.

15.2 Enabling the Correction Phase Indicator

When the payroll processing is in full swing, with the payroll due date looming the same day or the next day, an important step is to complete urgent master data updates as quickly as possible. This need is recognized by SAP, and thus, the correction phase indicator is provided to speed up the replication. Essentially, you can tell the system that, if the payroll area for any employee is in correction mode, then the system should perform the replication of changed employees immediately. In this way, the most up-to-date data is immediately available for payroll processing.

To turn this feature on, execute Transaction SM30 and enter "T77SO" in the **Table/View field** and click on **Maintain**, which will lead you to the screen shown in Figure 15.1.

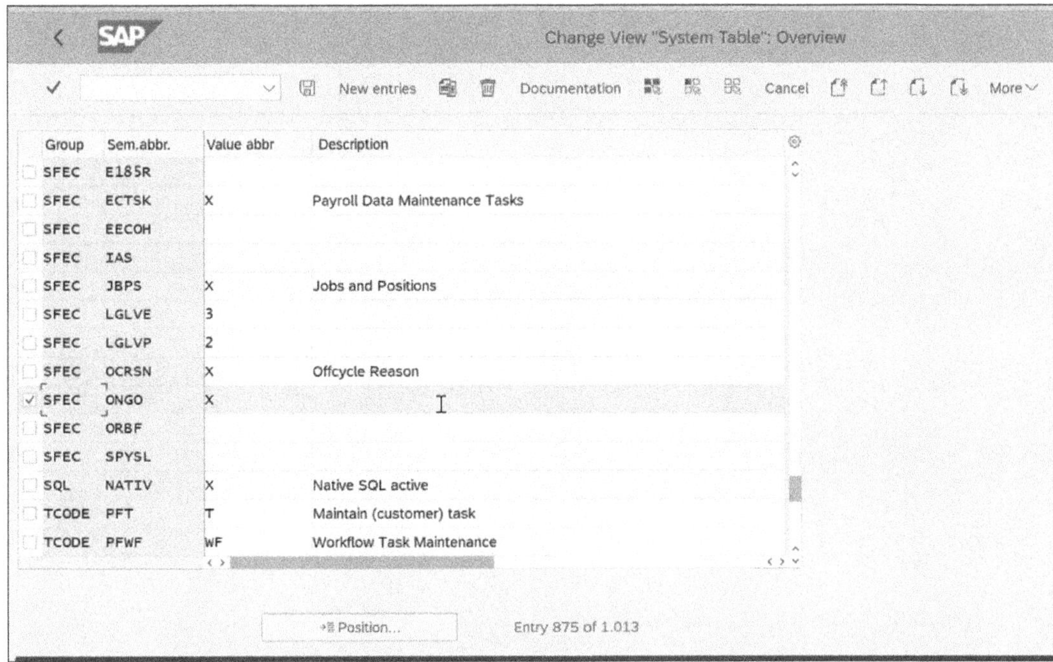

Figure 15.1 Enabling the Correction Phase Indicator

Select **Group SFEC** and **Sem.abbr. ONGO** and then enter the value "X" in the **Value abbr** field. After performing this step, the replication will trigger immediately after employee master data is changed.

15.3 Changing the Hire Date

A key date in Employee Central Payroll, the hire date is the date when the employee is eligible for pay and for other benefits. This data point is one of the first pieces of data that is replicated into Employee Central Payroll. The hire date, however, is also one of

15

the most difficult data items to change if a mistake has occurred. The mistake is generally on the Employee Central side, but the correction must be performed in both the Employee Central and Employee Central Payroll systems. Depending on the situation, the technique could be different and somewhat complicated. Follow the diagram shown in Figure 15.2 to fully understand the method. This flowchart describes the steps based on two events—whether payroll has been already run and whether the true hire date is in the past or in the future.

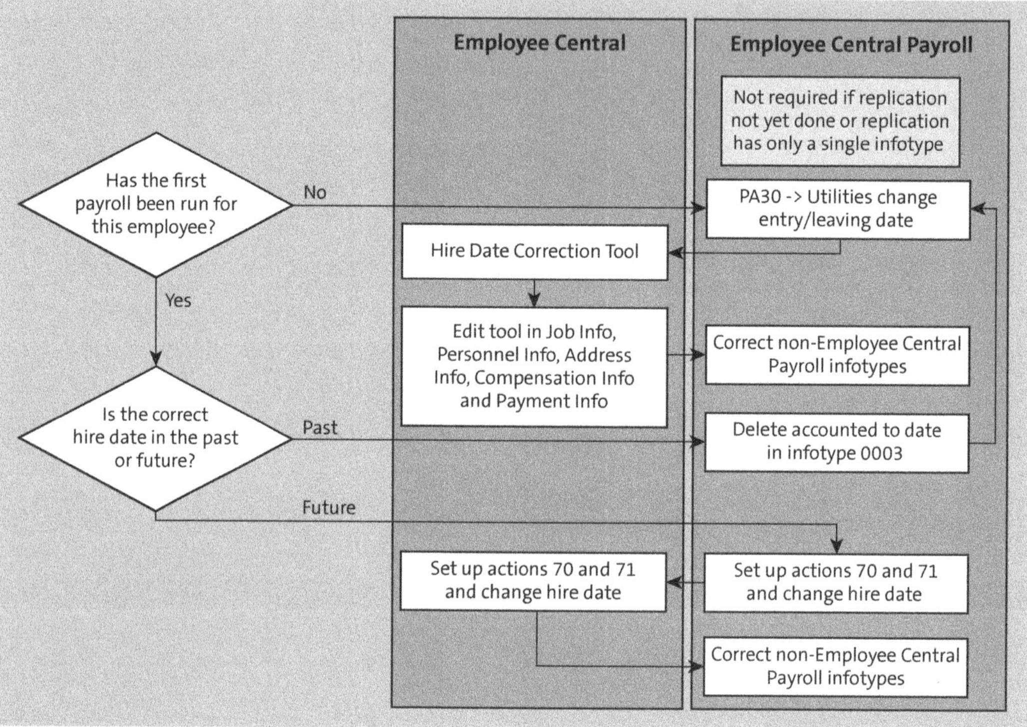

Figure 15.2 Hire Date Correction Process

In the following sections, we'll discuss how to change the hire date prior to payroll being run, after payroll has been run (with a hire date in the past), and after payroll has been run (with a hire date in the future).

15.3.1 Prior to Payroll Run

To change a hire date, if payroll has not been run yet, the first step is to make the change in Employee Central Payroll, using a utility that can correct the multiple infotypes that were created through replication. If replication has never been performed for the employee, this step can be skipped.

In Employee Central Payroll, use Transaction PA30 and select **Utilities • Change entry/ leaving date**, as shown in Figure 15.3.

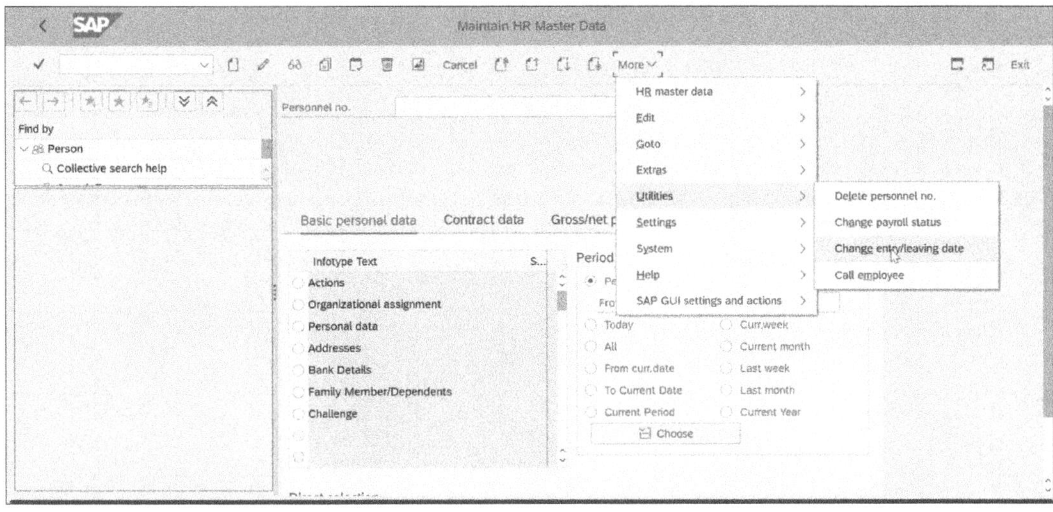

Figure 15.3 Change Entry/Leaving Date in Payroll

Change the hire date in the **Start** field and then save. When the **Postpone action** popup window opens, select all infotypes that you would like to change and click **Continue**.

The second step is in Employee Central. Enter "Hire Date Correction" into the search bar.

Enter a name in the field next to the **Hire Date Correction** field (in our example, we entered "test"), as shown in Figure 15.4. The original hire date appears.

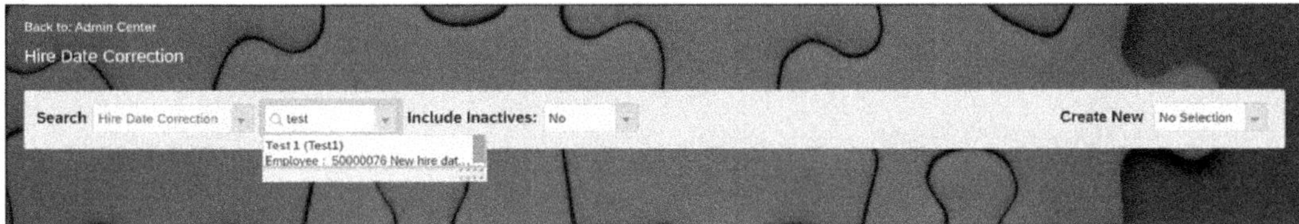

Figure 15.4 Hire Date Correction Tool in Employee Central: Identifying the Employee

From the **Take Action** dropdown, choose **Make Correction**, as shown in Figure 15.5.

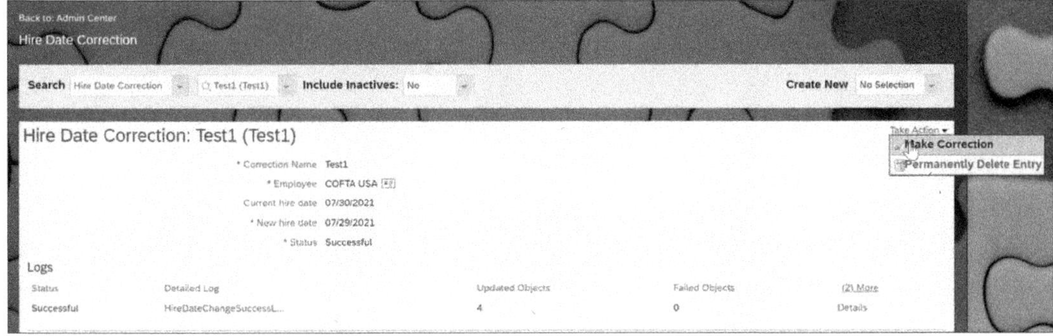

Figure 15.5 Take Action Menu

Maintain the **New hire date** field, as shown in Figure 15.6, and click **Save**.

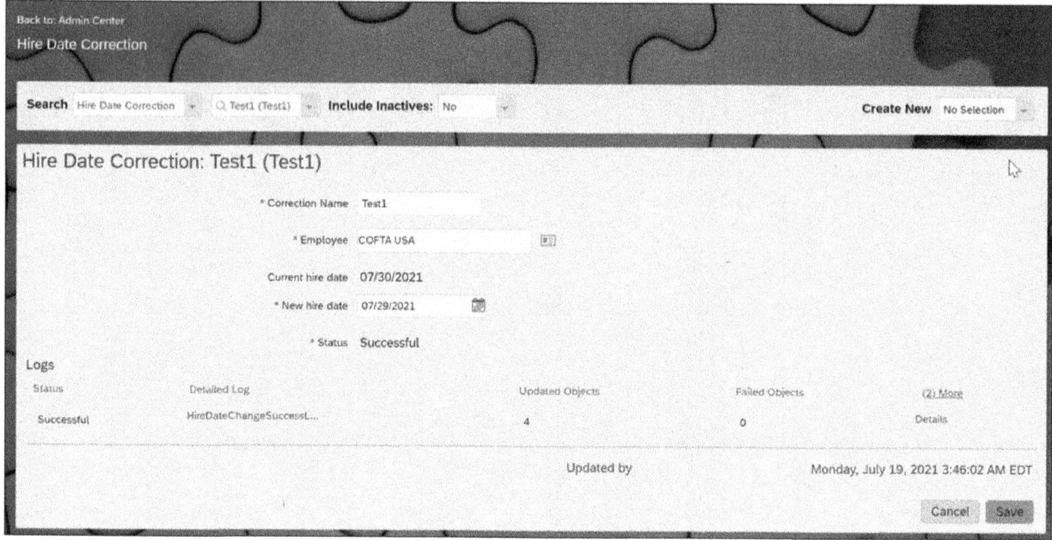

Figure 15.6 Changing Hire Date in the Hire Date Correction Tool in Employee Central

The following portlets will be automatically adjusted with the new hire date:

- **Employment Info**
- **Job Information**
- **Compensation Information**
- **Recurring Pay Components**
- **Job Relationships**
- **Personal Information**
- **Global Information**
- **Address Information**
- **Recurring Deductions**
- **Payment Information**
- **Dependents Information**

Although the hire date correction tool takes care of the hire date, in some portlets, history has already been collected. This tool does not correct the start date of previous transactions, which where the third step comes in. Use the **Edit** tool in Employee Central to change the effective dates in these portlets:

- **Personnel Information**
- **Address Information**
- **Compensation Information**
- **Payment Information**

To edit the history, click on the pencil icon next to the portlet and change the start date, as shown in Figure 15.7.

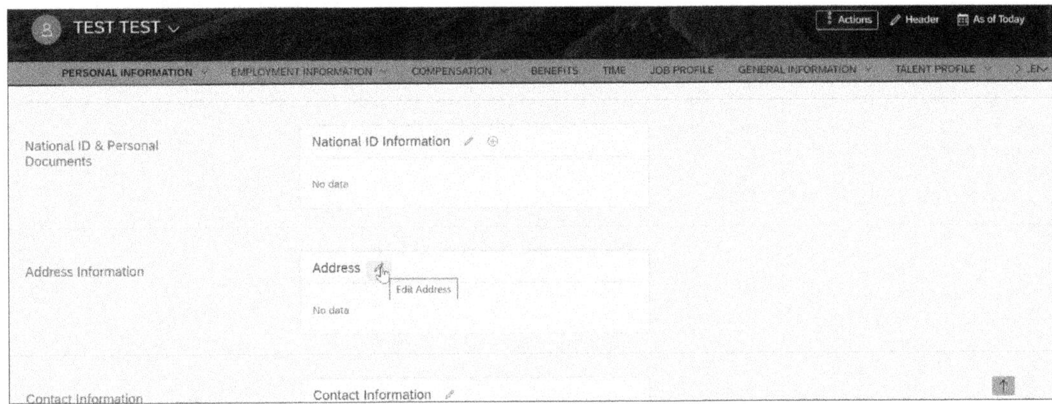

Figure 15.7 Editing the History of an Employee in Employee Central

The fourth step is to correct the start date of the infotypes that are not replicated from Employee Central, such as the tax or garnishment infotypes (for more information on these infotypes, see Chapter 2. For this task, use Transaction PA30, as shown in Figure 15.8 for a US tax infotype and then change the **Start** date.

15

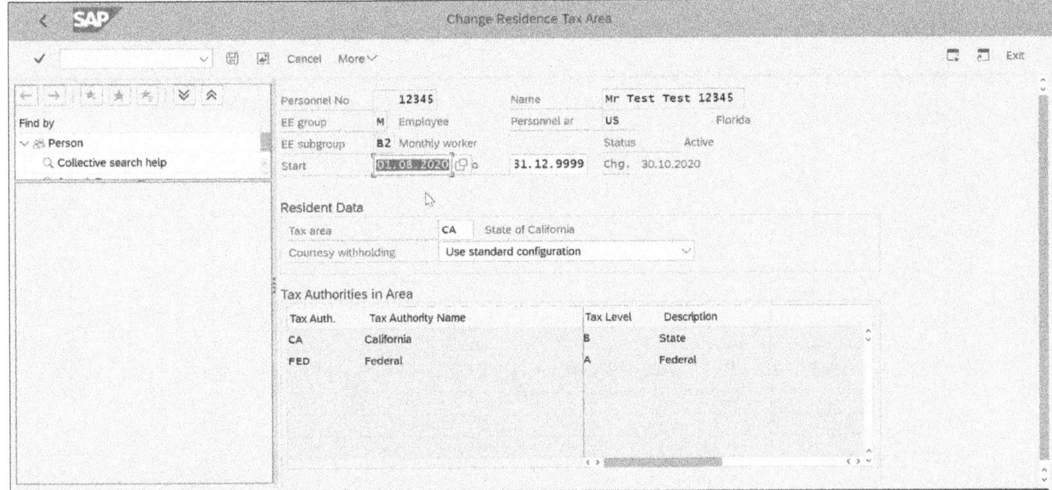

Figure 15.8 Correcting the Start Date in Payroll

15.3.2 After Payroll Is Run with Hire Date in the Past

To change the hire date after payroll has been run when the hire date is in the past, the first step is to update infotype 0003 in the Employee Central Payroll system, which records all payroll runs. Use Transaction PA30 and select **Utilities • Change payroll status**, as shown in Figure 15.9.

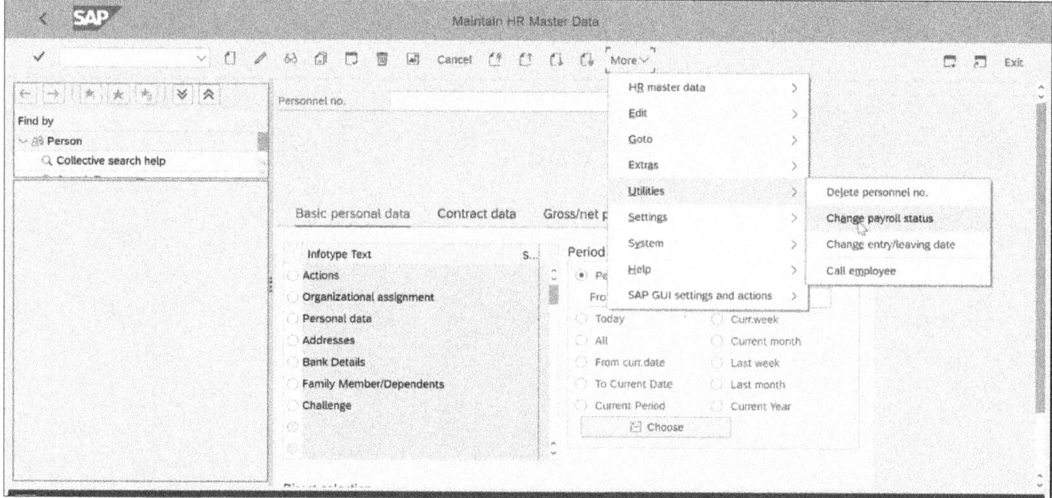

Figure 15.9 Changing Payroll Status for Employee Whose Payroll Is Already Run

Delete the entry in the **Accounted To** field and save.

The next steps are exactly as required for an employee whose payroll has not been run, as described in Section 15.3.1.

15.3.3 After Payroll Is Run with Hire Date in the Future

The steps required when the hire date is in the future are different from our previous discussion for when the hire date is in the past. In this scenario, changing the hire date is more simple. SAP provides two actions for Employee Central and Employee Central Payroll, actions 70 and 71, respectively, but these actions first must be configured.

The first step is to set these two actions up in Employee Central Payroll. Use Transaction SM30 and maintain table T529A, as shown in Figure 15.10.

Choose **New Entries** and set up action 70 with following values:

- **Name of Action Type**: Enter "Incorrect Entry."
- **FC**: Enter "1."
- **Employment**: Enter "0."
- **Spec Pymnt**: Enter "0."

Choose **New Entries** and set up action 71 with following values:

- **Name of Action Type**: Enter "Corrected Entry."
- **FC**: Enter "0."
- **Employment**: Enter "3."
- **Spec Pymnt**: Enter "1."

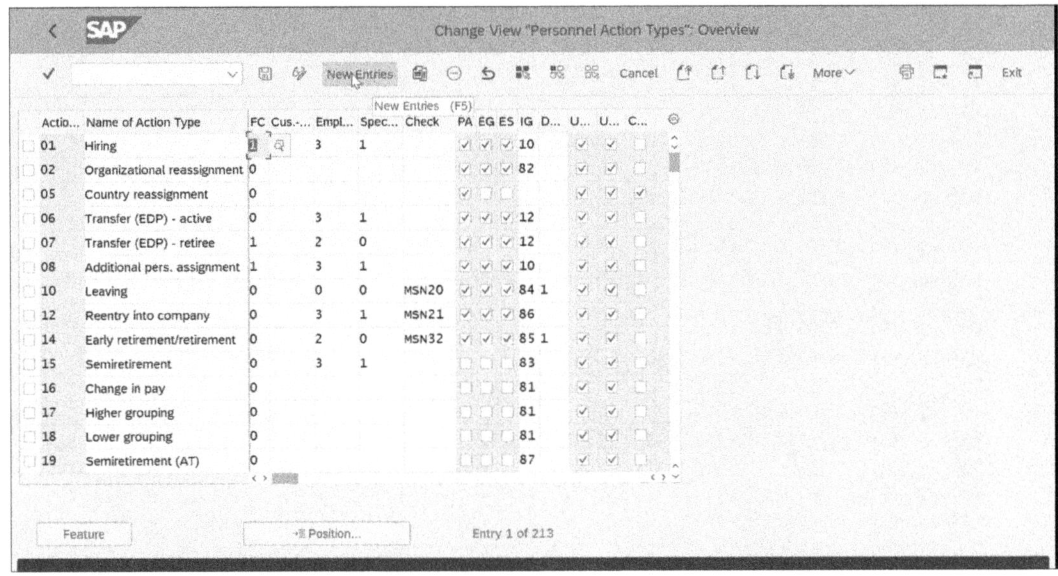

Figure 15.10 Setting Up Action 70 and 71 in Payroll

Save these settings, and then set up the action reasons by using the same transaction (Transaction SM30) with the view V_T530_ALL, as shown in Figure 15.11.

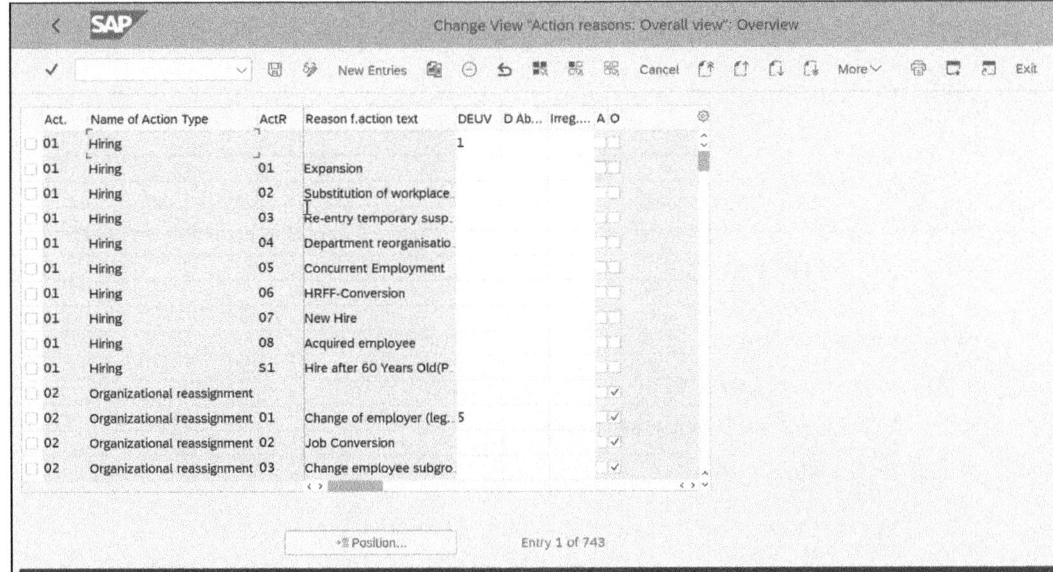

Figure 15.11 Setting Up Action Reasons in Payroll

Choose **New Entries** and set up reason 70 with following values:

- **Name of Action Type**: Enter "Incorrect Entry."
- **DEUEV**: Leave blank.

Choose **New Entries** and set up reason 71 with following values:

- **Name of Action Type**: Enter "Corrected Entry."
- **DEUEV**: Enter "1."

Save these settings, and now you're ready to move on to the second step—setting up the payroll mapping with Employee Central.

Use Transaction SPRO in Employee Central Payroll and follow the menu path **Personnel Management • Integration Settings for Employee Central Payroll • Configuration of Point-to-Point Replication • Assignment of Code Values • For Point-to-Point Replication • Maintain Code Value Mapping**. The screen shown in Figure 15.12 will appear.

Change View "Code value mapping (customer)": Overview

New Entries Cancel More

Code value mapping (customer)

Data type in Employee Central	ISO ...	Alt....	Employee Central Code Value	ERP Code Value
EVENT			10	BR
EVENT		01	12	16
EVENT		01	23	92
EVENT		01	26	20
EVENT		01	5	60
EVENT		01	6	06
EVENT		01	8	8L
EVENT		01	GA	05
EVENT		01	H	01
EVENT		01	R	12
EVENT	CA		12	16
EVENT	CA		23	92
EVENT	CA		26	20

Position... Entry 60 of 431

Figure 15.12 Mapping of Event Values from Employee Central to Payroll

Click **New Entries** and set up action 70 with following values:

- **Data type in Employee Central**: Enter "Event."
- **Employee Central Code Value**: Enter "70."
- **ERP Code Value**: Enter "70."

Click **New Entries** and set up action 71 with following values:

- **Data type in Employee Central**: Enter "Event."
- **Employee Central Code Value**: Enter "71."
- **ERP Code Value**: Enter "71."

The third step is to adjust the actions' infotype (infotype 0000). Use Transaction PA30 in Employee Central Payroll and copy the record with the hiring action by selecting **Edit • Copy**, as shown in Figure 15.13.

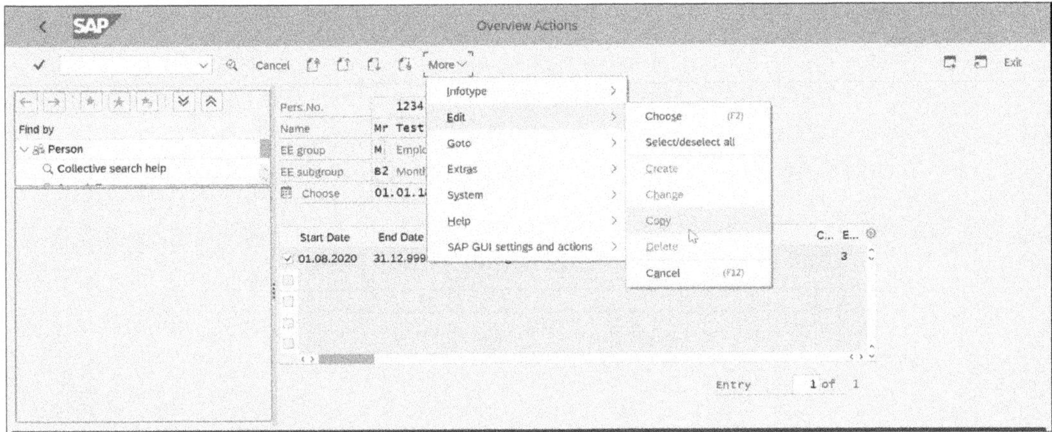

Figure 15.13 Copying a Hire Action Record in Transaction PA30

Enter "70" in the **Action** field and save without changing the date.

Now, copy the action again but enter "71" into the **Action** field. Change the validity start date to the correct hire date and save.

Now, go back to Employee Central. Enter "Manage Organization, Pay and Job Structures" into the search bar to pull up the screen to create two new event reasons (for events 70 and 71), as shown in Figure 15.14. From the **Create New** dropdown list, select **Event Reason** to open the screen shown in Figure 15.14. Then, fill in the details as shown. Be sure you enter "Corrected Entry" for payroll event 71 and "Incorrect Entry" for payroll event 70.

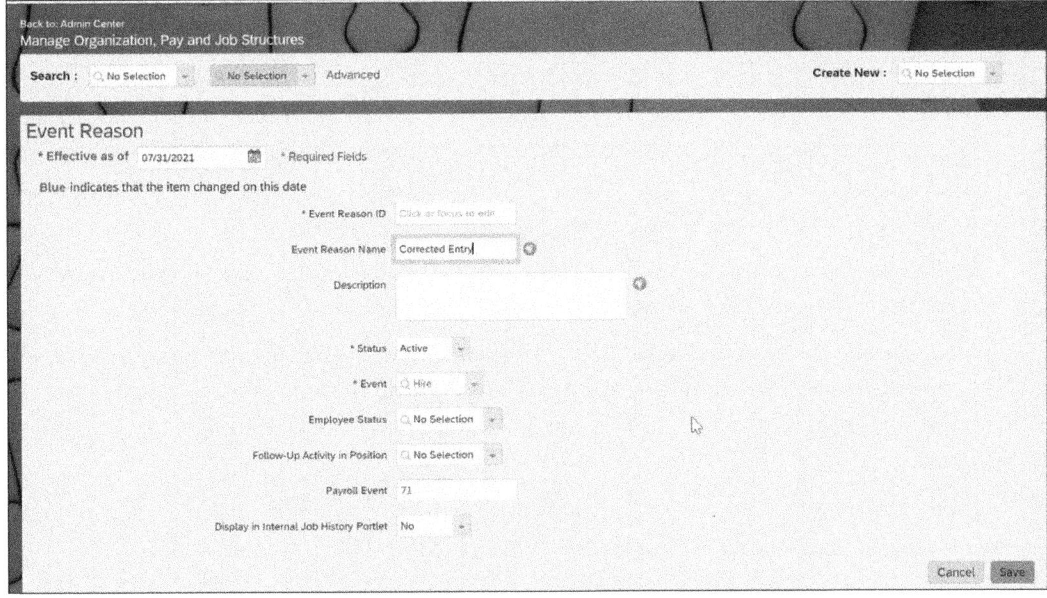

Figure 15.14 Creating New Event Reason in Employee Central for Hire Date Correction

Once done, edit the job information record by clicking the pencil icon, as shown in Figure 15.15:

- Add a new job record for the employee with the correct hire date and event 71 with the correct reason,
- Change the old job record with the event and event reason of 70.

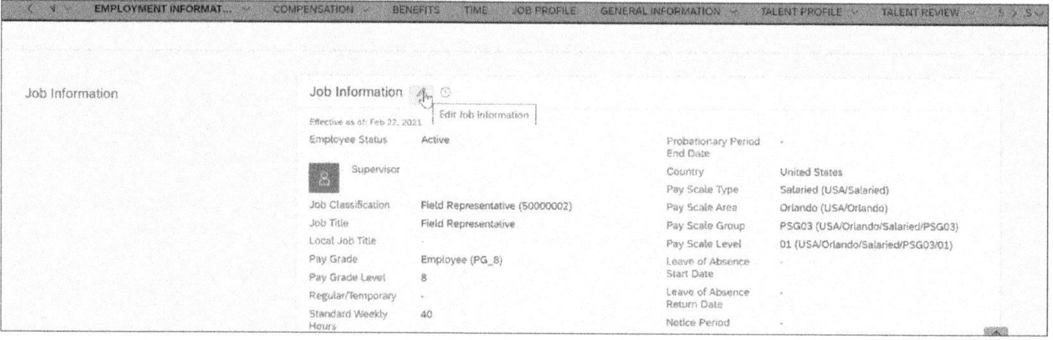

Figure 15.15 Edit Job Information

The final step is to correct the start date of the infotypes that are not replicated from Employee Central, such as the tax or garnishment infotypes. This process is the same as described in Section 15.3.2.

15.4 Summary

Employee data in SAP is completely event driven. The data that is stored in the system is driven by critical events that occur to an employee over the course of the employee's employment. This approach to data storage is a significant difference from how data is stored in other SAP modules, such as finance and logistics, where the relevant master data can be created independently of events. This data is required only to support the different events tracked in the system. In HR and payroll, the data is itself the result of a critical event and is thus tightly coupled with the event itself.

For this reason, the most troublesome issues with data center around the most critical events in an employee's employment lifecycle (hire, transfer job change, etc.).

Why did we focus on just replication errors in this chapter? Most other types of errors are easily detectable in the application itself: Missing fields result in prompts on infotype screens; incorrect syntax in schemas and rules cannot be saved until they are fixed.

This chapter provided guidance on some of the tough problems that you may encounter where the "instruction manual" is simply not sufficient for resolution.

Another type of error is tough to fix—errors that do not result in error messages. Unfortunately, no simple solution exists to resolve this type of error since members of this category show up as incorrect results in other areas of payroll. Missing wage type attributes for tax calculations will show up as incorrect tax results but only reveal itself to be a problem if you manually calculate the tax and find that the SAP result fails to meet expectations. Sometimes, a benefit deduction for a dependent is missed because the new dependent was not entered into the system on time. The only way to find and resolve these types of errors is through thorough and targeted testing. We'll turn to testing in the next chapter, with a focus on its role in implementation planning.

Chapter 16
Implementation Project Planning

Implementations of Employee Central Payroll, while similar to other IT implementations, have unique requirements. The purpose of this chapter is to define a payroll implementation plan that you can follow while we discuss some unique aspects to payroll implementations.

Implementations that include payroll, while complex, are run using a similar methodology as other SAP implementations. Cloud capabilities have introduced a more agile methodology, but key components of non-cloud implementations remain. In the past, SAP used the ASAP methodology but has since introduced SAP Activate, SAP's next-generation project methodology introducing an agile approach to implementing and delivering SAP solutions. While the methodologies may differ, the key components needed in an implementation are the same. A solid implementation approach is key to a successful delivery of the solution—without this approach, the implementation will not be successful. In this chapter, we'll describe the five phases of an Employee Central Payroll implementation—the discover, design, build, deploy, and hypercare phases—and explore key aspects of each phase.

Later in this chapter, we'll cover the importance of building a good delivery tracker or plan, of defining the conversion scope, and of attending to special testing considerations needed for payroll. We'll end this chapter by summarizing some key considerations for an implementation project.

16.1 Employee Central Payroll Implementation Phases

An Employee Central Payroll implementation starts before the project even begins with the discover phase where you'll select software and define the scope of the solution. These decisions continue to influence the implementation even after the project ends with the hypercare phase. Each phase has a set of activities and includes deliverables that are unique and critical to the implementation. Figure 16.1 shows a view of the implementation lifecycle. In this section, we'll cover each phase in more depth.

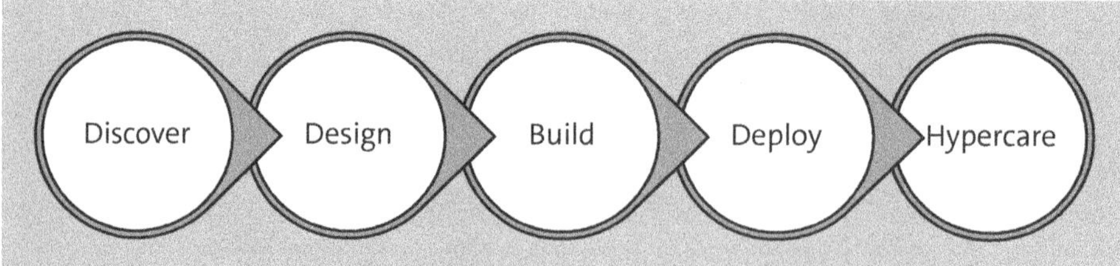

Figure 16.1 Payroll Implementation Lifecycle

16.1.1 Discover

The first phase of the implementation lifecycle, the *discover phase*, is the foundational starting point for your organization and when you'll define priorities and the path forward for success. Typical activities and deliverables include the following:

- Developing an understanding of the current landscape as well as areas of need or focus
- Selecting software
- Defining the scope of the implementation—functional, technical, and geographical
- Building a business case, including defining benefits and return on investment (ROI)
- Defining project governance
- Planning the implementation

The discover phase looks different for every company. Some companies go in knowing specifically the answers to these questions, while other companies spend months working through this process. However, the outcome is the same—a clear understanding of the path forward, the scope, and the timeline—to ensure planning is complete.

A key consideration in this phase, when determining the timeline, is defining the go-live date or specifically the first payroll period that will be run using Employee Central Payroll. We recommend going live at the beginning of a quarter so a clean break by period can be used for tax purposes. Ideally, going live would be planned for the beginning of the year. If you do not go-live at beginning of the year, you'll need to convert year-to-date (YTD) records for each employee's earnings, deduction records, as well as the taxes received for that year. This additional effort for conversions and validations also adds risk given the importance of converting this data correctly.

Another key consideration is the implementation strategy as it relates to implementing Employee Central and Employee Central Payroll. Table 16.1 outlines the possible scenarios.

Scenario	Pros	Cons
Employee Central and Employee Central Payroll together	▪ Design, build, and testing are conducted together.	▪ Lots of change for your organization ▪ Longer timeline for Employee Central to go live

Table 16.1 Deployment Plan Considerations

Scenario	Pros	Cons
Employee Central followed by Employee Central Payroll	■ Employee Central can be rolled out more quickly with no dependency on Employee Central Payroll.	■ Need to build a throwaway interface to the legacy payroll system ■ Design might need revisiting when Employee Central Payroll is implemented ■ Longer total timeline if run sequentially

Table 16.1 Deployment Plan Considerations (Cont.)

You'll need to look at your business requirements to determine which scenario works best for your company's priorities.

16.1.2 Design

During the design phase, the future state solution is defined, including process flows as well as the functional and technical solution itself. Workshops are held to design a solution that meets business requirements while also incorporating leading practices in process and design. Typical activities and deliverables include the following:

- Conducting workshops to understand requirements and to align to future design
- Designing future state processes
- Creating a functional requirements/requirements matrix
- Outlining interface specifications
- Determining data conversion requirements and specifications

During this phase, you should use standard functionality as much as possible instead of customizing the solution, which could cause long-term impacts.

16.1.3 Build

The build phase involves two key components—building the solution based on the design and testing the solution to ensure it meets your business requirements. In an agile methodology, this iterative process follows a sprint format that includes the build, review, and feedback processes. Typical activities and deliverables include the following:

- Delivering a configured solution
- Developing integrations
- Building conversion programs
- Creating configuration workbooks
- Defining user stories
- Building test scenarios
- Executing test scripts

16.1.4 Deploy

During the deploy phase, you would conduct a business and solution readiness check to validate that the project should move forward with implementation. Also part of this phase, the cutover process is rehearsed and then ultimately executed in the project environment. A critical step is carefully planning a cutover plan that incorporates steps and is coordinated across all parties—the project team, business stakeholders, and third-party vendors. Typical activities and deliverables include the following:

- Finalizing a cutover plan
- Rehearsing your cutover plan
- Validating the implementation through a readiness checklist
- End-user training
- Production run book
- Conducting knowledge transfer sessions
- Facilitating go/no-go meetings
- Executing the cutover process

16.1.5 Hypercare

The final phase of the project lifecycle, the hypercare phase, is about stabilizing and optimizing your new processes and solution. This phase begins at go-live (when the project team manages the system and the solution) and ends when the solution is fully handed off to the support team and end users. Twice a year, SAP SuccessFactors releases changes (H1 and H2 releases) that will need to be planned for. Typical activities and deliverables include the following:

- Creating a payroll run book
- Creating a release plan that includes H1/H2 release updates and required updates for payroll
- Formally handing off the solution
- Processing incidents

16.2 Maintaining a Good Delivery Tracker

As with any good project management methodology, tracking the status of deliverables is key. A good deliverables tracker is a log that tracks all deliverables that must be completed for a project. At a minimum, you should track the deliverable, its due date, and its completion percentage. Microsoft Excel can be used for this requirement since most project team members will have access to and understand this software, which can filter and search deliverables. You can also use Smartsheets or Microsoft Project.

Select the tool that best meets your company's needs because using a tool for tracking is the key step. Figure 16.2 shows an example of a clear and concise delivery tracker.

Deliverable Name	Owner	Due Date	Status	Completion Percentage
Workshops	Sue Toms	4/1/2021	Complete	100%
Future State Process Design	Riley Smith	6/15/2021	In Progress	75%
Functional Requirements	Heather Clark Kim Thomson	6/30/2021	In Progress	50%
Interface Specifications	Bob Engel	7/15/2021	Not Started	0%
Data Conversion Requirements	Rich Adams	7/30/2021	In Progress	25%

Figure 16.2 Sample Delivery Tracker in Microsoft Excel

16.3 Conversion Scope Consideration

Conversion is a process that extends along the full lifecycle of a project. During the discover phase, you'll try to understand your system landscape and what data will need to be converted based on scope. During the design phase, you'll complete the requirements for your data and map legacy system/fields so you can retrieve this data. During the build phase, you'll build out tools to load the data as well as test and validate the data conversion, and during the deploy phase, you'll execute the final cutover and validation of the data.

Three categories of data conversion exist for Employee Central Payroll implementations: human resources (HR), payroll, and YTD, as shown in Table 16.2. This list is only representative, and your needs will depend on what is in scope. For example, you may not need to convert benefits information.

Category	Description	Infotypes
HR	▪ This data is maintained in the HR system and can be replicated into, rather than converted into, Employee Central Payroll. ▪ For a new implementation of Employee Central and Employee Central Payroll, you would convert your data into Employee Central and use the prebuilt replication process to move all your employees from Employee Central to Employee Central Payroll.	▪ Infotype 0000 ▪ Infotype 0001 ▪ Infotype 0002 ▪ Infotype 0006 ▪ Infotype 0007 ▪ Infotype 0008 ▪ Infotype 0009 ▪ Infotype 0014 ▪ Infotype 0015

Table 16.2 Data Conversion Requirements for Payroll

Category	Description	Infotypes
Payroll	■ This data is not stored in the HR system but instead is created/updated through replication, directly into payroll, or has an interface for loading the data.	Tax: ■ Infotype 0207 ■ Infotype 0208 ■ Infotype 0209 ■ Infotype 0210 Benefits: ■ Infotype 0167 ■ Infotype 0168 ■ Infotype 0169 ■ Infotype 0170 Garnishments: ■ Infotype 0194 ■ Infotype 0195
YTD	■ If the go-live date is not the beginning of the calendar year, you must load YTD records for all associates that have received pay for that calendar year. This information is used for tax reporting, setting limits, etc. ■ This data must be converted into these tables using an upload program. ■ After the tables are loaded, the payroll driver (program RPCALCU0) must be run using schema ULK9 (or company-specific version) to load the YTD records into the employee's payroll results.	■ Table T558B ■ Table T558C ■ Table T5U8C

Table 16.2 Data Conversion Requirements for Payroll (Cont.)

Loading YTD records is a unique process to payroll. YTD records must be read during the payroll calculation process (described in detail in Chapter 2, Section 2.3) and are not stored in infotypes like the rest of the employee data, but in the payroll results. The payroll driver is used to read the three tables (T558B, T558C, and T5U8C) and update the results with the appropriate information needed.

Data validation is an important process in data conversions, particularly for the data stored in the YTD records, which are critical for payroll calculations. After the first production payroll is run, the data will be difficult to correct. This data is also compliance related; for example, tax deductions and taxable wages for the year are shared with tax authorities.

The scope of the data migration and the data migration process are important parts of the implementation and should be top of mind throughout.

16.4 Special Testing Considerations

As with all projects, testing is key. For payroll implementations, the standard types of testing still apply—unit testing, integration testing, and user acceptance testing. However, additional testing is needed. This additional testing and validation include what most other types of implementations do not—*payroll compare testing*, sometimes called payroll parallel testing or parallel payroll testing. This kind of testing consists of comparing legacy payroll results to SAP results to verify that the payroll function in SAP is working accurately. The results should be identical to legacy results, with the exception of some acceptable differences that we'll discuss later in this section.

A payroll compare test consists of three phases—design, plan, and execute—as shown in Figure 16.3.

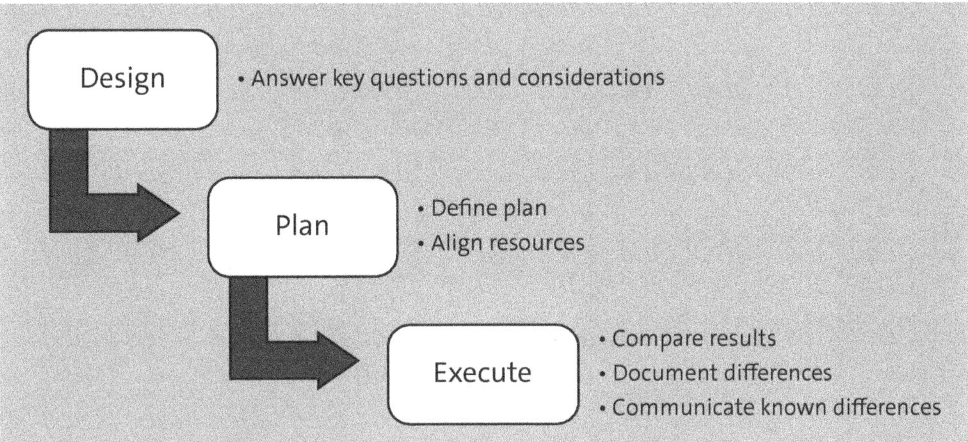

Figure 16.3 Payroll Compare Plan

During the design phase, you'll need to answer key questions to outline how you will execute payroll compare testing. Table 16.3 includes some of these key questions.

Key Questions/Considerations	Description
How many cycles should I test?	Running three cycles is recommended. Two runs could be sufficient for smaller, less complex companies, but three is the recommendation.
How long does a typical cycle take to validate?	The answer will depend on many factors, including the number of resources, tool availability, etc., but typically, they run 4 to 6 weeks per validation cycle. However, this time does not include the prep stage, only the validation or compare stage.

Table 16.3 Key Considerations in Payroll Compare Planning

Key Questions/Considerations	Description
Do I need a tool?	The goal of payroll compare testing is to validate that all payroll earnings, deductions, taxes, and accumulators are compared for 2 to 3 payroll cycles. Conducting this testing manually would be difficult for most companies. Having a tool to support this activity is essential to the timeliness of this test. You could use an in-house tool (which might be less economical and could impact timelines), or you can use one of the many third-party tools on the market.
When should I start this test?	Since payroll compare testing requires a clean converted environment with employees, you should start after a minimum of 2 mock cycles. (A mock conversion is the testing cycle that converts legacy data into Employee Central Payroll.) This approach allows for data defects to be corrected prior to the test. Also, to make the configuration stable, you would want a minimum of 1 integration cycle to be run prior to the start.
What periods should I use?	When selecting the periods, we recommend selecting periods that cover all possible scenarios—holidays, special union payroll calculations, minimum wage increases, etc. However, depending on timing of the test, this approach may not be possible.
What environment do I use, and what is required of this environment?	To execute a payroll compare test, you'll need converted master data, timesheet data, and YTD data, as well any other earnings and deductions in an Employee Central Payroll environment to execute the payroll. Typically, you would use one of the mock environments that has been validated and then load the other data needed specifically for payroll compare testing, for example, specific one-time payments and timesheets.
What threshold should I use?	This threshold usually changes with each cycle, each getting smaller. For the first test cycle, typically the threshold is $1.00, $.50 for the second, and $.10 for the third. The smaller the tolerance, the more time will be needed, but this approach reduces the risk that some problem will continue unnoticed. During planning, specify the threshold and tolerance levels for each cycle—each getting closer to 100% accuracy.

Table 16.3 Key Considerations in Payroll Compare Planning (Cont.)

Key Questions/Considerations	Description
Should other tests or validations be included in this testing cycle?	The goal of payroll compare testing is to validate the payroll results, but you can leverage this testing for other areas, for example, to validate financial postings. As you execute and validate the full employee population, you can also validate the resulting financial postings. This approach supplements integration testing but does not replace it.

Table 16.3 Key Considerations in Payroll Compare Planning (Cont.)

Now that you've answered key questions and created an overall design, you can now start planning for the test. This planning should happen early in the project to align these activities in your overall project plan. Key requirements and activities in this phase include the following:

- An environment with clean, converted data dedicated to the testing cycle should be set up. This environment should include the following components:
  - All employee data
  - Timesheet data
  - YTD loads (if applicable)
  - One-time earnings and deductions
  - Payroll executed for the defined payroll cycle
- Resources identified and onboarded. Some of the key resources needed for this activity include the following:
  - A lead to coordinate all the activities
  - Legacy payroll resources, with the ability to troubleshoot differences to perform the validation for this test
  - Data migration team to correct data defects
  - Project team to correct the system defects
- Tool setup, which is dependent on the tool. Common activities include the following:
  - Setting up the tool in the testing environment
  - Loading data into the tool, both legacy results and SAP results
  - Setting up tolerances
  - Setting up reports
- Defect management tool/process that must be defined and set up. Some considerations include the following:
  - Will I use the same tool for integration, or are any changes needed?
  - If following a manual approach, what is the template or process?
- Kick off for the payroll compare team

Now, you're ready for execution phase. Figure 16.4 shows the payroll compare process.

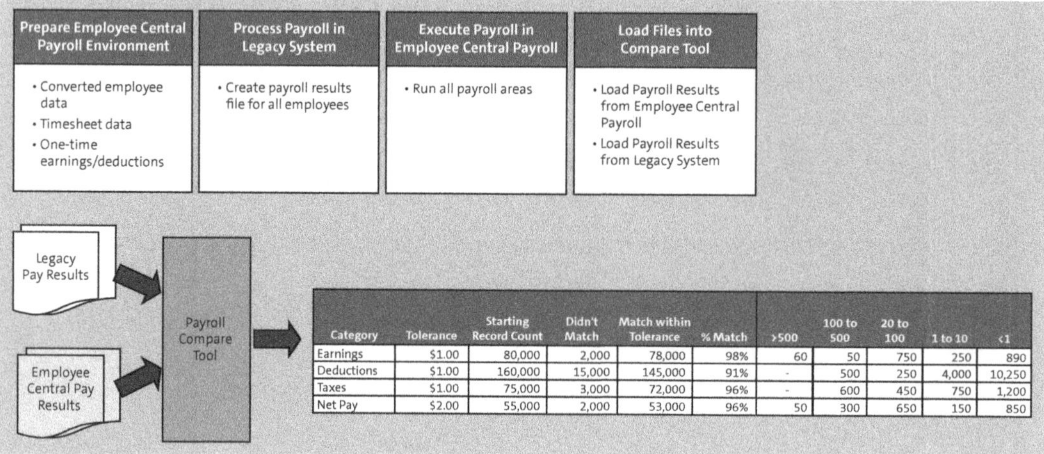

Category	Tolerance	Starting Record Count	Didn't Match	Match within Tolerance	% Match	>500	100 to 500	20 to 100	1 to 10	<1
Earnings	$1.00	80,000	2,000	78,000	98%	60	50	750	250	890
Deductions	$1.00	160,000	15,000	145,000	91%	-	500	250	4,000	10,250
Taxes	$1.00	75,000	3,000	72,000	96%	-	600	450	750	1,200
Net Pay	$2.00	55,000	2,000	53,000	96%	50	300	650	150	850

Figure 16.4 Payroll Compare Process

During this process, we're comparing the payroll results from the legacy system to payroll results in SAP and documenting the differences. You would start the analysis on earning differences, which impacts all the other categories. Once you feel comfortable with earnings, move on to deductions, followed by taxes and then net pay, as each are dependent on the other. For example, missing 20 hours of wages will impact an employee's 401k deductions, taxes, and net pay. You'll want to correct errors so you can continue to improve accuracy. Each difference must be documented as a defect and analyzed to determine the reason for the difference. A resolution must be found for each defect. Table 16.4 defines some common defects that you might encounter.

Common Defects	Resolutions
Data differences	Many differences will be caused by the data, even with the recommendation of running a few mock conversions before starting the testing. These defects should be forwarded to the data team for correction.
Missed requirements	During testing, you may find that some requirements were not raised and have not been solved, thus causing differences that must be addressed.
Compare tool errors	Errors may exist in the tool mapping, in files used for validation, etc., that could show a difference even though the calculations are correct.

Table 16.4 Common Defects

Common Defects	Resolutions
Defects with Employee Central Payroll configuration	Actual defects in the configuration are causing differences in results.

Table 16.4 Common Defects (Cont.)

During the testing process, you'll also encounter acceptable differences or differences in the new system that are not defects. Some examples are listed in Table 16.5. For each of these acceptable differences, a change or a communication plan must be put in place so the impacted employees can be informed and anticipate their changes in their next paychecks. You don't want your employees to be surprised when they receive their first paycheck from Employee Central Payroll, and the amount is different than what the legacy system had calculated. This inconsistency will raise many questions and concerns during hypercare that you must address. Better to address these issues beforehand. Communications could be broad or targeted to specific individuals or even request employee action, depending on the type of difference found. The key is to be proactive about these communications and potentially acting on acceptable differences.

Acceptable Differences	Description
Rounding	This common difference occurs when comparing payroll results due to how systems round numbers. Large differences should not occur, just differences of a few cents.
Local taxes	An error frequently encountered are differences in legacy and Employee Central Payroll calculations due to how the different payroll systems calculate local taxes. Many legacy payroll systems combine many of the taxes where Employee Central Payroll will split them. In the end, you'll likely get the same totals if combined, but subtotals may look different.
Incorrect legacy calculations	With almost every implementation, legacy defects will be found that, going forward, are corrected in SAP. These inconsistencies require communication with your employees. A recent example is the new W-4 form. If legacy payroll did not align with the new form, then tax calculations will be different, and nothing needs to be corrected in Employee Central Payroll.

Table 16.5 Common Acceptable Differences

16.5 Key Project Considerations

As discussed throughout this chapter, payroll implementations can be quite complex and require extra attention. A few key considerations important to all implementations, but specifically to payroll, are listed in Table 16.6.

Key Consideration	Description
Project plan	You'll need to create and maintain a good project plan that is actively used to track actual progress against the plan.
Business resources	Engaged/dedicated business resources are needed to share/confirm business requirements and should be involved in making key design decisions, testing the system, planning for rollout, and evaluating the impact to the organization.
Change management	Since payroll implementations impact every employee, a sound approach to change management is needed.
Support strategy	After the implementation is complete, you'll need support for ongoing changes, enhancements, and troubleshooting. Understanding what this support structure should be and engaging these resources early will allow for a smoother transition.

Table 16.6 Key Project Considerations

16.6 Summary

This chapter defined an implementation lifecycle and described how payroll compare testing can be leveraged for any payroll implementation. Payroll implementations are complex and require thorough planning and testing. You should expect differences between legacy payroll calculations and payroll calculations in Employee Central Payroll that need to be addressed. In this chapter, we also covered key considerations for payroll conversions and payroll compare testing.

The Authors

Deepankar Maitra has led payroll projects for more than 25 years for global businesses in industries like media and communications, oil and gas, defense, and manufacturing. Currently at Deloitte, he is responsible for architecting and implementing payroll solutions for several Fortune 500 companies, having delivered payroll for over 2 million employees in 50 states. Known for his solution development skills and dependable experience in human resources (HR) and payroll transformation projects, Deepankar has published numerous articles and presented at many conferences on topics that enhance the value customers receive from their investments in SAP.

Stacy Wilkins is a global SAP SuccessFactors project manager with more than 20 years of experience with SAP ERP HCM Payroll and SAP SuccessFactors Employee Central Payroll. She has been involved with all aspects of implementations, from sales through support. Her experience also includes serving as a payroll manager for a large, global entertainment company.

Index

D

E

F

G

H

I

Q

R

T

U

V

W

X

Z

- Find, hire, and onboard new talent with SAP SuccessFactors
- Implement Recruiting Marketing, Recruiting Management, and Onboarding in your HR landscape
- Apply best practices for successful talent acquisition projects

Amy Grubb, Kim Lessley

SAP SuccessFactors Recruiting and Onboarding

The Comprehensive Guide

This go-to guide for SAP SuccessFactors Recruiting and Onboarding has it all: functionality, implementation, and integration. Walk through essential HR processes such as creating job postings, evaluating candidates, and onboarding new colleagues, and optimize your SAP SuccessFactors system to accomplish them. With notes on implementation planning and integration with SAP ERP HCM, SAP SuccessFactors Employee Central, and third-party systems, this book is the key to your HR success.

672 pages, pub. 08/2017
E-Book: $79.99 | **Print:** $89.95 | **Bundle:** $99.99
www.sap-press.com/4290